Noise Coupling in System-on-Chip

Devices, Circuits, and Systems

Series Editor
Krzysztof Iniewski
Emerging Technologies CMOS Inc.
Vancouver, British Columbia, Canada

PUBLISHED TITLES:

Advances in Imaging and Sensing
Shuo Tang and Daryoosh Saeedkia

Analog Electronics for Radiation Detection
Renato Turchetta

Atomic Nanoscale Technology in the Nuclear Industry
Taeho Woo

Biological and Medical Sensor Technologies
Krzysztof Iniewski

Building Sensor Networks: From Design to Applications
Ioanis Nikolaidis and Krzysztof Iniewski

Cell and Material Interface: Advances in Tissue Engineering, Biosensor, Implant, and Imaging Technologies
Nihal Engin Vrana

Circuits and Systems for Security and Privacy
Farhana Sheikh and Leonel Sousa

Circuits at the Nanoscale: Communications, Imaging, and Sensing
Krzysztof Iniewski

CMOS: Front-End Electronics for Radiation Sensors
Angelo Rivetti

CMOS Time-Mode Circuits and Systems: Fundamentals and Applications
Fei Yuan

Design of 3D Integrated Circuits and Systems
Rohit Sharma

Diagnostic Devices with Microfluidics
Francesco Piraino and Šeila Selimović

Electrical Solitons: Theory, Design, and Applications
David Ricketts and Donhee Ham

Electronics for Radiation Detection
Krzysztof Iniewski

PUBLISHED TITLES:

Electrostatic Discharge Protection: Advances and Applications
Juin J. Liou

Embedded and Networking Systems: Design, Software, and Implementation
Gul N. Khan and Krzysztof Iniewski

Energy Harvesting with Functional Materials and Microsystems
Madhu Bhaskaran, Sharath Sriram, and Krzysztof Iniewski

Gallium Nitride (GaN): Physics, Devices, and Technology
Farid Medjdoub

Graphene, Carbon Nanotubes, and Nanostuctures: Techniques and Applications
James E. Morris and Krzysztof Iniewski

High-Speed Devices and Circuits with THz Applications
Jung Han Choi

High-Speed Photonics Interconnects
Lukas Chrostowski and Krzysztof Iniewski

High Frequency Communication and Sensing: Traveling-Wave Techniques
Ahmet Tekin and Ahmed Emira

High Performance CMOS Range Imaging: Device Technology and Systems Considerations
Andreas Süss

Integrated Microsystems: Electronics, Photonics, and Biotechnology
Krzysztof Iniewski

Integrated Power Devices and TCAD Simulation
Yue Fu, Zhanming Li, Wai Tung Ng, and Johnny K.O. Sin

Internet Networks: Wired, Wireless, and Optical Technologies
Krzysztof Iniewski

Introduction to Smart eHealth and eCare Technologies
Sari Merilampi, Krzysztof Iniewski, and Andrew Sirkka

Ionizing Radiation Effects in Electronics: From Memories to Imagers
Marta Bagatin and Simone Gerardin

Labs on Chip: Principles, Design, and Technology
Eugenio Iannone

Laser-Based Optical Detection of Explosives
Paul M. Pellegrino, Ellen L. Holthoff, and Mikella E. Farrell

Low Power Emerging Wireless Technologies
Reza Mahmoudi and Krzysztof Iniewski

PUBLISHED TITLES:

Magnetic Sensors and Devices: Technologies and Applications
Laurent A. Francis, Kirill Poletkin, and Krzysztof Iniewski

Medical Imaging: Technology and Applications
Troy Farncombe and Krzysztof Iniewski

Metallic Spintronic Devices
Xiaobin Wang

MEMS: Fundamental Technology and Applications
Vikas Choudhary and Krzysztof Iniewski

Micro- and Nanoelectronics: Emerging Device Challenges and Solutions
Tomasz Brozek

Microfluidics and Nanotechnology: Biosensing to the Single Molecule Limit
Eric Lagally

MIMO Power Line Communications: Narrow and Broadband Standards, EMC, and Advanced Processing
Lars Torsten Berger, Andreas Schwager, Pascal Pagani, and Daniel Schneider

Mixed-Signal Circuits
Thomas Noulis

Mobile Point-of-Care Monitors and Diagnostic Device Design
Walter Karlen

Multisensor Attitude Estimation: Fundamental Concepts and Applications
Hassen Fourati and Djamel Eddine Chouaib Belkhiat

Multisensor Data Fusion: From Algorithm and Architecture Design to Applications
Hassen Fourati

MRI: Physics, Image Reconstruction, and Analysis
Angshul Majumdar and Rabab Ward

Nano-Semiconductors: Devices and Technology
Krzysztof Iniewski

Nanoelectronic Device Applications Handbook
James E. Morris and Krzysztof Iniewski

Nanomaterials: A Guide to Fabrication and Applications
Sivashankar Krishnamoorthy

Nanopatterning and Nanoscale Devices for Biological Applications
Šeila Selimović

Nanoplasmonics: Advanced Device Applications
James W. M. Chon and Krzysztof Iniewski

Nanoscale Semiconductor Memories: Technology and Applications
Santosh K. Kurinec and Krzysztof Iniewski

PUBLISHED TITLES:

Noise Coupling in System-on-Chip
Thomas Noulis

Novel Advances in Microsystems Technologies and Their Applications
Laurent A. Francis and Krzysztof Iniewski

Optical, Acoustic, Magnetic, and Mechanical Sensor Technologies
Krzysztof Iniewski

Optical Fiber Sensors: Advanced Techniques and Applications
Ginu Rajan

Optical Imaging Devices: New Technologies and Applications
Ajit Khosla and Dongsoo Kim

Organic Solar Cells: Materials, Devices, Interfaces, and Modeling
Qiquan Qiao

Physical Design for 3D Integrated Circuits
Aida Todri-Sanial and Chuan Seng Tan

Power Management Integrated Circuits and Technologies
Mona M. Hella and Patrick Mercier

Radiation Detectors for Medical Imaging
Jan S. Iwanczyk

Radiation Effects in Semiconductors
Krzysztof Iniewski

Reconfigurable Logic: Architecture, Tools, and Applications
Pierre-Emmanuel Gaillardon

Semiconductor Devices in Harsh Conditions
Kirsten Weide-Zaage and Malgorzata Chrzanowska-Jeske

Semiconductor Radiation Detection Systems
Krzysztof Iniewski

Semiconductor Radiation Detectors, Technology, and Applications
Salim Reza and Krzysztof Iniewski

Semiconductors: Integrated Circuit Design for Manufacturability
Artur Balasinski

Smart Grids: Clouds, Communications, Open Source, and Automation
David Bakken

Smart Sensors for Industrial Applications
Krzysztof Iniewski

Soft Errors: From Particles to Circuits
Jean-Luc Autran and Daniela Munteanu

Solid-State Radiation Detectors: Technology and Applications
Salah Awadalla

PUBLISHED TITLES:

Structural Health Monitoring of Composite Structures Using Fiber Optic Methods
Ginu Rajan and Gangadhara Prusty

Technologies for Smart Sensors and Sensor Fusion
Kevin Yallup and Krzysztof Iniewski

Telecommunication Networks
Eugenio Iannone

Testing for Small-Delay Defects in Nanoscale CMOS Integrated Circuits
Sandeep K. Goel and Krishnendu Chakrabarty

Tunable RF Components and Circuits: Applications in Mobile Handsets
Jeffrey L. Hilbert

VLSI: Circuits for Emerging Applications
Tomasz Wojcicki

Wireless Medical Systems and Algorithms: Design and Applications
Pietro Salvo and Miguel Hernandez-Silveira

Wireless Technologies: Circuits, Systems, and Devices
Krzysztof Iniewski

Wireless Transceiver Circuits: System Perspectives and Design Aspects
Woogeun Rhee

FORTHCOMING TITLES:

3D Integration in VLSI Circuits: Design, Architecture, and Implementation Technologies
Katsuyuki Sakuma

Energy Efficient Computing: Devices, Circuits, and Systems
Santosh K. Kurinec and Krzysztof Iniewski

Nanoelectronics: Devices, Circuits, and Systems
Nikos Konofaos

Radio Frequency Integrated Circuit Design
Sebastian Magierowski

X-Ray Diffraction Imaging: Technology and Applications
Joel Greenberg and Krzysztof Iniewski

Noise Coupling in System-on-Chip

Thomas Noulis

CRC Press
Taylor & Francis Group
Boca Raton London New York

CRC Press is an imprint of the
Taylor & Francis Group, an **informa** business

CRC Press
Taylor & Francis Group
6000 Broken Sound Parkway NW, Suite 300
Boca Raton, FL 33487-2742

Printed on acid-free paper by CPI Group (UK) Ltd, Croydon, CR0 4YY

International Standard Book Number-13: 978-1-4987-9677-4 (Hardback)

Library of Congress Cataloging-in-Publication Data

Names: Noulis, Thomas, author.
Title: Noise coupling in system-on-chip / Thomas Noulis.
Description: Boca Raton : Taylor & Francis, a CRC title, part of the Taylor & Francis imprint, a member of the Taylor & Francis Group, the academic division of T&F Informa, plc, [2017] | Series: Devices, circuits, & systems | Includes bibliographical references.
Identifiers: LCCN 2017032132| ISBN 9781498796774 (hardback : alk. paper) | ISBN 9781315116693 (ebook)
Subjects: LCSH: Systems on a chip–Design and construction. | Electronic circuits–Noise. | Substrate noise–Prevention.
Classification: LCC TK7874.888 .N68 2017 | DDC 621.3815–dc23
LC record available at https://lccn.loc.gov/2017032132

Visit the Taylor & Francis Web site at
http://www.taylorandfrancis.com

and the CRC Press Web site at
http://www.crcpress.com

Contents

Preface

This book addresses noise coupling in modern system-on-chip design. Noise coupling is the root cause of the majority of systems-on-chip products' failure. As semiconductor technologies shrink below complementary metal-oxide semiconductor 28 nm to 16 m, 14 nm and 10 nm FinFET technology nodes, the dense integration and technology multisourcing to minimize area to cost ratio, renders it as the main design challenge. This becomes even more extreme in mobile communications, due to complete radio spectrum usage from 3 kHz to 30 GHz and performance specifications with signal/power below background noise and will become tighter to millimeter wave applications and moving from the 4G/LTE to the 5G mobile communications era. Currently, no noise coupling methodology is suitable enough as to address this phenomenon accurately. In particular, no methodology can be integrated in the current industry design flow, can capture mask (layout) level isolation performance trends, and provide accuracy from the baseband MHz system-on-chip applications (i.e., automotive semiconductor sector) to the mobile communications system-on-chips.

Especially for the mainstream 4G/LTE and the upcoming 5G mobile communications system-on-chip products, substrate crosstalk is a blocking point. In addition, substrate crosstalk is also crucial in automotive ICs and in BCD technologies where complicated coupling occurs through activation of parasitic devices. Critical limitations that are not addressed in any of the proposed methodologies are the large number of digital gates (>1 MGate), which are impractical in analog/RFMS simulations, the substrate extraction complexity, not only modeling wise but also model net list size and therefore simulation capacity/time wise and the controversy of the frequency region specification between the low-frequency digital versus the analog/RFMS. Finally with the available methodologies, noise coupling study can only be done after layout has been finalized, usually after receiving first silicon and the essential capability for pre-layout or early phase crosstalk simulation is missing.

As both IPs and design methods are migrated to the more advanced planar complementary metal-oxide semiconductor and 3D and FinFET smaller nodes below 28 nm, the necessity for a complete design infrastructure is demanded for first tape out silicon success. In this book, 2D and 3D system-on-chip coupling, analysis flow and modeling addressing the needs of the design community, are described. Academic and industry experts are contributing with the most up to date and modern scientific material to the topic of crosstalk. Using this book, designers can analyze and avoid complex noise coupling that degrades system-on-chip design performance, while reducing the need for overly conservative design practices.

Using this book, the design community will be able to use innovative coupling analysis flow and modeling, addressing all related needs, to analyze noise components, propagating not just through the substrate, but also through the parasitic interconnect and package and to identify substrate coupling noise contributors, levels, and transfer functions. In addition, technical analysis and methodologies are presented as to estimate and/or simulate coupling propagating via multiple paths, across the system-on-chip, and to avoid complex noise coupling that degrades silicon performance while reducing the need for conservative design practices.

Coverage includes novel methodologies to identify noise coupling in silicon, interconnect, and package, helping to locate potential noise issues, both before tape out and even earlier in the design process. The coupling mechanisms are addressed from silicon device level to package and printed circuit board level and from the kHz region until the millimeter-wave frequency region. Special focus is provided in 3D integration and through-silicon via coupling mechanisms. In addition, emerging coupling topics are addressed such as thermal and optical interconnects performance, power delivery networks, electro thermal considerations onto 3D integration, and 3D floor planning based on thermal interactions.

The book is written by a mixture of industrial experts and key academic professors and researchers. The intended audience is engineers with advanced system-on-chip and semiconductors background working in the semiconductor industry. This book can also be used as supplementary material in a graduate course curriculum and, in general, the intended audience is professionals working in the system-on-chip design field.

The book is a must for anyone involved in high performance system-on-chip design both in 2D and 3D ICs. I do believe you will enjoy reading this book as we have enjoyed writing it!

Thomas Noulis
August 31, 2016

Acknowledgments

Throughout the process of making this book happen, technical experts from the community have taken time out to help me. I would like to give special thanks to all the contributors for devoting their time and effort toward this book. A broad technical knowledge range is required, from silicon technology, device modeling, circuit design, package, and board development to electronic design automation and digital and analog/radio frequency design methodology, as to interpret the topic and create scientific value. This technical knowledge is contained in this book and I am grateful for all the contributions. I do believe that it will be a great asset to the community.

I initially got involved in this topic in 2008 during my employment in the R&D Department of HELIC, and from that time, this topic is not just a challenging technical field for me but it is more like a hobby. I would really like to thank HELIC Management team for providing me the opportunity to work in this field. I also thank the technology leader during my HELIC time, Dr. Sotiris Bantas, as well as the coauthor of Chapter 1 during my INTEL time and a good friend of mine, INTEL Principal Engineer, Dr. Peter Baumgartner, who also gave me the freedom to work on this topic. Both these technical experts believed in me and guided me through leadership and scientific excellence. Sotiris Bantas and Peter Baumgartner are highly skilled engineers; without them, this book may not have been written.

I also want to thank my teachers during my university studies and, in particular, my PhD supervisor, Prof. S. Siskos and Prof. T. Laopoulos. They have triggered me to focus in semiconductors and electronic circuits' development, and I am grateful for that.

Finally and foremost, I would like to thank my wife Magda for standing beside me throughout my career. I also thank my wonderful children, my daughter Georgia and our new addition, my son Apostolos, for always making me smile and for understanding on those nights and weekend mornings when I was working on this book instead of playing games. I would also like to thank my parents for enabling me to follow my ambitions throughout my childhood.

To my wife Magda,

To my kids, Georgia and Apostolos

Editor

Thomas Noulis is an assistant professor in the Department of Physics in Aristotle University, in the electronics laboratory. He was previously with INTEL Corp., as a staff RFMS engineer in the wireless platform R&D group, in Munich, Germany, specializing on 14 and 28 nm design platform enablement, in crosstalk, and in system-on-chip product area minimization. Before joining INTEL, Noulis was with HELIC Inc., initially as analog/radio frequency IC designer and then as R&D Engineer/Product Developer Lead, specializing in substrate coupling, signal, and noise integrity and analog/RFIC design.

Noulis holds a BSc in physics (2003) and an MSc in electronics engineering (2005) and a PhD in the "Design of signal processing integrated circuits" (2009) from Aristotle University of Thessaloniki, Greece and in collaboration with LAAS (Toulouse, France). During 2004 to 2009, he participated as principal researcher in multiple European and national research projects related to space application and nuclear spectroscopy IC design, while from 2004 to 2010, he also collaborated as visiting/adjunct professor with universities and technical institutes.

Noulis is in expert in sensor and detectors circuitries, analog/radio frequency integrated circuit design, noise modeling, simulation and optimization, substrate crosstalk, noise and power integrity, verification and simulation speed up, inductor flow, electromigration, and voltage drop methodology. He is focused in leading analog/radio frequency and electronic design automation developments in the technical area and in people management.

Dr. Noulis is the main author of more than 50 publications in journals, conferences, and scientific book chapters. He holds one French and World patent. His work received so far more than a 100 citations. He is an active reviewer of multiple international journals and a member of scientific journals editorial boards. He has given numerous invited presentations in European research institutes, conferences, workshops, and semiconductor companies on crosstalk and radiation detection IC design. He is the editor of the book *Mixed Signal Circuits* by CRC Press. He was awarded for his research activity by conferences and research organizations and can be reached at t.noulis@gmail.com.

Contributors

Peter Baumgartner
RF and Analog Devices
Intel Germany GmbH (Intel Corporation)
Neubiberg, Germany

Eric Beyne
3D and Optical I/O Technologies
IMEC
Leuven, Belgium

João Caldinhas Vaz
Instituto Superior Téncico
Universidade de Lisboa
Instituto de Telecomunicações
Lisbon, Portugal

Francis Calmon
Institut des Nanotechnologies de Lyon
Université de Lyon
Villeurbanne, France

J. R. Cárdenas-Valdez
Instituto Politecnico Nacional, (CITEDI-IPN)
Tijuana, Baja California, Mexico

Norocel D. Codreanu
POLITEHNICA University of Bucharest, Romania
Faculty of Electronics, Telecommunications and Information Technology
Department of Electronics Technology and Reliability
Center for Technological Electronics and Interconnection Techniques (UPB-CETTI)
Bucharest, Romania

Michael G. Dimopoulos
Dialog Semiconductor GmbH
Germering, Germany

James L. Drewniak
Department of Electrical and Computer Engineering
Missouri University of Science and Technology
Rolla, Missouri

Jun Fan
Department of Electrical and Computer Engineering
Missouri University of Science and Technology
Rolla, Missouri

Eby G. Friedman
Department of Electrical and Computer Engineering
University of Rochester
Rochester, NY

Vasileios A. Gerakis
Department of Electrical and Computer Engineering
Aristotle University of Thessaloniki
Thessaloniki, Greece

Christian Gontrand
Institut des Nanotechnologies de Lyon
Université de Lyon
Villeurbanne, France
and
Filière Génie Electrique
INSA Euro-Méditerranée
Université Euro-méditerranéenne de Fès
Fès, Maroc

Alkis A. Hatzopoulos
Department of Electrical and Computer Engineering
Aristotle University of Thessaloniki
Thessaloniki, Greece

Chulsoon Hwang
Department of Electrical and Computer Engineering
Missouri University of Science and Technology
Rolla, Missouri

Ciprian Ionescu
POLITEHNICA University of Bucharest, Romania
Faculty of Electronics, Telecommunications and Information Technology
Department of Electronics Technology and Reliability
Center for Technological Electronics and Interconnection Techniques (UPB-CETTI)
Bucharest, Romania

Swati Joshi
Department of Electronics and Communication Engineering
Indian Institute of Technology
Roorkee, India

Brajesh Kumar Kaushik
Department of Electronics and Communication Engineering
Indian Institute of Technology
Roorkee, India

Jingook Kim
School of Electrical and Computer Engineering
Ulsan National Institute of Science and Technology
Ulsan, South Korea

Joungho Kim
Department of Electrical Engineering
Korea Advanced Institute of Science and Technology
Daejeon, South Korea

Amit Kumar
Department of Electronics and Communication Engineering
Indian Institute of Technology
Roorkee, India

Errikos Lourandakis
Research and Development Department
HELIC
Santa Clara, California

Yue Ma
Institut des Nanotechnologies de Lyon
Université de Lyon
Villeurbanne, France

Luís Mendes
Instituto Politécnico de Leiria
Instituto de Telecomunicações
Lisbon, Portugal

Pedro Mendonça dos Santos
Academia Militar
Instituto de Telecomunicações
Lisbon, Portugal

Panayotis Merakos
Research and Development Department
HELIC
Santa Clara, California

Yiannis Moisiadis
Research and Development Department
HELIC
Santa Clara, California

Konstantinos Moustakas
Department of Physics, Electronics Laboratory
Aristotle University of Thessaloniki
Thessaloniki, Greece

Thomas Noulis
Department of Physics, Electronics Laboratory
Aristotle University of Thessaloniki
Thessaloniki, Greece

J. C. Núñez–Pérez
Instituto Politecnico Nacional, (CITEDI-IPN)
Tijuana, Baja California, Mexico

Henrique Quaresma
Instituto de Telecomunicações
NetThings Ltd
Edinburgh, Scotland

Martin Rack
Institute of Communication Technologies, Electronics and Applied Mathematics (ICTEAM)
Université catholique de Louvain
Louvain-la-Neuve, Belgium

Jean-Pierre Raskin
Institute of Communication Technologies, Electronics and Applied Mathematics (ICTEAM)
Université catholique de Louvain
Louvain-la-Neuve, Belgium

Stylianos Siskos
Department of Physics, Electronics Laboratory
Aristotle University of Thessaloniki
Thessaloniki, Greece

Emre Salman
Department of Electrical and Computer Engineering
Stony Brook University (SUNY)
Stony Brook, New York

Stefanos Stefanou
Research and Development Department
HELIC
Santa Clara, California

Xiao Sun
3D and Optical I/O Technologies
IMEC
Leuven, Belgium

Boris Vaisband
Department of Electrical and Computer Engineering
University of Rochester
Rochester, New York

Olivier Valorge
EASII-IC
Paris, France

Geert Van der Plas
3D and Optical I/O Technologies
IMEC
Leuven, Belgium

J. Verdier
Institut des Nanotechnologies de Lyon
Université de Lyon
Villeurbanne, France

1

*System-on-Chip Substrate Crosstalk Modeling and Simulation Flow**

Thomas Noulis and Peter Baumgartner

CONTENTS

1.1 Introduction

Substrate coupling in mixed-signal integrated circuits (ICs) and especially in wireless communications systems-on-chip (SoCs) create important performance degradation of the analog/RFMS (RF and mixed signal) circuits integrated on the same die as large digital systems. Couplings occur between noise transmitters, which, in most cases, are fast switching digital blocks, and a noise receiver, which, in most cases, is a sensitive analog, RF block or mixed signal architecture and takes place due to the capacitive and resistive nature of the substrate-devices interface (Figure 1.1). The present trend of CMOS technology shrink in technology nodes below 28 nm, the extremely dense integration to minimize the product area to cost ratio, and the new design concept of making analog/RF operations using digital architectures renders substrate crosstalk as one of the most crucial obstacles. Especially in the case of technology multi-sourcing (fabricating the same silicon product

* Part of the data presented in this chapter are Reprinted from *Electronic Letters*, Vol. 51, Issue 12, pp. 953–954, T. Noulis, P. Baumgartner, "Substrate cross-talk analysis flow for submicron CMOS system-on-chip," Copyright 2015," reproduced by permission of the Institution of Engineering & Technology, and from analog Integrated Circuits and Signal Processing, Vol. 90 Issue 2, pp. 477–485. T. Noulis, P. Baumgartner, "CMOS substrate coupling modeling and analysis flow for submicron SoC design," February 2017, reproduced With Permission of Springer.

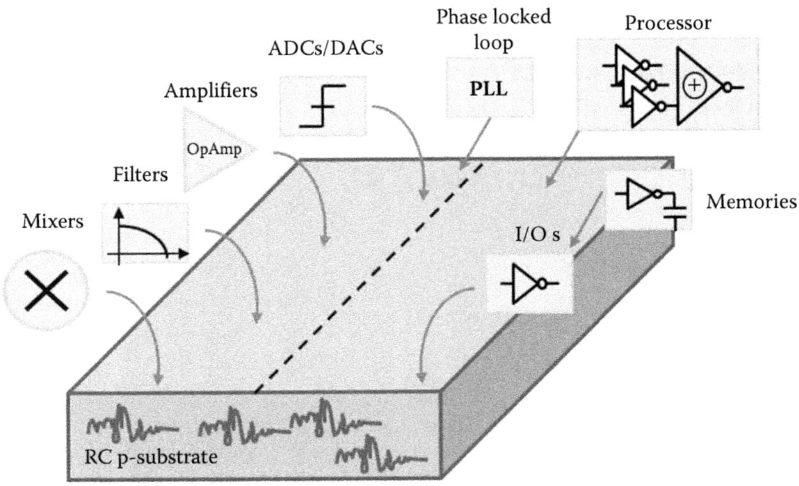

FIGURE 1.1
Representation of a full SoC substrate crosstalk with all the related analog, RF and digital, and mixed signal blocks integrated onto a common silicon substrate. (Reprinted from T. Noulis, P. Baumgartner, *Analog Integrated Circuits and Signal Processing*, 90, 477–485, February 2017, reproduced with permission of Springer.)

footprint in different CMOS technologies, but with the same minimum MOSFET transistor length, so as to minimize the cost between the silicon process vendors), and in particular in mobile communication SoC design. This becomes even more severe due to complete Radio Spectrum usage from 3 kHz to 30 GHz and specifications with signal/power below background noise. In addition, this will become a blocking point in reaching millimeter wave IC design and moving from the 4G/LTE mobile communications to the 5G mobile communications design era, and from 28 nm and 20 nm planar CMOS processes, to 14 nm and 10 nm FINFET [1,2].

All current injected into the substrate causes substrate voltage fluctuations. This is called substrate noise and is caused by the coupling of switching or noisy signals to the substrate. In digital CMOS circuits this noise is caused by three mechanisms: coupling from the digital power supply, coupling from transistor switching source-drain nodes, and impact ionization in the MOSFET channel. Noise on the digital power supply is caused by di/dt noise and resistive voltage drops due to the inductance and resistance in the power-supply connections to the chip. The combination of the inductance in the power-supply connection (package and printed circuit board [PCB]) and the on-chip capacitance between power and ground, also cause ringing in the power-supply voltage. These effects are also known as ground bounce or simultaneous switching noise. This noise coupling mechanism is the dominant cause of substrate noise, and in order to simulate substrate crosstalk, not only the substrate should be modelled (this is, in general, a missing item in the full integrated circuit design flow), but also the digital logic, the analog/RFMS part, and the full package and PCB.

Many methodologies and modeling approaches [2–14] to simulate substrate crosstalk have been suggested so far, but none of them are adequate to address this phenomenon to its full extent. Specifically, methodologies proposed in [3–6], were focused only on specific aspects of the problem and, in particular, only on the modeling of p-type substrates, such as low resistive bulk processes or outdated BiCMOS processes. Also, in the modeling of the digital logic switching activity, injected noise rather than the silicon substrate was addressed. While the substrate crosstalk phenomenon was addressed in product level ICs test-cases [7,14], the frequency region of interest was low (MHz region) and the processes used are currently totally outdated. Among the related methodologies, only Baharoglu et al. [13] and Nagata [14] have addressed the substrate coupling problem in a full SoC level, and provided methodologies that could confront the problem. However, these two methodologies cannot be integrated into the current industry analog/RFMS design flow and neither can capture the respective mask level isolation trends related to substrate crosstalk isolation and optimization. In addition, both are not suitable for use with high-frequency applications such as mobile communications SoCs due to the limitations imposed by their proposed models in high frequency regions. Other more recent analyses and methodologies were also presented [15–17], but none can capture the substrate isolation related performance degradation phenomena and, therefore, they cannot be practically used for substrate crosstalk simulation both in terms of accuracy and physical phenomenon behavioral trends.

There is no SoC substrate crosstalk modeling and simulation methodology that can be integrated into the current industry analog/RFMS design flow that can capture the respective mask level isolation trends related to substrate crosstalk isolation and optimization and also provide accuracy from the baseband MHz SoC applications (i.e., automotive semiconductor sector) to the most critical mobile communications SoCs. Especially for the mainstream 4G/LTE and the upcoming 5G mobile communications SoC products, substrate crosstalk is a blocking point. Other critical limitations that are not addressed in any of the proposed methodologies are the substrate extraction complexity, not only modeling wise, but in terms of size and therefore simulation capacity/time, and the need for early on substrate crosstalk estimation capability. In all the available methodologies, noise coupling analysis can be performed after layout has been finalized. This is usually done after receiving the first silicon where, obviously, capability for pre-layout or early phase crosstalk simulation is missing. Due to the absence of an accurate substrate crosstalk simulation flow, the SoC product vendors, especially in mobile communications related applications, have to fabricate a lot of "trial and error mode" test-microchips, implementing several tape-outs and, therefore, spending a lot of effort and resources to eventually be able to offer high performance working silicon products.

On the Electronic Design Automation (EDA) industry status and the respective software tool capabilities, only two solutions are commercially available [18,19], but both with limited accuracy and neither of them are seamlessly integrated into the industry standard IC design flow. Finally, in our recent letter communication [1], a preliminary substrate crosstalk analysis flow for SoC was briefly presented and in [20] a mixed-signal IC design guide to enhance substrate noise immunity was provided.

In this chapter, a novel and advanced substrate crosstalk modeling and analysis methodology is presented in detail. Advanced related topics are addressed, from the substrate RC modeling, to the respective data processing flow in the design environment, to the signal processing involved and the suitability of the simulation analysis type along with the respective obtained accuracy. This substrate crosstalk flow is fully compatible with the industry product level analog/RF and mixed signal design flow. It can be seamlessly integrated into the design environment and can model "on the fly" all the p-substrate guard ring substrate isolating structures used in mask design. It provides extreme flexibility in terms of the model netlist size and the substrate crosstalk aware respective simulation time. Its accuracy and superiority in capturing all the related effects is validated with measurement results in a CMOS 28-nm-ring oscillator substrate coupling vehicle.

1.2 Ring Oscillator Substrate Crosstalk Vehicle

An 11 stage CMOS CML differential ring oscillator [18] was designed and implemented in silicon. The process selected was a 28 nm CMOS, and the RF devices of the respective process design kit were used. A ring oscillator is a device composed of an odd number of NOT gates (inverters) in a ring whose output oscillates between two voltage levels representing true and false. The inverters are attached in a chain and the output of the last inverter is fed back into the first. In a physical device, no gate can switch instantaneously. In a device fabricated with MOSFETs, the gate capacitance must be charged before current can flow between the source and the drain. Thus, the output of every inverter in a ring oscillator changes a finite amount of time after the input has changed. Adding more inverters to the chain increases the total gate delay, reducing the frequency of oscillation [1,2].

The ring oscillator is a member of the class of time delay oscillators. A time-delay oscillator consists of inverting an amplifier with a delay element between the amplifier output and its input. The amplifier should have a gain larger than 1 at the intended oscillation frequency. Set the initial case where the amplifier input and output voltages are balanced at a stable point. A small amount of noise causes the amplifier output to rise slightly. After passing through the time-delay element, this small output voltage change will be presented to the amplifier input. The amplifier then has a negative gain of greater than one; therefore, the output will change in the direction opposite to the input voltage. It will change by an amount larger than the input value for a gain greater than 1. This amplified and reversed signal propagates from the output through the time-delay and back to the input where it is amplified and inverted again. The result of this sequential loop is a square-wave signal at the amplifier output with the period of each half of the square wave equal to the time delay. The square wave will grow until the amplifier output voltage reaches its limits, where it will stabilize.

The ring oscillator is a distributed version of the delay oscillator. The ring oscillator has an odd number of inverters to give the effect of a single inverting amplifier with a gain of greater than one. Rather than having a single delay element, each inverter contributes to the delay of the signal around the ring of inverters. Adding pairs of inverters to the ring increases the total delay and, thereby, decreases the oscillator frequency. Changing the supply voltage changes the delay through each inverter with higher voltages typically decreasing the delay and increasing the oscillator frequency.

In the specific architecture, the ring oscillator oscillation frequency was 670 MHz. The topology was comprised of a basic 11 stage CML and a five stage output buffer was implemented for extra amplification and for driving both the ring oscillator outputs onto the 50 Ω load of the Spectrum Analyzer. The Spectrum Analyzer used in the measurements had a 3 Hz to 50 GHz min–max frequency range and a resolution bandwidth of 1 Hz to 8 MHz. For the ring oscillator unit, core RF MOSFETs were used. For the buffer, thick gate devices were implemented for higher supply voltage usage and therefore higher allowed output signal swing. The supply of the core ring oscillator architecture was 1.2 V while the buffers supply was 1.8 V.

The physical implementation of the ring oscillator architecture is provided in Figure 1.2a and the silicon footprint microphotograph in Figure 1.2b. The substrate crosstalk vehicle is comprised of five identical ring oscillators related to the circuit architecture, but different substrate isolation structures are used in mask design level (layout). The different versions of the vehicle use different guard ring variations, as shown in Table 1.1.

A sinusoidal signal was injected into the substrate, for each ring oscillator, using a DC pad, as shown in Figure 1.2. This DC pad of each structure was tapped to the silicon substrate, 300 μm away from the ring oscillator. All DC biases—supplies and enable signals—and both the outputs were provided or measured with ground-signal-signal-ground (GSSG) on wafer-pad structures (Figure 1.2). The most specialized and modern commercial Fine-Pitch multi-contact probe was used to enable both DC and RF measurement capabilities to have minimum wafer level pad pitch.

The vehicles, described in Table 1.1, are depicted in Figure 1.2.

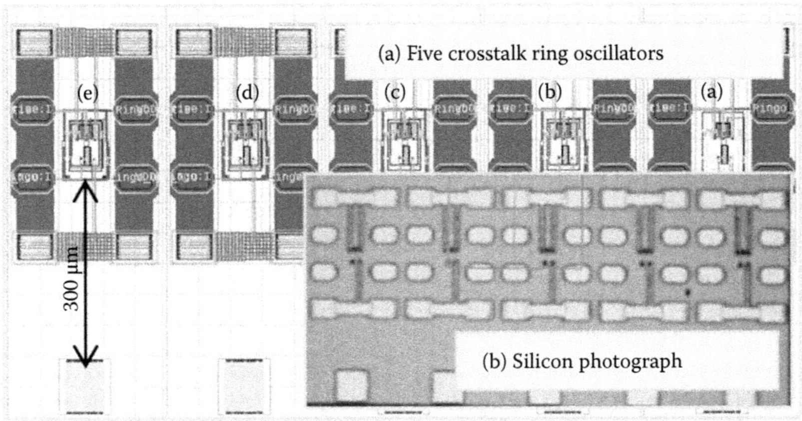

FIGURE 1.2
(a) Physical design of the ring oscillator crosstalk vehicles, (b) silicon microphotograph in magnification. (Reprinted from T. Noulis, P. Baumgartner, *IET Electronic Letters*, 51, 953–954, Copyright 2015, reproduced by permission of the Institution of Engineering & Technology, and from T. Noulis and P. Baumgartner, *Analog Integrated Circuits and Signal Processing*, 90, 477–485, February 2017, reproduced with permission of Springer.)

TABLE 1.1
Ring Oscillator Crosstalk Vehicles and their Mask Design Level Substrate Isolation Schemes

	First Guard Ring		Second Guard Ring		Third Guard Ring	
Ring Oscillator	**Connection**	**Width (μm)**	**Connection**	**Width (μm)**	**Connection**	**Width (μm)**
a						
b	gnd	1	float	4		
c	gnd	1	gnd	4		
d	gnd	1	gnd	1	float	4
e	gnd	1	gnd	1	gnd	4

Source: Reprinted from T. Noulis and P. Baumgartner, *Analog Integrated Circuits and Signal Processing*, 90, 477–485, February 2017, reproduced with permission of Springer.

1.3 Substrate Crosstalk Modeling and Analysis Flow

A substrate RC modeling approach, within the industry design flow was developed and a respective simulation analysis flow was followed. A block diagram of the model extraction flow is provided in Figure 1.3. As input data, the initial topology netlist and the original mask design (layout) are provided. The netlist is processed in order to define the substrate

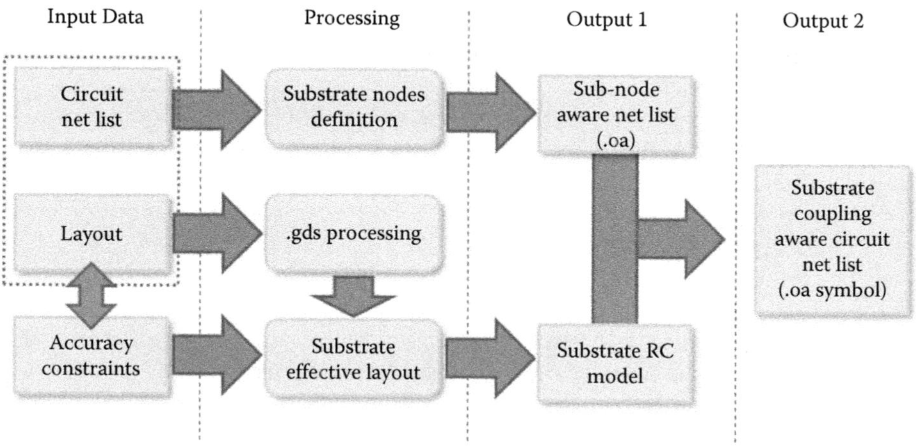

FIGURE 1.3
Silicon substrate model extraction flow. (Reprinted from T. Noulis and P. Baumgartner, *Analog Integrated Circuits and Signal Processing*, 90, 477–485, February 2017, reproduced with permission of Springer.)

effective nodes. On the layout side, the .gds is extracted and in .gds level it is manipulated as the only substrate effective layout to be extracted. In particular, as the effective structures to be identified, substrate contacts connecting metal layer segments to substrate layers, such as diffusions, are recognized. Metal segments of significant length are shorted to either p-type or n-type diffusion layers, effectively forming guard rings. A substrate effective "tap" is defined by a combination of layers and carries an additional definition (PSUB or NSUB) to differentiate it from a segment with only metal properties. The tap area is the area where the layers defined in the tap definition overlap. Vias are sized until they touch the boundary of the up and down layer overlap. During the processing of the layout, a new polygon is created; it has as layer—the device/effective area tap layer—and, as the net, the device bulk node/effective net. A pin is also created at the center of the polygon. With the above methodology all the substrate nodes—p-substrate or n-well/deep n-well contacts and devices coupling to substrate—are recognized. In particular, active devices are addressed as black boxes with a certain number of pins. They define a layout footprint, which is their effective substrate interface area. The area of the footprint may be either p (p-substrate bulk) or n (n-well bulk) depending on the bulk node of the device and the device type. The bulk node is where the substrate network will be connected. Full circuit blocks (which are not substrate crosstalk sensitive, in terms of injecting noise and receiving noise) can also be black boxed in the circuitry hierarchy so as to avoid unmanageable design complexity. Network reduction algorithms focus only on node reduction, but the simulator speed depends also on the number of elements (density of the matrix). Therefore, the merging of many small capacitors (<1 fF) to one or few capacitors per net-pair reduces the number of elements drastically without modifying the circuit operation significantly. To avoid the implementation of huge extracted netlists, accuracy constraints are set. Model extraction accuracy constraints are invoked according to the design phase (floor planning or prototype phase) and the desired model netlist size. In addition, multiple counted devices or effective layout objects can be merged into a single substrate interface area and all the bulk nodes can be connected into a single node. All the effective objects are connected by using trigonometry which serves to produce the smallest netlist size as possible. Simplifying and applying netlist reduction on the substrate RC mesh netlist has the penalty of lower simulation accuracy [1,2].

The model produced at output 1 stage in Figure 1.3 can describe the substrate network from block/domain level to device level. It is the model described in RC netlist-based form in order to model both the real and imaginary part, thus capturing low and high frequency coupling effects and ensuring validity from DC to RF operating regions. Accuracy constraints based on geometrical data invoke substrate contacts merging so as to avoid a huge RC substrate model netlist creation.

After the effective substrate RC network has been extracted and the circuit crosstalk defined netlist is available (Figure 1.3) both netlists can be merged into one .oa unified netlist where the circuit netlist is attached to a distributed substrate RC network. Substrate coupling aware simulations can be performed (output 2 on Figure 1.3).

On the substrate model, specific parameters are needed, as this is to be extracted. On the process parameters, these are the bulk resistivity (Ω * m) the bulk thickness, the well area junction capacitance (fF/μm^2), and the well-periphery junction capacitance (fF/μm). The participated objects (indicatively shown in Figure 1.4) during the substrate extraction flow are all the substrate taps (p+ contacts), the n-wells, all the active areas (active devices that couple either resistively, such as transistors or capacitively, such as inductors and metal capacitors), interconnects with the substrate. For highly accurate modeling, the metal also interconnects. The geometry data file contains the information of the location of the substrate effective objects. The technology file contains the substrate properties such as the dielectric constant, the thickness, and the resistivity for all layers.

A cross-section showing the implemented substrate modeling approach is provided in Figure 1.4 [21], where C_j is the well-junction capacitance and C_c are the metal interconnects capacitive coupling capacitances.

Symmetrical double-strip co-planar wave guide theory is utilized to calculate substrate coupling between all the interacting substrate effective objects. This methodology enables accurate model extraction from DC to the multi GHz frequency region. It calculates the capacitance between two lines on a substrate of finite thickness [21,22]. For the purposes of substrate model

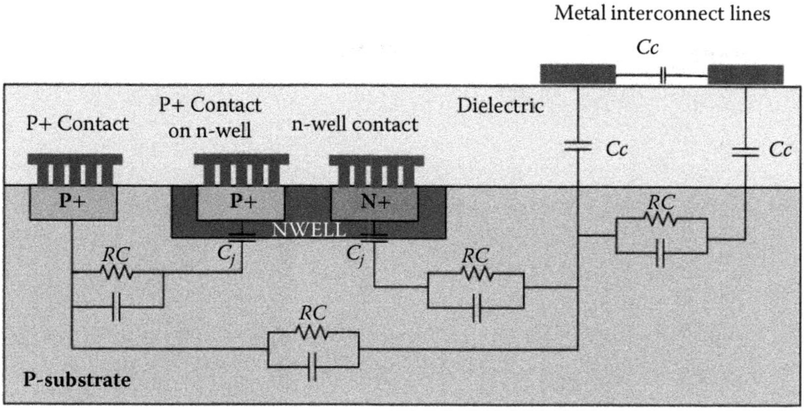

FIGURE 1.4
Silicon cross-section depicting the substrate modeling approach (Reprinted from T. Noulis and P. Baumgartner, *IET Electronic Letters*, 51, 953–954, Copyright 2015, reproduced by permission of the Institution of Engineering & Technology, and from T. Noulis and P. Baumgartner, *Analog Integrated Circuits and Signal Processing*, 90, 477–485, February 2017, with permission of Springer.)

extraction, the lines are considered to be substrate taps. The resistive and capacitive nature of the substrate is calculated by extracting the value of a resistor parallel to the capacitor using the fundamental relation between the two [21] given by

$$R \cdot C = \varepsilon_{Si} \cdot \varepsilon_0 \cdot \rho \tag{1.1}$$

where ε_{Si} is the dielectric constant of the silicon substrate, ε_0 is the electric constant and ρ is the electrical resistivity. All values in this approach are expressed per-unit length and therefore the total capacitance is derived my multiplying with the line length.

1.4 Crosstalk Simulation versus Silicon Measurements

An RC substrate model was extracted and the ring oscillator substrate coupling aware spectrum simulation is provided below together with the standard flow simulation result and the silicon measurement. The proposed flow (plotted in black in Figure 1.5) can capture the substrate coupling effect, simulating the intermodulation's spurs coming through the substrate +/−*f*(MHz) away from the carrier signal (where *f* is the frequency of Table 1.2). In Figure 1.5, the injected sinusoidal signal has a frequency of 10 MHz and amplitude of 1 V p-p and the addressed ring oscillator is

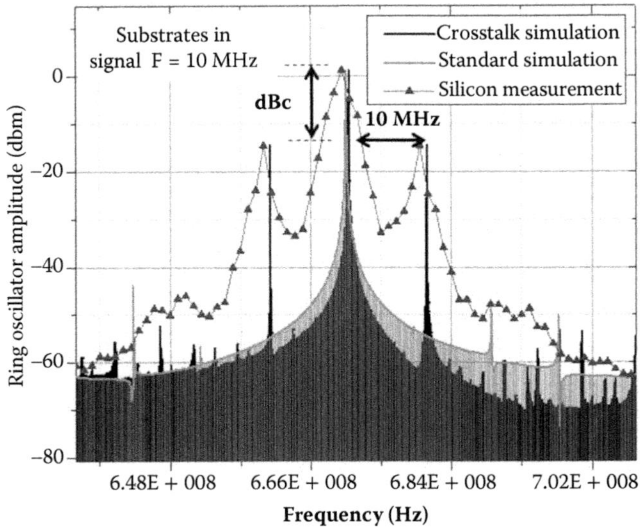

FIGURE 1.5
Ringo oscillator output-carrier signal spectrum: Standard flow simulation results versus the proposed substrate coupling aware analysis flow and silicon measurement. (Reprinted from T. Noulis and P. Baumgartner, *IET Electronic Letters*, 51, 953–954, Copyright 2015, reproduced by permission of the Institution of Engineering & Technology, and from T. Noulis and P. Baumgartner, *Analog Integrated Circuits and Signal Processing*, 90, 477–485, February 2017, with permission of Springer.)

TABLE 1.2
Silicon Measurement Results to Crosstalk Simulation Correlation

Ring Oscillator	Substrate Signal f (MHz)	Carrier Vf0 (dBm) Simulation	Carrier Vf0 (dBm) Measurement	Vspur (dBc) Simulation	Vspur (dBc) Measurement	Error (μV)
a	10	1.1772	1.1771	15.3569	15.3569	142.16
a	20	1.38	1.3786	21.1955	20.9941	532.03
b	10	1.3692	1.3586	15.4532	15.4446	−10.173
b	20	1.3892	1.3938	22.0604	22.19	−295.732
c	10	1.249	1.2515	16.344	16.2479	449.924
c	20	1.546	1.5457	22.776	22.4723	688.57
d	10	1.173	1.1715	15.001	14.7568	1288.45
d	20	1.295	1.2944	22.532	21.8867	1492.12
e	10	1.34	1.3338	16.751	16.1454	2709.57
e	20	1.328	1.3234	22.093	21.6606	1032.06

Source: Reprinted from T. Noulis and P. Baumgartner, *IET Electronic Letters*, 51, 953–954, Copyright 2015, reproduced by permission of the Institution of Engineering & Technology, and from T. Noulis and P. Baumgartner, *Analog Integrated Circuits and Signal Processing*, 90, 477–485, February 2017, with permission of Springer.

the one without any GR isolation, ring oscillator (a) of Figure 1.2. The ring oscillator output signal (carrier signal) power spectrum expressed in dBm (conversion of the output voltage signal into power onto the 50 V load of the Spectrum Analyzer) is provided in Figure 1.5.

The substrate crosstalk phenomenon trends and the respective proposed flow capturing capability in terms of the distance of the noise aggressor (substrate tap in our test case vehicles) to the noise victim (ring oscillator) and the use of guard rings is really crucial. In mask design level (layout) many guard rings are designed (double or even triple guard rings) to achieve extra substrate crosstalk isolation; thus, occupying silicon area and adding extra cost in the respective SoCs products. In addition, silicon area is also spent in order to increase the distance between sensitive substrate crosstalk blocks. Consequently the substrate modeling and analysis flow should be able to capture these performance trends.

Several substrate crosstalk aware simulations of the ring oscillators' carrier signal spectrum were performed with two different frequencies of 10 MHz and 20 MHz for the substrate injected signal and for all five ring oscillator structures. The results were correlated with silicon measurements and are provided in Table 1.2. The substrate crosstalk analysis flow accurately predicts the intermodulation spur amplitudes in the standard and the double guard ring oscillator structures. However, in the triple guard ring isolation structures, there is an overestimation resulting from larger errors, even though the general phenomenon performance behavior is captured.

The guard ring isolation trends are also taken into account as depicted in Figure 1.6a and b. The substrate isolation increases while moving from the

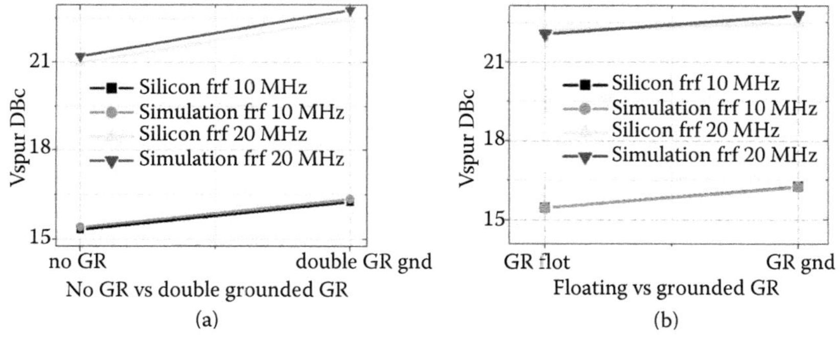

FIGURE 1.6
Silicon versus Simulation (a) guard ring biasing trend on substrate coupling isolation (b) no guard ring versus double grounded guard ring substrate coupling isolation trend. (Reprinted from T. Noulis and P. Baumgartner, *IET Electronic Letters*, 51, 953–954, Copyright 2015, reproduced by permission of the Institution of Engineering & Technology, and from T. Noulis and P. Baumgartner, *Analog Integrated Circuits and Signal Processing*, 90, 477–485, February 2017, with permission of Springer.)

floating outer guard ring to the grounded outer guard ring and from the no guard ring oscillator to the double guard ring isolation structure.

On the injected signal frequency trend, and in relation to the obtained substrate coupling isolation, again, the proposed modeling and analysis flow can capture the measurement results trend. In Figure 1.7, the isolation is depicted for all cases, from structure a to structure e, and for 10 MHz to 20 MHz injected signal frequency values. The proposed flow follows perfectly the input signal frequency trend in terms of the obtained isolation.

For all the cases, the accuracy metric is the difference between the amplitude of the carrier signal and the substrate intermodulation spur in dBc. The related errors after conversion of the dB values to rms voltage values are provided in Table 1.2. The mean error is 665 μV and the sigma is 765 μV. The substrate coupling dynamic range is 39.8 dB and the dynamic range sigma is 25.9 dB.

Regarding the related simulation time, different analysis types were studied in terms of the suitability for substrate crosstalk simulation. The main challenge is the size of the substrate extracted network netlist. Since each substrate tap (or family of taps, when merging is performed) is practically a node, the obtained netlist can be quite large. Four types of spice analysis were studied. Large signal analyses such as transient analysis (TRAN) and Quasi-Periodic Steady-State Analysis (QPSS) and small signal such as PSS driven Harmonic Balance (HB) and Envelope Following analysis (ENVLP). Quasi-Periodic Steady-State (QPSS) analysis is a large signal analysis than can be used for circuits with multiple large tones. QPSS analysis computes the steady-state responses of a circuit driven by two or more signals at

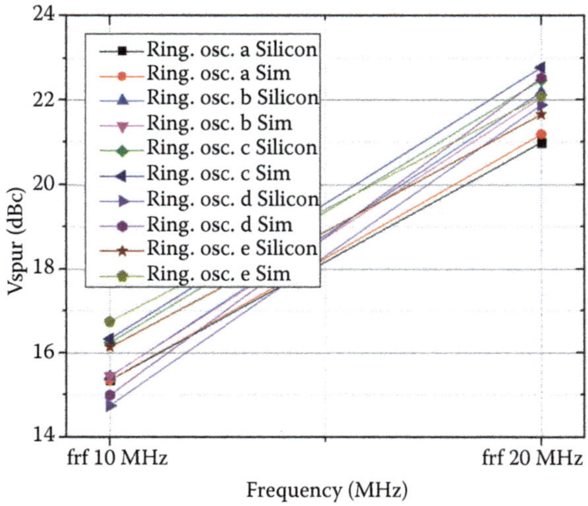

FIGURE 1.7
Silicon versus Simulation on substrate injected signal frequency trend in relation to isolation. (From T. Noulis and P. Baumgartner, *Analog Integrated Circuits and Signal Processing*, 90, 477–485, February 2017, reproduced with permission of Springer.)

unrelated frequencies. It is possible to select a sinusoidal or pulse signal as the large signal. Any additional signals, called moderate signals, must be sinusoids. The harmonic balance engine supports frequency domain harmonic balance analysis. It provides efficient and robust simulation for linear and weakly nonlinear circuits.

The ENVLP analysis is applied to efficiently and accurately analyze modulation signals in large circuits. ENVLP analysis reduces simulation time by exploiting the property that the behavior of a circuit in a given high frequency clock cycle is similar, but not identical, to its behavior in the preceding and following cycles. In particular, the envelope of the high frequency clock can be followed by accurately computing the circuit behavior over occasional cycle to try to capture the fast transient behavior. The slow varying modulation is accurately followed by a smooth curve. As a result, the spectrum of the circuit response can be obtained by combining the spectrum of the smooth curve and the spectrum of occasional clock cycles [23,24].

The results in terms of the obtained accuracy and simulation time are provided in Table 1.3. The TRAN simulation, although it provided accurate estimation of the phenomenon, required a quite large simulation time interval. The ENVLP was relatively faster, but the obtained results were not correct both in terms of the carrier and the crosstalk spur amplitude values. QPSS and HB could not provide the corresponding results in a reasonable simulation time.

The transient analysis simulation was performed using default values for the simulation tolerances *reltol, iabstol,* and *vabstol* and using the conservative-high accurate mode. The time window was selected from 0 to 2.2 μs and the time step was equal to 2 ps. The maximum time step was also set to 2 ps. Also, the respective frequency Fourier Transform (FFT) to translate the time domain signal result into the frequency spectrum, was performed setting *tstart* = 102.848 ns, *tstop* = 2.2 μs. This is the time window for which the oscillator's carrier signal amplitude is stable. The extracted parasitic ring oscillator attached to the RC substrate model was used in the simulations and the respective simulation time was equal to 6 h. This time interval was reduced to

TABLE 1.3

Simulation Analysis Type Suitability

Simulation Type (SPECTRE)	Substrate Coupling Simulation	Simulation Time	Accuracy
Transient (optimum)	✓	6 hours	✓
Quasi-Periodic Steady State	×	Impractical	**No result**
Harmonic Balance	×	Impractical	**No result**
Envelope	✓	1.5 hours	×

Source: From T. Noulis and P. Baumgartner, *Analog Integrated Circuits and Signal Processing,* 90, 477–485, February 2017, with permission of Springer.

43 min after applying the advanced parallel processing mode to the simulator. The respective bench test netlist size corresponds to 12,147 nodes.

Since the spur amplitudes that could be simulated are very low, the simulation accuracy is crucial. The most suitable analysis, the TRAN analysis, provides many options in its set up for simulation accuracy and simulation convergence. There are different options available in configuring transient analysis. In SPECTRE, *gear* is better than others, even though the help file says *"trapezoidal is used when we need high accuracy"* and *"gear can make systems more stable than they really are."* In particular, the "method" parameter specifies the integration method. The possible settings and their meanings are provided in Table 1.4 [23,24].

The trapezoidal rule is the most efficient when high accuracy is required. It exhibits point-to-point ringing, but that can be controlled by tightening the error tolerances. For this reason, if very loose tolerances are selected to obtain a quick answer, either backward-Euler or second-order Gear will probably give better results than the trapezoidal rule.

Several tolerances parameters determine the accuracy of the transient analysis. "reltol" and "abstol" control the discretized equation solution and specifically how well charge is conserved and how accurately steady-state or equilibrium points are computed. The integration error can be set or the errors in the computation of the circuit dynamics relative to "reltol" and "abstol" by setting the "lteratio" parameter. The parameter "relref" determines how the relative error is treated. The "relref" options are provided in Table 1.5.

1.4.1 FFT Processing Optimization

Since the selected optimum type of substrate coupling simulation analysis is the time domain simulation and FFT processing of the obtained waveform is afterwards used as to get the respective frequency domain (spectrum) response, important in the accuracy and in the dynamic range of the captured intermodulation products is the setup of the FFT processing. The selection of the FFT window type determines the obtained substrate noise dependent spectrum and the simulated noise spurs and carrier gain amplitudes. When using FFT analysis there are limits on resolution between different

TABLE 1.4

Transient Analysis Option Set Up Capabilities

Tran. Analysis Method	Meaning
euler	Backward-Euler is used exclusively
traponly	Trapezoidal rule is used almost exclusively
trap	Backward-Euler and the trapezoidal rule are used
gear2only	Gear's second-order backward-difference method is used almost exclusively
gear2	Backward-Euler and second-order Gear are used
Trapgear2	Allows all three integration methods to be used

TABLE 1.5
Transient Analysis Error Parameter "Relref" Options

Error Parameter "Relref" Option	Meaning
pointlocal	Compares the relative errors in quantities at each node to that node alone
alllocal	Compares the relative errors at each node to the largest values found
sigglobal	Compares relative errors in each of the circuit signals to the maximum for all signals at any previous point in time
allglobal	Same as "relref=sigglobal" except that it also compares the residues for each node to the maximum of that node's past history

Source: Virtuoso Spectre Circuit Simulator RF Analysis User Guide, Product Version 6.2, June 2007, Cadence; SpectreRF Theory, Product Version 5.0, July 2002, Cadence.

frequencies and on the detectability of a small signal in the presence of a large one. There are two problems: the signal can be measured for a limited time and the FFT only calculates results for certain discrete frequency values, the so called "FFT bins." The limit on measurement time is fundamental to any frequency analysis technique [25,26].

The FFT implements a Fourier transform at a discrete set of frequencies and from a time-domain waveform sampled at discrete times over a finite interval of time. Because of the finite interval, the FFT tends not to be very frequency-selective. When using the FFT to measure the frequency content, the analysis should be based on a finite set of data. The FFT assumes that the finite data set is one period of a periodic signal. For the FFT, both the time domain and the frequency domain are circular topologies, so the two endpoints of the time waveform are interpreted as though they were connected together. Therefore, the finiteness of the sampling record results in a truncated waveform with different spectral characteristics. The finiteness can also introduce sharp transition changes into the measured data. The sharp transitions are discontinuities. To minimize this effect, we apply a Window Function to the measured signal in the time domain. This makes the endpoints of the waveform meet and therefore result in a continuous waveform without sharp transitions. Applying a window function to the acquisition that connects the waveform endpoints in a smoother fashion before computing the FFT results in better spectral resolution. This technique is also referred to as "applying a window" or simply "windowing". There are different types of window functions available, each with their own advantage and preferred application.

Table 1.6 contains all the available "windowing" types of FFT processing. This table is considered to be the "windowing" SNA optimum selection bench-mark in relation to each SNA test-case, since the requirement of the signal separation specification is greatly dependent on the application.

TABLE 1.6
FFT "Windowing" Type in Relation to Application

Signal Type	Window
Transients whose duration is shorter than the length of the window	Rectangular
Transients whose duration is longer than the length of the window	Exponential, Hanning
General-purpose applications	Hanning
Spectral analysis (frequency-response measurements)	Hanning (random excitation), Rectangular (pseudorandom excitation)
Separation of two tones with frequencies very close to each other but with widely differing amplitudes (optimum)	Kaiser-Bessel
Separation of two tones with frequencies very close to each other but with almost equal amplitudes	Rectangular
Accurate single-tone amplitude measurements	Flat top
Sine wave or combination of sine waves	Hanning
Closely spaced sine waves	Uniform, Hanning
Unknown content	Hanning
Narrowband random signal (vibration data)	Hanning
Broadband random (white noise)	Uniform

1.5 Conclusion and Discussion

Substrate crosstalk modeling and simulation flow, seamlessly integrated to the analog/RFMS design methodology was developed. A substrate RC mesh-modeling concept was presented that enabled accuracy from DC to multi-GHz and validity in mobile communications applications. This flow was validated using measurement results in a 28 nm CMOS vehicle and in terms of achieved accuracy in capturing the substrate crosstalk of different mask design level isolation structures. All the available simulation analysis types were examined and time domain transient analysis was designated as the most suitable, both in terms of the obtained results and netlist capacity. The particular substrate crosstalk modeling and extraction flow, enables, for the first time, everything from substrate crosstalk estimations and "what if" early phase simulations analyses to full detailed product release crosstalk simulations. It also enables progressive substrate crosstalk optimization and increases predictability, while, lowering SoC performance risk significantly.

There are a lot of extra items that should be addressed as part of substrate crosstalk in order to estimate accurately in all design phases. More than 50% of the wireless communications SoC product failures [27], especially in the case of technology multi-sourcing, are related to substrate crosstalk. While

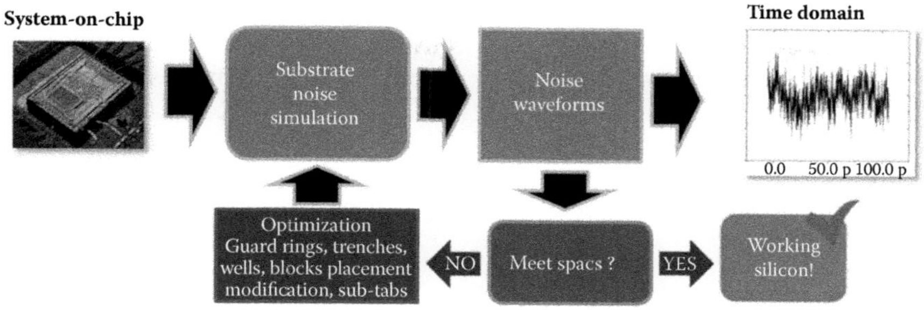

FIGURE 1.8
Substrate coupling aware design flow.

an SoC prototype-level fully detailed crosstalk simulation flow is needed, it's even more significant to be able to make preliminary substrate crosstalk simulation in the early design phase, since, by the time the full product mask design (layout) and PCB and package layouts are ready, it is already too late to make extreme modifications with risking substrate coupling performance degradation. Currently, the absence of an accurate substrate crosstalk simulation flow, forces SoC product vendors, especially those in mobile communications applications, to fabricate a lot of "trial and error mode" test-microchips, implementing several tape-outs and, therefore, spending extreme amount of efforts and resources to eventually high performance working silicon products.

This complex problem can only be faced with a novel breakthrough and "easy to use" substrate crosstalk analysis flow addressing all practical needs of the design community. A crosstalk verification flow, such as the one in Figure 1.8, provides the capability to simultaneously analyze substrate noise propagating not just through the substrate and through the parasitic interconnect and the package.

The substrate noise contributors, levels and transfer functions between all the SoC blocks can be clearly identified and displayed in all the design phases, from the floor-planning all the way to the prototype and the product release phase. A solid and unified modeling toolset is required, with simulation capacity that can enable IC integrators, architects and analog/RF IP designers to estimate and/or simulate (according to the design phase) noise propagating via multiple paths, across the entire SoC and optimize/simulate the SoC design, placing guard rings, trenches, and applying alternative block placement. Since noise coupling in RFMS designs has been, for many years, a blocking point, and considering that both IPs and design methods are migrating to the smaller nodes, and, particularly in the mobile communications ICs, the necessity for a complete and accurate design infrastructure for high performance/high speed RFMS designs is demanding, there is a need for a first tape out silicon success. Following a substrate aware design flow as depicted in Figure 1.8, designers are able to analyze and avoid complex noise coupling that degrades RFMS design performance, while reducing the need for overly conservative design practices. Unique capabilities can be provided to quickly identify noise coupling not only in silicon, but also other critical design aspects of interconnect and package/PCB analysis, all of which are significant in helping locate potential noise issues, both before tape-out and even earlier in the design process.

Acknowledgment

The contribution of Hui Zhang, Antonius Koller, and Reinhard Golly of INTEL—Mobile Communications Group—RFAD Characterization Laboratory, during the silicon measurements, is greatly appreciated.

A small part of this work was financed by Aristotle University Research Committee and the Special Account for Research Grants, in the framework of the "Strengthening Research Activity in Aristotle University of Thessaloniki—2015" action, and under the project "SoCs NOISE & SIGNAL INTEGRITY"—Project Code Number 93296.

References

1. T. Noulis and P. Baumgartner, Substrate cross-talk analysis flow for submicron CMOS system-on-chip, *IET Electronic Letters*, vol. 51, no. 12, 953–954, 2015.
2. T. Noulis and P. Baumgartner, CMOS substrate coupling modeling and analysis flow for submicron SoC design, *Analog Integrated Circuits and Signal Processing*, vol. 90, no. 2, 477–485, 2017.
3. A. Muhtaroglou, G. Taylor, T. Rahal-Arabi, K. Callahan, On-die droop detector for analog sensing of power supply noise, *IEEE Journal of. Solid-State Circuits*, vol. 39, no. 4, 651–660, 2004.
4. M. Felder and J. Ganger, Analysis of ground-bounce induced substrate noise coupling in a low resistive bulk epitaxial process: Design strategies to minimize noise effects on a mixed signal chip. *IEEE Transaction on Circuits and Systems II*, vol. 46, no. 11, 1427–1436, 1999.
5. E. Charbon, P. Miliozzi, L. P. Carloni, A. Ferrari, and A. Sangiovanni-Vincentelli, Modeling digital substrate noise injection in mixed signal IC's, *IEEE Transactions on Computer Aided Design of Integrated Circuits and Systems*, vol. 18, no. 3, 301–310, 1999.
6. R. Gharpurey and G. Meyer, Modeling and analysis of substrate coupling in integrated circuits, *IEEE Journal of Solid state circuits*, vol. 31, no. 3, 125–128, 1996.
7. B. R. Stanisic, N. K. Verghese, R. A. Rutenbar, L. R. Carley, and D. J. Allstot, Addressing substrate coupling in mixed-mode IC's: Simulation and power distribution synthesis, *IEEE Journal of Solid State Circuits*, vol. 29, no. 3, 226–238, 1994.

8. R. Senthinathan and J. L. Prince, Simultaneous switching ground noise calculation for packaged CMOS devices, *IEEE Journal of Solid State Circuits*, vol. 26, no. 11, 1724–1728, 1991.
9. K. Joardar, A simple approach to modeling cross-talk in integrated circuits, *IEEE Journal of Solid State Circuits*, vol. 29, no. 10, 1212–1219, 1994.
10. A. L. L. Pun, Substrate noise coupling through planar spiral inductor, *IEEE Journal of Solid State Circuits*, vol. 33, no. 6, 877–884, 1998.
11. J. M. Casalta, X. Aragonès, and A. Rubio, Substrate coupling evaluation in BiCMOS technology, *IEEE Journal of Solid State Circuits*, vol. 32, no. 4, 598–603, 1997.
12. A. Helmy and M. Ismail, *Substrate Noise Coupling in RFICs*, New York, NY: Springer Science, 2008.
13. M. Baharoglu, M. van Heijningen, V. Gravot, S. Donnay, H. De Man, G. Gielen, M. Engels, and I. Bolsens, Modeling and experimental verification of substrate noise generation in a 220k-gates WLAN system-on-chip with multiple supplies, *IEEE Journal of Solid State Circuits*, vol. 38, no. 9, 1250–1260, 2003.
14. M. Nagata, "On-Chip measurements complementary to design flow for integrity in SoCs," In *Proceedings of the 44th annual Design Automation Conference*, San Diego, California, June 04–08, pp. 400–403, 2007.
15. N. Azumam, S. Shimazaki, N. Miura, M. Nagata, T. Kitamura, S. Takahashi, M. Murakami, K. Hori, A. Nakamura, K. Tsukamoto, M. Iwanami, E. Hankui, S. Muroga, Y. Endo, S. Tanaka, M. Yamaguchi, "Measurements and Simulation of Substrate Noise Coupling in RF ICs with CMOS Digital Noise Emulator," *In 9th International Workshop on Electromagnetic Compatibility of Integrated Circuits (EMC Compo)*, Nara, Japan, pp. 42–46, 2013.
16. S. Bronckers, G. Van der Plas, G. Vandersteen., and Y. Rolain, Substrate noise coupling mechanisms in lightly doped CMOS transistors, *IEEE Transations on Instrumentation and Measurement*, vol. 59, no. 6, 1727–1733, 2010.
17. M. Shen, J. Mikkelsen, O. K. Jensen, and T. Larsen, "A Compact P+ Contact Resistance Model for Characterization of Substrate Coupling in Modern Lightly Doped CMOS Processes," In *7th European Microwave Integrated Circuits Conference (EumIC)*, Amsterdam, Nederland, pp. 29–30, 492–495, 2012.
18. F. Clement, "Method and Apparatus for Assisting Integrated Circuit Designing with a Substrate Coupling," Patent No. 8306803, Assignee: Coupling Wave Solutions CWS, November 6, 2012.
19. Y. Bando, D. Kosaka, G. Yokomizo, K. Tsuboi, Y. S. Li, S. Lin, M. Nagata, "Power Supply and Substrate Noise Analysis; Reference Tool Experience with Silicon Validation," In *Design Automation Conference*, San Francisco, CA, July 26–31, 2009.
20. O. Valorge, F. Calmon, C. Andrei, P. Dautriche, C. M. Gontrand, Mixed-signal IC design guide to enhance substrate noise immunity in bulk silicon technology, *Analog Integrated Circuits and Signal Processing*, vol. 63, no. 2, 185–196, 2010.
21. R. K. Hoffmann, *Handbook of Microwave Integrated Circuits*, Norwood, MA: Artech House, Inc., 1987.
22. A. J. Van Genderen, N. P. Van der Meijs, and T. Smedes, "Fast Computation of Substrate Resistances in Large Circuits," In *Proceeding of EDTC '96, European Conference on Design and Test*, Paris, France, March 11–14, pp. 560–566, 1996.
23. Virtuoso Spectre Circuit Simulator RF Analysis User Guide, Product Version 6.2, June 2007, Cadence

24. https://wenku.baidu.com/view/57255d020740be1e650e9a21.html
25. https://en.wikipedia.org/wiki/Fast_Fourier_transform
26. http://www.labbookpages.co.uk/audio/firWindowing.html
27. http://www.eetimes.com/document.asp?doc_id=1280687

2

Substrate Induced Signal Integrity in 2D and 3D ICs

Emre Salman

CONTENTS

2.1 Introduction

Silicon substrate represents a common medium for large numbers of transistors that are integrated to form a monolithic die. Since silicon is a semiconducting material, the signals injected into the substrate propagate throughout the die and can disturb the operation of highly sensitive circuits through multiple mechanisms [1,2]. This phenomenon, typically referred to as substrate noise coupling, should be considered during the design and verification stages of a traditional flow to avoid substrate noise induced failures or performance degradations. In small-scale circuits with transistors numbering in the range of several hundreds, substrate coupling noise can be analyzed by extracting the substrate and simulating the extracted substrate with the switching circuit. As the complexity of the circuit grows, however, the extraction and simulation processes start to be computationally prohibitive [3]. More efficient analysis methods are, therefore, required to understand the substrate effects on large-scale circuits. Efficient analysis of substrate noise coupling is significant for mixed-signal ICs for two primary reasons: (1) to ensure (through simulations) that the circuit operates correctly and required design objectives (such as gain, signal-to-noise ratio, harmonic distortion, and bandwidth) are satisfied and (2) to determine the required noise isolation between analog and digital components. The second reason is important, since noise isolation methodologies typically exhibit overhead in circuit area and power consumption. Thus, the required isolation should be quantified to minimize this overhead while ensuring that the substrate noise is within the tolerable range. This chapter first provides an overview of substrate noise coupling by summarizing noise analysis/modeling methodologies and noise mitigation techniques. Next, figures-of-merit are introduced to evaluate the importance of substrate noise in analog circuits. These figures-of-merit rely on frequency dependent comparisons of input-referred equivalent device noise and input-referred equivalent substrate noise [4]. Finally, the opportunities and challenges provided by through-silicon via (TSV) based three-dimensional (3D) integration are discussed.

2.2 Overview of Substrate Noise Coupling in 2D ICs

In mixed-signal ICs, substrate coupling noise has long been a critical issue due to dense and monolithic integration of sensitive analog/RF and aggressive digital components on the same die. Switching (also referred to as induced) noise caused by a transition of a digital signal is injected into the substrate through multiple mechanisms and propagates through the substrate, finally reaching an analog/RF circuit, as shown in Figure 2.1. On the receiver side, substrate noise can degrade important performance specifications such as signal-to-noise ratio, gain, and bandwidth [5,6].

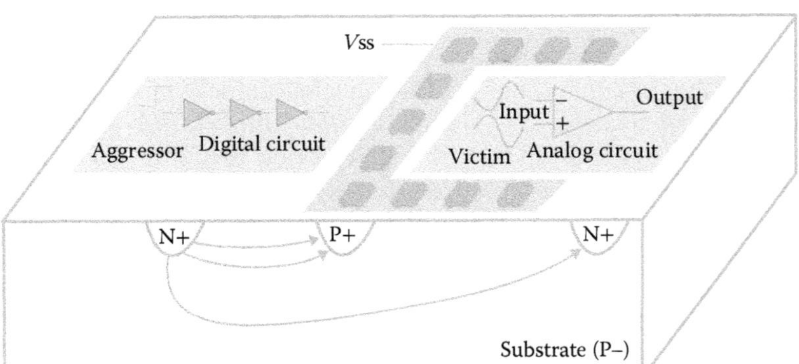

FIGURE 2.1
Switching noise coupling from a digital block to a sensitive analog circuit in the presence of guard rings.

Noise is injected into the substrate through three primary mechanisms [7], as depicted in Figure 2.2, where a cross-section of a CMOS inverter with power and ground networks is shown. The ground network has a substrate contact, resistively connecting the substrate with the ground distribution network. Alternatively, the power network has an n-well tie, which capacitively connects the substrate to the power network due to n-well capacitance. According to this figure, the three noise injection mechanisms are: (1) power and ground coupling through, respectively, the substrate contact and n-well tie, (2) source drain junction coupling through reverse-biased pn junction capacitance, and (3) impact ionization (not shown in the figure) due to the high electric field within the depletion region. Since the power and ground distribution networks typically exhibit resistive and inductive characteristics, non-negligible power supply and ground noise (*IR* drop and *Ldi/dt* noise) exist on, respectively, power and ground networks [8,9]. The ground noise resistively couples into the substrate through the substrate contacts; whereas, the power noise first resistively couples into n-well through the n-well tie and then capacitively couples into the substrate through the n-well capacitance. Thus, a capacitive isolation exists between power network and substrate. When the digital signals on the source/drain terminals switch, noise couples into the substrate through the junction capacitances. Of these three mechanisms, impact ionization is typically negligible as compared to the first two mechanisms. The dominance of power/ground versus source/drain coupling depends on multiple parameters such as the number of switching gates and power/ground network impedances [7,10]. For large-scale circuits, power/ground coupling is the typically the dominant noise

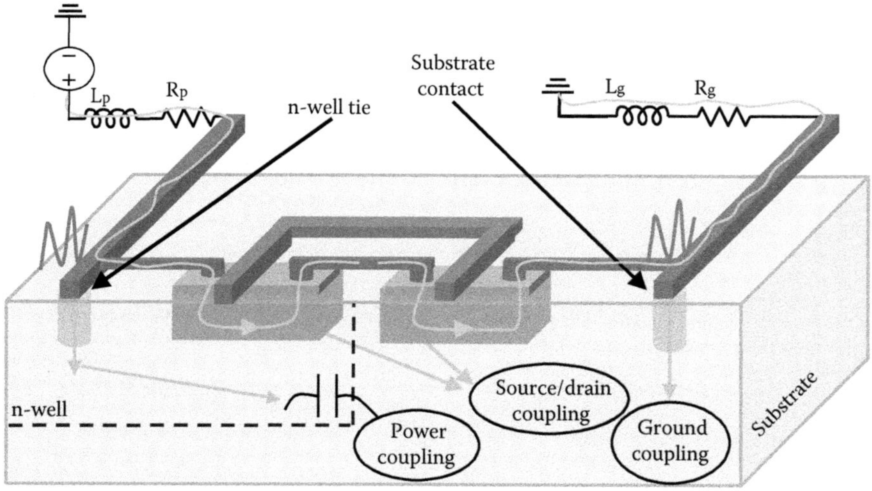

FIGURE 2.2
Substrate noise injection mechanisms.

injection mechanism into the substrate. Thus, any method that reduces the power supply noise has a positive impact on substrate noise coupling.

2.2.1 Substrate Noise Analysis and Modeling

The analysis of switching noise and substrate coupling is challenging due to prohibitive computational complexity. Accurate estimation of the substrate noise at the bulk node of an analog/RF transistor requires simultaneous consideration of the digital switching activity, power/ground networks, and the substrate network [11]. Since the substrate can have deleterious impact on the operation of sensitive blocks in mixed-signal circuits, it is highly desirable to include substrate noise analysis within the conventional design flow, as shown in Figure 2.3.

Significant effort has been made to characterize and efficiently analyze substrate noise coupling. For example, in [12,13], experimental circuits have been used to model substrate noise in a mixed-signal environment and several isolation strategies have been introduced. In [14], a simplified equivalent circuit has been developed to model switching noise and its effects on analog-to-digital converter and voltage-controlled oscillator. Specific logic gates have been used in [15] to model and detect switching noise in digital circuits. In [16], a voltage comparator has been designed as a noise detector to measure the equivalent substrate noise waveforms. The uniformity of the voltages on the ground distribution network is exploited in [3] to efficiently analyze substrate noise coupling.

In general, existing methods consist of two primary approaches: (1) high-level substrate noise analysis and (2) substrate modeling techniques. Chip-level analysis techniques using SPICE are typically not feasible due to complexity requirements, as depicted in Figure 2.4.

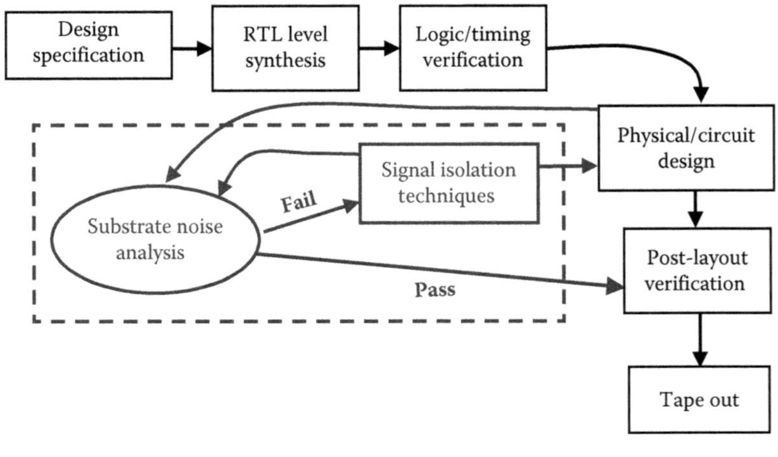

FIGURE 2.3
Integration of substrate noise analysis into a conventional design flow for mixed-signal integrated circuits.

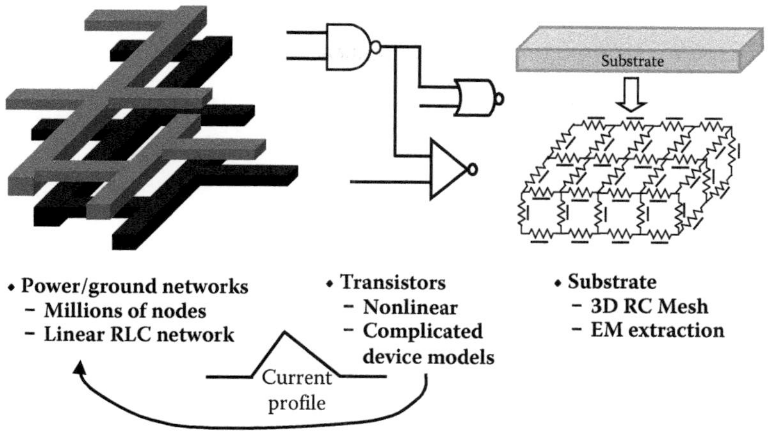

FIGURE 2.4
Complexity requirements of large scale networks consisting of power/ground networks, nonlinear devices, and substrate represented as a 3D *RC* mesh.

2.2.1.1 High-Level Simulation Methods

A schematic based analysis methodology has been proposed in [17] to reduce the number of elements obtained from the post-layout extraction process. Substrate impedances are obtained through compact models and back annotated to the schematic. The transistor level simulation of a large scale circuit including the back annotation of the substrate resistance of every port, however, is not feasible for large scale circuits due to the nonlinear nature of the device models.

A methodology is proposed in [18] for accurately estimating the switching current drawn by a digital block. Two different techniques are introduced: an input pattern dependent scheme for high accuracy and a pattern independent scheme for high computational efficiency. Current profiles are used to analyze the substrate noise. Efficient modeling of the substrate network of a large scale circuit, however, remains as the primary issue.

A high-level simulation methodology is provided in [19] by generating a linear macro model for each standard cell in the circuit, as depicted in Figure 2.5. The switching gates are represented with current sources. The proposed approach is, however, challenging, particularly for bulk type substrates, where the substrate cannot be represented by a single equipotential node. Shorting all of the substrate contacts to a single node, as suggested in [19], is not a valid approach for cases where power/ground noise (and therefore the substrate noise) exhibits large spatial variation.

A methodology is described in [3] based on extracting only those regions of the substrate where there is significant current flow. The interaction between the ground and substrate networks is exploited to significantly reduce the computational complexity of the substrate extraction process. Specifically, small

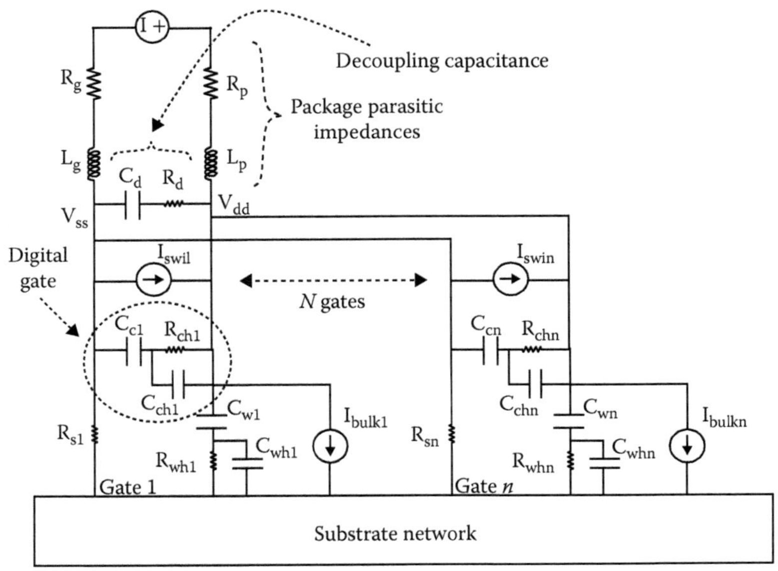

FIGURE 2.5
High-level simulation methodology that generates a linear macro model for each standard cell in the circuit.

spatial voltage differences along the ground network are utilized to determine *voltage domains* on the substrate. Note that these voltage domains on the substrate have approximately the same voltage. As illustrated in Figure 2.6, these regions of the substrate are short-circuited by the ground network; therefore, no major current flow exists in these domains. Since these regions are short-circuited, a coarse extraction is sufficient for these regions [20]. This coarse extraction is achieved by reducing the number of substrate ports to only one. Note that a fine extraction is performed for those regions where there is significant current flow. The proposed heterogeneous extraction of the substrate significantly reduces the computational complexity while maintaining a reasonable accuracy [21].

To evaluate the error in the estimated noise voltage and the improvement in computational complexity, the methodology is compared with a full extraction of the substrate achieved by a commercial tool. An aggressor digital core located close to a sensitive block in an industrial transceiver circuit with a bulk type substrate is used for the analysis. Full extraction of the substrate using the vendor tool produces 312,096 resistors and requires approximately six hours to complete. Alternatively, the proposed heterogeneous extraction methodology reduces the number of substrate resistances to 15 under the same computation environment, achieving more than four orders of magnitude reduction and requires negligible time to perform. The accuracy of the methodology in estimating the substrate noise voltage is shown in Figure 2.7. As illustrated in this figure, the peak noise voltage

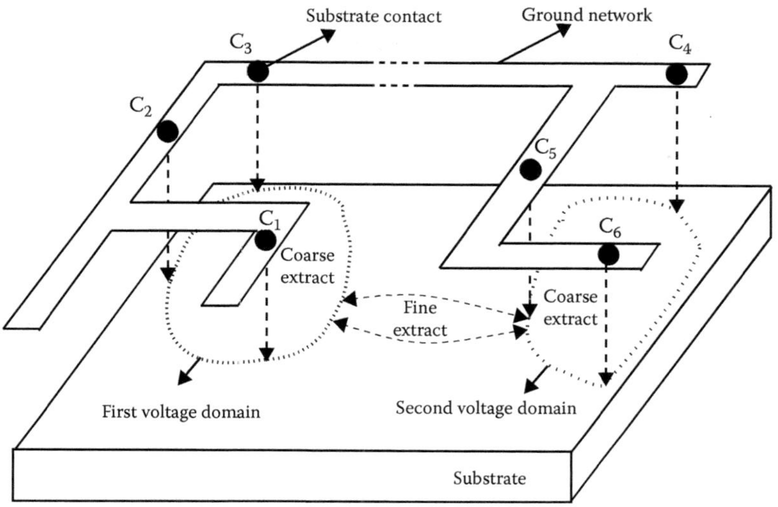

FIGURE 2.6
Identifying voltage domains on the substrate. C_1, C_2, and C_3 produce the first domain assuming $V_{C1} \approx V_{C2} \approx V_{C3}$. Similarly, C_4, C_5, and C_6 produce the second domain assuming $V_{C4} \approx V_{C5} \approx V_{C6}$.

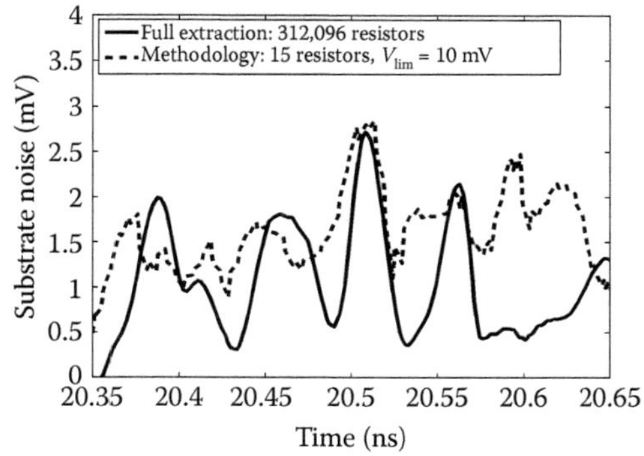

FIGURE 2.7
Comparison of the transient substrate noise at the victim node by simulating the fully extracted circuit and application of the methodology.

is accurately estimated with an error less than 5%. The overall agreement between the two waveforms is also reasonable where the rms error over several clock cycles is less than 10%. Note that even though the error is relatively higher at specific time instances, the methodology is still significantly useful due to the great improvement in computational efficiency. The limitation of the methodology in terms of run time is the requirement to pre-characterize each cell in the library for various input switching patterns and to perform a gate level simulation of the circuit to extract the required timing information.

2.2.1.2 Substrate Modeling

Current approaches to model the substrate can be divided into two classes. The first class includes those techniques that discretize the substrate into a 3D *R*(*C*) mesh to determine the impedances, such as the finite difference method (FDM) [22,23] and the boundary element method (BEM) [24,25]. Neglecting magnetic field effects on the substrate, a simplified Maxwell's equation can be derived as

$$\frac{1}{\rho}\Delta \bullet E + \in \frac{\partial}{\partial t}(\Delta \bullet E) = 0 \tag{2.1}$$

where ρ and $\in$ represent, respectively, the sheet resistivity and the permittivity of the semiconductor, and *E* is the electric field.

Equation 2.1 can discretize the substrate volume in *differential* form using FDM, resulting in a huge, sparse matrix. Although the non-uniformities distributed throughout the substrate can be included using FDM [26], the overall accuracy is a strong function of the resolution of the discretization process, making the extraction of bulk-type substrates challenging [27].

Alternatively, Equation 2.1 can be discretized in *integral* form using BEM with an appropriate Green's Function [28]. For BEM, the size of the resulting matrix is significantly smaller, yet highly dense, as compared to FDM, since BEM only discretizes the ports into the substrate. As such, BEM does not consider the non-uniformity of the substrate, such as channel stop implants.

Several different techniques have been proposed to obtain a more efficient solution of the algebraic equations produced by FDM or BEM to reduce an *RC* network, such as moment matching techniques [29,30], a fast Fourier transform algorithm [25], a fast eigendecomposition technique [26], a numerically stable Green Function [31], and a combination of BEM and FEM techniques [32]. The primary limitation of these approaches (FDM and BEM), however, is the increase in computational complexity with the size of the circuit, prohibiting the efficient analysis of large scale mixed-signal circuits.

The second class of substrate modeling methods is the use of macromodels to represent the impedance between two ports on a substrate [33–35]. Although computationally more efficient, as compared to FDM and BEM, only limited accuracy can be achieved. Other limitations of these macromodels are the requirement to use process-dependent fitting parameters obtained through empirical data and the need to scale these models for smaller geometries.

2.2.2 Substrate Noise Reduction

Various approaches exist to reduce switching noise or alleviate the effect of switching noise on sensitive circuits. These approaches can be classified under three primary categories [5]: (1) to reduce the input noise magnitude of the circuit, (2) to modify the noise transfer medium, and (3) to reduce the sensitivity of the analog/RF circuit to substrate coupling noise.

Since power supply noise is a primary source of substrate noise, any design technique that mitigates power/ground noise also helps in reducing the substrate coupling noise. These methods fall under the first category and include reducing the parasitic inductance (through, e.g., flip-chip packages with low impedance C4 bumps), using separate power/ground networks for analog and digital circuits (a standard design method for mixed-signal ICs), using decoupling capacitors to reduce power supply noise, on-chip voltage regulators [36,37], and more radical design choices, such as asynchronous circuit design methodology [38]. Skew and slew rate control have also been investigated to reduce, respectively, the peak supply current and change of rate in supply current, thereby lowering *Ldi/dt* noise [1].

Typical examples for the second category include utilizing guard rings, a higher resistivity substrate, physical separation, and triple-well isolation. Physical separation is a simple, but effective method for lightly-doped substrates that exhibit a relatively uniform current profile throughout the substrate. Alternatively, for Epi-type substrates where a high resistivity Epi layer is deposited over the low resistivity bulk, physical distance is only partially effective in reducing substrate coupling noise. More specifically, peak noise observed at a victim node is reduced as the distance between the victim and aggressor is increased until a critical distance is reached. This critical distance is characterized as four times the thickness of the Epi layer [12]. Thus, beyond this critical distance, physical separation has negligible effect on reducing the substrate coupling noise.

Higher resistivity substrate (also referred to as p-well blockage) and triple-Well (also referred to as deep n-well) isolation are process dependent techniques to mitigate substrate noise [39]. The use of triple-Well isolation is illustrated in Figure 2.8. Both the aggressor and victim are placed within

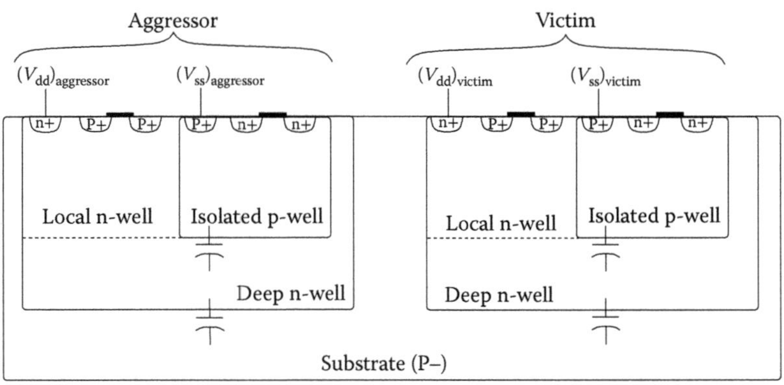

FIGURE 2.8
Triple-Well (deep n-well) isolation where the aggressor and victim are surrounded with a deep n-well and the nMOS devices are fabricated within isolated p-wells.

deep n-wells. The nMOS devices are fabricated within isolated p-wells. Thus, additional capacitive isolation (effective at low to medium frequencies) exists between the noisy substrate and sensitive devices. A useful design practice is to divide a large, deep n-well into smaller sections to reduce the junction capacitances, thereby enhancing the capacitive isolation. Thus, each victim block should have a separate deep n-well [40].

Finally, guard rings represent one of the most common substrate noise mitigation techniques within the second category [41,42]. In most of the practical situations, either the aggressor or the victim blocks are surrounded by a guard ring, as depicted in Figure 2.9, where a ring is placed around the perimeter of the aggressor. The primary objective of a guard ring is to filter substrate noise by ensuring a low resistive path to ground. A guard ring consists of a metal line with a large number of substrate contacts connected to a low impedance ground network. As illustrated in Figure 2.9a, part of the current injected into the substrate is picked by the guard ring. Guard rings, however, cannot fully prevent substrate noise from propagating. Part of the current can bypass the contacts within the guard ring and can

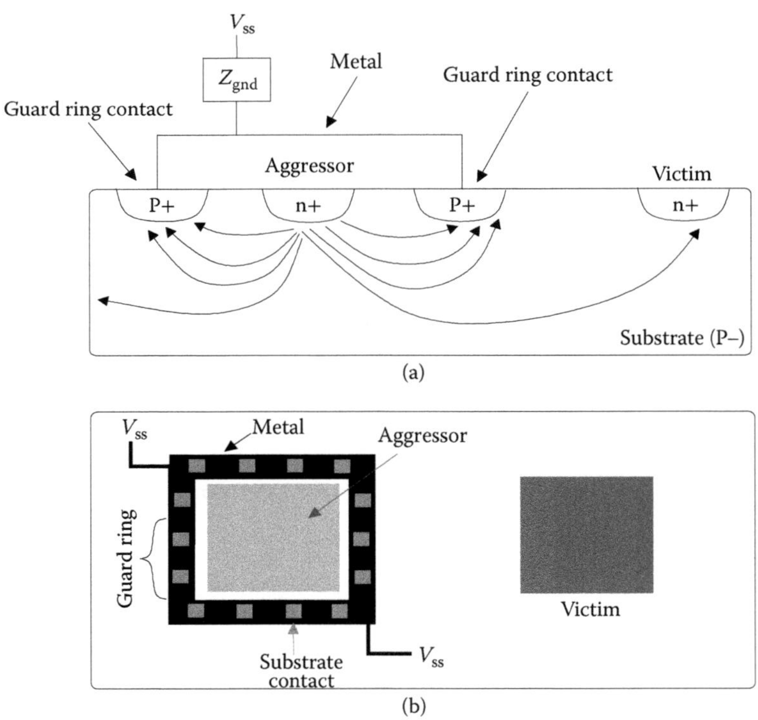

FIGURE 2.9
Guard rings (placed around an aggressor) providing a low impedance path for the injected substrate noise current, (a) cross-section view illustrating the current flow and (b) top view illustrating the metal layer and contacts.

reach the victim device. To minimize this portion of the substrate current, the guard ring and the ground network it is connected to should have the least impedance. For example, if the guard ring is connected to the ground network of the noisy digital circuit, significant noise can be injected into the substrate through contacts. To avoid this situation, guard rings typically possess a separate ground network where ground bounce is minimized.

A common method to reduce the sensitivity of the analog/RF circuit to substrate noise (last category) is differential signaling. A differential analog circuit achieves higher common mode and power supply rejection ratios, thereby alleviating the effects of substrate noise. Note that this design technique is highly dependent on the symmetry of the circuit's physical layout, since noise behaves as a common mode only if both pairs are fully symmetric.

2.3 Figures-of-Merit to Characterize the Significance of Substrate Coupling Noise

Analog/RF circuits suffer not only from induced substrate noise, but also from intrinsic device noise such as thermal [43], flicker [44], and shot noise [45].Traditional analog design flows typically focus on intrinsic device noise, since extensive analysis and simulation methods exist [46,47]. Alternatively, the analysis of switching noise and substrate coupling can be challenging due to prohibitive computational complexity, as described in the previous section. Accurate estimation of the substrate noise at the bulk node of an analog/RF transistor requires simultaneous consideration of the digital switching activity, power/ground networks, and the substrate network [11].

For analog design flows, it is highly important to evaluate the significance of substrate coupling noise and understand the conditions under which substrate noise starts to be dominant over device noise. This evaluation is important since conventional noise mitigation techniques (such as guard rings, deep n-well, slew/skew control, and power network optimization) typically aim at reducing switching noise amplitude at the expense of area and power consumption. It is, therefore, critical to determine the required reduction in switching noise amplitude to minimize the overhead.

2.3.1 Concept of Input Referred Switching Noise

Input-referred switching noise was introduced as a figure-of-merit to determine the significance of induced noise [4]. Specifically, input-referred switching noise can be compared with equivalent input device noise to identify dominance regions as a function of time domain switching noise characteristics, such as the peak amplitude, period, oscillation frequency within each period, and damping coefficient. Time domain peak amplitude that leads to equal input-referred switching and device noise in the frequency domain is characterized and used as the second figure-of-merit. This analysis flow is summarized in Figure 2.10.

The time domain substrate noise profile at the bulk node of the victim transistors is modeled as a decaying sine wave, as shown in Figure 2.11.

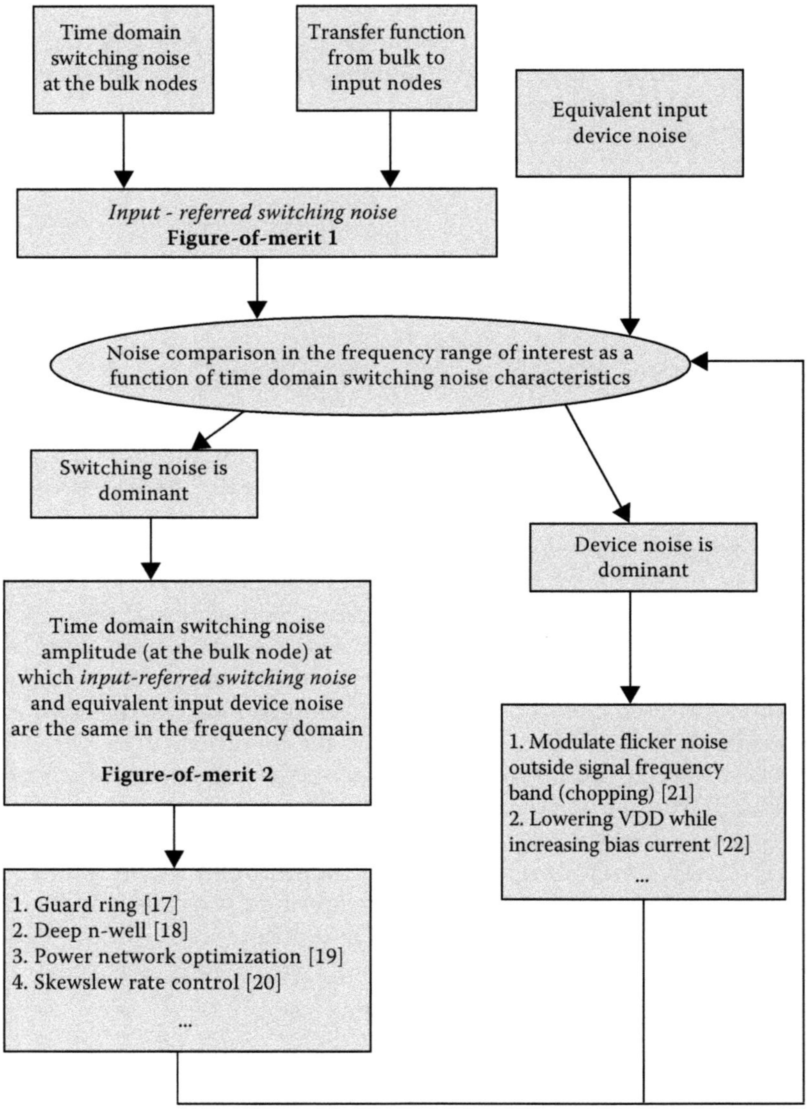

FIGURE 2.10
Analysis flow and the concept of *input-referred switching noise* to determine the significance of induced substrate noise in analog circuits.

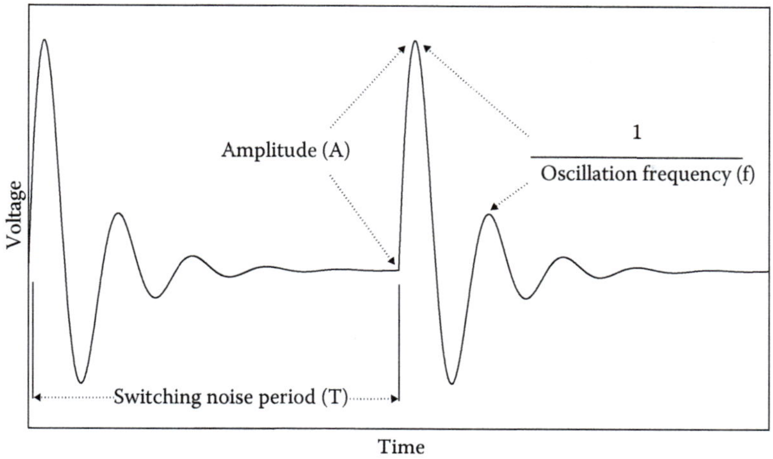

FIGURE 2.11
Characteristics of a decaying sine wave used to model time domain substrate noise at the bulk node of a transistor.

Some of the important characteristics of the substrate noise profile include the amplitude, noise period, and oscillation frequency. Existing techniques can also be utilized to estimate the substrate noise profile for a specific circuit (see the previous section). Corresponding frequency dependent transfer functions are used to transfer substrate noise from the bulk node to the input node of the circuit. This noise is referred to as input-referred switching noise and used as a primary figure-of-merit to determine the significance of substrate noise. By comparing the input-referred switching noise with the equivalent input device noise, the dominant noise source in the frequency range of interest is determined. Note that the noise analysis at the input eliminates the effect of gain, providing a more fair comparison framework.

Once the dominant noise is determined, applicable noise reduction techniques can be applied. Furthermore, time domain substrate noise amplitude (at the bulk node) that leads to equal switching and device noise at the input node (in the frequency domain) is characterized and used as a guideline, while applying substrate noise reduction techniques. As an example, this analysis flow is applied to a two-stage amplifier with a DC gain of 72 dB. The peak substrate noise at the bulk terminals of the transistors within the amplifier is 30 mV (1% of the V_{DD}), which is comparable to the measured substrate noise in [13] and [48]. Input-referred device (thermal and flicker) and substrate noise are compared to identify the dominant noise source as a function of multiple parameters [49].

These noise sources are plotted as a function of frequency in Figure 2.12. According to Figure 2.12, substrate noise is dominant at low frequencies (until approximately 1 kHz) despite high flicker noise. Also note that at DC, input-referred switching noise is 30 dB higher than equivalent input device noise.

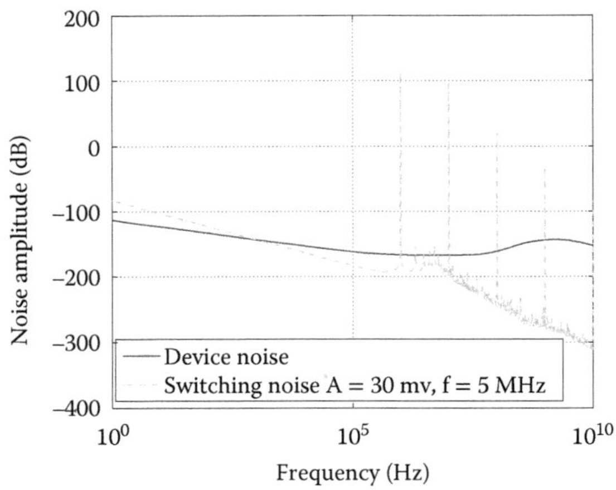

FIGURE 2.12
Comparison of input-referred switching noise and equivalent input device noise for a two-stage amplifier.

Both noise sources decrease with increasing frequency, but substrate noise decreases at a faster pace. Thus, device noise starts to be dominant after the crossover point, except for the fundamental and harmonic frequencies that are determined by the period of the substrate noise (modeled as a decaying sine wave) in time domain. Note that the dominant noise analysis illustrated in Figure 2.12 is not significantly affected by temperature variations since flicker noise dominates thermal noise at low frequencies. Alternatively, at high frequencies, device noise proportionally increases with temperature due to thermal noise. This increase (approximately 5 dB at 10 MHz when temperature rises from 27°C to 165°C), however, is negligible since the overall noise is sufficiently low.

To better investigate the effect of the substrate noise period, the frequency domain dominance region is illustrated in Figure 2.13 for the two-stage amplifier. The y axis represents the frequency, while the switching noise period varies from 1 μs to 10 μs (x axis). The peak amplitude of the noise is 30 mV. The black line with square markers and dotted black lines represent the operating points where equivalent input device noise and input-referred switching noise are equal. Substrate noise is dominant in the shaded region; whereas, the blank region represents the operating points where device noise is dominant.

As the period of the switching noise increases in the time domain, the dominance region of the substrate noise is reduced. Note that at constant period, substrate and device noise become equal at multiple frequencies, as indicated by the black line with square markers and the black dots at higher frequencies.

These crossover points can also be observed in Figure 2.12 at a constant period of 1 μs. In the immediate vicinity of the black dots, substrate noise

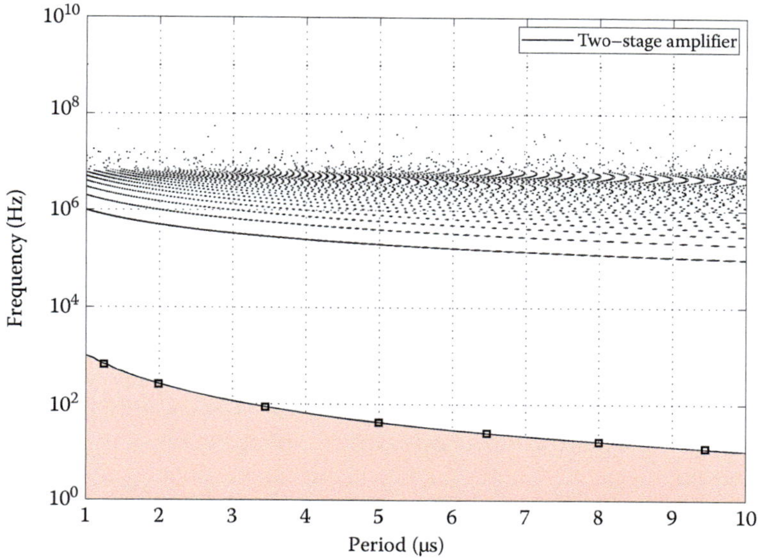

FIGURE 2.13
Noise dominance regions for the two-stage amplifier.

dominates due to fundamental and harmonic frequencies. Thus, if the amplifier bandwidth of interest coincides with these points, substrate noise significantly affects circuit operation. Alternatively, in the blank region, emphasis should be placed on reducing device noise.

The noise dominance region illustrated in Figure 2.13 is obtained at a constant substrate noise amplitude (A = 30 mV) in the time domain. Existing noise mitigation techniques typically aim at reducing this amplitude. It is however difficult to determine an acceptable level of substrate noise since no reliable figures-of-merit exist. Thus, the concept of input referred switching noise can be utilized to numerically solve for the peak amplitude A of substrate noise (in the time domain) that makes the input-referred switching noise and equivalent input device noise the same in the frequency domain.

For example, when the period of the switching noise is 1 µs at 100 Hz, the peak substrate noise amplitude in the two-stage amplifier should be 9.5 mV to satisfy equal input-referred switching and device noise. Note that if the period of the switching noise is equal to 10 µs, the fundamental frequency is 100 kHz. At this frequency, the substrate noise amplitude that produces equal input-referred switching and device noise is sufficiently small since the effect of switching noise is very strong at the fundamental frequency. This figure-of-merit provides a guideline on the acceptable level of substrate noise, assuming that the time domain characteristics of the substrate noise (at the victim bulk node) are known.

2.3.2 Reverse Body Biasing to Alleviate Substrate Noise

As mentioned previously, existing substrate noise mitigation techniques (such as guard rings, deep n-well, and power network optimization) typically focus on reducing the peak noise amplitude. An alternative approach is to reduce the magnitude of the transfer function from a bulk node (where substrate noise is present) to the input node of a victim circuit, thereby reducing input-referred switching noise even though the substrate noise at the bulk node remains the same. This method falls under the third substrate noise reduction category described in Section 2.2.2. Note that in low voltage operational transconductance amplifiers, the bulk node has been properly biased and utilized as the input node, since threshold voltage has not scaled proportionally with the power supply voltage [50,51]. Alternatively, in this method, the bulk node is reverse biased to reduce noise coupling from the bulk node (where substrate noise is present) to the input node.

Decreasing bulk transconductance g_{mb}, while maintaining constant transconductance g_m, reduces the magnitude of the transfer function, thereby alleviating the effect of substrate noise at the input of the two-stage amplifier.

According to the following expression, bulk transconductance g_{mb} of the input nMOS transistors can be decreased by applying reverse body bias to these transistors [52],

$$g_{mb} = g_m \frac{\gamma}{2\sqrt{2\phi + V_{SB}}}, \tag{2.2}$$

where γ is the body effect coefficient and ϕ is the surface potential. When body bias decreases, the threshold voltage increases and the current flowing through input nMOS transistors decreases due to a higher threshold voltage. Less current causes the source voltage of input transistors to decrease. However, the change in the source voltage is relatively small as compared to the body bias. Thus, V_{SB} increases and g_{mb} decreases. Since both V_{GS} and threshold voltages simultaneously increase, the g_m of the input nMOS transistor remains approximately constant, which is highly important for maintaining the primary design objectives of the amplifiers.

The effect of reverse body bias on g_m and g_{mb} is illustrated in Figure 2.14 for the two-stage common source amplifier. As the body bias changes from 0 to -3 V (up to -V_{DD},) g_m changes only from 1.73 mA/V to 1.601 mA/V (7.5% reduction); whereas, g_{mb} decreases from 403.8 µA/V to 135.8 µA/V (more than 66% reduction). Thus, as shown in Figure 2.14b, the ratio g_{mb}/g_m significantly decreases as the body bias changes from 0 to –3 V, thereby weakening the bulk-to-input transfer function and reducing input-referred switching noise.

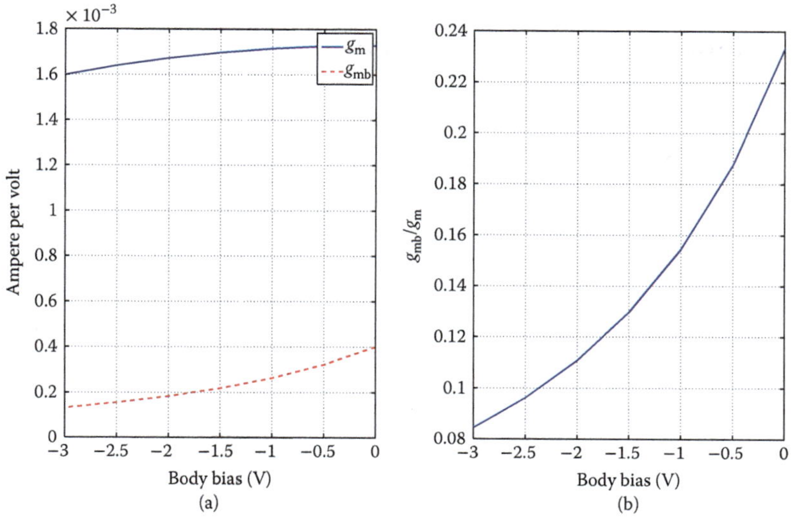

FIGURE 2.14
Effect of reverse body bias on (a) transconductance and bulk transconductance and (b) the ratio of bulk transconductance to transconductance for two-stage common source amplifier.

TABLE 2.1

The Effect of Reverse Body Biasing on Primary Design Objectives for Both Amplifiers

Two-Stage Common Source Amplifier					
V_B (V)	I_D (μA)	DC $_{gain}$ (dB)	g_{mb}/g_m	V_{GS} (V)	V_{th} (V)
0	89.58	72.3	0.23	941.6 m	807.6 m
−0.8	87.73	73.4	0.17	1.072	941.1 m
−1	87.23	73.2	0.15	1.099	969 m
−2	84.23	68.6	0.11	1.213	1.085
−3	79.75	59.1	0.08	1.298	1.172

The effect of reverse body biasing on primary design objectives is listed in Table 2.1 for a two-stage common source amplifier. DC gain, output swing, phase margin, and bandwidth remain approximately the same until a body bias of −1 V. Note that the increase in the threshold voltage is compensated by an increase in the gate-to-source voltage (due to a reduction in the drain current). The peak substrate noise amplitude that satisfies having equal noise at the input increases with body biasing. For example, when the period of the substrate noise is 1 μs at 1 kHz the peak amplitude increases from 30 mV to 42.1 mV for a two-stage amplifier, indicating that approximately 40% more switching noise can be tolerated thanks to reverse body biasing.

2.4 Implications of 3D Technology on Substrate Noise Coupling

Through-silicon via (TSV) based 3D technology exhibits both opportunities and challenges for managing substrate induced signal integrity. Heterogeneous integration is widely considered as one of the most significant advantages of vertical integration [53]. Since each plane on a 3D chip has a separate substrate, partitioning aggressor and victim blocks into distinct planes alleviates the issue of substrate coupling. Several practical issues, however, limit this advantage, as discussed in this section.

Despite the high integration density achieved by TSV based 3D ICs, the physical size of the TSVs have not scaled as fast as the devices [54]. For example, the diameter of practical TSVs is in the several micrometers range; whereas, modern transistors have dimensions in the range of 10–16 nm. From an electrical point of view, this large difference (orders of magnitude) in physical size produces TSV capacitance (tens of fF) into substrate that is much greater than device related capacitances (sub fF). Furthermore, TSVs transmit signals having rail-to-rail voltages, possibly with fast transitions. As a result of these factors, significant noise can couple into the substrate from the TSVs, producing non-negligible shift in threshold voltages [55,56]. Due to this electrical impact and mechanical stress, TSVs typically have a specific keep-out zone where device placement is avoided [57]. This keep-out zone, however, can be insufficient for sensitive analog/RF circuits, since the noise current that couples into the substrate can propagate throughout the plane.

A case study was performed to better understand TSV related substrate noise coupling in a two-plane 3D IC [58]. A counter is used as an aggressor circuit while a sense amplifier is used as the analog victim block. The aggressor and victim are placed on separate planes. Specifically, the top plane that is closer to the I/O pads is dedicated to the analog/RF circuitry to minimize the impedances between package and analog devices. Alternatively, the aggressor can be placed within the bottom plane. It is assumed that the two planes are bonded with a face-to-face bonding technology. Highly distributed 3D electrical models based on the transmission line matrix method are utilized to analyze signal integrity. Specifically, both the substrate and TSVs are discretized using unit cells consisting of *RLC* impedances and an entire electrical model is generated in both 2D and 3D technologies. A conceptual representation of the electrical model for the 3D technology is depicted in Figure 2.15. The bulk nodes in the schematic are connected to the corresponding nodes on the substrate. Two separate substrates exist: (1) the substrate of the bottom plane where the bulks of the digital aggressor are connected and (2) the substrate of the upper plane where the bulks of the victim sense amplifier are connected. TSVs pass through the upper (analog) substrate and reach the metal layers of the analog plane. The top most metal layer of the analog plane is connected to the top most metal layer of the digital plane

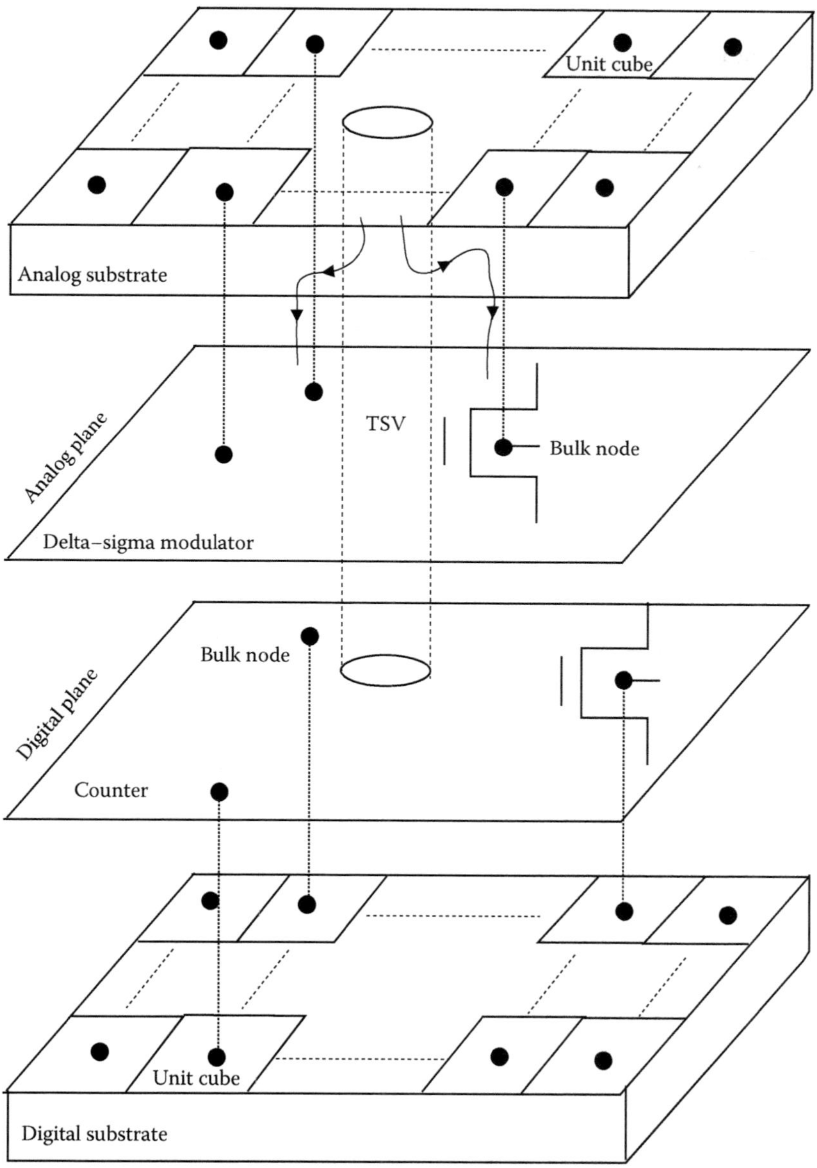

FIGURE 2.15
Conceptual representation of the highly distributed electrical model used to analyze signal integrity in a two-plane 3D IC.

using bumps. There are eight TSVs: five for the clock signals (each at 1 MHz), two for power supply voltage (3 Volts), and one for data signal.

Signal integrity results demonstrate that the peak substrate noise at the bulk node of the victim device exceeds 6 mV for the 3D technology; whereas for 2D technology, the substrate noise is approximately 3 mV. If TSVs are assumed to be ideal with zero capacitance to substrate, the peak substrate noise at the victim device is below 2 mV. This analysis demonstrates the significant noise that couples into the substrate from the TSVs. The same conclusion is validated by observing the rms noise over a long period of time. The rms noise is reduced by more than 50% only if the TSVs are ideal, that is, TSV-to-substrate noise coupling is prevented. Thus, having separate substrates for aggressor and victim blocks can be an important advantage in heterogeneous 3D ICs, provided that TSVs are sufficiently shielded from the victim devices.

2.5 Summary and Conclusions

An overview of substrate noise coupling was provided with emphasis on high-level simulation methods, computationally efficient compact models, and primary techniques to mitigate substrate noise. Figures-of-merit were described to quantify the significance of substrate noise in analog circuits. Specifically, the concept of input-referred switching noise was discussed and compared with input-referred equivalent device noise (thermal and flicker) to identify the dominance regions as a function of multiple parameters. From this comparison, peak substrate noise (at the bulk of a victim transistor) that leads to equal input-referred switching and device noise can be determined, providing useful guidelines for substrate noise reduction methodologies. This analysis is important since it is typically difficult to determine the required degree of noise isolation techniques in mixed-signal circuits. Reverse body biasing was also proposed as an effective method to weaken the bulk-to-input transfer function, thereby reducing the effect of substrate noise without significant changes in primary performance characteristics. Finally, the implications of TSV based 3D technology on substrate noise coupling were discussed. Specifically, the noise that couples into substrate due to large TSV capacitance was demonstrated. Separating aggressor and victim blocks into different planes is an effective design strategy only if sufficient shielding is achieved to protect victim blocks from TSV related substrate noise coupling.

References

1. E. Salman and E. G. Friedman, *High Performance Integrated Circuit Design*, 1st edition, New York, NY: McGraw-Hill, September 2012.
2. E. Salman, "Switching Noise and Timing Characteristics in Nanoscale Integrated Circuits," PhD Thesis, University of Rochester, 2009.

3. E. Salman, R. Jakushokas, E. G. Friedman, R. M. Secareanu, and O. L. Hartin, Methodology for efficient substrate noise analysis in large scale mixed-signal circuits, *IEEE Transactions on Very Large Scale Integration (VLSI) Systems*, vol. 17, no. 10, 1405–1418, October 2009.
4. Z. Gan, E. Salman, and M. Stanacevic, Figures-of-merit to evaluate the significance of switching noise in analog circuits, *IEEE Transactions on Very Large Scale Integration (VLSI) Systems*, vol. 23, no. 12, 2945–2956, December 2015.
5. A. Afzali-Kusha, M. Nagata, N. K. Verghese, and D. J. Allstot, Substrate noise coupling in SoC design: Modeling, avoidance, and validation, *Proceedings of the IEEE*, vol. 94, no. 12, 2109–2138, December 2006.
6. S. Donnay and G. Gielen, *Substrate Noise Coupling in Mixed-Signal ASICs*, Berlin: Springer, 2003.
7. E. Salman, E. G. Friedman, R. M. Secareanu, and O. L. Hartin, Identification of dominant noise source and parameter sensitivity for substrate coupling, *IEEE Transactions on Very Large Scale Integration (VLSI) Systems*, vol. 17, no. 10, 1559–1564, October 2009.
8. H. Wang and E. Salman, Closed-form expressions for I/O simultaneous switching noise revisited, *IEEE Transactions on Very Large Scale Integration (VLSI) Systems*, vol. 25, no. 2, 769–773, February 2017.
9. E. Salman, E. G. Friedman, R. M. Secareanu, and O. L. Hartin, Worst case power/ground noise estimation using an equivalent transition time for resonance, *IEEE Transactions on Circuits and Systems I: Regular Papers*, vol. 56, no. 5, 997–1004, May 2009.
10. E. Salman, E. G. Friedman, R. M. Secareanu, and O. L. Hartin, "Dominant Substrate Noise Coupling Mechanism for Multiple Switching Gates," in *Proceedings of the IEEE International Symposium on Quality Electronic Design*, Santa Clara, CA, USA, pp. 261–266, March 2008.
11. E. Salman, E. G. Friedman, and R. M. Secareanu, "Substrate and Ground Noise Interactions in Mixed-Signal Circuits," in *Proceedings of the IEEE International SoC Conference*, Austin, TX, USA, pp. 293–296, September 2006.
12. D. K. Su, M. J. Loinaz, S. Masui, and B. A. Wooley, Experimental results and modeling techniques for substrate noise in mixed-signal integrated circuits, *IEEE Journal of Solid-State Circuits*, vol. 28, no. 4, 420–430, April 1993.
13. M. Xu, D. K. Su, D. K. Shaeffer, T. H. Lee, and B. A. Wooley, Measuring and modeling the effects of substrate noise on the LNA for a CMOS GPS receiver, *IEEE Journal of Solid-State Circuits*, vol. 36, no. 3, 473–485, March 2001.
14. G. Boselli, G. Trucco, and V. Liberali, "Effects of Digital Switching Noise on Analog Circuits Performance," in *Proceedings of the IEEE European Conference on Circuit Theory and Design*, Sevilla, Spain, pp. 160–163, August 2007.
15. M. Van Heijningen, J. Compiet, P. Wambacq, S. Donnay, M. Engels, and I. Bolsens, "Modeling of Digital Substrate Noise Generation and Experimental Verification Using a Novel Substrate Noise Sensor," in *Proceedings of the IEEE European Solid-State Circuits Conference*, Duisburg, Germany, pp. 186–189, September 1999.
16. K. Makie-Fukuda, T. Anbo, T. Tsukada, T. Matsuura, and M. Hotta, Voltage-comparator-based measurement of equivalently sampled substrate noise waveforms in mixed-signal integrated circuits, *IEEE Journal of Solid-State Circuits*, vol. 31, no. 5, 726–731, May 1996.

17. P. Birrer, S. K. Arunachalam, M. Held, K. Mayaram, and T. S. Fiez, Schematic-driven substrate noise coupling analysis in mixed-signal IC designs, *IEEE Transactions on Circuits and Systems I: Regular Papers*, vol. 53, no. 12, 2578–2587, December 2006.
18. S. M. Reddy and R. Murgai, "Accurate Substrate Noise Analysis Based on Library Module Characterization," in *Proceedings of the IEEE International Conference on VLSI Design*, Hyderabad, India, January 2006.
19. M. Badaroglu, G. Van der Plas, P. Wambacq, S. Donnay, G. G. E. Gielen, and H. J. De Man, SWAN: High-level simulation methodology for digital substrate noise generation, *IEEE Transactions on Very Large Scale Integration (VLSI) Systems*, vol. 14, no. 1, 23–33, January 2006.
20. E. Salman, R. Jakushokas, E. G. Friedman, R. M. Secareanu, and O L. Hartin, "Input Port Reduction for Efficient Substrate Extraction in Large Scale IC's," in *Proceedings of the IEEE International Symposium on Circuits and Systems*, Seattle, WA, USA, pp. 376–379, May 2008.
21. E. Salman, R. Jakushokas, E. G. Friedman, R. M. Secareanu, and O L. Hartin, "Contact Merging Algorithm for Efficient Substrate Noise Analysis in Large Scale Circuits," *Proceedings of the ACM/IEEE Great Lakes Symposium on VLSI*, Boston, MA, USA, pp. 9–14, May 2009.
22. K. Fukahori and P. R. Gray, Computer simulation of integrated circuits in the presence of electrothermal interaction, *IEEE Journal of Solid-State Circuits*, vol. 11, no. 6, 834–846, December 1976.
23. B. R. Stanisic, N. K. Verghese, R. A. Rutenbar, R. Carley, and D. J. Allstot, Addressing substrate coupling in mixed-mode IC's: Simulation and power distribution synthesis, *IEEE Journal of Solid-State Circuits*, vol. 29, no. 3, 226–238, March 1994.
24. T. Smedes, N. P. van der Meijs, and A. J. van Genderen, Boundary element methods for 3D capacitance and substrate resistance calculations in inhomogeneous media in a VLSI layout verification package, *Advances Engineering Software*, vol. 20, no. 1, 19–27, 1994.
25. R. Gharpurey and R. G. Meyer, Modeling and analysis of substrate coupling in integrated circuits, *IEEE Journal of Solid-State Circuits*, vol. 31, no. 3, 344–352, March 1996.
26. J. P. Costa, M. Chou, and L. M. Silveria, Efficient techniques for accurate modeling and simulation of substrate coupling in mixed-signal IC's, *IEEE Transactions on Computer-Aided Design of Integrated Circuits and Systems*, vol. 18, 1597–607, May 1999.
27. N. K. Verghese, D. J. Allstot, and M. A. Wolfe, Verification techniques for substrate coupling and their application to mixed-signal IC design, *IEEE Journal of Solid-State Circuits*, vol. 31, no. 3, 354–365, March 1996.
28. T. Smedes, N. P. van der Meijs, and A. J. van Genderen, "Extraction of Circuit Models for Substrate Crosstalk," in *Proceedings of the IEEE/ACM International Conference on Computer-Aided Design*, San Jose, CA, USA, pp. 199–206, November 1995.
29. N. K. Verghese, D. J. Allstot, and S. Masui, "Rapid Simulation of Substrate Coupling Effects in Mixed-Mode ICs," in *Proceedings of the IEEE Custom Integrated Circuits Conference*, San Diego, CA, USA, pp. 18.3.1–18.3.4, May 1993.
30. L. T. Pillage and R. A. Rohrer, Asymptotic waveform evaluation for timing analysis, *IEEE Transactions on Computer-Aided Design of Integrated Circuits and Systems*, vol. 9, no. 4, 352–366, April 1990.

31. A. M. Niknejad, R. Gharpurey, and R. G. Meyer, Numerically stable green function for modeling and analysis of substrate coupling in integrated circuits, *IEEE Transactions on Computer-Aided Design of Integrated Circuits and Systems*, vol. 17, no. 4, 305–315, April 1998.
32. E. Schrik and N. P. van der Meijs, "Combined BEM/FEM Substrate Resistance Modeling," in *Proceedings of the IEEE/ACM Design Automation Conference*, New Orleans, LA, USA, pp. 771–776, June 2002.
33. D. Ozis, T. Fiez, and K. Mayaram, "A Comprehensive Geometry-dependent Macromodel for Substrate Noise Coupling in Heavily Doped CMOS Processes," in *Proceedings of the IEEE Custom Integrated Circuits Conference*, Orlando, FL, USA, pp. 497–500, September 2002.
34. H. Lan, T. W. Chen, C. O. Chui, P. Nikaeen, J. W. Kim, and R. W. Dutton, Synthesized compact models and experimental verifications for substrate noise coupling in mixed-signal ICs, *IEEE Journal of Solid-State Circuits*, vol. 41, no. 8, 1817–1829, August 2006.
35. R. Jakushokas, E. Salman, E. G. Friedman, R. M. Secareanu, O. L. Hartin, and C. Recker, "Compact Substrate Models for Efficient Noise Coupling and Signal Isolation Analysis," in *Proceedings of the IEEE International Symposium on Circuits and Systems*, Paris, France, pp. 2346–2349, May/June 2010.
36. Y. Park and E. Salman, "On-Chip Hybrid Regulator Topology for Portable SoCs with Near-Threshold Operation," in *Proceedings of the IEEE International Symposium on Circuits and Systems*, Montreal, Canada, pp. 786–789, May 2016.
37. C. Yan, Z. Gan, and E. Salman, "In-Package Spiral Inductor Characterization for High Efficiency Buck Converters," in *Proceedings of the IEEE International Symposium on Circuits and Systems*, Baltimore, MD, USA, pp. 2396--2399, May 2017.
38. J. Sparso, *Asynchronous Circuit Design: A Tutorial*, Technical University of Denmark, Lyngby, Denmark, 2006.
39. K. W. Chew, J. Zhang, K. Shao, W. Boon Loh, and S.F. Chu, "Impact of Deep N-well Implantation on Substrate Noise Coupling and RF Transistor Performance for Systems-on-a-Chip Integration," in *Proceedings of the IEEE European Solid-State Device Research Conference*, Firenze, Italy, pp. 251–254, September 2002.
40. A. Gothenberg, E. Soenen, and H. Tenhunen, "Modeling and Analysis of Substrate Coupled Noise in Pipelined Data Converters," in *Proceedings of the Southwest Symposium on Mixed-Signal Design*, San Diego, CA, USA, pp. 125–130, February 2000.
41. E. Salman and E. G. Friedman, "Methodology for Placing Localized Guard Rings to Reduce Substrate Noise in Mixed-Signal Circuits," in *Proceedings of the IEEE Conference on PhD Research on Microelectronics and Electronics*, Istanbul, Turkey, pp. 85–88, June 2008.
42. E. Salman, E. G. Friedman, R. M. Secareanu, and O L. Hartin, "Substrate Noise Reduction Based on Noise Aware Cell Design," in *Proceedings of the IEEE International Symposium on Circuits and Systems*, New Orleans, LA, USA, pp. 3227–3230, May 2007.
43. A. van der Ziel, Thermal noise in field-effect transistors, *Proceedings of the IRE*, vol. 50, no. 8, 1808–1812, August 1962.
44. Y. Nemirovsky, I. Brouk, and C. G. Jakobson, 1/f noise in CMOS transistors for analog applications, *IEEE Transactions on Electron Devices*, vol. 48, no. 5, 921–927, May 2001.

45. A. van der Ziel, Theory of shot noise in junction diodes and junction transistors, *Proceedings of the IRE*, vol. 43, no. 11, 1639–1646, November 1955.
46. W. Leach Jr., Fundamentals of low-noise analog circuit design, *Proceedings of the IEEE*, vol. 82, 1515–1538, 1994.
47. F. Bizzarri, A. Brambilla, and G. Gajani, Steady state computation and noise analysis of analog mixed signal circuits, *IEEE Transactions on Circuits and Systems*, vol. 59, 541–554, 2012.
48. M. Badaroglu, S. Donnay, H. J. De Man, Y. A. Zinzius, G. G. E. Gielen, W. Sansen, T. Fonden, and S. Signell, Modeling and experimental verification of substrate noise generation in a 220-K gates WLAN system-on-chip with multiple supplies, *IEEE Journal of Solid-State Circuits*, vol. 38, no. 7, 1250–1260, July 2003.
49. Z. Gan, E. Salman, and M. Stanacevic, "Methodology to Determine Dominant Noise Source in a System-on-Chip Based Implantable Device," in *Proceedings of the IEEE International SoC Conference*, Buffalo, NY, USA, pp. 115–119, September 2012.
50. J. Rosenfeld, M. Kozak, and E. G. Friedman, "A Bulk-Driven CMOS OTA with 68 dB DC Gain," in *Proceedings of the IEEE International Conference on Electronics, Circuits and Systems*, Tel Aviv, Israel, pp. 5–8, December 2004.
51. Y. Haga, H. Zare-Hoseini, L. Berkovi, and I. Kale, "Design of a 0.8 Volt Fully Differential CMOS OTA Using the Bulk-Driven Technique," in *IEEE International Symposium on Circuits and Systems*, Kobe, Japan, pp. 220–223, May 2005.
52. B. Razavi, *Design of Analog CMOS Integrated Circuits*, New York, NY: McGraw-Hill, 2001.
53. V. F. Pavlidis and E. G. Friedman, *Three-Dimensional Integrated Circuit Design*, Burlington, MA: Morgan Kaufmann, 2009.
54. C. Yan, S. Kontak, H. Wang, and E. Salman, "Open Source Cell Library Mono3D to Develop Large-Scale Monolithic 3D Integrated Circuits," in *Proceedings of the IEEE International Symposium on Circuits and Systems*, Baltimore, MD, USA, pp. 2581–2584, May 2017.
55. H. Wang, M. H. Asgari, and E. Salman, Compact model to efficiently characterize TSV-to-transistor noise coupling in 3D ICs, *Integration, the VLSI Journal*, vol. 47, no. 3, 296–306, June 2014.
56. E. Salman, "Noise Coupling Due to Through Silicon Vias (TSVs) in 3-D Integrated Circuits," in *Proceedings of the IEEE International Symposium on Circuits and Systems*, Rio de Janeiro, Brazil, pp. 1411–1414, May 2011.
57. H. Wang, M. H. Asgari, and E. Salman, "Efficient Characterization of TSV-to-Transistor Noise Coupling in 3D ICs," in *Proceedings of the ACM/IEEE Great Lakes Symposium on VLSI*, Paris, France, pp. 71–76, May 2013.
58. E. Salman, M. H. Asgari, and M. Stanacevic, "Signal Integrity Analysis of a 2-D and 3-D Integrated Potentiostat for Neurotransmitter Sensing," in *Proceedings of the IEEE Biomedical Circuits and Systems Conference*, San Diego, CA, USA, pp. 17–20, November 2011.

3

TSV-to-Substrate Noise Coupling in 3D Systems

Boris Vaisband and Eby G. Friedman

CONTENTS

3.1 Introduction

A three-dimensional (3D) structure is an effective platform for integrating heterogeneous circuits within a single system, as illustrated in Figure 3.1. Each layer of a 3D integrated circuit (IC) is typically independently fabricated using different substrate materials for different applications [1]. The 3D structure is, therefore, a natural platform for modern diverse applications including high performance computing and mobile and wearable devices.

Noise coupling is a key issue in integrated circuits [2,3]. This issue is aggravated in 3D circuits since multiple layers (each with an individual substrate) are placed above each other, allowing signals to propagate throughout the 3D structure. Through substrate vias (TSVs) connect the different layers within the 3D system, carrying power, clock, and data. TSVs, a seminal component of 3D technology, are short vertical interconnections between the different

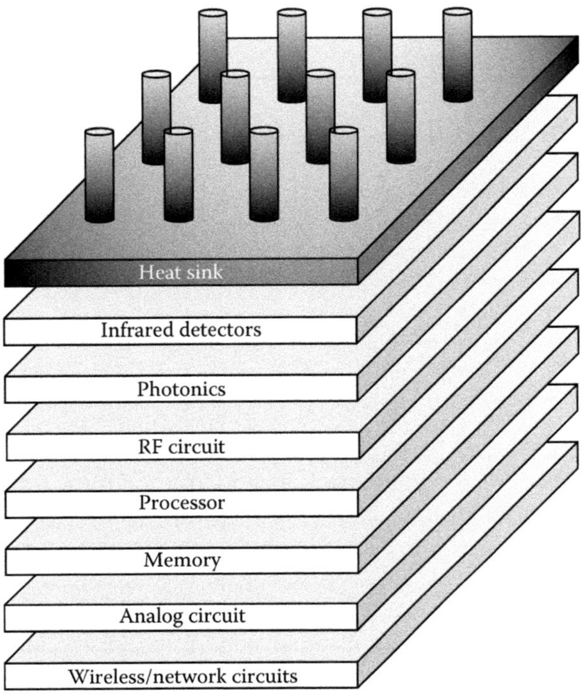

FIGURE 3.1
Heterogeneous 3D integrated circuit.

layers that can support global signaling requirements [1]. The TSVs penetrate the substrate of a layer and connect to either the first or last metal within that layer [4]. Typical TSV dimensions are 20 µm in length and 2 µm in diameter [5]. Similar to two-dimensional (2D) substrate coupling, the signals within the TSVs can couple capacitive and inductive noise into the substrate, affecting nearby victim circuits or other signals within a TSV.

The rest of the chapter is composed of the following sections. Heterogeneous substrate and circuit characteristics are reviewed in Section 3.2. Substrate noise models are offered in Section 3.3. The proposed models are evaluated in Section 3.4. Noise mitigation techniques are reviewed in Section 3.5. Some conclusions are offered in Section 3.6.

3.2 Heterogeneous Substrate and Related Circuit Characteristics

Previous work has addressed noise coupling from TSVs to the substrate in homogeneous circuits (processor/memory stacks), typically on a silicon substrate [6,7]. The purpose of the work described in this chapter is to consider

TABLE 3.1

Common Circuits and Compatible Substrate Types

Circuits	Substrate Material	Electrical Resistivity Ω·cm
Processor/memory	Silicon (Si)	1–10
RF/analog	Gallium arsenide (GaAs)	$4 \cdot 10^7$
Photonics	Germanium (Ge)	$1.3 \cdot 10^{-3}$
Space applications/detectors	Mercury cadmium telluride (HgCdTe)	2

coupling noise in heterogeneous 3D systems composed of different substrate materials.

The electrical resistivity and thermal conductivity of different commonly used substrate materials are listed in Table 3.1. Important examples of these commonly used materials in modern integrated circuits are silicon (Si), gallium arsenide (GaAs), germanium (Ge), and mercury cadmium (MerCad) telluride (HgCdTe) [8–10]. The electrical resistivity of the substrate materials is a key issue in noise coupling. Therefore, due to the wide range of resistivities, as listed in Table 3.1, an individual noise coupling model for each of the substrate materials is necessary.

Each of the substrate materials listed in Table 3.1 is appropriate for a specific circuit application. Si is typically lower cost and more mature as compared to the other materials and therefore used for mainstream processor and memory applications. The superior electron mobility of GaAs makes GaAs attractive for high performance analog devices. Ge is a favorable substrate material for photovoltaic and photodetector applications due to the high absorption coefficient of Ge. Military and space applications that require high quality infrared detectors commonly use HgCdTe [11], which has a tunable bandgap ranging from 0.1 eV to 1 eV. This property of HgCdTe supports detection of long wavelength light.

3.3 Substrate Noise Models

Heterogeneous models for Si, Ge, and GaAs are proposed in this section. Each model is based on the electrical properties of the substrate material.

3.3.1 Si Model

The noise coupling model proposed by Salman [6] is shown in Figure 3.2. A distributed *RC* model composed of four sections is used to characterize the TSV impedance and capacitive coupling into the silicon substrate. The substrate is modeled using distributed lateral and vertical resistors. The ground network is modeled as a resistive-inductive (RL) impedance [6]. The resistance of a TSV [1] is

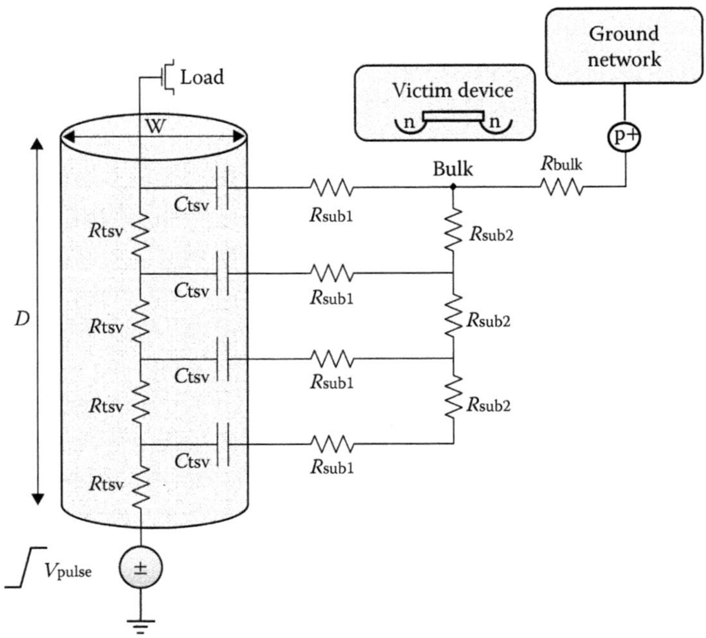

FIGURE 3.2
Distributed model for noise coupling from a TSV to a victim through a silicon substrate. (From E. Salman, "Noise Coupling Due To Through Silicon Vias (TSVs) in 3-D Integrated Circuits," In *Proceedings of the IEEE International Symposium on Circuits and Systems*, pp. 1411–1414, May 2011.)

$$R_{tsv} = (1/N_{tsv}) \cdot \rho_c D / \left(\pi \cdot (W/2)^2\right) \tag{3.1}$$

where the number of distributed sections of the TSV is N_{tsv}, resistivity of the conductive material within the TSV is ρ_c, and depth (length) and diameter of the TSV are, respectively, D and W. With a copper resistivity of 2.8 μΩ·cm [12], depth of 20 μm, and diameter of 2 μm [5], the resistance is 0.18 Ω for $1/N_{tsv} = 1$. This resistance is relatively small as compared to the resistance of a typical digital buffer. A lumped RC model for the TSV is therefore used [2,13], as shown in Figure 3.3. Another important issue is the model of the ground network. The victim device is commonly connected to the ground network through a bulk contact; the inductive behavior of this network, therefore, also has to be considered. Note that the electrical resistivity of MerCad telluride is similar to silicon. The same model shown in Figure 3.2 can therefore be used.

3.3.2 Ge Model

For Ge, a "short-circuit" model (shown in Figure 3.4) is proposed. The resistance of the substrate is negligible and the model, therefore, only exhibits a coupling capacitance from the TSV to the substrate. Ge is particularly

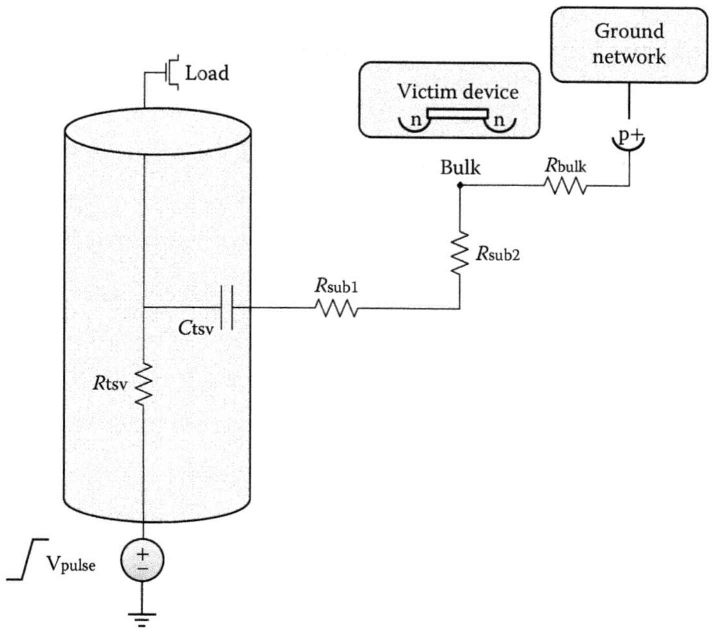

FIGURE 3.3
Lump model of coupling noise from a TSV to a victim through a silicon substrate.

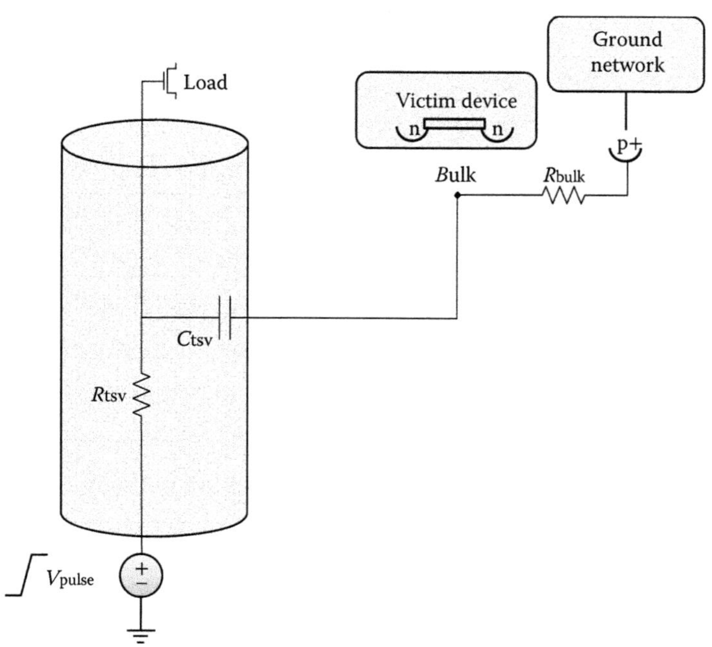

FIGURE 3.4
Coupling noise from a TSV to a victim, short-circuit Ge substrate model.

susceptible to substrate noise coupling due to the low resistivity of the substrate. Special emphasis on noise mitigation techniques is therefore necessary for this substrate material.

3.3.3 GaAs Model

GaAs exhibits significantly higher resistivity as compared to the other evaluated substrates. The proposed model for GaAs is therefore an "open-circuit" model that ignores capacitive coupling, as illustrated in Figure 3.5. This model characterizes a noise coupling system as an electrical open circuit.

3.4 Model Evaluation

The proposed models have been verified in the time domain using COMSOL Multiphysics [14] and SPICE. The finite element method (FEM) evaluation in Section 3.4.1 is used here to validate the proposed models for each substrate material. The validated models are evaluated for different ground network inductances and compared to a distributed model in Section 3.4.2.

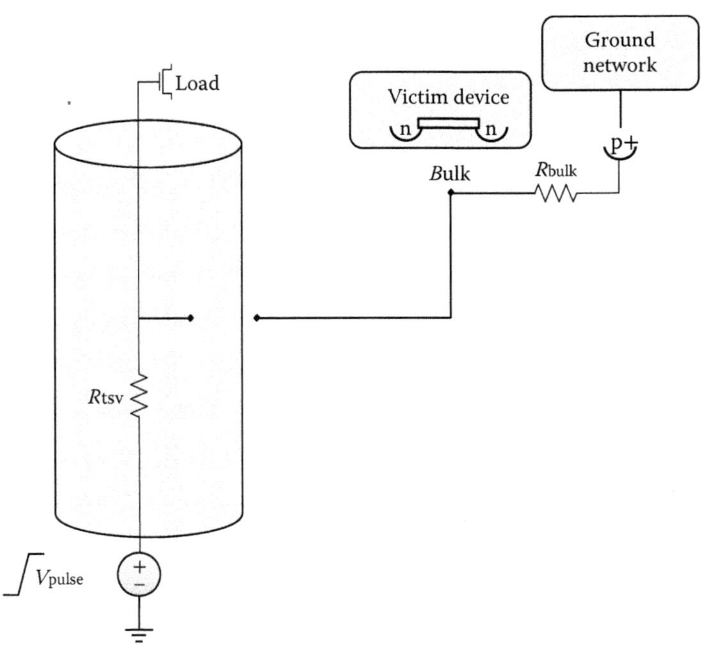

FIGURE 3.5
Coupling noise from a TSV to a victim, open circuit GaAs substrate model.

3.4.1 COMSOL Evaluation

COMSOL is used to validate the proposed heterogeneous models and evaluate the performance characteristics of each model. The physical structure is shown in Figure 3.6. A schematic of the circuit is shown in Figure 3.7. The parameters used for the evaluation are listed in Table 3.2.

The voltage at the victim node, obtained from the numerical evaluation, for each substrate material is shown in Figure 3.8. The peak noise at the victim is 130 mV, 760 mV, and 27 nV for, respectively, Si, Ge, and GaAs. The COMSOL results are compared to the SPICE evaluation presented in the following subsection.

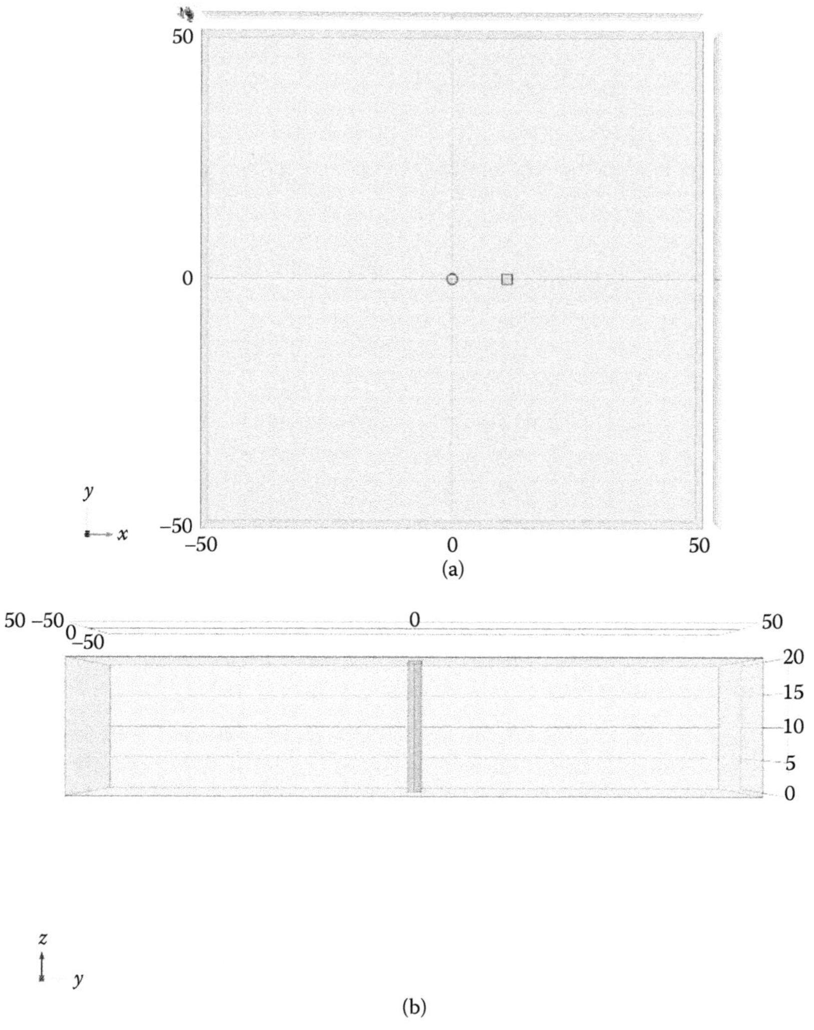

FIGURE 3.6
Structure evaluated in COMSOL. (a) Top, and (b) side views.

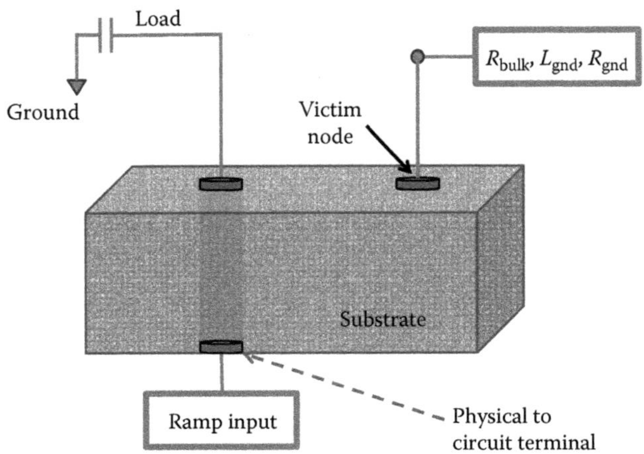

FIGURE 3.7
Physical and circuit structure. The terminals connect the physical and circuit domains.

TABLE 3.2
Parameters for Model Evaluation

Parameter	Value	Description
R_{tsv}	1 μm	TSV radius
H_{tsv}	20 μm	TSV height
V_{pulse}	1 to 0 volts	Ramp input
t_{rise}	10 psec	Input rise time
L	10 μm	Distance between aggressor TSV and victim
R_{gnd}	2 Ω	Ground network resistance
L_{gnd}	1 nH	Ground network inductance

3.4.2 SPICE Evaluation

A comparison of a lumped model versus a distributed model with three sections is listed in Table 3.3 for Si, GaAs, and Ge. The models have been evaluated using SPICE. A 10 psec input ramp (V_{pulse}) is applied to simulate switching of the aggressive digital circuits. The voltage is evaluated at the victim device node. Both the peak noise voltage and settling time (2% of the final value) are recorded for three different ground network inductances.

The error of the lumped model as compared to the distributed model for Si is 1.2%. A lump model can therefore be used to accurately characterize a silicon substrate. As observed from the results listed in Table 3.3, the inductance of the ground network can significantly affect the peak noise voltage. In the worst case (an inductance ranging from 0.1 to 10 nH), a difference of 26.5 mV (14.2% error) is noted.

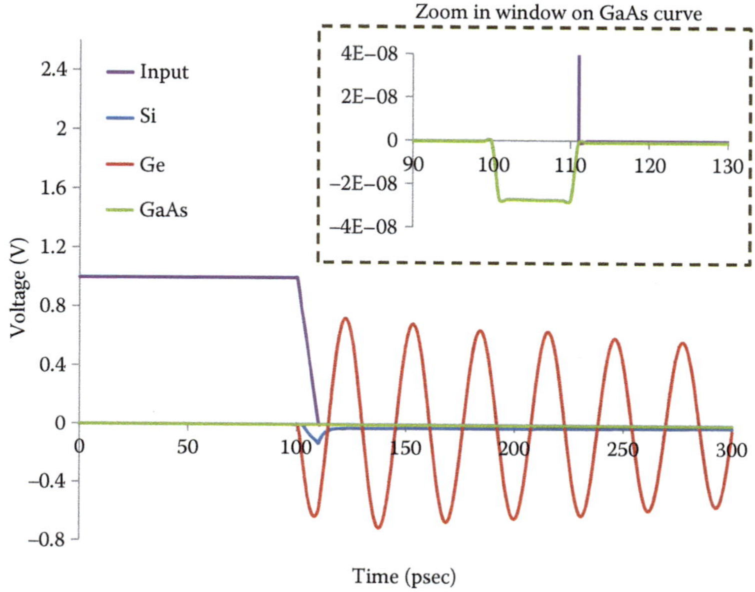

FIGURE 3.8
Voltage at victim node for Si, Ge, and GaAs substrate materials, based on COMSOL.

The peak noise voltage for both a lumped and distributed model of a GaAs substrate is in the range of picovolts and is, therefore, negligible in most applications. It is also observed from Table 3.3 that the inductance of the ground network has no effect on the peak noise voltage. This behavior is due to the resistivity of the substrate, which is sufficiently large to shunt the inductance of the ground network.

Ge is highly dependent on the inductance of the ground network. Comparing lumped and distributed models, a distributed model provides negligible improvement in accuracy as compared to a lumped model. The worst case difference in the peak noise voltage is 0.2 mV (2.3% error), while the settling time is similar. A lump model, which incorporates fewer nodes, is therefore preferable. The short-circuit model deviates from the lump model by 2.6 mV (23.4% error) for a ground network inductance of 0.1 nH, and 2 nsec (25% error) for a ground network inductance of 10 nH, for, respectively, the peak noise voltage and settling time. A lump model, similar to the model for silicon as shown in Figure 3.3, should therefore be used. If the circuit specifications are not particularly strict (a higher peak noise voltage and longer settling times are allowed), a short-circuit model can be used to reduce the computational time.

The peak noise listed in Table 3.3 for each substrate material ($L_{gnd} = 1$ nH) is compared to the noise levels obtained from COMSOL. The percent error for each substrate material is listed in Table 3.4. The difference in error between the models is due to the finer structural model used in COMSOL. The finite element model (FEM) evaluates the structure in all directions. The SPICE

TABLE 3.3

Comparison of Lumped, Distributed, and Short-Circuit Models for Si, GaAs, and Ge Substrates for Different Ground Network Inductances

		Si		Ge		GaAs	
Model	**Ground Inductance (nH)**	**Peak Noise (mV)**	**Settling Time (nsec)**	**Peak Noise (mV)**	**Settling Time (nsec)**	**Peak Noise (mV)**	**Settling Time (nsec)**
Short-circuit	0.1	-	-	11.1	0	-	-
	1	-	-	645.5	1.46	-	-
	10	-	-	954.4	8	-	-
Lumped	0.1	159.8	1.57	8.5	~ 0	$3.8 \cdot 10^{-8}$	~ 0
	1	162.4	1.57	638.5	1.1	$3.8 \cdot 10^{-8}$	~ 0
	10	186.3	1.57	950.8	6.3	$3.8 \cdot 10^{-8}$	~ 0
Distributed	0.1	161.8	1.55	8.7	~ 0	$3.9 \cdot 10^{-8}$	~ 0
(three	1	164.5	1.55	637.5	1.08	$3.9 \cdot 10^{-8}$	~ 0
sections)	10	188.6	1.55	950.1	6.26	$3.9 \cdot 10^{-8}$	~ 0

TABLE 3.4

Comparison of Peak Noise at the Victim Between COMSOL and SPICE for Si, Ge, and GaAs Substrate Materials

Substrate Material	COMSOL	SPICE	Percent Error (%)
Si	134.1 mV	162.4 mV	17.4
Ge	724 mV	638.5 mV	11.8
GaAs	27 nV	38 nV	28.9

model only considers the electrical components between the aggressor and victim. Note that the difference in the peak noise between COMSOL and SPICE, as listed in Table 3.4, is not large (particularly for the GaAs substrate), ranging from 11 nV to 85.5 mV.

3.5 Noise Mitigation Techniques

Reducing the noise coupled through the substrate is important. Several noise mitigation techniques—guard rings, distance between aggressor and victim, and epitaxial (epi) layer within the bulk—are discussed in Sections 3.5.1, 3.5.2, and 3.5.3, respectively. The effectiveness of these two techniques in reducing the peak noise at the victim is evaluated by COMSOL. Note that the noise coupled to the victim in a GaAs substrate is negligible; mitigation techniques for this type of substrate are therefore unnecessary.

3.5.1 Guard Rings

Guard rings are grounded substrate contacts diffused into the bulk substrate [15]. In guard rings, a low impedance path is formed between the aggressor and the guard ring (much less than the impedance between the aggressor and the victim), thereby reducing the flow of charge to the victim. This system is depicted in Figure 3.9. The guard ring surrounds the aggressor TSV and is connected to ground. In the example structure, the width and length of each side are, respectively, 2 μm and 14 μm. The electrical conductivity of the guard ring is 10^4 $(\Omega{\cdot}\text{m})^{-1}$.

A key design parameter for guard rings is the depth of the diffusion. The peak noise at the victim as a function of the depth of the guard ring for Si and Ge substrate materials is depicted in Figure 3.10. The peak noise is inversely proportional to the depth of the guard ring. The reduction in peak noise at the victim due to the guard rings is listed in Table 3.5. Note in Table 3.5 that the guard ring is highly effective in reducing the peak noise to negligible levels in both Si and Ge substrates. Due to the low resistivity of the Ge substrate, the path between the aggressor TSV and the guard ring exhibits significantly lower resistance than the path to the victim. The coupling noise therefore terminates at the guard ring, lowering the peak noise at the victim to approximately 140 μV. The guard ring however requires additional fabrication steps to place a ground contact in the substrate.

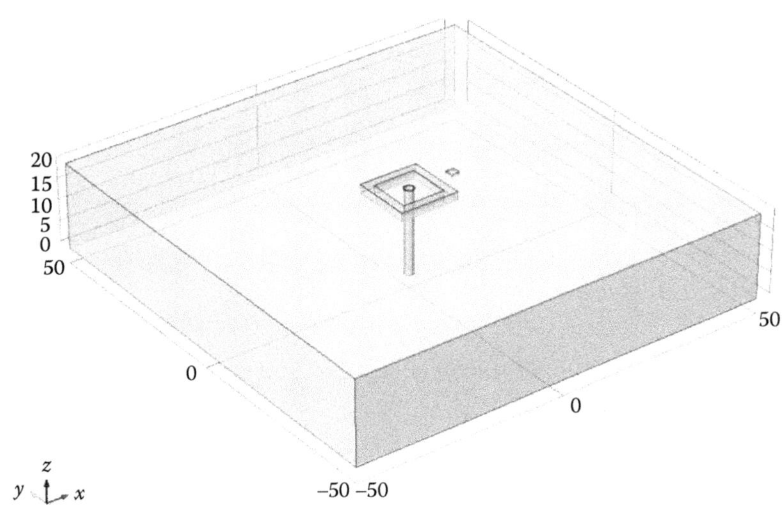

FIGURE 3.9
Physical aggressor-victim structure evaluated in COMSOL, including a guard ring around the aggressor TSV.

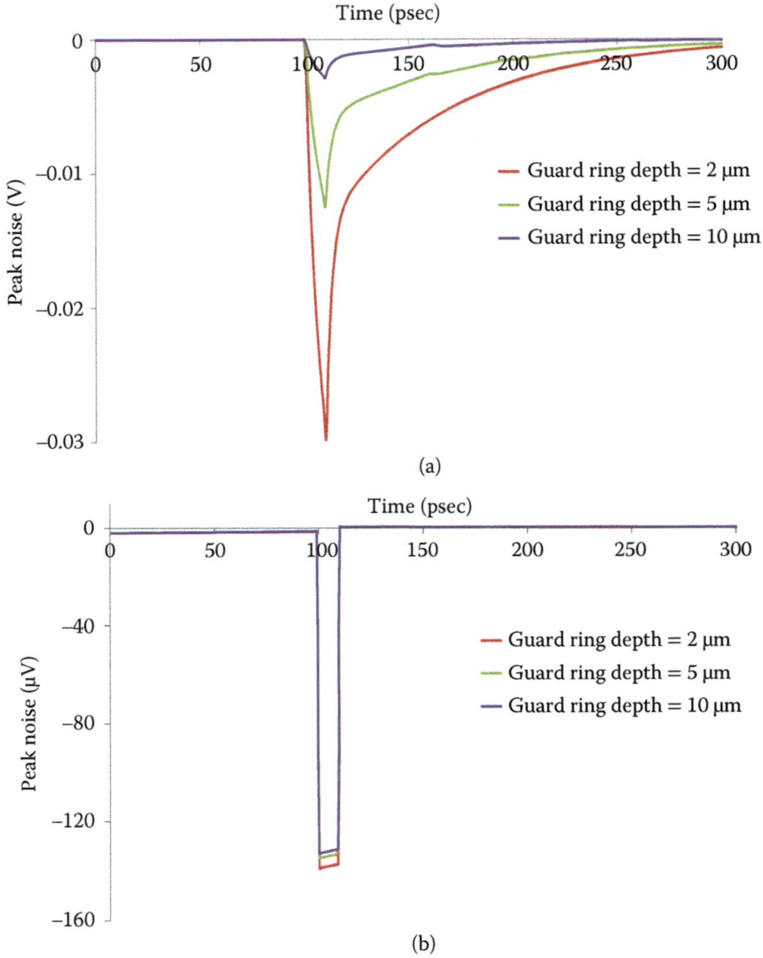

FIGURE 3.10
Peak noise at the victim for different depths of a diffused guard ring, (a) Si substrate, and (b) Ge substrate.

TABLE 3.5
Effectiveness of Guard Rings in Terms of Peak Noise at the Victim for Si and Ge Substrate Materials

	Si Substrate		Ge Substrate	
Guard Ring Depth (μm)	**Peak Noise at Victim Protected by Guard Ring (mV)**	**Percent Reduction of Peak Noise (%)**	**Peak Noise at Victim Protected by Guard Ring (μV)**	**Percent Reduction of Peak Noise (%)**
2	29.7	77.9	138.6	~ 100
5	12.4	90.8	134.5	~ 100
10	2.8	97.9	132.5	~ 100

3.5.2 Distance between Aggressor and Victim

A useful mitigation technique to reduce substrate coupling noise is to increase the distance between the aggressor TSV and the victim circuit [16]. This technique requires a keep out region around the aggressor TSV. Outside the keep out region, the substrate noise is below a pre-set threshold as long as the victim circuits are placed within this region. COMSOL evaluation of the peak noise at the victim for different distances between the aggressor TSV and victim is shown in Figure 3.11. The effectiveness of increasing the distance in terms of the peak noise at the victim as compared to the peak

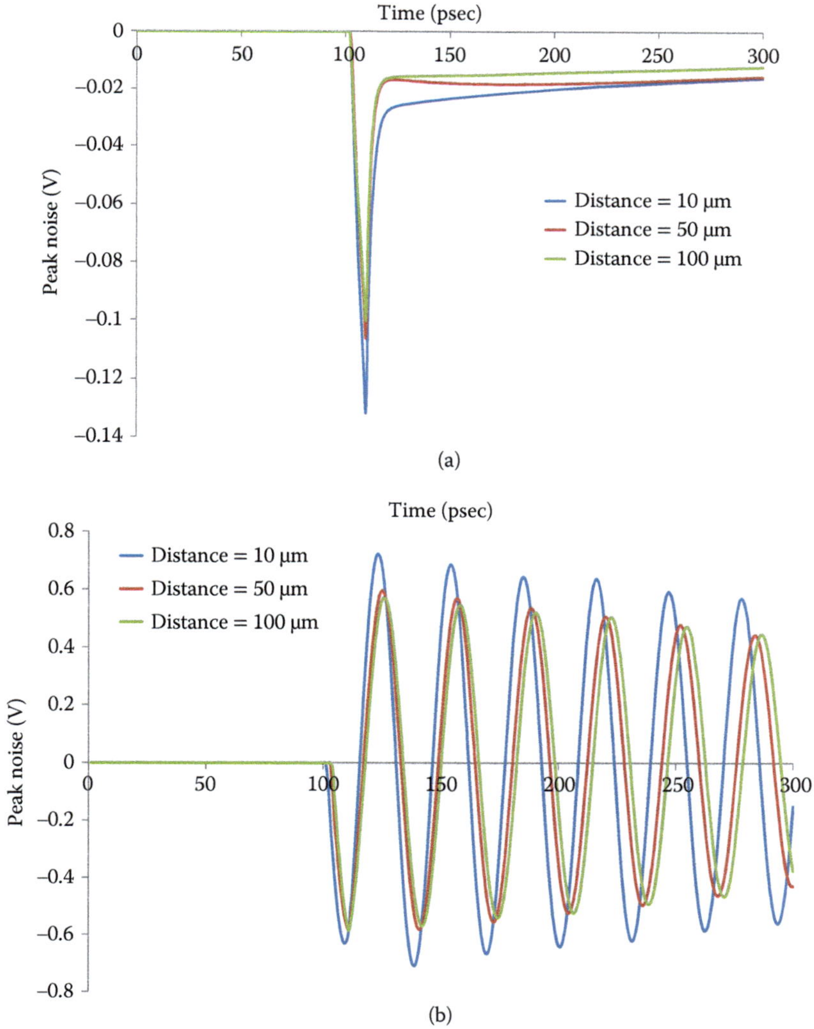

FIGURE 3.11
Peak noise at the victim for increasing distance between the aggressor TSV and victim circuit, (a) Si substrate, and (b) Ge substrate.

noise without applying any noise mitigation techniques (see Table 3.4) is listed in Table 3.6. Increasing the distance between the aggressor and victim is therefore an effective technique to protect the victim circuits, albeit requiring greater area.

3.5.3 Epitaxial Layer

An epitaxial layer is a thin resistive layer typically grown on top of the bulk substrate. The epi layer is lightly doped, exhibiting a resistivity approximately 300 times greater than the resistivity of bulk substrate [17]. One purpose of the epi layer is to force the coupling noise to the lower part of the substrate, which is less resistive, that is, creating a low resistive path farther from the devices fabricated within the epi layer. The coupling noise in an epi substrate is shown in Figure 3.12. Additional fabrication steps are required to form the epi layer, making this technology more expensive.

TABLE 3.6

Effectiveness of Increasing Distance Between the Aggressor and Victim in Terms of the Peak Noise at the Victim for Si and Ge Substrate Materials

Distance between Aggressor and Victim (μm)	Si Substrate		Ge Substrate	
	Peak Noise (mV)	Percent Reduction of Peak Noise (%)	Peak Noise (mV)	Percent Reduction of Peak Noise (%)
50	105.5	21.3	584.4	19.3
100	99.5	25.8	579.5	20

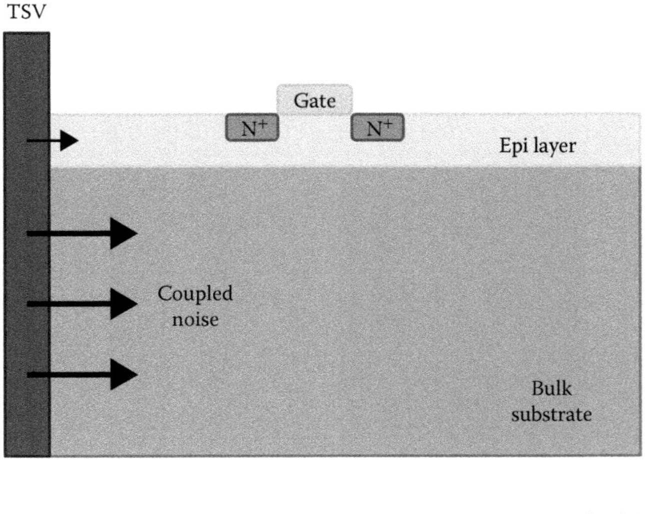

FIGURE 3.12
Noise coupling in epitaxial substrate. (From D. K. Su et al., *IEEE Journalof Solid-State Circuits*, 28, 420–430, April 1993.)

TABLE 3.7
Effectiveness of Epitaxial Layer in Lowering the Peak Noise at the Victim for Si and Ge Substrate Materials

	Si Substrate		Ge Substrate	
Epi Layer Thickness µm	Peak Noise (mV)	Percent Reduction of Peak Noise (%)	Peak Noise (mV)	Percent Reduction of Peak Noise (%)
0.5	41	69.4	391.5	45.9
5	40.6	69.7	91.3	87.4

COMSOL evaluation of the peak noise at the victim within an epi layer is listed in Table 3.7. The epi layer is effective in reducing the peak noise at the victim. For an epi layer with a thickness of 0.5 µm, the peak noise is reduced by 69.4% and 45.9% in, respectively, Si and Ge substrate materials. Increasing the thickness of the epi layer can further reduce the peak noise. For an epi layer with a thickness of 5 µm, the peak noise is reduced by 69.7% and 87.4% in, respectively, Si and Ge substrate materials.

3.6 Conclusions

Substrate noise coupling is a primary concern in TSV-based 3D ICs. TSVs, a seminal component of 3D ICs, create additional noise coupling paths into the substrate of adjacent layers within a 3D structure.

A general noise coupling model for different types of semiconductor materials is insufficient for heterogeneous 3D ICs where each layer often consists of different substrate materials.

Individual noise coupling models are suggested for different common substrate materials. The effectiveness of each noise coupling model is validated using COMSOL. The effects of the ground network inductance on the peak noise at the victim circuit is determined for each model. The effectiveness of different noise mitigation techniques is described only for the Si and Ge substrates, since the GaAs substrate exhibits low coupling noise due to the high resistivity of the substrate. The proposed models for Si and Ge exhibit up to 17.4% error as compared to numerical analysis. The open circuit model for GaAs exhibits 28.9% error as compared to numerical analysis; however, the actual deviation is only 11 nV. The suggested models are therefore useful for noise coupling analysis in heterogeneous 3D ICs.

A shallow guard ring reduces substrate noise by 77.9% and 100%, respectively, in Si and Ge substrates. Increasing the distance between the aggressor TSV and the victim circuit is less effective than using guard rings, providing a reduction of 25.8% and 20% for, respectively, the Si and Ge substrates for a distance of 100 μm. An epi layer grown on top of the bulk substrate can also reduce the coupling noise at the victim. An epitaxial layer reduces the peak noise by up to 69.7% and 87.4% in, respectively, Si and Ge substrate materials.

References

1. V. F. Pavlidis, I. Savidis, and E. G. Friedman, *Three-Dimensional Integrated Circuit Design*, 2nd edition, Burlington, MA; Morgan Kaufmann, 2017.
2. E. Salman and E. G. Friedman, *High Performance Integrated Circuit Design*, New York, NY; McGraw-Hill Publishers, 2012.
3. R. M. Secareanu, S. Warner, S. Seabridge, C. Burke, J. Becerra, T. E. Watrobski, C. Morton, W. Staub, T. Tellier, I. S. Kourtev, and E. G. Friedman, Substrate coupling in digital circuits in mixed-signal smart-power systems, *IEEE Transactions on Very Large Scale Integration (VLSI) Systems*, vol. 12, no. 1, 67–78, January 2004.
4. S. M. Satheesh and E. Salman, Effect of TSV fabrication technology on power distribution in 3D ICs, In *Proceedings of the ACM/IEEE Great Lakes Symposium on VLSI*, pp. 287–292, May 2013.
5. I. Savidis and E. G. Friedman, Closed-form expressions of 3-D via resistance, inductance, and capacitance, *IEEE Transactions on Electron Devices*, vol. 56, no. 9, 1873–1881, September 2009.
6. E. Salman, Noise coupling due to through Silicon Vias (TSVs) in 3-D integrated circuits, In *Proceedings of the IEEE International Symposium on Circuits and Systems*, pp. 1411–1414, May 2011.
7. C. Xu, R. Suaya, and K. Banerjee, Compact modeling and analysis of through-si-via-induced electrical noise coupling in three-dimensional ICs, *IEEE Transactions on Electron Devices*, vol. 58, no. 11, 4024–4034, November 2011.
8. L. Shifren, R. Aitken, A. R. Brown, V. Chandra, C. Binjie, C. Riddet, C. L. Alexander, B. Cline, C. Millar, S. Sinha, G. Yeric, and A. Asenov, Predictive simulation and benchmarking of Si and Ge pMOS FinFETs for future CMOS technology, *IEEE Transactions on Electron Devices*, vol. 61, no. 7, 2271–2277, July 2014.
9. G. K. Dalapati, C. K. Chia, C. Mahata, S. Krishnamoorthy, C. C. Tan, H. R. Tan, C. K. Maiti, and C. Dongzhi, Impact of buffer layer on atomic layer deposited TiAlO alloy dielectric quality for epitaxial-GaAs/Ge device application, *IEEE Transactions on Electron Devices*, vol. 60, no. 1, 192–199, January 2013.
10. A. Rogalski, New trends in infrared and terahertz detectors, In *Proceedings of the IEEE Optoelectronic and Microelectronic Materials Devices Conference*, pp. 218–220, December 2014.
11. A. L. Betz and R. T. Boreiko, Space applications for HgCdTe at FIR wavelengths between 50 and 150 μm, In *Proceedings of the SPIE Materials for Infrared Detectors*, pp. 1–9, November 2001.
12. M. G. Farooq, T. L. Graves-Abe, W. F. Landers, C. Kothandaraman, B. A. Himmel, P. S. Andry, C. K. Tsang, E. Sprogis, R.P. Volant, K.S. Petrarca, K. R. Winstel, J. M. Safran, T. D. Sullivan, F. Chen, M. J. Shapiro, R. Hannon, R. Liptak,

D. Berger, and S. S. Iyer, 3-D copper TSV integration, testing and reliability, In *Proceedings of the IEEE International Electron Devices Meeting*, pp. 7.1.1–7.1.4, December 2011.

13. T. Sakurai, Approximation of wiring delay in MOSFET LSI, *IEEE Journal of Solid-State Circuits*, vol. 18, no. 4, 418–426, August 1983.
14. "COMSOL Multiphysics 5.0," 2013. Available at: http://www.comsol.com/.
15. E. Salman and E. G. Friedman, Methodology for placing localized guard rings to reduce substrate noise in mixed-signal circuits, In *Proceedings of the Fourth Conference on PhD Research on Microelectronics and Electronics*, pp. 85–88, June 2008.
16. B. Vaisband and E. G. Friedman, Noise coupling models in heterogeneous 3-D ICs, *IEEE Transactions on Very Large Scale Integration (VLSI) Systems*, vol. 24, no. 8, 2778–2786, August 2016.
17. D. K. Su, M. J. Loinaz, S. Masui, and B. A. Wooley, Experimental results and modeling techniques for substrate noise in mixed-signal integrated circuits, *IEEE Journal of Solid-State Circuits*, vol. 28, no. 4, 420–430, April 1993.

4

3D Interconnects with IC's Stack Global Electrical Context Consideration

Yue Ma, Olivier Valorge, J. R. Cárdenas-Valdez, Francis Calmon, J. C. Núñez–Pérez, J. Verdier, and Christian Gontrand

CONTENTS

4.1 Introduction

Three-dimensional (3D) Si integration seems like the right way to compete with Moore's law (more than Moore versus more Moore). However, it is still a long way to go.

In 2010, the question was: Why 3D? Today, the questions are: When to use 3D and how to use 3D? The 3D chip stacking is known to overcome conventional 2D-IC issues, using in-depth contacts or through-silicon for signal transmission.

Mixed Signal ICs design has become key for systems-on-chip (SoCs) VLSIs. Functional analog blocks such as LNA, filters, and AD/DA converters, are placed on a die with high speed digital processing elements composed of a few million digital gates. Parasitic phenomenon, commonly called crosstalk, occurs between the noisy digital and the sensitive analog part of the device. Due to many parasitic coupling mechanisms, there is a distinct possibility that the transient regimes in the digital circuitry of this kind of device will corrupt low-level analog signals and seriously compromise the achievable performance [1]. One of the main coupling phenomena is the substrate coupling. The chip substrate acts as a collector, integrator, and distributor of coupled noise on chip. Many developments describe methods to simulate this perturbing kind of noise [2–4]. However, the majority of the models presented only considers the package bonding, whereas many different parts of the supply lines are important in substrate noise generation [5,6]: PCB, package, padring, on chip metal lines, and other parasitic elements of supply lines.

Nowadays, micro/nanotechnology and the development of semiconductor technology enable designers to integrate multiple systems into a single chip, not only in two-dimensional (2D) (planar), but also in 3D (in the bulk). This design technology reduces cost, while improving performance and makes the possibility of a system to be on chip [1–4].

The kind of silicon substrate in which 3D interconnections are processed is an important parameter and has a strong impact on 3D interconnection electrical behavior. Electrical modeling methods are dependent on the substrate type of the application. Two different kinds of P silicon substrates are typically used in CMOS/BiCMOS processes: uniform lightly doped substrate (>1Ω.cm) and heavily doped substrate (<1Ω.cm) with lightly thinned doped (>1Ω.cm) epitaxial layer. Other semiconductor substrates are used in micro/nano technologies and could embed 3D interconnections (SOI technologies, MEMS technologies, sensor technologies).

Advancements in the field of V(U)LSI have led to more compact ICs having higher clock frequencies and lower power consumptions. The paradigms of these integrated technologies are SoCs and Systems-in-Package (SiPs). Now, these conventional 2D planar technologies face several strategic challenges: physical limits, processing complexity, fabrication costs, and so on. Consequently, technological approaches other than scaling are investigated to continue following or get over Moore's Law. Currently, 3D integrated systems can be obtained by stacking vertically 2D integrated circuits using mature, controlled micro and nano technologies. The electrical connections are ensured by new pre- or post-processed metallic structures (Figure 4.1): redistribution layers (RDL), which distribute power and high speed signals on die top or backside surfaces; copper pillars and mostly through-silicon via (TSV) are keys enabling technology for 3D integration, by propagating signals through the silicon layers.

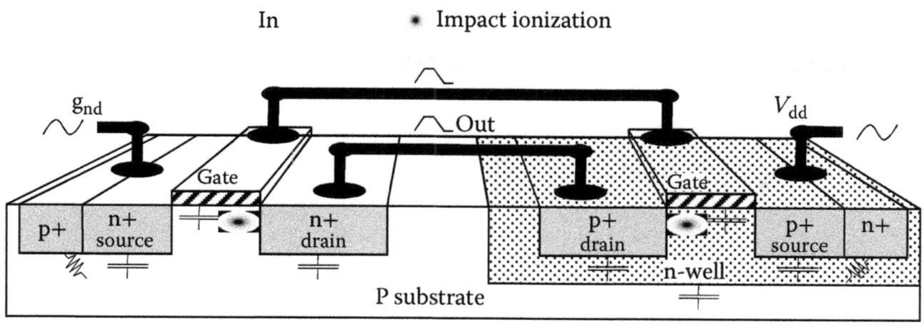

FIGURE 4.1
Schematic diagram showing the main mechanisms of substrate noise generated by a CMOS inverter.

Chips can also be stacked vertically using wire-bonding and flip-chip techniques. However, the flip-chip technique solely provides an interconnection between two chips; whereas, wire-bonding only enables the connection of chip input/output pads located at their perimeter. Moreover, wire-bonding presents disadvantages in terms of surface and propagation delays depending on the application frequency clock and the wire-bond lengths. Finally, solder balls are put on the 3D systems bottom layers backsides to ensure connection with their environment. Compared with 2D classical schemes, 3D integration potential benefits are numerous: performance improvement, flexible heterogeneous system combinations (logic CMOS, RF analog function, memories, and sensors), significantly shortened interconnect line networks that decrease interconnect line delays and power consumption (which become significant obstacles in 2D VLSI systems relative to delays in transistor switching), smaller form factors, and reduced fabrication costs [1,2].

Many challenges are encountered with 3D integration due to its status as an emerging technology, notably in properly characterizing and electrically modeling the 3D interconnects. Few present-day CAD tools can design 3D architectures. Many works have been mainly dedicated to advanced TSV electrical modeling.

Different types of via play a central role in realizing high-density integrated 3D systems. This enables a large variation in shape, dimension (radius, length), dielectric thickness, and filling material. The difficulty in TSV modeling is that the overall 3D interconnection electrical context must be considered. For example, current paths must be modeled if the environment comprises top/back redistribution metal lines. As with local metal layer interconnect line structures, the 3D interconnect global environment requires the modeling of the substrate coupling effects which can no longer be neglected at high frequency. Several substrate extraction techniques have been proposed in the literature but most of them cannot be employed for SoCs with realistic dimensions or are only suitable for a particular type of structure or technology process [7,8].

4.2 Substrate Noise Generators and Propagation

In this section, we describe the different injection mechanisms in a digital device. Most CMOS logic elements can be reduced into simple CMOS inverters. Figure 4.1 shows a CMOS inverter and its different substrate noise sources.

The main substrate noise source is due to the power and ground bounces that are transmitted into the substrate through its biasing contacts [6]. In a CMOS inverter, the NMOS source transmits the ground bounce and the PMOS source transmits the power fluctuations. The power and ground bounces are generated by large transient-current spikes pulled by the simultaneous switching of multiple digital gates. Those spikes and the parasitic elements of the package, PCB, and pad ring can introduce resonant pathways and so power and ground ringing. The drain depletion capacitance of each MOS also transmits voltage variations into the bulk. In the case of a basic inverter, the output voltage is directly coupled to the substrate through the drain depletion capacitance of the NMOS and the PMOS. Additionally, the gate electrode is coupled to the substrate through its oxide and the channel capacitance. Another noise generator is the well-known impact ionization. When a MOS transistor is biased in the saturation regime, a high electric field develops in the depleted region of the channel, near the drain. This high field creates hot carriers that dissipate their energy by generating some electron-hole pairs. For a NMOS, the holes created are swept to the substrate.

Other mechanisms exist and are described in different works: capacitive coupling with signals interconnects, gate induced drain leakage, photon-induced current and diode leakage current.

4.2.1 ICEM Standard Approach and Its Extension

In this part, we introduce a basic model to simulate the substrate voltage. We will consider the power and ground bounces as the unique source of substrate noise. This approach drastically simplifies the modeling methodology and will give, as shown hereafter, correct results. The ICEM describes the electromagnetic behavior of an integrated circuit [6,7]. Two coupling mechanisms of the internal activities for emission are proposed in this model: the conducted emissions or through input/output lines and the direct radiated emissions. The basic architecture of the ICEM model is shown in the Figure 4.2. It consists in two sub-models: a Passive Distribution Network sub-model and an Internal Activity sub-model.

The use of the ICEM allows an accurate modeling of the integrated circuit power and ground bounces [6,7]. To simulate parasitic substrate voltage induced by the supply bounces, a Substrate Network sub-model is added to the classical ICEM (Figure 4.2). The IA sub-model describes the internal

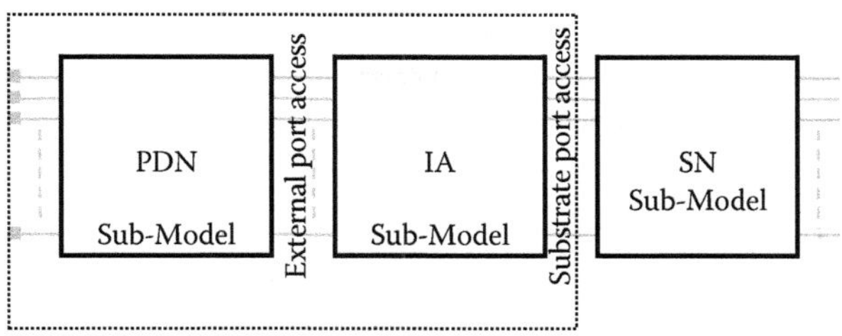

FIGURE 4.2
Basic architecture of extended ICEM.

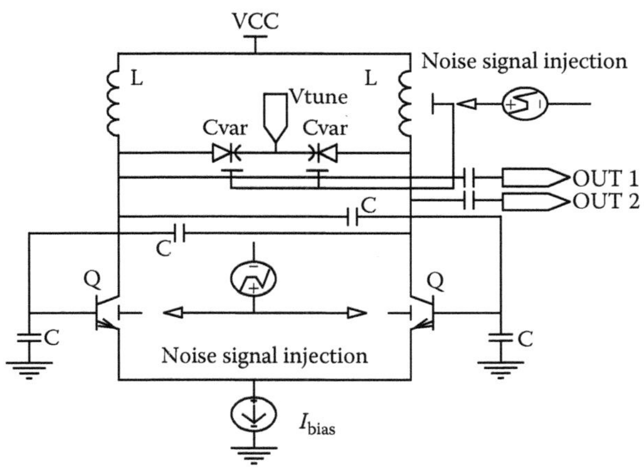

FIGURE 4.3
Simplified VCO core with injection contacts [7].

activity of the digital circuit and is represented by the transient current consumption and by the capacitance of the digital circuitry.

The PDN is the sub-model used to describe the electrical structure for powering the chip from the PCB to the on-chip power metal lines. The added module, SN, is an RC network that models the substrate noise propagation. From the existing ICEM and an additional substrate sub-model, a new methodology of electronics design optimization is shown: substrate noise characterization.

4.2.1.1 Analog Part and Substrate Coupling

The design of mixed signal ICs supposes the resolution of coupling problems between analog and digital blocks placed on the same die. The switching events of digital gates involve variations of supply voltage function in digital signal transitions that are not compensated by decoupling capacitors. Isolation between analog and digital parts with buried layers, guard rings, or separation of the supply nets are methods to reduce coupling mechanisms [7–9].

However, substrate parasitic capacitors and resistors allow the noise signal generated by the digital part to pass through the substrate and to reach the analog and radio frequency (RF) blocks. The literature is quite poor concerning the impact on analog and RF blocks, many papers being focused on noise generation and propagation into digital CMOS circuits [10–12].

This chapter deals with the impact of low frequency substrate noise onto a fully integrated VCO. We have investigated the way the substrate noise (digital noise or injected noise by means of substrate taps inside VCO core) is converted close to the carrier frequency and impacts VCO spectral purity. Simple calculations and measurements without phase noise formalism are possible as we quantify the coupling mechanism by evaluating the side-bands power relative to the carrier (in dBc). Compared to previous work [11,13], this chapter aims to locally analyze coupling mechanisms between substrate noise and the VCO spectrum as two injected points are placed inside the VCO core. We do not use a global approach by considering only power supply bounces due to coupling between substrate and power supply metal rail. We accurately analyze coupling mechanism in order to determine which devices are sensitive to bulk noise in our specific VCO test-chip, including integrated injection taps. This chapter is organized as follows. Section 4.2 presents the VCO cell structure and its layout implementation. The VCO oscillation frequency sensitivity functions of tuning voltage and bias current are measured and analyzed as the VCO static sensitivities in Section 4.3. Measured VCO spurious side-bands under injected noise perturbation are also presented. Then, simulated varactor and bipolar transistor contributions to the carrier frequency variations are analyzed in Section 4.4.

4.2.1.2 VCO Structure and Layout

The VCO was designed into a 0.35 μm BiCMOS high-ohmic process technology. Figure 4.3 shows the simplified VCO schematic. The main part of the VCO is the LC-tank. The negative resistance is obtained by a cross-coupled differential pair of hetero-junction bipolar transistors biased by an I_{bias} current.

In order to inject a parasitic signal directly into the substrate with an external generator, we have placed several contacts near the differential transistors and the inductors (4). The circuit supply voltage is 3.3 V. The tuning voltage can vary from 0 V to 4 V to obtain a carrier frequency from 4.25 GHz to 4.60 GHz. The output signals pass through buffers formed by a common collector circuit configuration. The power magnitude delivered to a 50 Ω load under this current is closed to 0 dBm. The phase noise measured at 100 kHz from carrier varies from –90 dBc/Hz up to –100 dBc/Hz (depending on chip dispersion and biasing). To avoid frequency variations due to supply voltage noise, many decoupling capacitors are placed in the VCO layout (Figure 4.2). For the same purpose, substrate contacts and guard rings around the devices that have a large surface in contact with the bulk have been added. Other substrate contacts with different areas and locations are placed on the test-chip to allow us to estimate the contact and substrate resistance.

Substrate contacts for perturbation injection have different areas and locations on the chip that could allow us to estimate the contact and substrate resistance.

4.2.1.3 Static Sensitivity Functions

VCO carrier frequency variations depend on bias current and tuning voltage.

Two sensitivity functions can be used to describe the variations of the VCO carrier function of the bias current or tuning voltage. These are, respectively, K_{bias} and K_{tune} (see below). A change in the I_{bias} current can modulate the carrier frequency in the same manner that a modification of V_{tune} voltage can simultaneously change the varactor capacitance and the output frequency. In this framework, consider a sine wave perturbation superimposed on the bias current or the tuning voltage:

$$v_m(t) = A_m \cos(\omega_m t) \tag{4.1}$$

The carrier frequency changes according to the expression:

$$f = f_c + K \cdot v_m(t) \tag{4.2}$$

where f_c is the VCO carrier frequency and K the sensitivity function.

After frequency integration to determine the phase, the output signal of the oscillator is determined by the following equation:

$$s(t) = A \cos\left(\omega_c t + \frac{A_m K}{f_m} \sin(\omega_m t)\right) \tag{4.3}$$

The modulation index ($A_m K / f_m$) depends on a sensitivity function. For a supply voltage perturbation, the modulation index has a low value, so we can assume in our calculations that the modulation band is placed in a narrow range. We refer to the carrier in order to compute the perturbation level in the spectral domain. With these assumptions, the spectral power relative to the carrier in dBc units is given by

$$P_{\text{sbc}} = 20 \cdot \log\left[\frac{K \cdot A_m}{2 f_m}\right] \tag{4.4}$$

It measures the gap between the spurious side-band power and the carrier magnitude and the parasitic side band one (Figure 4.5). A perturbation superimposed on the bias current or tuning voltage generates side-bands that have the power magnitude directly proportional to K sensitivity (Equation 4.4). In the same time, K sensitivity is function of the bias current and of the tuning voltage. Thus, the noise power below the carrier varies when biasing or tuning are modified.

Figure 4.4 shows the plot of the measured VCO frequency as a function of the tuning voltage and the bias current.

In order to compare the sensitivity functions and their effects on the power below carrier when a substrate perturbation is applied, we represent the frequency evolution versus the bias current and tuning voltage on Figure 4.4. The sensitivity function is obtained by derivation of the frequency curve (see Figures 4.5 and 4.6).

Figure 4.5 represents the simulated and measured sensitivity function of the bias current. Simulations of the sensitivity functions have been made with the Cadence-SpectreRF simulator. Perturbation effect on bias current results in a small variation of the carrier frequency. The voltage applied to the varactors generates a much more important sensitivity level than the bias current sensitivity. Roughly, the sensitivity function magnitude for bias

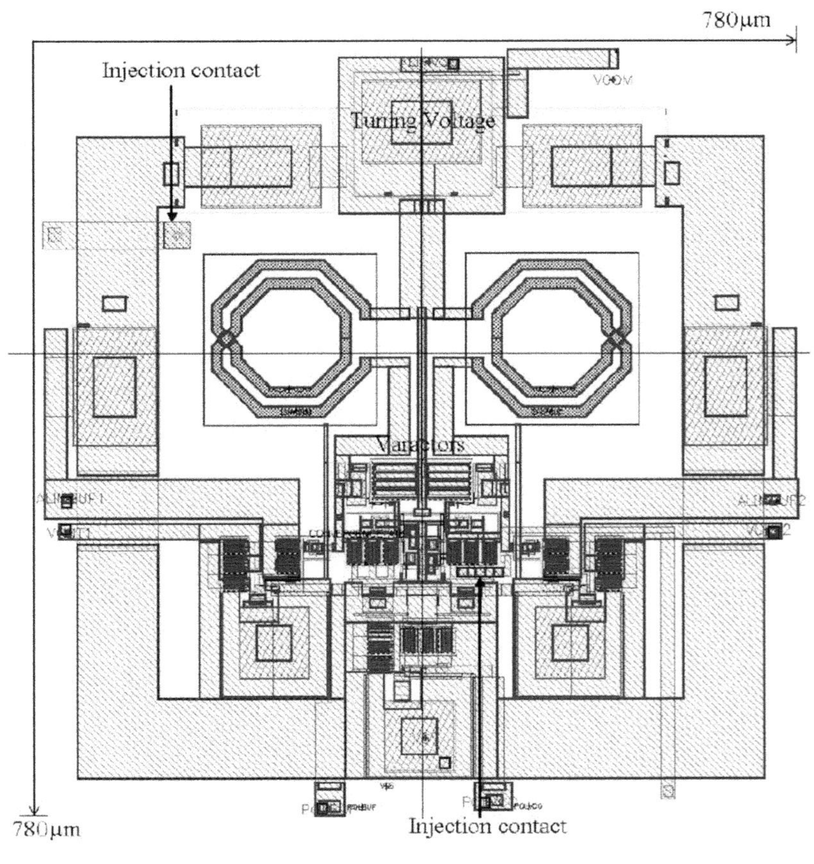

FIGURE 4.4
VCO layout.

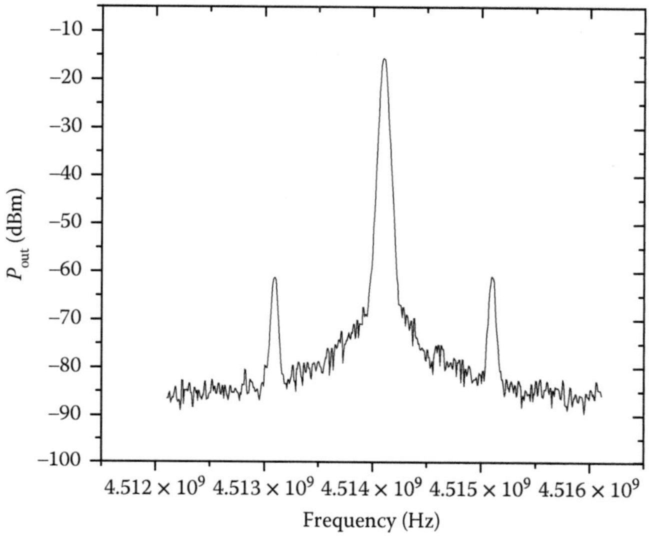

FIGURE 4.5
P_{sbc} for a small sine wave applied to the contact near inductor (50 mV peak at 1 MHz).

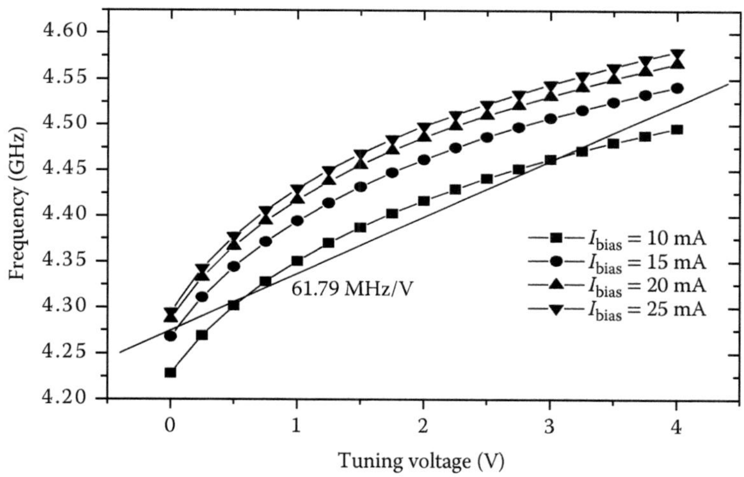

FIGURE 4.6
VCO carrier frequency as a function of the tuning voltage for different bias currents.

current K_{bias} is ten times lower compared to the tuning voltage sensitivity K_{tune} (Figures 4.5 and 4.6).

This method lets us establish the VCO sensitivities as functions of bias circuit conditions by measurements and/or simulations. From Equation 4.4, the impact of a perturbation on bias source or tuning voltage can be easily evaluated. By these assumptions, the impact of supply noise could be reduced. VCO bias has to be properly chosen in order to decrease the VCO frequency sensitivity.

4.2.1.4 VCO Spurious Side-Bands Due to Substrate Perturbation

Experimentally, when sine wave voltage signal is injected into substrate, we measure VCO spurious side-bands closed to carrier frequency f_c. The VCO sensitivity due to injected perturbation (level, location, frequency, etc.) is quantified by P_{sbc} values (Figures 4.3 and 4.7).

Substrate noise induces V_{ss} and V_{dd} power supply bounces. In our specific test-chip—considering the high-ohmic substrate (noise attenuation with distance, substrate cannot be considered as a single node), numerous on-chip decoupling capacitors, and injection nodes inside VCO—we, obviously, have a direct path between injected nodes and passive or active device bulk (varactor, inductor, HBT's, etc.). Studying components sensitivity can be used to analyze the contribution of each one to the output noise. Simulated transfer functions between a source directly applied to the device substrate pins and VCO output have been studied (Table 4.1). The P_{sbc} simulated data shows that an important contribution to output noise is given by the devices that have a large n-well and/or n-buried layer area (varactors and transistors).

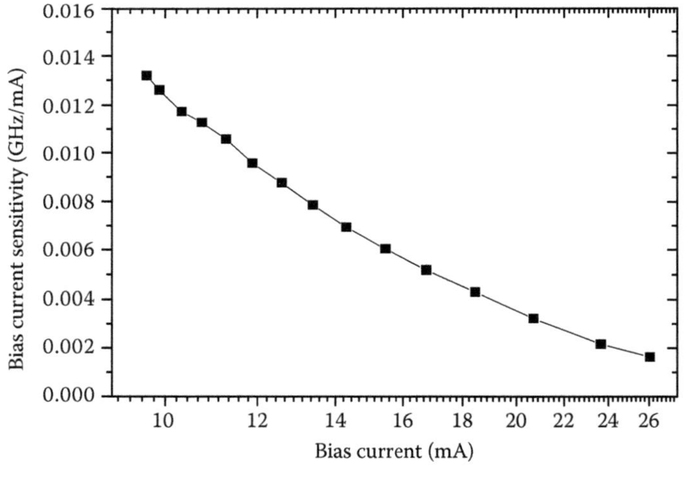

FIGURE 4.7
Bias current sensitivity function (K_{bias}).

TABLE 4.1

Component Contributions to the Output Power

Device	Area[a] (μm²)	P_{sbc} (dBc)
Varactors	2800	–35.93
Inductors	-	–104.25
Transistors	4800	–5.54
Varactors and transistors	7600	–5.65
All	7600	–5.91

[a] n-well and/or n-buried area.

A correlation between sensitive components and P_{sbc} values has been established. Injection of the sine wave noise close to the bipolar transistor substrate generates important spurious side-bands. An equivalent impact is obtained when the same sine wave is injected into both bipolar transistor and varactor substrate. For similar injection conditions, bipolar substrate noise impact surpasses varactor contribution. A deep study of this correlation will be made in a future work.

To accurately simulate the impact of a substrate sine wave perturbation, some of the substrate elements based on measurements between injection contacts and grounds have been determined. The passive and active device models used for our simulations contain equations for the parasitic coupling elements. In this way, to compare measured and simulated data, it is necessary to implement elements describing coupling paths between the injection point and VCO components. The contacts used to inject noise into the substrate have different areas and are placed in different locations. The circuit devices are surrounded by substrate isolations (guard ring resistance) of different resistance values connected to the ground. Due to parasitic elements, the resistance of injection substrate contact to the ground depends on its layout location. Thus, the injected perturbation will be more attenuated when the substrate isolation (guard ring resistance) of a device has a low resistance value. With these remarks and our measurement interpretations, we created a basic substrate model by adjusting resistances from guard ring to ground and from substrate to ground for injection into the transistor, inductor and varactor substrates. Figures 4.8 and 4.9 shows the tuning voltage sensitivity and the power of the side band below carrier P_{sbc} respectively as a function of the tuning voltage sensitivity. Measurements and simulations of P_{sbc} have been obtained using 150 mV injection peak voltage. This sine wave has been injected near one of the inductors (Figure 4.2), at 150 kHz. These measurements show that there is a direct link between the VCO tuning voltage sensitivity and the side band power (P_{sbc}).

Circuit simulations in our basic substrate model show that two main devices can convert substrate noise into the output power: transistors and varactors. Converted noise by transistors versus tuning voltage sensitivity (Figure 4.10) is a roughly constant function and has an important contribution to P_{sbc} level. On the other hand, converted noise by varactors leads to an important decrease of the P_{sbc} with tuning voltage sensitivity. The comparison between simulated data obtained with simple substrate model and measured data proves that P_{sbc} values are dominated by varactor behavior. (Increase in P_{sbc} depends on tuning voltage sensitivity level). With the injection close to one of the inductors and considering the substrate attenuation, VCO side-bands are mainly introduced by the varactor sensitivity to the low frequency harmonic substrate noise. With the injection in the vicinity of transistors of the cross-coupled pair, measured P_{sbc} values are higher compared to P_{sbc} for injection close inductor (for the same injected perturbation) and do not depend on the tuning voltage sensitivity. This remark agrees with the simulated behavior obtained with our simple substrate model (Figure 4.8).

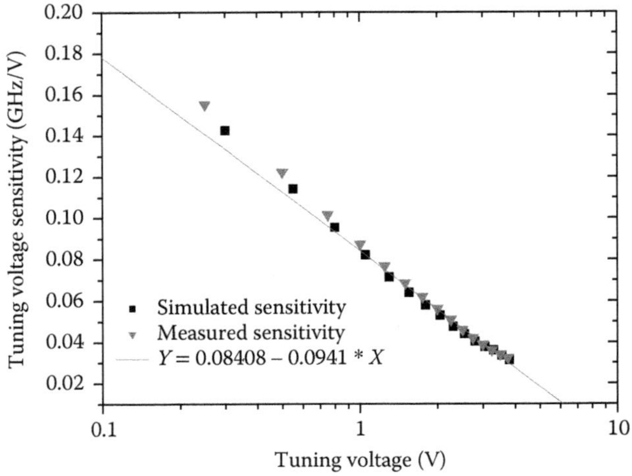

FIGURE 4.8
Tuning voltage sensitivity function (K_{tune}).

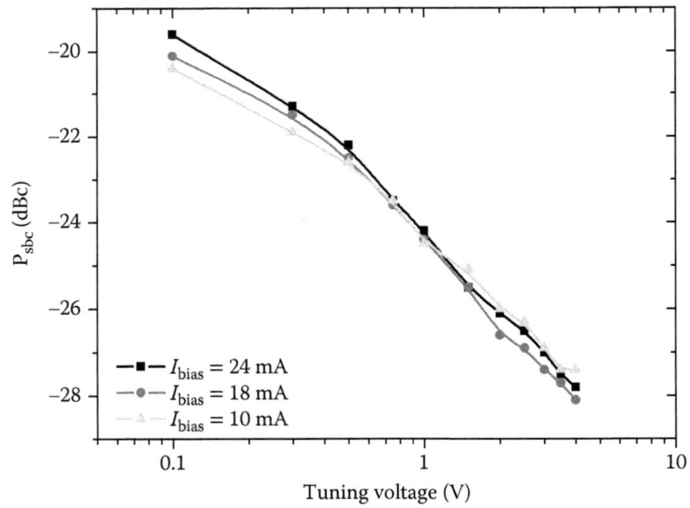

FIGURE 4.9
Measured P_{sbc} function of tuning voltage (50 m Vpk at 100 kHz perturbation is injected close to inductors).

4.2.2 Test Chips Presentation

To confirm our approach, we use a standard BiCMOS process with 5 metal layers and a high resistive substrate. This allows us to characterize substrate noise generated by the digital part of a mixed signal integrated circuit. Figure 4.11 shows a chip microphotograph.

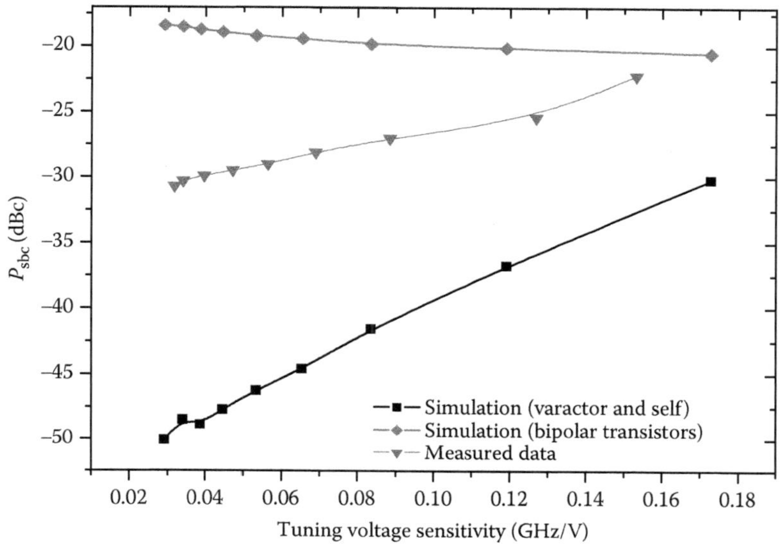

FIGURE 4.10
P_{sbc} versus tuning voltage sensitivity.

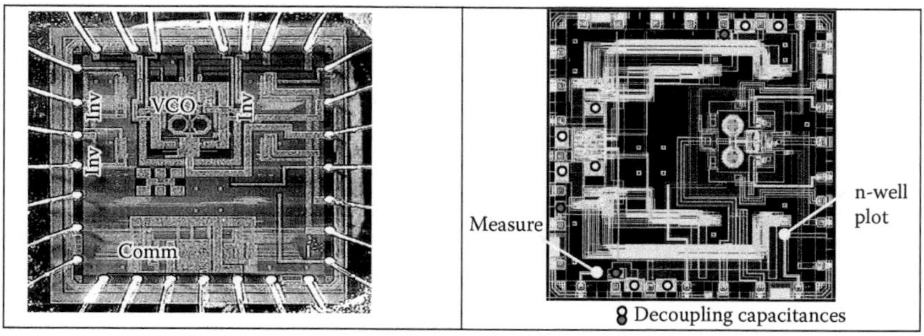

FIGURE 4.11
Left: Optimized circuit layout. Right: Microphotograph of evaluation chip.

Four inverter networks are placed to generate noise into the substrate. They can all be driven, with many configurations, by a programmable logical command. We can switch to different numbers of inverters according to their drive or their location on the layout. This command is isolated in the substrate from the other part of the circuit by a P+ guard ring connected to a dedicated ground point. This isolation reduces the incidence of the command switching events on the measured substrate voltage. All the noisy inverters can switch in a very large frequency range, from 0 to more than 500 MHz, depending on the external clock frequency. On the same die, lies a high frequency Voltage Controlled Oscillator that should be disturbed by the digital switching noise. Many bulk taps, all over the system, allow substrate voltage measurements and also external noise injection; different configurations can be programmed. All these configurations make it possible to have many test cases for the comparison between measurements and simulations.

The package of the circuit is a QFN (quad-flat no-leads) with very short bonding wires and a backside or slug contact. This kind of package is currently used for low-noise or RF applications.

4.2.3 Extractions, Simulations, and Experimental Results

4.2.3.1 ICEM Model Parameters Extraction

The extended-ICEM model of our test-chip is composed of five main parts: the PCB supply lines, the socket and the package, the supply lines of the chip, the core of each digital part of the circuit, and the simplified substrate model. The parasitic inductances, capacitances and resistances of supply lines are extracted from layouts and 3D models with different dedicated tools [14–17]. Those tools provide us an equivalent RLC network of each part. Figure 4.12 gives an equivalent RLC lumped network that can be used to model two adjacent bonding wires. This schematic is a good example of how the different parts of the supply lines can be modeled. Typical values of such model parasitic elements are: $L_{bond1} = L_{bond2} = 1.45$ nH, $C_{bond1} = C_{bond2} = 5.1$ fF, $C_{bond12} = 67$ fF and $K_{12} = 0.26$.

The core consumption current is a waveform obtained with SPICE transient simulations of each block. The core capacitance is also provided from AC Spice simulations. This capacitance can also be approximated by adding the capacitances of all the logical gates included in the digital design. Lastly, the substrate model is extracted from a simplified layout with a CAD tool substrate extractor [14–17]. We obtain, after a long step of extraction, a SPICE netlist of the substrate that can be linked in the global simulation schematic. The measurement lines have also been modeled in order to consider the different parasitic elements that should be disturbing for measurements.

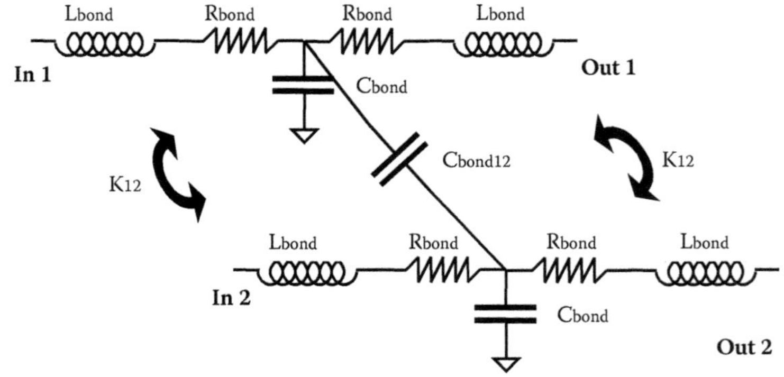

FIGURE 4.12
Adjacent bonding wires RLC model.

4.2.3.2 Measurements and Simulations

The comparisons confirm the effectiveness of the proposed simulation method. Different configurations are tried by changing some important parameters in substrate noise generation or propagation: the power magnitude of injected noise, the transfer function of the package, and the location of the measurement substrate contact.

When high current spikes are requested by the inverter networks, the simulation results are more pessimistic than the measurements. This is mainly due to the consumed current characterization step. During this step, power supply and ground lines are supposed to be perfect; thus, the voltage drop effects are not simulated.

Another way to validate this simple model is to change the transfer function of the package by cutting off some power bond wires. The chip has three dedicated power bond wires in the classical configuration. To show the effectiveness of the extended-ICEM model, we have kept only one power bonding wire. Figures 4.13 and 4.14 still show good correlations.

Levels measured are quite similar between both bonding configurations. As expected, frequencies decrease when only one power supply bond wire is

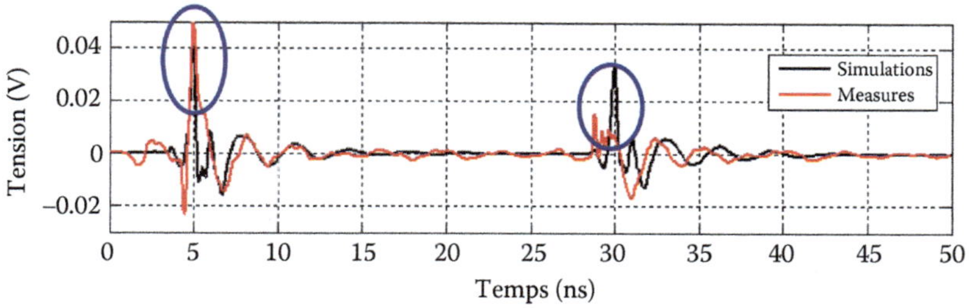

FIGURE 4.13
Comparisons between measurements and simulations of the substrate voltage generated by the networks of reversers "1" and "2" on the extnoise4 measurement plot to show defects of the simulated model.

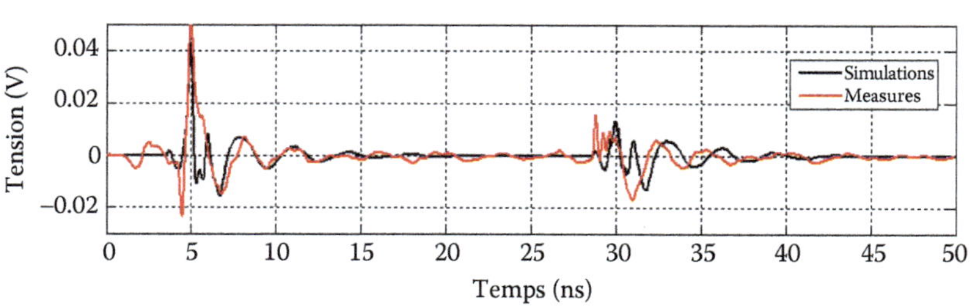

FIGURE 4.14
Comparisons between measurements and simulations of the substrate voltage generated by the networks of reversers "3" and "4" on the extnoise4 measurement plot to show defects of the simulated model.

kept. This is due to the increase of the power supply line parasitic inductance that involves the down-shifting of the resonance frequency.

Finally, we can also measure the substrate voltage on different P+ contacts at different points of the die. The resistance between the ground and the substrate tap depends on its location. The first contact is closer to the system ground than the second one; as a consequence, the peak to peak substrate voltage measured on the first contact is higher than the second substrate contact measurement. These simulations follow measurements and still confirm the effectiveness of the test-chip simulation schematic.

4.3 Overview of the 3D Interconnect and Substrate Modeling Approach

4.3.1 The Different Steps of the Modeling Approach

3D interconnects require the consideration of the global electrical context to correctly evaluate system performances. We propose physics-based compact models derived from semiconductor and electromagnetic theory. At first, the models are derived from parametrical extractions performed on test structures. The *RLCG* parametrical extractions are performed on wafer through S-parameters measurements with a Vector Network Analyzer (VNA) and ground-signal-ground probes (SOLT calibration) for a frequency sweep from 70 kHz to 40 GHz. The Thru-Reflect-Line (TRL) de-embedding technique is applied to remove the contact and access lines effects from the S-parameters measurements and, thus, obtain the test structures actual RF behaviors [12]. A more physical modeling approach relies on analyzing the test structures afterwards by means of: (1) a 3D full-wave electromagnetic tool using the Finite Element Method (FEM) to understand the physical and electrical phenomena occurring in 3D interconnects during the propagation of a signal, and (2) the Transmission Line Method (TLM).

TSVs cannot always be considered as perfect cylinders. Diameter increases toward the TSV top surface (BEOL [back end of line] level) while the thicknesses of the copper and oxide layers decrease. For a W1 configuration (e.g., Figure 4.15), the TSV diameter, including the copper layer and the polymer material, varies from 60 µm to 59 µm. In a W2 configuration, the diameter varies from 40 µm to 37 µm. At each extremity, for both configurations, the TSV copper and SiO_2 layers thicknesses vary respectively from 1 µm to 3 µm and from 0.2 µm to 0.5 µm (W1 and W2 configurations).

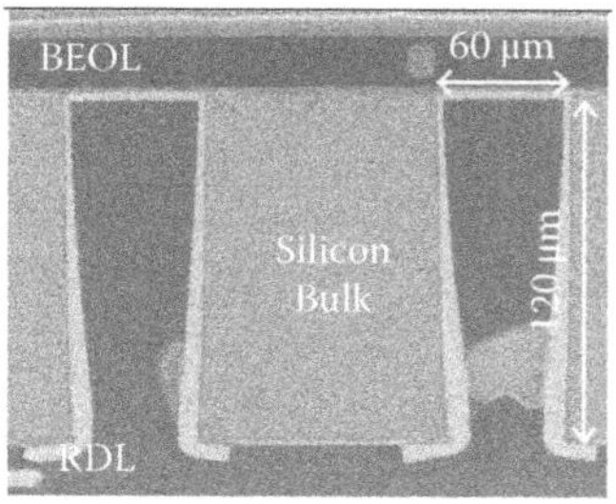

TSV geometrical variations	BEOL level	RDL level
TSV diameter (polymer material + copper layer)	60 µm(W1)	59 µm(W1)
	40 µm(W2)	37 µm(W1)
TSV copper layer thickness	1 µm	3 µm
TSV oxide layer thickness	0.2 µm	0.5 µm

FIGURE 4.15
SEM cross section of a medium-density TSV used in the TSV chains (W1 configuration).

4.3.1.1 Compact Models of the Medium-TSV (and Coplanar Line)

As in planar technologies, 3D interconnects can be built with *RLCG* passive components described in Π or T networks. It does not matter which type of network is chosen as long as the interconnect length is much smaller than the length of the propagated signal wavelength.

Otherwise, both have to be distributed in a certain number of RLCG elementary cells to give an equivalent response. We consider the T network in the rest of this chapter. Each of the 3D interconnect *RLCG* networks is modeled with serial elements (resistances or partial inductances) to model the signal propagation, and parallel elements (capacitance or conductance) and to model the interconnect environment. The proximity (or coupling) effects are also included in the compact model description. The coplanar line resistances and the self- and mutual- inductances are calculated from some well-known formula [14].

The resistances are calculated for a DC value (the skin effect could be considered later) and the inductances are calculated depending on the partial inductance analytical expressions [18–21].

The compact models that we propose works in low and medium frequencies for medium density TSV. We have chosen to calculate resistances with DC values.

$$R = \rho \frac{L}{W \cdot T} \tag{4.5}$$

$$L_{\text{self}} = \mu_c \frac{L}{2\pi}\left[\ln\left(\frac{2L}{W+T}\right) + 0.5 + \frac{0.447(W+T)}{2L}\right] \tag{4.6}$$

$$M = \mu_c \frac{L}{2\pi}\left[\ln\left(\frac{2L}{p}\right) - 1 + \frac{p}{L}\right] \tag{4.7}$$

where ρ is the metal resistivity (copper: $1.72 \times 10^{-8}\Omega{\cdot}m$); μ_c the copper permeability; *W*, *L*, *T* and *P*, respectively, the line width, length, thickness and the pitch between two interconnect lines (Figure 4.7).

The interline capacitance C_{inter} (or coupling capacitance) is calculated according the set of equations defined as function of the concerned line surfaces and the fringe capacitances taken at the extremities (*Cf*) and the middle (*Cf'*) of the coupled-line system. The electrical parameter *Cp* corresponds to the coupling capacitance between the ground plane and the concerned line surface (Figure 4.16). It is referred in the literature as the line's self-capacitance.

$$Cf = \varepsilon_{ox}\left(0.075\frac{W}{H} + 1.4\left(\frac{T}{H}\right)^{0.222}\right) \tag{4.8}$$

$$Cf' = \frac{Cf}{1 + (H/S)} \tag{4.9}$$

$$Cp = \varepsilon_0\varepsilon_{ox}\frac{W \cdot L}{H} \tag{4.10}$$

$$C_{inter} = Cf - Cf' + \frac{\varepsilon_{ox}L}{100}\left(3\frac{W}{H} + 83\frac{T}{H} - 7\left(\frac{T}{H}\right)^{0.222}\right)\cdot\left(\frac{H}{S}\right)^{1.24} \tag{4.11}$$

where ε_{ox} is the silicon oxide permittivity and S is the inter-line gap. An average value of the *W* width in the formula is used, the signal and ground lines having different widths.

We propose our own analytical expressions for the calculation of the medium-density TSV compact model resistance and oxide capacitance. The

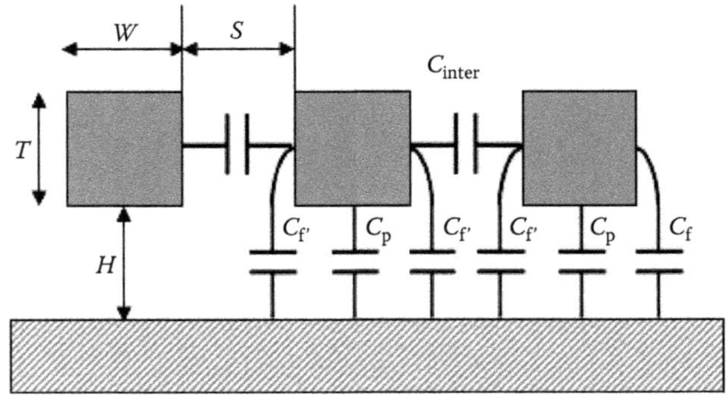

FIGURE 4.16
Illustration of the capacitive couplings between parallel lines located over a ground plane.

TSV being conical, the expressions are derived from the formula defining a perfect cylindrical conductor, for which we have applied, along the conductor length, linear variations for the thickness and radius.

$$R_{TSV} = \frac{H}{\pi\sigma}\frac{1}{2(R_{mt}\cdot T_{md} - R_{md}\cdot T_{mt})}\ln\left(\frac{T_{md}}{2R_{md} - T_{md}}\cdot\frac{2R_{mt} - T_{mt}}{T_{mt}}\right) \quad (4.12)$$

$$C_{ox(TSV)} = \frac{\varepsilon_0\varepsilon_{ox}\cdot 2\pi\cdot H}{T_{ot} - T_{od}}\left[R_{ot} - R_{od} + \frac{R_{ot}\cdot T_{od} - R_{od}\cdot T_{ot}}{T_{ot} - T_{od}}\ln\left(\frac{T_{od}}{T_{ot}}\right)\right],$$

$$R_{ot} = R_{mt} + T_{ot},\quad R_{od} = R_{md} + T_{od} \quad (4.13)$$

where R_{mt} and T_{mt} are the copper layer radius and thickness of the TSV top surface, R_{ot} *and* T_{ot} *are the oxide layer radius and thickness of the TSV top surface,* R_{md} *and* T_{md} *are the copper layer radius and thickness of the TSV bottom surface, and* R_{od} *and* T_{od} *are the oxide layer radius and thickness of the TSV bottom surface.*

According to the geometrical data, supposing the W1 configuration: R_{mt} = 30 μm, R_{md} = 29.5 μm, T_{mt} = 1 μm, T_{md} = 3 μm, T_{ot} = 0.2 μm, T_{od} = 0.5 μm.

The polymer material filling the TSV does not significantly affect its RF behaviour. Besides, the variations of the TSV radius and the copper layer thickness are very low. Consequently, the analytical expressions of a filled cylindrical conductor are still used in order to calculate the self and mutual inductances.

$$L_{TSV} = \mu_0\mu_r\frac{H}{2\pi}\left[\sinh^{-1}\left(\frac{H}{R_a}\right) + \frac{R_a}{H} + 0.15 + \sqrt{\left(\frac{R_a}{H}\right)^2 + 1}\right] \quad (4.14)$$

$$M_{TSV} = \mu_0\mu_r\frac{H}{2\pi}\left[\sinh^{-1}\left(\frac{H}{p}\right) + \frac{p}{H} - \sqrt{\left(\frac{p}{H}\right)^2 + 1}\right] \quad (4.15)$$

where p is the pitch between two nearby TSVs and R_a the TSV average radius.

4.3.1.2 Global Electrical Context Modeling

The modeling of the global electrical context is illustrated for TSV chains and for CPWs (coplanar wave guides), the latter of which will not be presented here. In the TSV chain, the currents propagate vertically through the epitaxial and the oxide layers to reach the ground planes corresponding to the BEOL level ground lines (Figure 4.15). The epitaxial and oxide layers electrical parameters ($R_{epi(TSV)}$, $C_{epi(TSV)}$, and $C_{ox(GND)}$) are calculated in the same way as the coplanar waveguide. The couplings between the substrate and the TSVs are taken into account with the geometrical capacitance $C_{ox(TSV)}$. Only the proximity effects between TSVs (TSV/TSV couplings) are considered. The TSV chain electrical context modeling involves dividing the BEOL ground lines into serial blocks, by dissociating, on the overall surface, the parts reached by the current coming from the TSVs.

These surface parts have a length corresponding to the TSVs contact pads length L_{pad}. The epitaxial layer parameters ($R_{epi(TSV)}$, $C_{epi(TSV)}$) and the oxide capacitance ($C_{ox(GND)}$), which model the current propagation to the ground line through the TSV, must therefore be calculated for $L = L_{pad}$.

The ground lines surfaces parts involved in the current paths are simply modeled by *RL* networks.

4.3.1.3 To a Simulation Platform

4.3.1.3.1 Substrate Modeling Approach

The main difficulty with a priori exploration is that the choice of architectural block parameters must be made according to non-ideal effects (implementation technology, sensitivity to noise and temperature variations, dispersion), which are not known here. It is therefore essential to determine the application-defined design space and to map this to a technology-defined (see, e.g., Figure 14.8) design space for each architectural variant to be considered. The development of this approach will thus rely heavily on high-level models, as well as on design database exploitation (data mining techniques and the suitability of Pareto fronts has to be investigated to optimize the search speed and database size) and estimation methods. This in

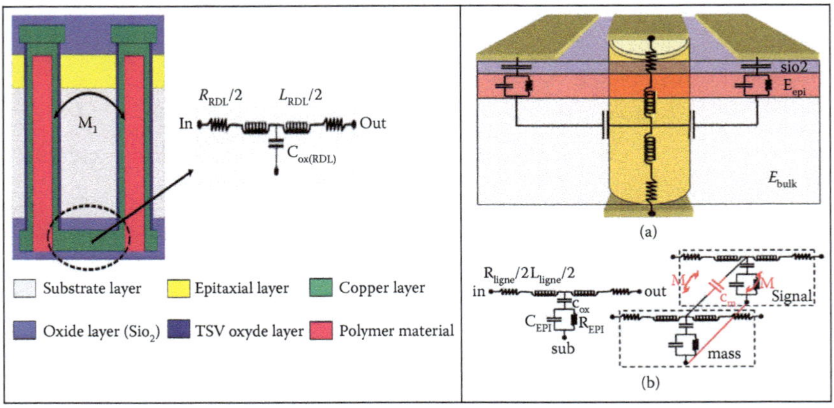

FIGURE 4.17
Left: Medium-density TSV front view. Right: *RLCG* compact modeling of the medium-density TSV including the current path.

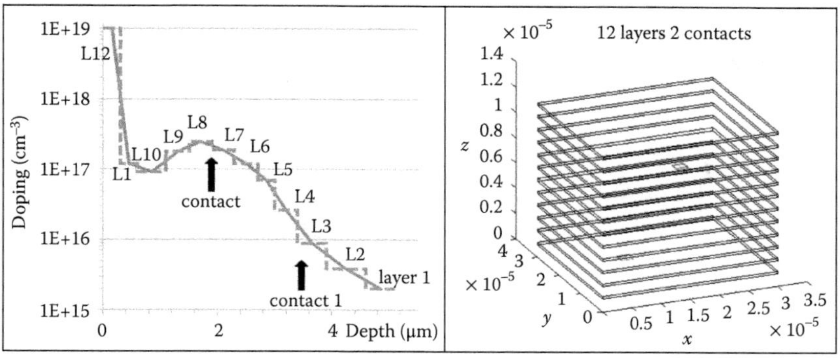

FIGURE 4.18
Left: Specific depth profile (p^+/p-region). Right: 3D schematic of the structure.

combination with bottom-up performance space models as a way to improve the overall efficiency of the flow will be investigated.

Commercially available EDA (Electronic Design Automation) tools are required to allow a flexible floor plan and better vertical and horizontal place and route steps, including better thermal modeling and specific DFY (Design for Yield) constraints. Design rules have to be derived and provided to the designer using simulation results and, when possible, measurement data. These rules are related to optimize the thermal management of the structure and minimize crosstalk, signal delays, electro-thermal interactions, losses at the RF domain, etc. Parametric simulation studies will also be carried out to derive the rules. Among others, parameters are geometry, materials used, and material properties.

Currently, EDA tools supporting IC package, module and SiP design are architected in a way that supports and assumes all connections between dies going through IC package fabrics (e.g., bumps, wire bonds, traces and vias, or package-on-package [PoP] solder balls).

With emerging request for 3D die stacking, this assumption is no longer valid. In order to support 3D architectures, EDA tools will have to be improved to handle direct die-to-die connections within the IC layout tool, including the support of stacked chip configuration, connectivity management, parasitic extraction, analysis, and simulation of the stacked IC.

Methodology, flow, and tools ought to be developed to allow flexible stacked IC connectivity between different ICs and also the final package. Different styles of connectivity will be defined and supported to allow greater cost-efficiency and better electrical and thermal results.

The definition and implementation of new methods for flexible floor plan generation, vertical placing and complex models for interconnects, power grid and clock routing are needed.

So we will develop an EDA flow for the following:

1. Define TSV and contacts placement and automatic alignment and update driven from top/bottom die.
2. Define the power network based on TSV and contact direct connection to the power grid.
3. Provide timing analysis of the interconnections through various configurations, such as TSV, stacked dies, and face to face interconnection. This will be based on standard RC parasitic extractions and static timing analysis.
4. Generate a thermal model of a system composed of several dies to be used later in package and PCB simulation.
5. Generate an electrical model of a system composed of several dies to be used later in package and PCB simulation.

4.3.1.3.2 General Design Flow

TSV-based stacked IC design methodology requires a comprehensive development of both the existing flow approach and a totally new development of the design flow and its associated tools.

The TSV design flow development shall initially be developed according to the baseline approach (Figure 4.19).

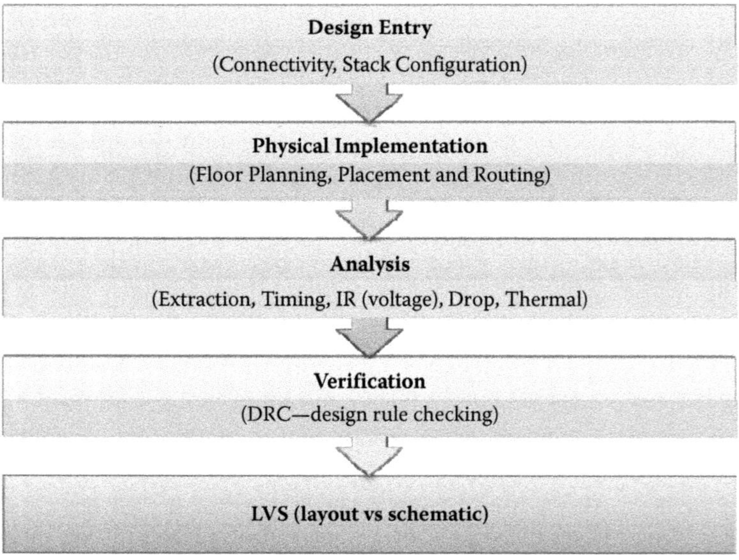

FIGURE 4.19
Design Flow synopsis.

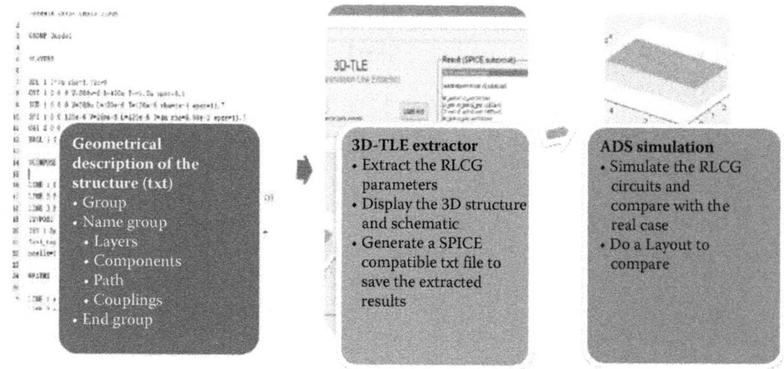

FIGURE 4.20
Our global platform.

4.3.1.3.3 Transmission Line Extractor 3D-TLE

When we consider the global electrical context (e.g., substrate impedance) and the compact model's distribution, the total compact model can become very complex. So, we develop a 3D extraction tool, 3D-TLE (3D Transmission Line Extractor), based on our proposed approach, which integrates the substrate extraction method algorithms.

The modeling of a conductive substrate as a simple node is only viable for low and medium frequency domains. At high frequency, the substrate coupling effects must be added to the system electrical modeling by representing the substrate as a RLCG network to more accurately evaluate the IC's losses. The modeling can become rapidly complex since the coupling effects must be considered between of all the components of the chip. Moreover, the substrate can be non-uniform with different doping values, that is, having different resistivity and permittivity throughout the considered volume. The substrate is therefore modeled as a stack of parallel heterogeneous dielectric layers, because the doping profile evolves in the perpendicular direction to the Si–SiO_2 surface. The extraction tool generates, from this text file, a SPICE net list containing the system RLCG electrical description, which can be exported toEDA tools such as Agilent Technologies ADS®. The viability of 3D-TLE was checked through comparisons with COMSOL (FEM) simulations.

4.3.1.3.4 Input Txt File Geometrical Description

The file is composed of four classes: the layers, components, paths, and couplings. The layers represent the basic information about the layers, the components define every single component, the paths represent the interconnections, and the couplings define the coupling relationship.

TABLE 4.2

A Sample of the Input Txt File

```
*Modele 2xTSV CHAIN 240um
GROUP Umodel
#LAYERS
RDL 1 T=7u rho=1.72e-8
OXI 1 0 0 0 W=280e-6 L=420u T=-1.2u epsr=6.1
SUB 1 0 0 0 W=280u L=420e-6 T=120e-6 rho=1e-4 epsr=11.7
EPI 1 0 0 120e-6 W=280e-6 L=420e-6 T=4u rho=9.09e-2 epsr=11.7
OXI 2 0 0 124.5e-6 W=280e-6 L=420e-6 T=1e-6 epsr=5.8
BEOL 1 T=0.8u rho=1.72e-8
#COMPONENTS
LINE 1 RDL 1 W=90u L=240u ncells=20 inter OXI(1)->subnode(1)
LINE 2 BEOL 1 W=80u L=240u ncells=20 inter OXI(2)->EPI(1)->subnode(1)
LINE 3 RDL 1 W=90u L=90u ncells=1
TSVPORT 1 BEOL 1 W=80u L=90u ncells=1
TSV 1 Rmet_top=30e-6 Rmet_down=29e-6 Tmet_top=1e-6 Tmet_down=3e-6 +
Toxi_top=0.2e-6 Toxi_down=0.5e-6 h=120e-6 rho=1.72e-8 epsr=6.5 +
ncells=1 updown BEOL(1)->RDL(1) inter TSVPORT(1)->OXI(2)->EPI(1)
#PATHS
LINE 1 signal 95e-6 90e-6 -8.2e-6 11 12
LINE 2 gnd1 0 90e-6 125e-6 3 4 coupled
LINE 2 gnd2 200e-6 90e-6 125e-6 5 6 coupled
LINE 3 rldin 95e-6 0 -8.8e-6 1 11
LINE 3 rdlout 95e-6 330e-6 -8.8e-6 12 2
TSVPORT 1 pad1 0 0 125e-6 0 3 101 coupled
TSVPORT 1 pad2 0 330e-6 125e-6 4 0 201 coupled
TSVPORT 1 pad3 200e-6 0 125e-6 0 5 102 coupled
TSVPORT 1 pad4 200e-6 330e-6 125e-6 6 0 202 coupled
TSV 1 tsv1 140e-6 45e-6 0 in 1 101 102 coupled
TSV 1 tsv2 140e-6 375e-6 0 2 out 201 202 coupled
#COUPLINGS
LINE gnd1 LINE gnd2
TSVPORT pad1 TSVPORT pad3
TSVPORT pad2 TSVPORT pad4
TSV tsv1 TSV tsv2
end group
end
```

4.3.1.3.5 Layers

We can see substrate layers definition in Table 4.3. In this table, the type of layers can be SUB (substrate), OXI (oxide), EPI (epitaxial), RDL (redistribution layer), and BEOL (back-end-of line). But for RDL and BEOL, they have a different definition, as shown in Table 4.4.

TABLE 4.3

Substrate Layers Description

SUB	**1**	**0**	**0**	**0**	**W = 280 e − 6**	**L = 420u**	**T = −1.2u**	**epsr = 6.1**
Types of layers	Number of layers of the same type	X coordinate of Begin point	Y coord. of Begin point	Z coord. of Begin point	Width	Length	Thick.	Dielec.

TABLE 4.4

Redistribution Layer Description

RDL	**1**	**T = 7u**	**rho = 1.72e−8**
Types of layers	Number of layers of the same type	Thickness	Density

4.3.1.3.6 Devices

We can see devices description in Table 4.5.

TABLE 4.5

Devices Description

LINE	**1**	**RDL**	**1**	**W = 90μ**	**L = 240μ**	**n_{cells} = 20**	**Inter OXI(1)->subnode(1)**
Type of components	Number of components of the same type	Type of layer	Layer number	Width	Length	Thickness	Dielectric

Here the type of components can be LINE, TSV, and TSVPORT. But for TSV, the description is a little heavier.

4.3.1.3.7 Paths

For paths, we can see a definition in Table 4.6.

TABLE 4.6

Interdevices Path Description [7]

LINE	**1**	**Signal**	**95e−6**	**90e−6**	**−8.2e−6**	**11**	**12**
Type of components	Number of paths of the same type	Type of path	X coordinate	Y coord.	Z coord.	Port 1	Port 2

Here, the type of components can be LINE, TSV, and TSVPORT, which are predefined in the former part. Type of paths can be signal, GND, RDL, and pad for TSVPORT, TSV for TSV. If the component is coupled with other components in the path's description, it will be added as "coupled" at the end. The coupling relationship will be defined in the following part "coupling."

4.3.1.3.8 Coupling

TABLE 4.7

Devices Coupling Description [7]

LINE	gnd1	LINE	gnd2
Type of component	The exact coupling component	Type of component	The exact coupling component

3D-TLE will automatically generate a 3D structure in the geometry block, a schematic in the schematic block, and a SPICE-compatible txt file. Its content will be displayed in the result block (Figure 4.21).

4.3.1.3.9 Substrate Extractor

The layered substrate is often treated as a simple node, but, at high frequency, the substrate coupling effects must be added to the system electrical modeling.

Our extractor is developed to analytically extract the RC network and to more accurately evaluate the IC' losses. With an affordable speed, the substrate extractor can be divided into five parts. The upper-left corner block defines the general study information, which includes the study frequency domain, the number of study frequency points per decade, the dimension

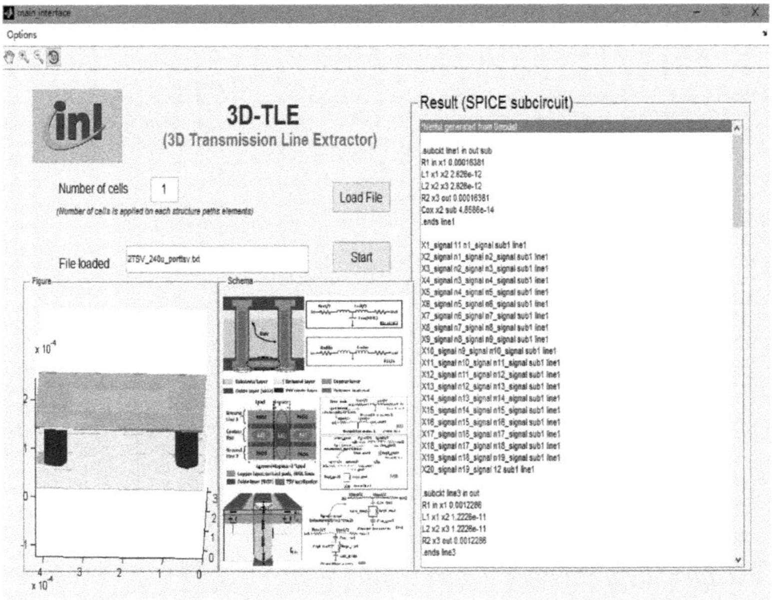

FIGURE 4.21
3D TLE simulation platform.

of the substrate, the mesh scale with 1 μm as reference, and the study mode—that is, contact–contact mode, TSV–contact mode, and TSV–TSV mode (contact–contact mode studies the cases where the contacts are embedded in the substrate. TSV–contact mode extracts the RC parameters between an insertion TSV and a contact embedded in the substrate and TSV–TSV mode performs the RC extraction of 2 TSVs inserted in the substrate [22]. The upper-middle block defines the TSV or contacts geometries. The lower-left bloc defines the layers' physical characteristics, like the conductivity, thickness, and relative permittivity [23]. Once we define the geometries, we can click on the draw figure button in the middle of the interface and it will automatically generate the 2D section image. The extracted RCZ result will be shown in the right section. Here is a TSV–contact extraction result.

4.4 Conclusion

We propose a compact and pragmatic modeling approach to the entanglement of 3D substrate and its interconnects. For interconnects, the modeling approach is applied to redistribution lines and medium-density TSVs. It is based on parametrical extractions performed on realistic test structures and analytical formulas. Moreover, it includes the system global electrical context with the current paths and the modeling of proximity and/or substrate effects. Complete equivalent electrical models are illustrated for TSV chains structures. The modeling approach gives finally some equivalent electrical models. However, while it implies the modeling of all the substrate coupled effects existing between the elements sharing this substrate, it still requires an external substrate substrate coupling simulation method.

This chapter proposes a substrate coupling simulation method suitable for execution in a conventional CAD environment. A standard approach called ICEM (Integrated Circuit Emissions Model) has been extended to allow fast substrate noise simulations. Currently, this model considers only the noise generated in the substrate by the power supply and ground bounces. A more complex model is needed to integrate other injection phenomena, such as impact ionization or drain coupling, for instance. Moreover, to apply this approach to a realistic mixed signal design several challenges have to be grasped. The first one concerns the way to obtain the transient consumption current of an entire large digital part. Some CAD tools offer the possibility to simulate these currents. Finally, it is necessary to analyze influenced analog device of interest, such as VCOs disturbed by parasites coming often from digital block *via* the substrate.

The impact of low frequency substrate noise on a typical RF analog device, a voltage-controlled oscillator, is studied. At first, we presented the test-chip structure with fully integrated VCO and two substrate taps inside the VCO core for noise injection or measurement (one close inductor/varactor and the other in vicinity of the HBT cross-coupled pair). VCO characterization is performed through measured static sensitivity curves that represent the frequency variations function of tuning voltage or bias current. An analytical approach to harmonic perturbation superimposed on bias current or tuning resumes the classical frequency modulation mechanism with observed spurious side-bands. Then measured VCO side-bands are analyzed when harmonic voltage noise is injected.

The next step was the study of the sensitive devices. Many simulations were carried out with a simple substrate model, leading to the conclusion that mainly varactors and bipolar transistors are affected by substrate noise. These simulations agree with the measurements: side-band power below carrier is more important when the injection is located close to the HBT's of the cross-coupled pair and does not depend on the tuning voltage. On the other hand, side-band power below carrier decreases with tuning voltage when the injection is located close inductor and varactor. Finally, we demonstrate that there is a direct link between the VCO tuning voltage sensitivity and the side-band power. Output noise power is a function of circuit sensitivities and device isolations. Furthermore, we explicitly show, in this chapter that a global approach, since it only considers resistive and/or capacitive coupling between substrate and power supply rails, is not sufficient for precise quantification of noise impact on the radio frequency oscillator.

Moreover, the ICEM model has been tested for contact pads, coplanar waveguides and TSV structures. This simple approach has the advantage of being easily automated. Additionally, the model can be improved by including skin effect and eddy current. On-going works will consist of extending the modeling approach to other types of 3D interconnects, such as copper pillars, and to different types of 3D interconnect matrices.

Last of all, we begin to build a simulation platform dedicated to substrates with their connectics. The user describes his system in a text file by means of specific hierarchical syntax and statements. He declares the system as a subcircuit with a certain number of input/output ports (there is no limitation in number). The subcircuit contains components connected with internal nodes. Instead of defining *RLCG* components with resistance/inductance/capacitance/conductance values, the user defined structures such as lines, pillars, TSVs, or CPWs with associated parameters corresponding to the technological and or/geometrical data. He can specify whether the structures in the subcircuit are coupled or not, possible current paths, and so on.

Once the user has finished its system/subcircuit description, he loads it in our 3D-TLE simulator and launches the extraction tool. Then, 3D-TLE provides, as an output file, a SPICE-like format netlist of the subcircuit, in which all the components are described electrically with *RLCG* parameters. These electrical *RLCG* structures correspond to the compact model that we proposed. The I/O ports of the subcircuit are the ones declared by the user in his text file. If necessary, in addition to the internal node indicated by the user in his subcircuit description, 3D-TLE generates automaticallyextra internal nodes to connect the structures to the I/O ports in the SPICE netlist.

The netlist is exportable to CAD tools like ADS [24] or Cadence Virtuoso [25]. In the case of ADS, by means of different options, the user can create from the SPICE netlist a black box containing the netlist and for which the I/O ports are the subcircuit ones.

When descending into the black box, the user is able to visualize the SPICE description and can also modify it. After generating the black box, the user instantiates it in his top level schematic. The viability of 3D-TLE is checked by comparing, for a given system, the simulated S-parameters of its 3D-TLE SPICE netlist with the ones of its electrical schematic designed under ADS®. Obviously, the expected results show some good fits describing the system, since using 3D-TLE seems easier and faster than designing it using a standard CAD tool.

Acknowledgments

This work is supported by INFIERI (Intelligent Fast Interconnected and Efficient Devices For Frontier Exploitation In Research And Industry) European Program and by UPM (Union Pour la MEDITERRANEE), and One of Us (Yue Ma). Thanks to the China Scholarship Council (CSC).

References

1. X. Aragonès, J. L. Gonzalez, F. Moll, and A. Rubio, Noise generation and coupling mechanisms in deep-sub micron IC's, *IEEE Design & Test of Computers*, vol. 19, no. 5, 27–35, September–October 2002.
2. D. K. Su, M. J. Loinaz, S. Masui, and B. A. Wooley, Experimental results and modeling techniques for substrate noise in mixed-signal integrated circuits, *IEEE Journal of Solid-State Circuits*, vol. 28, no. 4, 420–430, April 1993.
3. M. Xu, D. K. Su, D. K. Shaeffer, T. H. Lee, and B. A. Wooley, Measuring and modeling the effects of substrate noise on the LNA of a CMOS GPS receiver, *IEEE Journal of Solid-State Circuits*, vol. 36, no. 3, 473–485, March 2001.
4. M. van Heijningen, M. Baradoglu, S. Donnay, G. G. E Gielen, and H. J. De Man Substrate noise generation in complex digital systems: Efficient modeling and simulation methodology and experimental verification, *IEEE Journal of Solid-State Circuits*, vol. 35, no. 7, 1065–1072, July 2000.
5. J. Briaire and K. S. Krisch, Principles of substrate crosstalk generation in CMOS circuits, *IEEE Transactionson Computer-Aided Design of Integrated Circuits and Systems*, vol. 19, no. 6, June 2000.
6. M. Felder and J. Ganger, Analysis of ground-bounce induced substrate noise coupling in a low resistive bulk epitaxial process: design strategies to minimize noise effects on a mixed-signal chip, *IEEE Transactionson Circuits and Systems*, vol. 46, no. 11, November 1999.
7. O. Valorge, F. Sun, J.-E. Lorival, F. Calmon, and C. Gontrand, Mixed-signal IC design to enhance substrate noise immunity in bulk silicon technology, *Analog Integrated Circuits and Signal Processing*, Vol. 63, no. 2, 185–196, 2010.
8. O. Valorge, F. Sun. J. E. Lorival, M. Abouelatta-Ebrahim, F. Calmon, and C. Gontrand, Analytical and numerical model confrontation for transfer

impedance xtraction in three dimensional radiofrequency circuits, *Circuits and Systems*, vol. 3, 126–135, 2012. Available at: http://www.SciRP.org/journal/cs/.
9. F. Hertzel and B. Razavi, A study of oscillators jitter due to supply and substrate noise, *IEEE Transaction on Circuit and Systems II: Analog and Digital Signal Processing*, vol. 46, no. 1, 56–62, January 1999.
10. T. Blalack, Y. Leclercq, and C. P. Yue, On-Chip RF isolation techniques, In *IEEE Bip./BiCMOS Circuit and Technology Meeting Conference, Monterey*, CA, October 2002.
11. C. Soens, G. van der Plas, P. Wambacq, and S. Donnay, Performance degradation of LC-Tank VCO by impact of digital switching noise, In *Solid-State Circuits Conference, ESSCIRC*, Belgium, Europe, 2004.
12. Y. Zinzius, G. Gielen, and W. Sansen, Modelling impact of digital substrate noise on embedded regenerative comparators, In *IEEE European Solid-State Circuit Conference*, Lisbon, Portugal, September 2003.
13. H. Jacquinot, J. Majos, and P. Penn, "5 GHz low-noise bipolar and CMOS Monolithic VCOs," In *European Solid-State Circuits Conference*, Stockholm, Sweden, September 2000.
14. M. Kamon, M.J. Tsuk, and J.K. White, FASTHENRY: a multipole-accelerated 3-D inductance extraction program, *IEEE Transactions on Microwave Theory and Techniques*, vol. 42, no. 9, 1994.
15. HFSS by Ansoft Corporation. Available at: http://www.ansys.com/products/electronics/ansys-hfss
16. FastCAP Systems Corporation. Available at: https://fastcapsystems.com/
17. Y. Ma, O. Valorge, J.R. Cárdenas-Valdez, J.C. Núñez-Perez, J. Verdier, F. Calmon, and C. Gontrand, Electro-thermal considerations dedicated to 3-D integration; noise coupling (Chap 15), In *Noise Coupling in System-on-Chip*, T. Noulis (ed.), Taylor and Francis, CRC Press (this book).
18. Y. Eo and W. R. Eisenstadt, High-speedVLSI interconnect modeling based on S-parameter measurements, *IEEE Transactions on Components, Hybrids, and Manufacturing Technology*, vol. 16, no. 5, 555–562, August 1993.
19. E. B. Rosa, The self and mutual inductances of linear conductors, *U. S. Bulletin of Standards*, 1907.
20. A. Weisshaar, H. Lan, and A. Luoh, Accurate closed-form expressions for the frequency-dependent line parameters of on-chip interconnects on lossy silicon substrate, *IEEE Transactions on Advanced Packaging*, vol. 25, no. 2, 288–296, May 2002.
21. C. Xu, V. Kourkoulos, R. Suaya, and K. Banerjee, A fully analytical model for the series impedance of through-silicon vias with consideration of substrate effects and coupling with horizontal interconnects, *IEEE Transactions on Electron Devices*, vol. 58, no. 10, 3529–3540, October 2011.
22. D. H. Kim, S. Mukhopadhyay, and S. K. Lim, Fast and accurate analytical modeling of through-silicon-via capacitive coupling, *IEEE Transactions on Components, Packaging and Manufacturing Technology*, vol. 1, no. 2, 168–180, 2011.
23. T. Brandtner and R. Weigel, "Hierarchical Simulation of Substrate Coupling in Mixed-Signal ICs Considering the Power Supply Network," In *Design, Automation and Test in European Conference and Exhibition*, Paris, France, March 2002.
24. Advanced Design System–ADS, Keysight EEs of EDA, 2018. Available at: http://www.home.agilent.com.
25. Cadence Virtuoso Spectre Circuit Simulator, Available at: http://www.cadence.com.

5

Modeling of On-Chip Power Distribution Network

Chulsoon Hwang, Jingook Kim, Jun Fan, Joungho Kim, and James L. Drewniak

CONTENTS

5.1 Introduction

The main goal of a power distribution network (PDN) is to provide a low impedance path (ideally 0 ohms) from the source, i.e., a voltage regulator module, to the transistors on the chip. However, the parasitics of the PDN cause the voltage to fluctuate, resulting in voltage noise on power and ground nets. The voltage noise goal is typically ±10%; for example, a system with nominal V_{dd} = 1.0 V may guarantee the actual supply remains within 0.9–1.1 V. Two fundamental sources of power supply noise are IR drops and L d*I*/d*t* noise. As current flows through the resistance of the PDN, it causes a voltage drop proportional to IR. As the time-varying current flows through the inductance of the PDN, it causes a voltage drop proportional to the rate of current change, L d*I*/d*t*. The resistance of the PDN includes the resistance of the on-chip wires and vias, the resistance of the bond wires or solder bumps, and the resistance of the package and printed circuit board (PCB). Because the package and PCB typically use copper that is much thicker and wider than on-chip wires, the on-chip network dominates the IR drop. The inductance of the power supply is typically dominated by the inductance of the bond wires or solder bumps connecting the die to the package. Modern packages devote many of their pins or bumps to power and ground to minimize the inductance. L dI/dt noise is becoming a major bottleneck for high speed system design as package inductance remains unchanged while the speed and current continues to increase [1].

The PCB and package PDN are relatively well understood and can be modeled very accurately [2,3]. Because of its nature, the approach to an on-chip PDN is quite different from that of an off-chip PDN. In this chapter, on-chip PDN modeling is discussed. We begin in Section 5.2 by modeling on-chip capacitors. Section 5.3 extracts the electrical characteristics of the on-chip wires forming the PDN. Section 5.4 is devoted to an off-chip PDN incorporating PCB and vias. Section 5.5 investigates power supply-induced jitter. Trends regarding signal output contaminated by supply voltage fluctuation are analyzed at the transistor level and physically understood.

5.2 On-Chip Capacitors

Chips need a substantial amount of capacitance between power and ground to provide instantaneous current demands. The decoupling capacitors, also known as bypass capacitors, are distributed across the chip so that a local spike in current can be supplied from nearby decoupling capacitance rather than the decoupling capacitors on the package and/or PCBs through an inductive path, reducing the L dI/dt noise. The decoupling capacitance is typically realized as a gate capacitance in large MOS transistors. In a similar manner, a through-silicon via can provide a significant amount of capacitance as a supplementary method in 3D ICs [4]. In systems with mixed memory and logic, however, the decoupling capacitance can also be realized as a trench capacitance [5,6].

5.2.1 MOS Capacitor

An MOS (metal-oxide-semiconductor) capacitor is formed by the gate electrode on one side of the oxide layer and the inverted channel beneath the oxide layer. The most common substrate in contemporary CMOS technology is the p-type substrate and electrons (the minority carriers) form the conductive channel. However, electron generation cannot keep up with voltage changes on the gate in high frequencies (> 100 Hz) and so electrons remain in the depletion region, resulting in smaller capacitance [7]. In an NMOS capacitor, the n+ diffusions (drain and source) provide electrons (majority carriers) to the channel and they can remain in the inversion region in high frequencies. Although an NMOS capacitor is widely used for the decoupling purpose, which is, strictly speaking, different from an MOS capacitor, it is commonly called an MOS capacitor for the sake of simplicity and the term MOS capacitor will be used hereafter.

Because the channel is highly resistive compared to conductors such as the gate metal or metal wires, the resistance of the channel dominates the effective series resistance (ESR) of the MOS capacitor. Although a smaller ESR is usually preferred, it is not always the case as it is beneficial in damping the resonance peak (on-die capacitance—package inductance). An example of the resonance is demonstrated in Section 5.4. The distributed nature of the channel resistance/capacitance can be modeled as a distributed RC model as shown in Figure 5.1 [8]. The current path through the substrate is negligible and excluded in the model.

Writing the total channel resistance and capacitance as R and C, the circuit in Figure 5.1 can be seen as a transmission line with the propagation constant $\gamma = \sqrt{j\omega RC}/l$ and the characteristic impedance $Z_0 = \sqrt{R/(j\omega C)}$. l is the length of the transistor. Since the circuit is symmetrical with respect to the center, the right-hand half of the RC line can be folded onto the left-hand, giving $R' = R/2$, $C' = 2C$, $l' = l/2$. The input impedance can then be given as

$$Z_{\text{in}}(\omega) = \sqrt{\frac{R/4}{j\omega C}} \coth\left(\sqrt{\frac{j\omega RC}{4}}\right) \tag{5.1}$$

The impedance of a single lumped resistor in series with a capacitor is

$$Z_{\text{lump}}(\omega) = R_{\text{lump}}(\omega) + \frac{1}{j\omega C_{\text{lump}}(\omega)} \tag{5.2}$$

The frequency-dependent lumped capacitance and inductance can be obtained by comparing Equations 5.1 and 5.2. From a series of expansions of

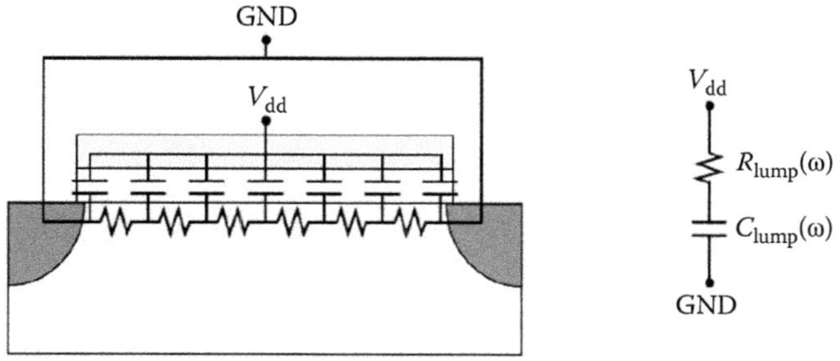

FIGURE 5.1
Distributed RC line model of MOS transistor and RC equivalent model.

Equation 5.1 and by mapping a distributed RC circuit onto a lumped model, we can also obtain $R = 12 * R_{lump}(0)$ [8].

The model is compared with measurement in Figure 5.2. Frequency dependent R and C values of a 10 μm × 10 μm transistor are measured by a high-frequency microprobe and a network analyzer. The parasitics of the measurement pad are excluded by using a de-embedding technique [9]. *R* and *C* values for the model are determined from the measured respective *R* and *C* values in low frequency.

With less resistance, less capacitance degeneration occurs and a larger W/L ratio is preferred for the high frequency decoupling purpose. When extremely low ESR is required, for example, on-chip electromagnetic bandgap structures [10], an accumulation-mode MOS capacitor can be used instead, which has been used for high frequency circuits such as a VCO. When the p-type substrate of a typical NMOS capacitor is replaced with the n-type substrate, that is, an N-well, the NMOS operates in an accumulation region at the nominal power supply voltage, resulting in lower channel resistance. In [10], an accumulation-mode MOS capacitor showed an ESR value six times smaller than that of a typical MOS capacitor.

5.2.2 Through-Silicon Via

In the last decade, through-silicon via (TSV)-based 3D ICs have emerged as a promising technology to extend Moore's Law. TSV interconnections have shorter lengths and a higher density compared to the wire bonds, thereby enabling smaller form factors and wider bandwidths. In addition to signal routings, TSVs are also used for PDNs. The contribution of TSVs on PDN impedance has been studied [11]. Although the capacitance per area of TSVs is usually less than that of transistors, capacitance from thousands of TSVs would reduce the burden of on-die decoupling capacitors. However, as with

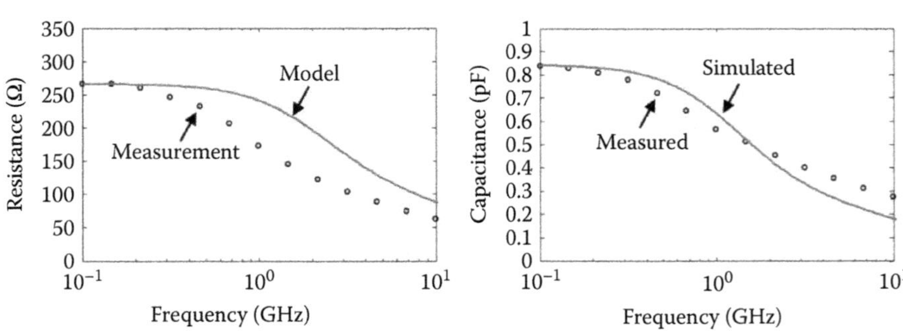

FIGURE 5.2
Measured and simulated equivalent resistance and capacitance of a W/L = 10 μm/10 μm transistor.

the MOS capacitor, the capacitance of TSV also degenerates at higher frequencies. This problem can be resolved by placing an n+ doping contact on the top surface of silicon substrates surrounding the oxide as shown in Figure 5.3, connecting to ground [4]. By forming the n+ doping contact next to the oxide layer, the electrons are pulled in from the n+ region (majority carriers) instead of the p-type silicon substrate (minority carriers) so that the inversion layer can be formed even at high frequencies. Consequently, the capacitance of the power TSV structure becomes significantly bigger than that of a conventional TSV structure with depletion capacitance.

Because the maximum capacitance available from the power TSV structure is eventually limited to the oxide capacitance, the total capacitance from the power TSVs used in the 3D IC is limited, but it can still help increase the on-chip capacitance without additional cost rather than solely provide the on-chip capacitance sufficient for a PDN impedance requirement.

A distributed RC model can also be used for a TSV structure. One end of the TSV structure can be considered open, thus the input impedance of the open-ended RC transmission line can be calculated as

$$Z_{\text{in}} = \sqrt{\frac{R_w}{j\omega C_w}} \coth\left(\sqrt{j\omega R_w C_w} l\right) \tag{5.3}$$

where R_w and C_w are the per-unit-length channel resistance and capacitance, respectively, and ω is angular frequency. l represents thickness of the silicon substrate. The resistance of the metal TSV body is much less than that of the channel resistance and thus can be neglected. The per-unit-length capacitance is the same as the oxide capacitance. The cylindrical capacitance and the per-unit-length channel resistance are given by [12]:

$$C_w = C_{ox} = \frac{2\pi\varepsilon_{ox}}{\ln\left(r_{ox}/r_{\text{metal}}\right)} \tag{5.4}$$

$$R_w = \frac{1}{\mu_n C_{ox}(V_{GS} - V_{TH})} \tag{5.5}$$

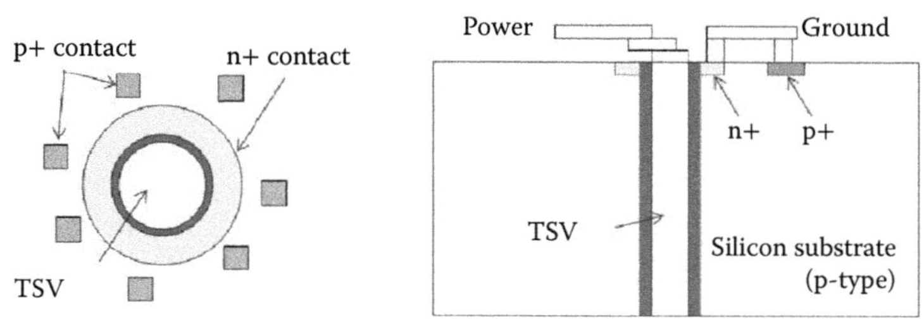

FIGURE 5.3
Layout of power TSV structures embedded in silicon substrates.

where

$$V_{TH} = V_{FB} + 2\phi_F + \frac{qN_a(r_{max}^2 - r_{ox}^2)}{2\varepsilon_{ox}} \ln\left(\frac{r_{ox}}{r_{metal}}\right)$$

$$V_{FB} = (\phi_M - \phi_S) = \phi_M - \chi - \frac{E_q}{2q} - \phi_t \ln\left(\frac{N_a}{n_i}\right)$$

r_{metal} is the metal radius; r_{ox} is the oxide outer radius; μ_n is the electron mobility; ϕ_F is the Bulk potential $\phi_F = \phi_t \ln(N_a/n_i)$; ϕ_t is the KT/q (0.026 V); ϕ_M is the Work function of metal (Al = 4.1 V); q is the electronic charge (1.6 × 10–19 C); n_i is the Intrinsic carrier concentration of Si (1.18 × 1010 cm^{-3}); N_a is the doping concentration of acceptor ions (1.3 × 1015 cm^{-3}); χ is the affinity of Si (4.05 V); E_q is the Bandgap energy of Si (1.12 eV) and r_{max} is maximum depletion radius.

It is known that the value of the electron mobility can be evaluated only through measurement, but, roughly half of the bulk mobility [13] is used (7.5 × 10^{-2} m^2/V·s). The maximum depletion radius, r_{max}, can be evaluated from the equation in [12]:

$$\frac{qN_a}{4\varepsilon_{si}}\left(r_{ox}^2 - 2r_{max}^2 \ln(r_{ox}) + r_{max}^2(2\ln(r_{max}) - 1)\right) - 2\ln\left(\frac{N_a}{n_i}\right) = 0 \tag{5.6}$$

For a TSV structure (r_{metal}: 5 µm, r_{ox}: 5.1 µm) the calculated equivalent capacitance values are compared with the numerical simulation in Figure 5.4. Simulations were conducted using Silvaco ATLAS [14].

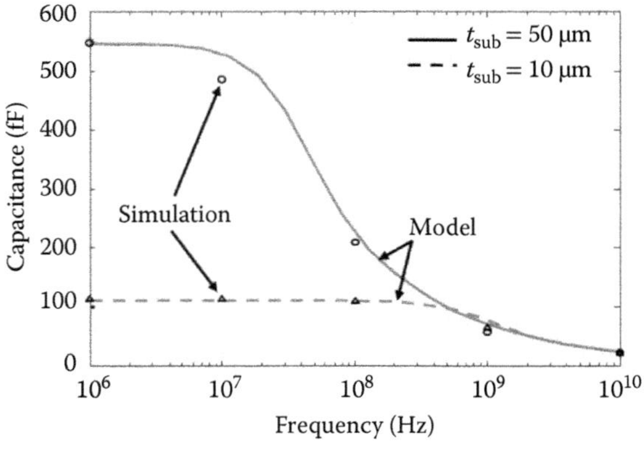

FIGURE 5.4
Equivalent capacitance of a TSV.

5.3 On-Chip Interconnects

5.3.1 Geometry and Equivalent Circuit

Accurate characterization of power/ground wires in environments such as silicon interposers and integrated circuits was traditionally unessential because the circuits operated at relatively low frequencies and could be modeled sufficiently by simple lumped-element RC models. However, as the operating frequencies continue to increase, more accurate modeling is necessary, especially, since the wire characteristics predominate the PDN performance in silicon interposers where a MOS capacitor does not exist. The two power distribution types are illustrated in Figure 5.5. On-chip PDN consist of power and ground wires within the cells and more wires connecting the cells together. Most cells contain internal power and ground busses routed on metal 1 or metal 2. These rows can be strapped together with wider vertical metal wires, this is called as "ladder type." The power and ground lines can also be interdigitated in both vertical and horizontal directions and equipotential lines across different layers are connected through vias, which is called a "grid type." For modeling and analysis perspectives, the geometries can be split into two uniform transmission line sections, coplanar waveguide and microstrip. Modeling each transmission line section is discussed in subsequent sections.

Characteristics of transmission lines on lossy silicon substrates are quite different from those on organic substrates used in PCBs such as FR4 epoxy. Because of the conductivity of the silicon substrate, propagation characteristics change in different frequency regions and with the different resistivity

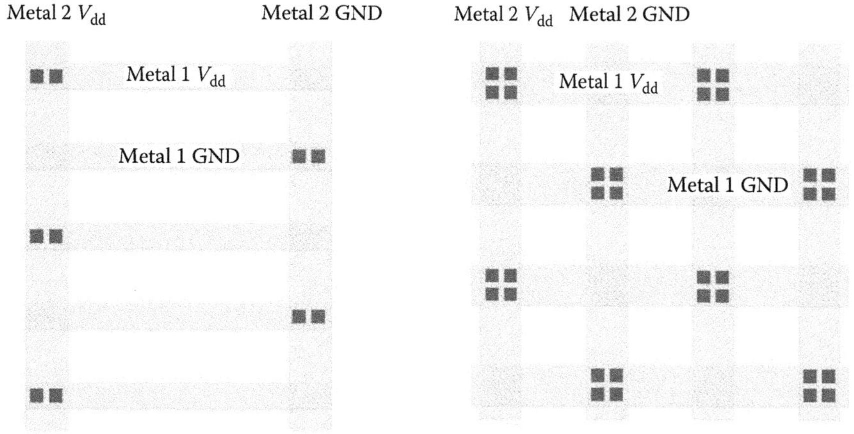

FIGURE 5.5
Power distribution for on-chip: ladder and grid types.

of the silicon substrate. For that reason, a slow-wave mode mechanism occurs in low frequency and a quasi-TEM mechanism when the frequency exceeds the silicon relaxation frequency. Because of the low impedance of silicon substrate, most of the electrical energy is confined to the lossless insulating layer below the conductors. However, the magnetic field freely penetrates the substrate, because the quasi-static range of the magnetic field is much less than the silicon's skin depth. Accordingly, the magnetic field distribution is nearly that of normal transmission line on an insulating substrate (TEM). This separation of electric and magnetic energies results in slow-wave mode propagation [15].

Kwon showed the slow-wave model to be a quasi-TEM mode by adequately describing the transmission lines with a single distributed resistance, inductance, conductance, and capacitance (RLGC) model. A general equivalent circuit based on the physical parameters is shown in Figure 5.6, where the relaxation between two modes is taken into account in terms of C_{sub} and G_{sub} [16]. Capacitors C_{pg} and C_{pgs} represent the power-to-ground and power/ground to silicon substrate capacitances, respectively. The conductance G_{sub} is due to transverse current induced in the silicon substrate by the electric field, while C_{sub} similarly represents transverse displacement currents. The effects of the longitudinal currents and transverse magnetic field are modeled in terms of R and L. To account for the skin-depth effect, L is further divided into internal and external inductance, L_{int} and L_{ext}. The shunt capacitors C_{pgs}, C_{sub}, and C_{pg} model the transition from the slow-wave mode at lower frequencies to the quasi-TEM mode at higher frequencies. Namely, in the low-frequency limit, the conductance G_{sub} effectively shunts C_{sub} and the total capacitance is mainly C_{pgs}. In the limit of frequency equal to infinity, the capacitances make up the total dielectric quasi-TEM limit capacitance.

For the modeling of micro-strip and strip lines, the same approach can be applied. Starting with quasi-TEM mode analysis, the lossy silicon substrate effect for the slow-wave mode is then included. First, air-filled capacitance C_0 can be calculated for each cross section by using conformal mapping which is an effective method for capacitance calculation for a pair of wires [17–20]. The capacitance, including multilayer dielectrics, depends on geometry and capacitance calculation for each transmission line given in subsequent sections.

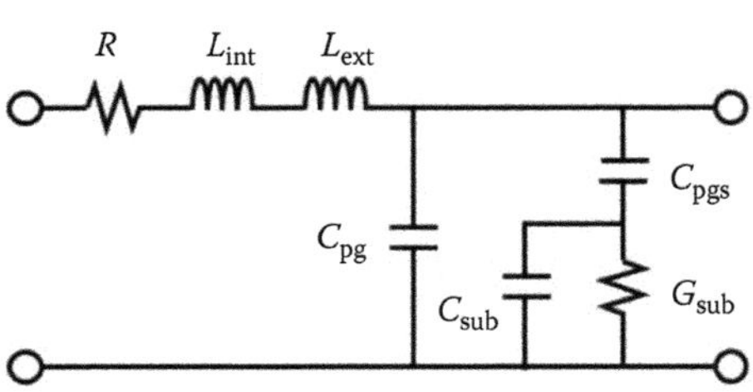

FIGURE 5.6
Equivalent RLGC model for on-chip transmission lines.

Then, assuming the TEM mode propagation, the external inductance of a lossless line L_{ext} can be represented as

$$L_{ext} = \frac{1}{c^2 C_0} \quad (5.7)$$

where c is the free-space phase velocity.

Next, the internal impedance is included, the resistance R and the internal inductance L_{int} caused by the internal current distribution. The phenomenological loss equivalence method (PEM) introduced in [21] was used. In the PEM method, a planar quasi-TEM transmission line having a finite conductor thickness on the order of the penetration depth is approximated by an equivalent single strip which has the same conductor loss as the transmission line. Derivations for each cross section are given in subsequent sections.

5.3.2 Coplanar Waveguide Structure

When the power and ground wires are interdigitated without overlapping wires in different layers, the structure can be seen as a coplanar waveguide, as shown in Figure 5.7. In this section, RLGC components for the coplanar waveguide structure were extracted.

As the power and ground wires are interdigitated, a zero tangential H-field on the side boundaries and, therefore, a perfect magnetic conductor (PMC) boundary can be set. Furthermore, the structure is symmetrical with respect to the center power line; the shunt components are the same for both directions. First, we need to calculate the air-filled capacitance, C_i^a, caused by the electric fields between the power and ground lines bounded by PMC. Applying the Schwarz-Christoffel transformation, which is used to transform the upper half-plane onto the interior of a simple polygon, the capacitance can be estimated as Equations 5.8 through 5.12.

In Figure 5.8, the dotted lines represent the PMC boundaries. The capacitance in the open boundary and the capacitance confined in the dielectric layer are calculated, taking the finite thickness of the dielectric layer into

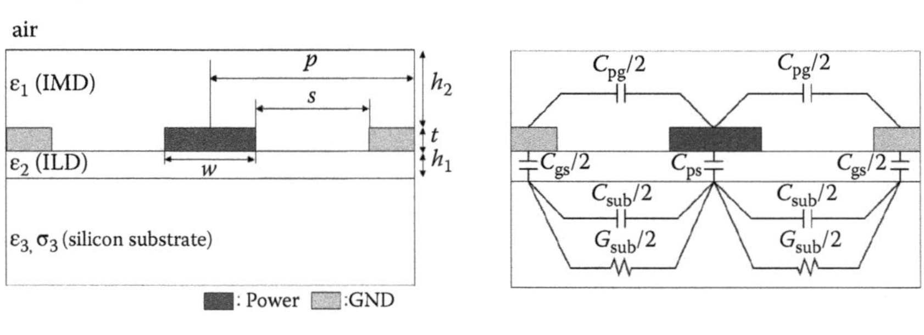

FIGURE 5.7
Coplanar waveguide structure and its equivalent shunt components.

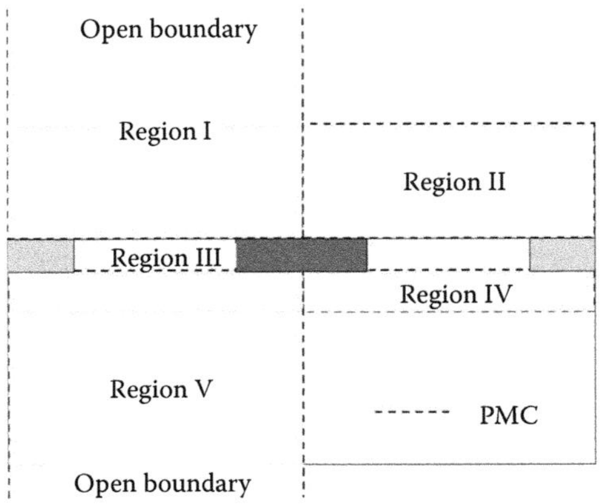

FIGURE 5.8
Boundaries for air-filled capacitance estimation; the boundaries of the dielectric layers are also replaced with PMC walls.

account; the boundary between the air and the dielectric layer is modeled as a magnetic wall and the dielectric layer is replaced by an air-filled capacitance. Using the conformal mapping, each region can be transformed into an intermediate t-plane and then converted to a final w-plane. Finally, the strips are converted to a parallel-plate capacitor so that capacitance can be represented as closed-form expressions. Since the typical range of the silicon substrate is in the order of several hundreds of μm while the space is in an order of several tens of μm, for simplicity, the open boundary is assumed for the silicon substrate. If the silicon substrate becomes comparable to the space, for example, grinded die for stacking, the finite thickness effect can be taken into account by bounding the silicon substrate with a PMC wall.

The air-filled capacitance for each region can be represented in terms of the ratio of the complete elliptic integral of the first kind and its complement, $K(k)$ and $K(k')$ as [18]:

$$C_i^a = \frac{\varepsilon_0}{2}\frac{K(k_i')}{K(k_i)} \quad i = \text{I, II, IV,V} \tag{5.8}$$

$$C_{\text{III}}^a = \varepsilon_1 \frac{t}{s} \tag{5.9}$$

where

$$k_{\text{I,V}} = \cos\left(\frac{w\pi}{2p}\right)$$

$$k_{\text{II,IV}} = sn\left(\frac{s}{2w+s}K(\zeta_i), \zeta_i\right)$$

$$\zeta_{\mathrm{II}} = \left(\left(e^{\frac{(2w+s)\pi}{2h_2}} - 2 \right) / \left(e^{\frac{(2w+s)\pi}{2h_2}} + 2 \right) \right)^2$$
$$\zeta_{\mathrm{IV}} = \left(\left(e^{\frac{(2w+s)\pi}{2h_1}} - 2 \right) / \left(e^{\frac{(2w+s)\pi}{2h_1}} + 2 \right) \right)^2 \tag{5.10}$$

$K(k)$ and $K(k')$ are the complete elliptic integral of the first type and its complement, respectively. k is the elliptic modulus and $k' = \sqrt{1-k^2}$. sn is the Jacobi elliptic function. As the conformal mapping assumes a zero-thickness conductor, Region III is introduced to take the thickness of the conductor [19]. Accurate expression for the ratio $K(k')/K(k)$ are available [22] as

$$\frac{K(k')}{K(k)} = \frac{1}{\pi} \ln\left(2\left(1+\sqrt{k'}\right)/1-\sqrt{k'}\right) \quad \text{for } 0 \le k \le \frac{1}{\sqrt{2}}$$
$$\frac{K(k')}{K(k)} = \frac{\pi}{\ln\left(2\left(1+\sqrt{k}\right)/1-\sqrt{k}\right)} \quad \text{for } \frac{1}{\sqrt{2}} \le k \le 1 \tag{5.11}$$

Next, the dielectric constants are taken into account. C_{pg} includes the electric fields confines in air and lossless dielectric layer (both upper and lower). C_{sub} corresponds to the electric fields penetrating the silicon substrate in a quasi-TEM mode. From the air-filled capacitance in Equation 5.8 and permittivity defined in Figure 5.7, the capacitance can be calculated. The calculations (C_{pg} + C_{sub}) are compared with full EM simulation results in Figure 5.9, assuming the silicon substrate is lossless. The error rate is less than 10% in most physical dimensions.

$$C_{pg} = 2\left(C_{\mathrm{air}} + C_{\mathrm{dielectric}}\right)$$
$$C_{\mathrm{sub}} = 2\left(C_{\mathrm{sub}}\right) \tag{5.12}$$

where,

$$C_{\mathrm{air}} = \left(C_{\mathrm{I}}^a - C_{\mathrm{II}}^a\right)$$

$$C_{\mathrm{dielectric}} = \varepsilon_1 C_{\mathrm{II}}^a + \varepsilon_1 C_{\mathrm{III}}^a + \varepsilon_2 C_{\mathrm{IV}}^a$$

$$C_{\mathrm{sub}} = \varepsilon_3\left(C_{\mathrm{V}}^a - C_{\mathrm{IV}}^a\right) \;\; C_0 = 4C_{\mathrm{I}}^a + 2C_{\mathrm{III}}^a$$

Then, the conductivity of the silicon substrate needs to be included for a slow-wave mechanism. The conductance can be calculated from the air-filled capacitance, as shown in Equation 5.13. We assume that C_{sub}/C_{pgs} = C_{sub} as

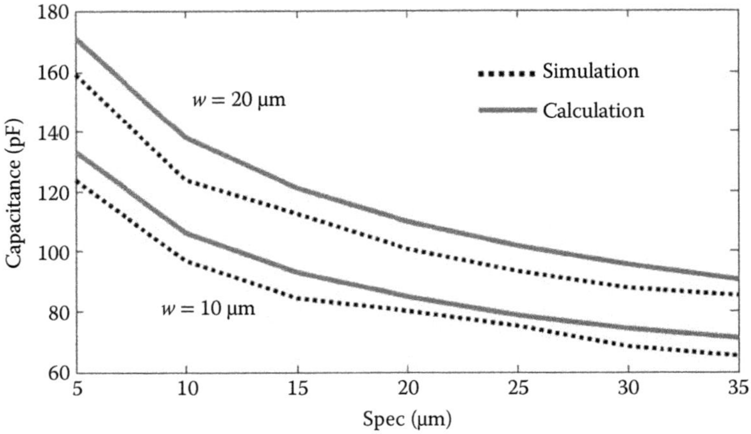

FIGURE 5.9
Capacitance comparisons between calculations and simulations ($h_1 = h_2 = 2$ μm, t = 1 μm).

the C_{pgs} is much bigger than C_{sub}, so the C_{pgs}—the capacitance between the power/ground wires and the conductive silicon substrate—are represented based on conformal mapping between a strip and infinitely large ground.

$$G_{sub} = 2\frac{\sigma_3}{\varepsilon_o}\left(C_{V}^{a} - C_{IV}^{a}\right) \tag{5.13}$$

$$C_{pgs} = \frac{1}{2}C_{ps} = \frac{1}{2}C_{gs} = \frac{1}{2}\varepsilon_2\varepsilon_0\left(\frac{w}{h_1} + \frac{4\ln 2}{\pi}\right) \tag{5.14}$$

Finally, the internal impedance was modeled, which is especially important for the ESR estimation.

The internal impedance can be calculated using the equivalent strip width ($W_e = 1/G_f$) equivalent thickness ($t_e = AG_f$) and the surface impedance of a flat-plane conductor with finite thickness, as represented by [21]:

$$Z_{int} = R + j\omega L_{int} = \frac{Z_s}{W_e}\coth\left(\frac{(1+j)t_e}{\delta}\right) = Z_s G_f \coth\left(\frac{1+j}{\delta}AG_f\right) \tag{5.15}$$

Z_s is the surface impedance of the conductor medium. G_f and A are the geometric factor and cross-sectional area of a transmission line section, respectively. The geometric factor G_f is obtained from the derivative of external inductance, L_{ext}, as presented in Equation 5.16.

$$G_f = \frac{1}{\mu}\sum_m\left(\frac{\partial L_{ext}(t,s)}{\partial n_m}\right) = \frac{1}{\mu}\left[\left(-\frac{\partial L_{ext}}{\partial t}\right)_{(1)} + \left(-\frac{\partial L_{ext}}{\partial t}\right)_{(2)} + \left(\frac{\partial L_{ext}}{\partial s}\right)_{(3)} + \left(\frac{\partial L_{ext}}{\partial s}\right)_{(4)}\right] \tag{5.16}$$

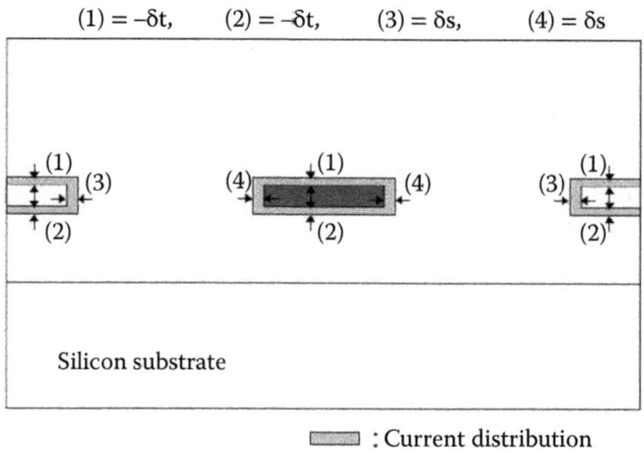

FIGURE 5.10
Current distribution caused by the field penetration and physical parameter variation according to the current distribution.

where the number m is the number of metal faces in which the shallow penetration was considered, as depicted in Figure 5.10.

Using the geometric factor, we can calculate the internal impedance of power/ground lines as

$$Z_{i,pwr} = Z_s G_f \coth\left(\frac{1+j}{\delta} A_{pwr} G_f\right) \quad Z_{i,gnd} = Z_s G_f \coth\left(\frac{1+j}{\delta} A_{gnd} G_f\right) \tag{5.17}$$

where

$$Z_s = (1+j)\frac{1}{\sigma\delta} \quad A_{pwr} = wt \quad A_{gnd} = \frac{w}{2}t$$

From Equation 5.17, the resistance and internal inductance can be calculated as:

$$R = \mathrm{Re}\left(Z_{i,pwr} + \frac{1}{2} Z_{i,gnd}\right) \quad L_{\mathrm{int}} = \frac{1}{\omega}\mathrm{Im}\left(Z_{i,pwr} + \frac{1}{2} Z_{i,gnd}\right) \tag{5.18}$$

5.3.3 Microstrip

When the power and ground wires are overlapped in different layers, the structure can be seen as a supported microstrip, as shown in Figure 5.11. Then, on the condition that the space (s) is much greater than the height (h_1) which is usually the case, a supported coplanar waveguide can be considered to be a microstrip line. In this section, RLGC components for the microstrip structure were extracted.

First, the air-filled capacitance is calculated using Equation 5.19, which is a well-known equation obtained from the conformal mapping-based empirical approach [23]:

$$C_0 = \varepsilon_0 \left(\frac{w}{h_1} + 0.77 + 1.06 \left(\frac{w}{h_1} \right)^{0.25} + 1.06 \left(\frac{t}{h_1} \right)^{0.5} \right) \tag{5.19}$$

Assuming that the partial capacitance confined in the air is negligible, the capacitance is then calculated by

$$C_{pg} = \varepsilon_1 C_0 \tag{5.20}$$

Figure 5.12 compares the capacitance from Equation 5.20 to numerical simulation results. Physical dimensions are $t = 0.5$ μm, $\varepsilon_1 = 4$, and $h_1 + h_2 = 5.5$ μm.

The next step is the extraction of the internal impedance based on a PEM approach. The geometric factor (G_f) which is associated with the external inductance, should be calculated.

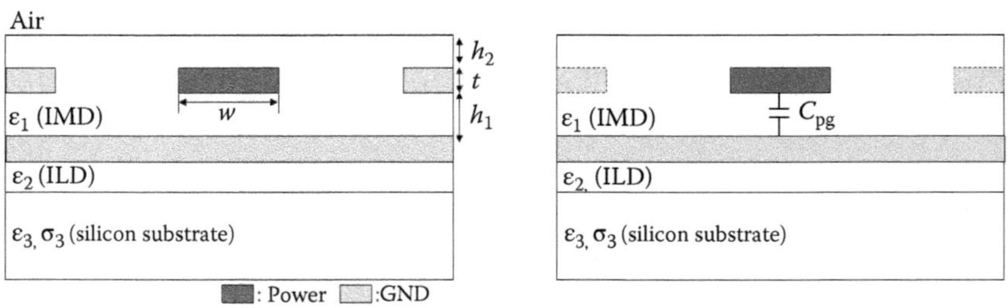

FIGURE 5.11
Microstrip structure and its equivalent shunt component.

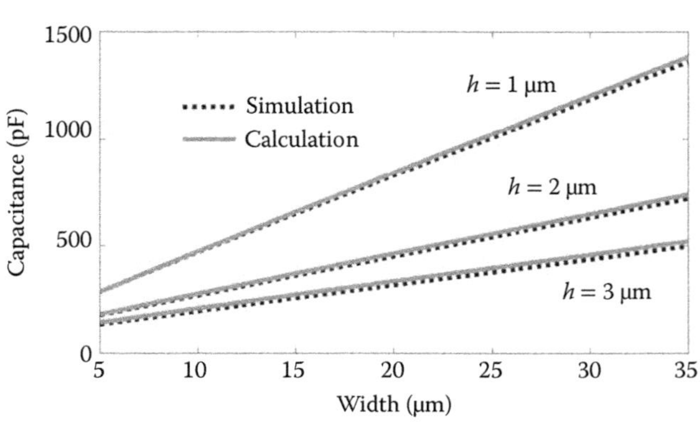

FIGURE 5.12
Capacitance comparisons between calculations and simulations.

$$G_f = \frac{1}{\mu}\sum_m \left(\frac{\partial L_{\text{ext}}(w, h_1, t)}{\partial n_m}\right)$$
$$= \frac{1}{\mu}\left[\left(\frac{\partial L_{\text{ext}}}{\partial h_1}\right)_{(1)} + \left(-\frac{\partial L_{\text{ext}}}{\partial t}\right)_{(2)} + \left(\frac{\partial L_{\text{ext}}}{\partial h_1} - \frac{\partial L_{\text{ext}}}{\partial t}\right)_{(3)} + \left(-2\frac{\partial L_{\text{ext}}}{\partial w}\right)_{(4)}\right] \quad (5.21)$$

where the number m is the number of metal faces in which we consider the shallow penetration, as depicted in Figure 5.13.

Then, the internal impedance ($Z_{i,pwr}$ and $Z_{i,gnd}$) of the power/ground lines is calculated as

$$Z_{i,pwr} = Z_s G_f \coth\left(\frac{1+j}{\delta} A_{pwr} G_f\right) \quad Z_{i,gnd} = Z_s G_f \coth\left(\frac{1+j}{\delta} A_{gnd} G_f\right) \quad (5.22)$$

where

$$Z_s = (1+j)\frac{1}{\sigma\delta} \quad A_{pwr} = wt \quad A_{gnd} = 10h_1 t$$

In the above equation, $10h_1$ is used as the width of the current density because the current predominantly flows in the area $A_{gnd}(= 10h_1 t)$. From these internal impedances, the resistance (R) and internal inductance (L_{int}) of the cross section can be calculated, as presented in Equation 5.23:

$$R = \text{Re}(Z_{i,pwr} + Z_{i,gnd}), \quad L_{\text{int}} = \frac{1}{\omega}\text{Im}(Z_{i,pwr} + Z_{i,gnd}). \quad (5.23)$$

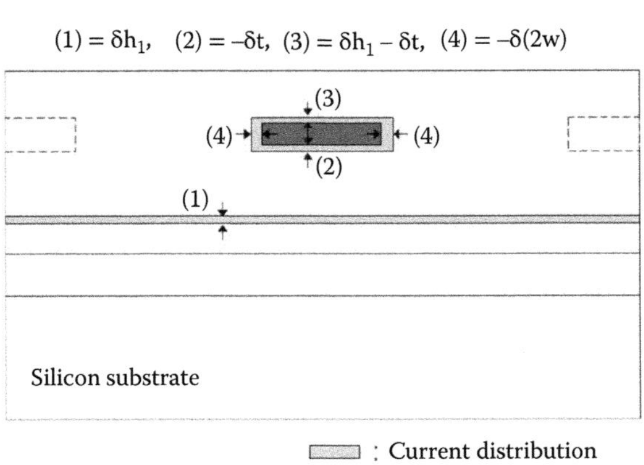

FIGURE 5.13
Current distribution caused by the field penetration and physical parameter variation according to the current distribution.

Because of the existence of the ground conductor, the effect of the silicon substrate can be ruled out. Once a model of the unit cell is obtained, the whole geometry can be modeled through the connection between the face-to-face ports of the adjacent unit cell models based on the segmentation method [24]. The segmentation method is discussed in the subsequent section.

5.3.4 Segmentation

In order to calculate the impedance of an on-chip PDN or chip-package hierarchical PDN, the segmentation method is useful to combine the segments into one network parameter [24]. Two independent structures, called structure 1 and structure 2, have rectangular-shaped PDNs with arbitrary dimensions. Port a and b, which are used for PDN impedance observations at locations a and b, are called external ports. Meanwhile, ports p at structure 1 and q at structure 2, which are used for assigning interconnections between the two independent PDNs, are referred to as internal ports.

As shown in Figure 5.14, the total PDN structure is composed of two independent, rectangular-shaped structures.

When the internal port of each structure is interconnected to complete the impedance matrix of the total structure, new boundary conditions are generated. Using these new boundaries conditions and the impedance matrices of the independent structures, structures 1 and 2, the impedance matrix of the total structure can be derived by

$$\begin{bmatrix} V_a \\ V_p \end{bmatrix} = \begin{bmatrix} Z_{aa} & Z_{ap} \\ Z_{pa} & Z_{pp} \end{bmatrix} \begin{bmatrix} I_a \\ I_p \end{bmatrix} \quad \begin{bmatrix} V_b \\ V_q \end{bmatrix} = \begin{bmatrix} Z_{bb} & Z_{bq} \\ Z_{qb} & Z_{qq} \end{bmatrix} \begin{bmatrix} I_b \\ I_q \end{bmatrix} \tag{5.24}$$

$$\begin{bmatrix} V_a \\ V_b \end{bmatrix} = \begin{bmatrix} Z_{aa} - Z_{ap}(Z_{pp} + Z_{qq})^{-1} Z_{pa} & Z_{ap}(Z_{pp} + Z_{qq})^{-1} Z_{qb} \\ Z_{ap}(Z_{pp} + Z_{qq})^{-1} Z_{qb} & Z_{bb} - Z_{bq}(Z_{pp} + Z_{qq})^{-1} Z_{qb} \end{bmatrix} \begin{bmatrix} I_a \\ I_b \end{bmatrix} \tag{5.25}$$

Consequently, the impedance matrix of the total structure can be derived using the impedance matrices of the independent structures and the segmentation method.

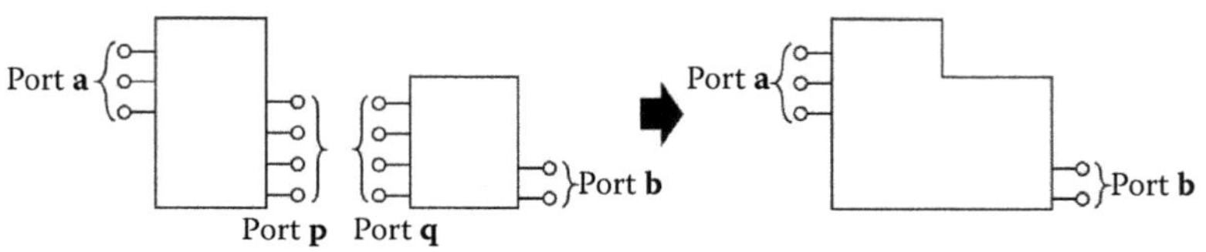

FIGURE 5.14
Segmentation method to combine two networks.

In this section, as an example, a grid-type PDN, illustrated in Figure 5.15, which provides a low impedance path in integrated circuits and silicon interposers, is considered. The grid structure is periodic such that a unit cell analysis can be applied to model the entire structure. The repeated pattern is defined as a unit cell, as shown in Figure 5.15. Each unit cell is combined into a single network using the segmentation method, which is a matrix calculation using Equation 5.25. Then, by connecting all the ports of a unit cell with the face-to-face ports of the adjacent unit cell based on the segmentation method, the whole PDN structure can be modeled.

In similar manner, the unit cell is also modeled by connecting the transmission line sections based on a segmentation method, as depicted in Figure 5.16.

For validation, a simple grid-type interposer PDN pattern with dimensions of 2 mm × 2 mm was tested. The estimated PDN impedance curves are compared with full EM simulations in Figure 5.17. Two ports are defined in

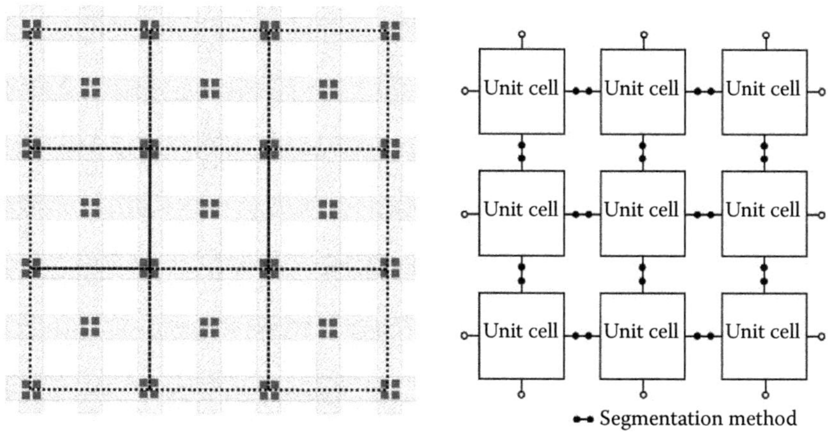

FIGURE 5.15
An example of grid-type PDN and unit cells defined from the periodic structures.

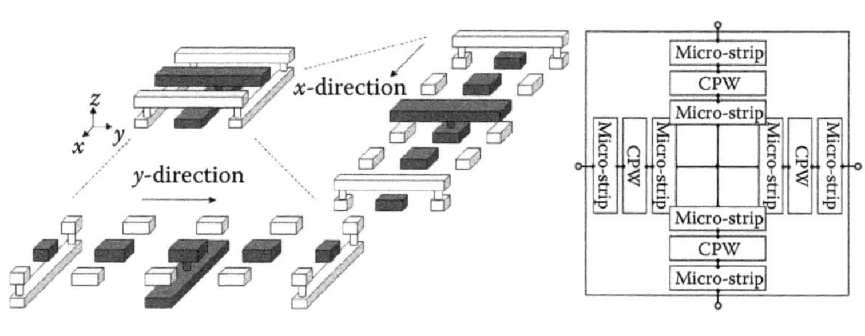

FIGURE 5.16
A unit cell is modeled by connecting the RLGC-lumped models of all transmission line sections based on a segmentation method.

the edge of the PDN (port 1: corner, port 2: middle). While the 3D EM tools take a long time (usually more than a day) to simulate the interposer PDN, the equations-based method only takes a few seconds [19].

It is quite interesting that the mode resonances appear. On-chip/silicon interposer exhibit similar behaviors to package and/or PCB PDNs where parallel plates form the mode resonances. It can be understood that the on-chip PDN can be seen as a transmission line supporting quasi-TEM mode in high frequencies and interaction with open boundaries results in the mode resonances.

The variation in the size of the interposer PDN was investigated. As shown in Figure 5.18, there are three different sizes, 2 mm × 2 mm, 4 mm × 4 mm, and 6 mm × 6 mm. Each PDN is composed of the same unit cells. The PDNs have the same port position (0 mm, 0.1 mm) in the x- and y-axis. Along with PCB-type PDN (where the PDNs are square) the inductance, L_{PDN}, and the capacitance, C_{PDN} increase proportionally to the increase in the length. As shown in Figure 5.18, the mode resonance occurs at a lower frequency range as the PDN size increases. However, Q factors at the resonance frequencies are reduced because of the resistive silicon. Remember that the slow-wave

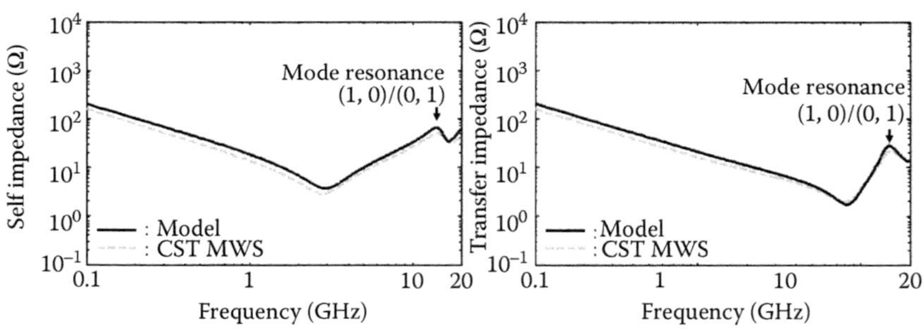

FIGURE 5.17
Self- and transfer-impedances of an interposer PDN on the silicon subtrates.

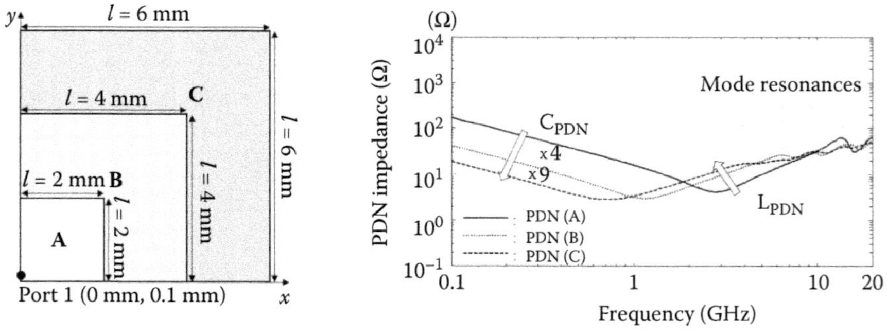

FIGURE 5.18
Interposer PDNs with different sizes and corresponding impedance curves.

mode dominates in low frequency and the resistive silicon reduces the Q factor, thereby dampening the resonance peaks.

5.4 Off-Chip PDN Structures

This section focuses on the off-chip PDN and how to combine it with an on-chip PDN. The board level PDN, ordinarily parallel planes, can be easily modeled based on a cavity resonant model. The impedance matrix between ports at anti-pads of vias in a pair of parallel planes is solved from Maxwell's equations, whose solution is the resonant cavity model as [25]

$$Z_{ij}(\omega)=\frac{j\omega\mu d}{ab}\sum_{m=0}\sum_{n=0}\frac{\varepsilon_m^2\varepsilon_n^2}{k_{mn}^2-k^2}f(x_i,y_i,x_j,y_j) \tag{5.26}$$

where

$$f(x_i,y_i,x_j,y_j)=\cos\left(\frac{m\pi x_i}{a}\right)\operatorname{sinc}\left(\frac{m\pi t_{xi}}{2a}\right)\cos\left(\frac{n\pi y_i}{b}\right)\operatorname{sinc}\left(\frac{n\pi t_{yi}}{2b}\right)$$
$$\cos\left(\frac{m\pi x_j}{a}\right)\operatorname{sinc}\left(\frac{m\pi t_{xj}}{2a}\right)\cos\left(\frac{n\pi y_j}{b}\right)\operatorname{sinc}\left(\frac{n\pi t_{yj}}{2b}\right)$$

a and b are the plane dimensions in the x- and y-directions, respectively; d is dielectric thickness or the separation between the two planes; (x_i, y_i) and (x_j, y_j) are the coordinates of the center of the ith and jth ports, respectively; (t_{xi}, t_{yi}) and (t_{xj}, t_{yj}) represent the widths of the ith and jth ports, respectively; $k=k'-jk''$, $k'=\omega\sqrt{\mu\varepsilon}$ and $k''=\omega\sqrt{\mu\varepsilon}(\tan\delta+r/d)/2$; $k_{mn}^2=(m\pi/a)^2+(n\pi/b)^2$ with m, n as the mode numbers along the x and y directions, respectively; μ, ε, and $\tan\delta$ are the permeability, permittivity, and the loss tangent of the dielectric material; $r=\sqrt{2/(\omega\mu\sigma)}$ is the skin depth, where σ is the conductivity of the metal planes; and the constant $\varepsilon_{m,n}=1$ if $m=n=0$, $\varepsilon_{m,n}=\sqrt{2}$ if $m,n\neq 0$.

Impedance Equation 5.26 can be further expressed as a parallel-plate capacitance C_P for the first cavity mode with $(m, n) = (0, 0)$ and an inductance $L_{ij}(\omega)$ for the summation of all other cavity resonant modes as [26]:

$$Z_{ij}(\omega)=\frac{1}{j\omega C_P}+j\omega L_{ij}(\omega) \tag{5.27}$$

where $C_P=\varepsilon\frac{ab}{d}$ and $L_{ij}(\omega)=\frac{\mu d}{ab}\sum_{m=0}^{\infty}\sum_{n=0}^{\infty}\frac{\varepsilon_m^2\varepsilon_n^2}{k_{mn}^2-k^2}f(x_i,y_i,x_j,y_j)\Big|_{(m,n)\neq(0,0)}$

The frequency dependent inductance $L_{ij}(\omega)$ is nearly constant at the frequencies below 60% of the first cavity resonance frequency, and, in the low frequency range, it can be approximated to a single value at DC L_{ij} when k is equal to zero. The infinite summation for m and n can be truncated in practice as soon as convergence is achieved. Equation 5.27 can be directly represented with an L-C lumped circuit model using DC inductances, L_{ij}, as shown in Figure 5.19.

The impedance parameters extracted from the measurement are compared with those from the cavity model and the L-C lumped circuit model in Figure 5.19. Additional parasitic inductance, accounting for the measurement port (SMA connector, 424 pH from a 3D full-wave solver) is included at both ports. The resonant cavity model agrees very well with measurement up to 8 GHz, whereas the circuit model agrees with the measurement only below the first cavity resonance. However, the PDN impedance seen at the IC looking into PCB is usually dominated by the IC PDN and the inductance of IC-PCB connection at the high frequency range, so the lumped circuit model is practically acceptable.

Next, the overall PDN modeling of the IC assembled with the PCB is built from the impedance directly measured on the IC. The measurement pads at the IC are connected to the PCB though the IC metals and wire-bonding and via connection, as shown in Figure 5.20. The circuit topology for IC PDN modeling can be estimated from the physics of the IC metallization. Ladder-type

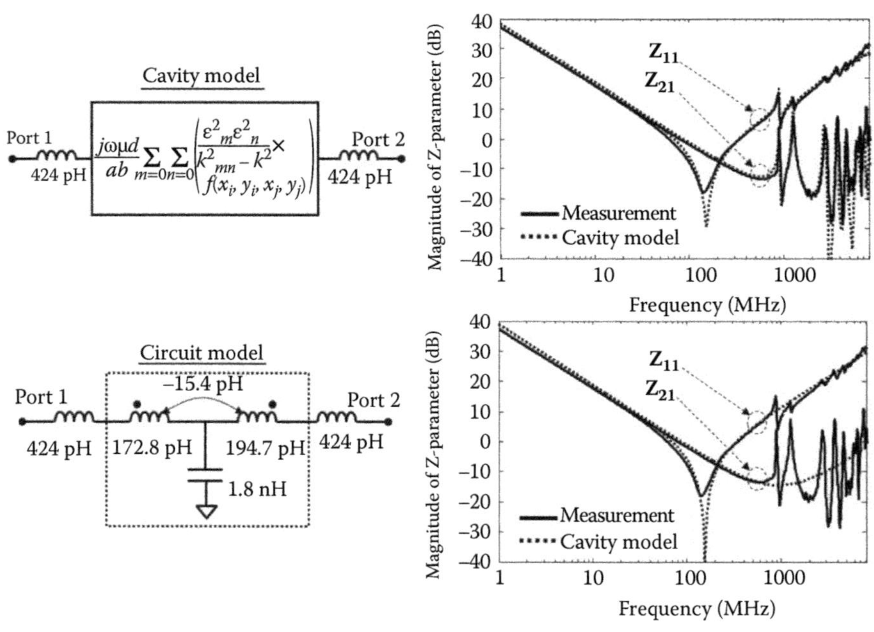

FIGURE 5.19
Simplified circuit model of the cavity model and comparisons with measurements.

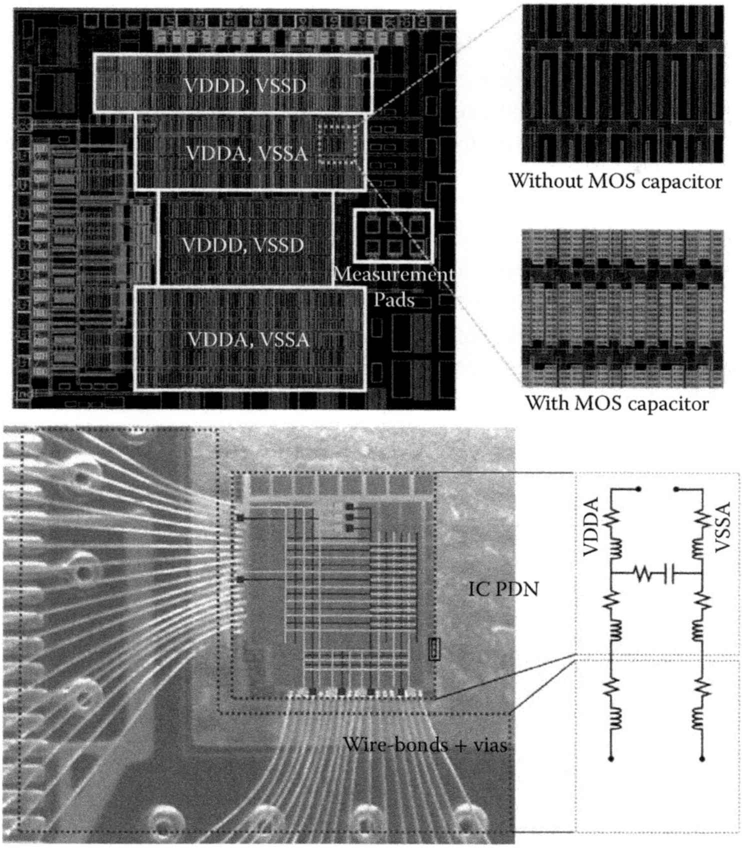

FIGURE 5.20
Layout of IC with measurement pad and circuit modeling topology for IC PDN and wire-bonding connection.

PDN was used for VDDA and GNDA rails in Metal 1 and Metal 2. Each segment can be modeled by lumped RLGC components and then combined through a segmentation method. In addition, when MOS on-chip decoupling capacitors are employed, the capacitance between VDDA and GNDA is increased significantly by the gate-channel capacitance of MOSFET. The parasitic series resistance at the silicon substrate is also decreased by the channel formed at the MOS devices. For all cases, with or without the MOS capacitor, the PDN model from the IC pads to the PCB can be approximately modeled by the circuit topology in Figure 5.20.

The circuit model of the overall PDN with the IC is built from the impedance measurement using the probing pads on the IC. In order to reduce measurement uncertainty, the transfer impedance was measured at the IC using two measurement pads in close proximity, since unwanted series contact resistance between the microprobe and the IC pad can be minimized by two-port input impedance measurement [27]. For the model of the PCB, a lumped circuit model was implemented using Equation 5.27 based on the cavity model. The modeling parameters for the IC, wire-bonding, and power supply were fitted to measured results using the physics-based circuit topology. The models and measurements for the cases without and with on-chip MOS decoupling capacitor are shown in Figures 5.21 and 5.22, respectively. The impedances from the physics-based models agree well with those from measurements. As discussed earlier, the on-chip capacitance is increased significantly and the series resistance with the capacitor is decreased when MOS device capacitors are employed. The presence of the capacitors also has the effect of reducing the impedance a few hundred MHz. Also, after the second resonance peak at hundreds of MHz, the PDN impedance seen at the IC is dominated by the impedance of the IC and wire-bonding, which means the lumped circuit model of the PCB is acceptable in terms of power integrity of the IC, as discussed.

The measurement for the IC with on-chip MOS decoupling capacitor was performed with 1.8 V and 0 V bias between VDDA and GNDA, respectively. The fitted on-chip capacitance value with 1.8 V bias is 2.0 nF, whereas that with 0 V bias is only 0.5 nF, as shown in Figure 5.22. The extracted on-chip capacitance values are significantly different for different biases, since the

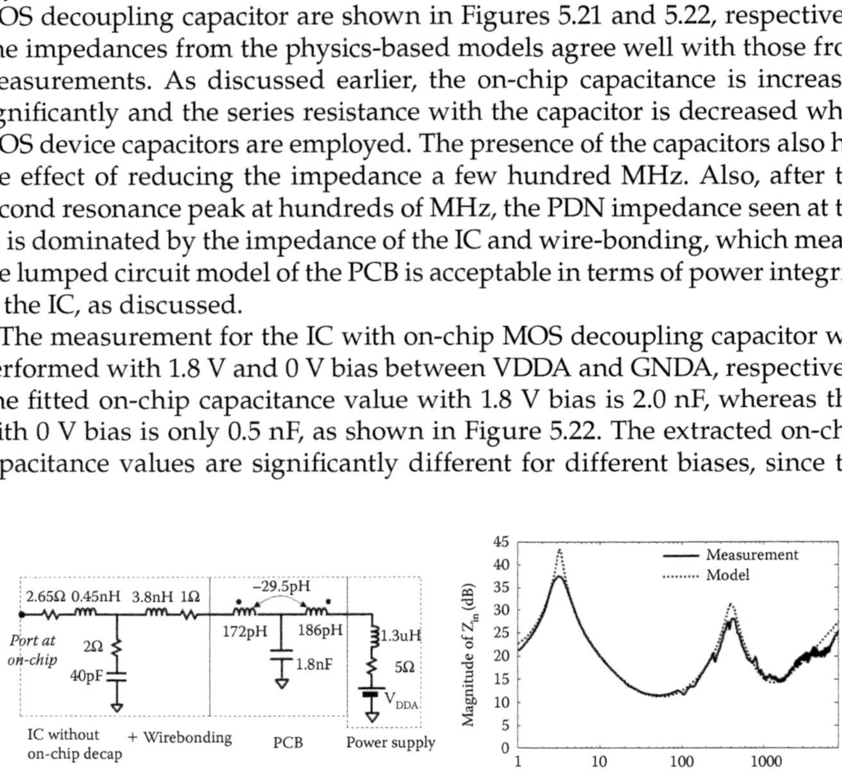

FIGURE 5.21
Lumped circuit model of overall PDN with IC not including the MOS decoupling device.

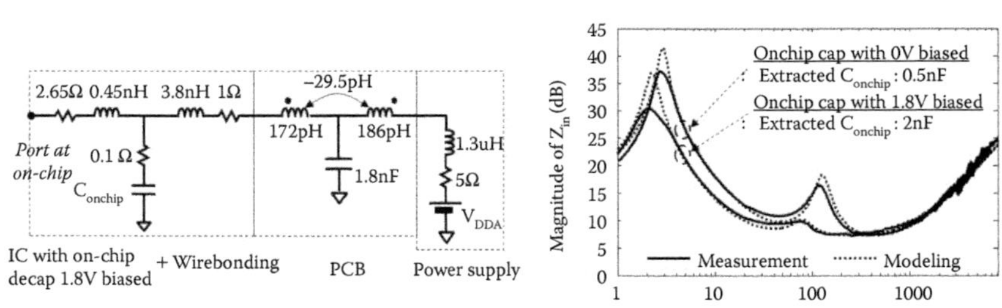

FIGURE 5.22
Lumped circuit model with IC including MOS decoupling device.

channel at MOSFET is not generated between the source and drain with 0 V gate bias, which is critical to the MOSFET device decoupling capacitor. The VDDA should be biased to a power level in the impedance measurement in order to measure actual PDN impedance for operating the IC.

5.5 Power Supply-Induced Jitter (PSIJ)

Jitter is one of the most important challenges in today's I/O interfaces as the I/O speed increases up to multi-gigabit data rates. Allowable jitter for a link has continuously decreased in order to satisfy the required bit error rate (BER) with a decreasing unit interval (UI). Jitter is commonly classified into deterministic jitter (DJ) and random jitter (RJ), which can be further split into data dependent jitter (DDJ), bounded uncorrelated jitter (BUJ), periodic jitter (PJ), Gaussian jitter, and so on [28]. Each type of jitter has corresponding root causes and the supply voltage fluctuation is one of the main causes of DJ.

Supply voltage fluctuations are mainly generated from the switching current flowing through logic and buffer circuits and the non-ideal impedance of the power distribution network (PDN). This has been an important issue in system integration for high performance and reliability [18,29–31]. Supply voltage fluctuations can be electromagnetically coupled to signal lines or cause significant delay changes, resulting in power supply-induced jitter (PSIJ) and amplitude uncertainty in the output voltage of the transmitting or receiving buffers [32–34].

Jitter due to supply voltage fluctuations has been actively studied. The delay change at a single-ended inverter buffer due to a dc level shift of supply voltage is analytically solved in [33,34]. However, the dynamic effect of supply voltage fluctuations during the buffer transition has not been considered. In [35], the power supply-induced jitter sensitivity (PSIJS) has been studied based on an abstract model that a buffer chain is assumed to be a high bandwidth homogenous medium through which the clock wave propagates at a velocity that is proportional to the power supply. Also, the jitter sensitivity of a general interface system due to supply voltage fluctuations is defined and simulated in [36,37]. The jitter sensitivity as a transfer function relating supply voltage fluctuations to jitter is demonstrated and proven to be very useful in practical engineering applications for jitter estimation.

5.5.1 Analytical Approaches for PSIJ Modeling

Analytical or semi-analytical approaches for PSIJ modeling have lots of advantages because they could provide closed-form expressions, physical insights, and the possibility of leading to significant speed-up after being combined with other link analytical methods. In [36,37], the jitter sensitivity was defined and simulated by a circuit solver, but analytical expressions were not provided.

Analytical closed-form expressions are obtained by detail circuit analyses. The expressions for the output waveform of a single-ended inverter buffer with supply voltage fluctuations can be solved from a piecewise linear buffer model [38,39]. The modeling parameters can be extracted from a SPICE device library or from measurements. As an example, the I-V curves of the PMOS and NMOS in an inverter buffer are plotted from a SPICE device library in Figure 5.23. The levels of V_{dd} and V_{SS} are 1.2 V and 0 V, respectively. Several curves of the drain current for small changes of Dv_{GS} are plotted as a function of the drain-source voltage, v_{DS}, in solid lines.

The I-V curves are divided into two regions according to the magnitude of v_{DS}. In the region where the absolute magnitude of v_{DS} is sufficiently large and the MOSFETs are in the saturation region, the drain current can be approximately represented in linear functions of the gate-source and drain-source voltages as

$$I_D(t) = G_{mp}V_{GS} + g_{mp}\Delta v_{GS}(t) + \lambda_p v_{DS}(t) \quad \text{for PMOS} \tag{5.28}$$

$$I_D(t) = G_{mn}V_{GS} + g_{mn}\Delta v_{GS}(t) + \lambda_n v_{DS}(t) \quad \text{for NMOS} \tag{5.29}$$

where G_{mp} and g_{mp} are the large- and small-signal transconductances of the PMOS; G_{mn} and g_{mn} are the corresponding parameters of the NMOS; and λ_p and λ_n are the output conductances of the PMOS and NMOS relative to the drain-source voltage, respectively. v_{DS}(t) represents the drain-source voltage and V_{GS} and Δv_{GS}(t) represent the dc level and fluctuation portion of the gate-source voltage, respectively.

In the triode region with relatively small v_{DS}(t), the PMOS and NMOS can be replaced by turn-on resistors, r_{onp} and r_{onp} respectively. The approximated linear curves for each region are plotted in dotted lines. Using the piecewise

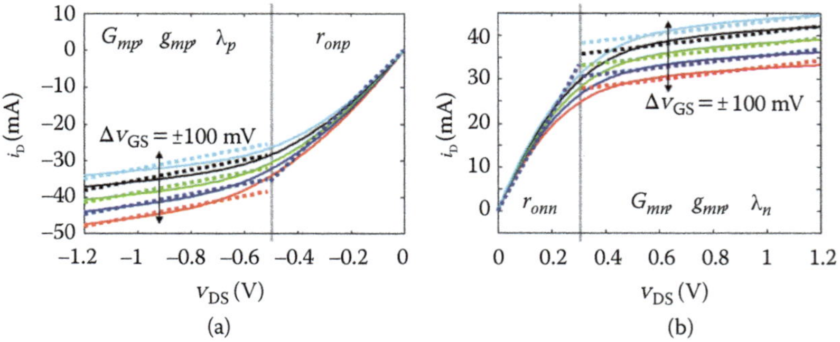

FIGURE 5.23
Piecewise linear models and modeling parameters for the total I-V curve of (a) a PMOS for pull-up transitions; and, (b) an NMOS for pull-down transitions.

modeling parameters of the MOSFETs, the expressions for a single-stage buffer output transition can be analytically calculated with given load conditions.

For a reference example, the final stage buffer driving the parallel connection of *C* and *R* through *L* is calculated, as shown in Figure 5.24. *C*, *R*, and *L* represent the load capacitance, termination resistance, and parasitic inductance of an off-chip channel, respectively. When the fluctuating voltages on the power, ground, and input nets of the buffer are denoted as $\Delta v_{DD}(t)$, Δv_{SS} (t), and $\Delta v_{in}(t)$, each equivalent MOSFET model in pull-down and pull-up transitions is expressed, as shown in Figure 5.25a and b for saturation and triode regions, respectively. The input voltage fluctuation, $\Delta v_{in}(t)$, may be caused by the output of the previous buffers in the multi-stage buffers. The reference of the load capacitance and termination resistor is assumed as zero potential.

In the pull-down transition, the NMOS is in the saturation region first because the magnitude of $v_{DS}(t)$ is relatively large. As the output rises, $|v_{DS}(t)|$ gets smaller. After $|v_{DS}(t)|$ exceeds the boundary voltage between the triode and saturation regions, NMOS moves to the triode region. Similarly, the operation condition of the PMOS is changed from the saturation to the triode region in the pull-up transition.

The piecewise linear models of the MOSFETs in the dotted blue box of Figure 5.25a and b can be universally expressed using the Thevenin equivalent voltage source and resistor, $v_{Thev}(t)$ and R_{Thev}, as shown in Figure 5.26. The expressions of $v_{Thev}(t)$ and R_{Thev} for the pull-up and pull-down transitions includes the power-ground fluctuation waveforms, $\Delta v_{DD}(t)$, $\Delta v_{SS}(t)$, as summarized in [40].

Using the universal equivalent circuit of Figure 5.26, the output voltage in the time domain is obtained as

$$v_{out}(t) = v_{out}(0) + \left(v_{Thev}(t) - \left(\frac{R_{Thev}}{R} + 1 \right) v_{out}(0) \right) u(t) * h(t) + LCv'_{out}(0)h(t) \quad (5.30)$$

where *h*(*t*) represents the transient impulse response of the circuit model in Figure 5.26. The initial conditions, v_{out} (0) and v'_{out} (0), for a triode region are

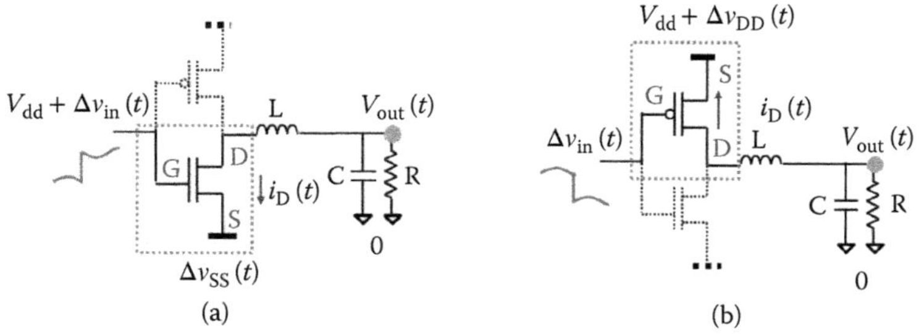

FIGURE 5.24
Buffer configurations used in the analysis for (a) pull-down and (b) pull-up transitions.

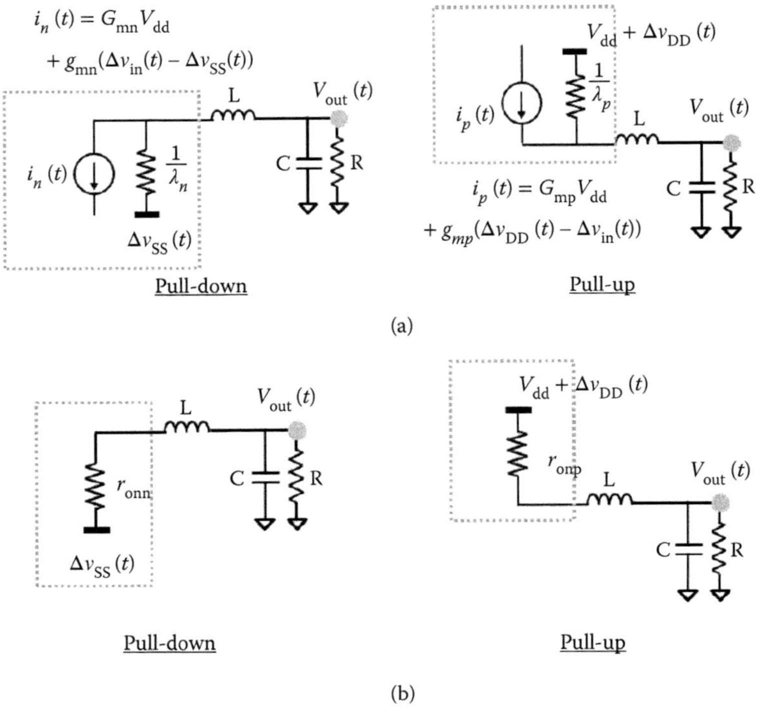

FIGURE 5.25
The equivalent MOSFET models (a) in the saturation region and (b) in the triode region.

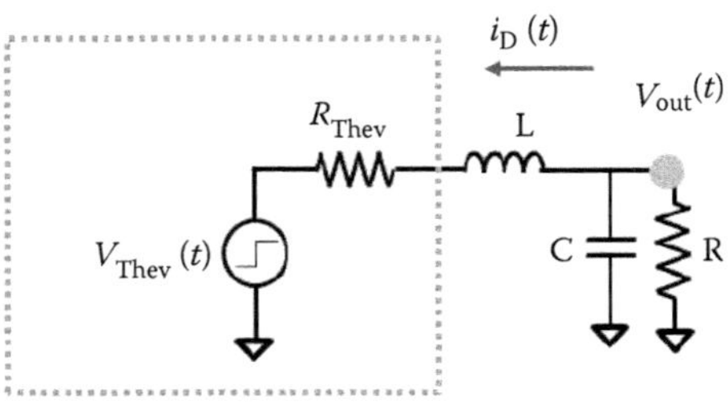

FIGURE 5.26
The universal buffer model using Thevenin equivalent circuits.

the value and slope of the output voltage at the end of the previous saturation region. Also, the initial conditions for a saturation region can be obtained from the output voltage at the triode region in the previous opposite transition. As a result, the output voltage of the buffer for both pull-up and pull-down transitions for given power supply fluctuations can be calculated directly in the time domain using Equation 5.30.

The supply voltage fluctuations on the victim buffer can be extracted from the measurement or SPICE simulations with a proper model. The on-chip power and the on-chip ground voltages at the victim buffer should be obtained separately with regard to the reference of the receiving load. For a certain supply voltage fluctuation, the output voltage waveforms of the multi-stage buffers can be calculated by applying Equation 5.30. When a buffer in the multi-stage buffers makes a transition, the gate input of the next stage buffer behaves as a capacitor load. As an example, three-stage buffers can be regarded as separated three single-stage buffers with appropriate load capacitances, as shown in Figure 5.27a. The step responses for each single-stage buffer are individually calculated. While changing the switching time of the first-stage input with regard to the fluctuating supply voltages, the step responses of the victim multi-stage buffer are piled up as "multiple step responses" in Figure 5.27b. However, since the multiple step responses

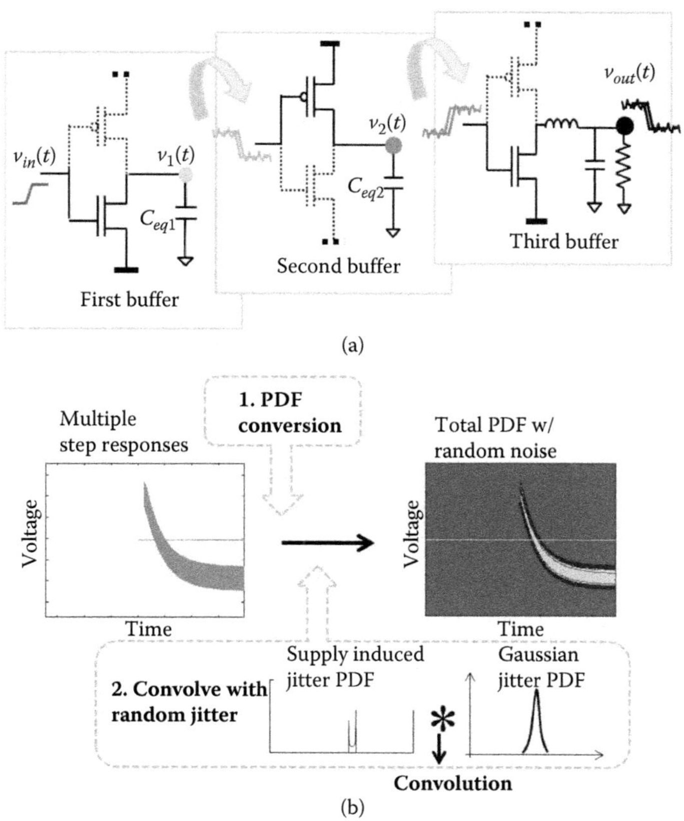

FIGURE 5.27
(a) Procedure to calculate the output voltage of three-stage buffers considering the delay variation, (b) Procedures to calculate the total PDF including supply fluctuations and random noise.

do not provide sufficient information on how often the transitions arise, they should be converted into PDFs.

At every time step after the initial switch to first-stage input, the variations in the multiple step responses according to the initial switching time can be calculated. The initial switching input can occur at any arbitrary time with regard to the supply and ground fluctuations. The channel response, due to the power supply fluctuations, is also a function of the input switching time. Mathematically, when a random variable Y is given as a function of a random variable X, the PDF of Y, $f_Y(y)$, can be calculated from the PDF of X, $f_X(x)$, as [41]:

$$f_Y(y)=\frac{f_X(x_1)}{|dy/dx|_{x=x_1}}+\frac{f_X(x_2)}{|dy/dx|_{x=x_2}}+\ldots+\frac{f_X(x_k)}{|dy/dx|_{x=x_k}} \tag{5.31}$$

where x_1, x_2, …, and x_k represent the value of X when the random variable Y has the value of y.

Then, the value of the output fluctuation voltage is swept increasing y from the smallest to the largest, while finding the corresponding input switching time, x_1, x_2, …, and x_k, for the value of y. The derivative of y at each value of x_k is also calculated. Assuming a uniform PDF of the buffer switching event, $f_X(x)$, the PDFs of the output fluctuations, $f_Y(y)$, can be numerically computed using Equation 5.32.

After calculating the PDF of step responses due to the supply voltage fluctuations, the Gaussian jitter PDF can be convolved with the supply-induced PDF to include the random noise effect. The PDFs and jitters of two-stage and three-stage output buffers are calculated and plotted in Figure 5.28a. The measured output waveforms and jitter histograms at a fabricated IC assembled with PCB are also plotted in Figure 5.28b. The values of the deterministic jitter (DJ) are approximately obtained from the time interval between the two peaks in the jitter histogram, as the tails of the histograms should correspond to the random jitter. The results from calculation and measurement show that the deterministic jitter of two-stage output is smaller than that of three-stage output. Since the detail characteristics of the off-chip output channel are lost in the lumped circuit modeling of the manufactured PCB, the calculated waveforms do not accurately capture the small bumps in the measured output waveforms; however, the overall envelope of the output waveforms and the effect of supply voltage fluctuation are sufficiently captured.

5.5.2 Analytical Derivations of PSIJ Transfer Function

For estimating the jitter caused by dynamic supply voltage fluctuations, it is very helpful to use the continuous time interval error (CTIE) concept as an alternative to the time interval error (TIE) [38]. The TIE of a signal is defined as the timing difference between the ideal edges and the actual edges, as shown in Figure 5.29. The total jitter of the signal is then determined by the

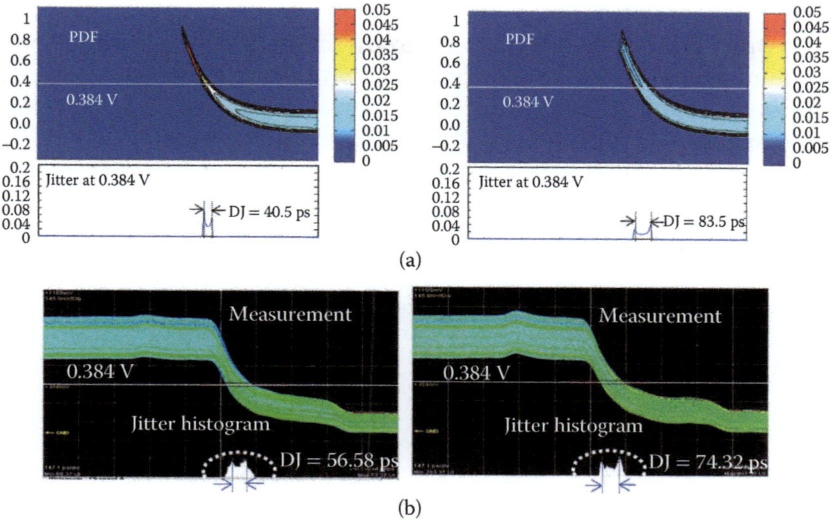

FIGURE 5.28
(a) Calculated PDFs and jitters for two- and three-stage buffers, (b) measured output voltage waveforms and jitter histogram for two- and three-stage buffers.

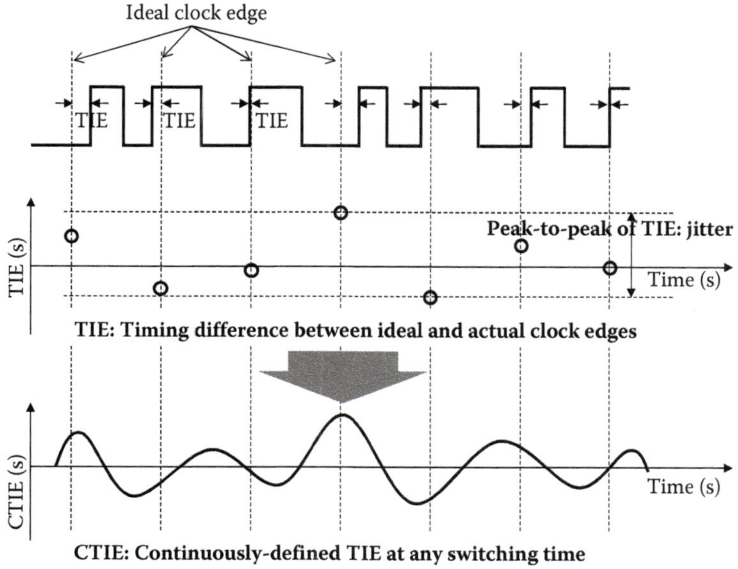

FIGURE 5.29
Comparison of jitter, TIE, and continuous TIE (CTIE).

peak-to-peak value of the TIE. In CTIE, the continuously-defined time difference between the ideal and the actual edges at any time is defined. Sampling of the CTIE at the actual switching time is then equivalent to the TIE. The

CTIE can be analytically calculated from the transfer functions relating supply voltage fluctuations to jitter, and the TIE and jitter are then determined by sampling the CTIE at the buffer input switching time. Based on the proposed concept of CTIE, the jitter caused by arbitrary supply voltage fluctuations can be easily calculated using the jitter transfer functions.

Utilizing the CTIE definition, the transfer functions relating the power and ground voltage fluctuations to jitter for a low-to-high output transition in the frequency domain can be derived by

$$H_{LH}^{ground}(\omega) = \frac{C}{G_{mp} + \lambda_p} \frac{1}{\sqrt{C^2 r_{onn}^2 \omega^2 + 1}} \frac{1}{V_{DD}} \angle(\pi - \text{atan2}(Cr_{onn}\omega, 1)) \tag{5.32}$$

$$\begin{aligned} H_{LH}^{power}(\omega) = &\frac{C}{0.5\lambda_p + G_{mp}} \frac{(\lambda_p + g_{mp})}{\lambda_p^2 + C^2\omega^2} \frac{1}{V_{DD}} \\ &\sqrt{A(\omega)^2 + B(\omega)^2} \angle(\pi - \text{atan2}(B(\omega), A(\omega))) \end{aligned} \tag{5.33}$$

where the MOSFET parameters of G_m, g_m, λ, and r_{on} are the same as shown in Figure 5.23, and

$$A(\omega) = \lambda_p \cos \omega t_{pLH0} + C\omega \sin \omega t_{pLH0} - \lambda_p \exp\left(-\frac{\lambda_p}{C} t_{pLH0}\right)$$

$$B(\omega) = -\lambda_p \sin \omega t_{pLH0} + C\omega \cos \omega t_{pLH0} - C\omega \exp\left(-\frac{\lambda_p}{C} t_{pLH0}\right)$$

$$t_{pLH0} = \frac{C}{\lambda_p} \ln \frac{G_{mp} + \lambda_p}{G_{mp} + 0.5\lambda_p}$$

Similarly, the jitter transfer functions for a high-to-low output transition are derived as

$$H_{\text{HL}}^{\text{power}}(\omega) = \frac{C}{G_{mn} + \lambda_n} \frac{1}{\sqrt{C^2 r_{onp}^2 \omega^2 + 1}} \frac{1}{V_{DD}} \angle(-\text{atan2}(Cr_{onp}\omega, 1)) \tag{5.34}$$

$$\begin{aligned} H_{\text{HL}}^{\text{ground}}(\omega) = &\frac{C}{0.5\lambda_n + G_{mn}} \frac{(\lambda_n + g_{mn})}{\lambda_n^2 + C^2\omega^2} \frac{1}{V_{DD}} \\ &\times \sqrt{A(\omega)^2 + B(\omega)^2} \angle(-\text{atan2}(B(\omega), A(\omega))) \end{aligned} \tag{5.35}$$

where $A(\omega)$ and $B(\omega)$ represent the N-MOS counterparts.

For arbitrarily given power and ground voltage fluctuations, the spectral components of the CTIE for each output transition are then calculated, respectively, as

$$CTIE_{\text{LH}}(\omega) = H_{\text{LH}}^{\text{ground}}(\omega)V_g(\omega) + H_{\text{LH}}^{\text{power}}(\omega)V_p(\omega) \tag{5.36}$$

$$CTIE_{\text{HL}}(\omega) = H_{\text{HL}}^{\text{ground}}(\omega)V_g(\omega) + H_{\text{HL}}^{\text{power}}(\omega)V_p(\omega) \tag{5.37}$$

where $V_g(\omega)$ and $V_p(\omega)$ are the spectra of power and ground voltage fluctuations, respectively. The CTIE(ω) in Equations 5.36 and 5.37 are reconstructed in the time domain, and then the total jitter is determined by the peak-to-peak value of the total propagation delay waveforms in the time domain as

$$\begin{aligned} Jitter &= \max(t_{pLH}(t_s), t_{pHL}(t_s)) - \min(t_{pLH}(t_s), t_{pHL}(t_s)) \\ t_{pLH}(t_s) &= t_{pLH0} + CTIE_{LH}(t_s) \\ t_{pHL}(t_s) &= t_{pHL0} + CTIE_{HL}(t_s) \end{aligned} \tag{5.38}$$

where t_s represents the switching time of the buffer input, while t represents the time passed after the buffer input.

In Equation 5.38, the jitter caused by the propagation delay skew, $|t_{pHL0} - t_{pLH0}|$, is also included. The propagation delay skew is considered here to take into account possible propagation delay unbalance between the low-to-high transition and the high-to-low transition. The overall procedure for estimating jitter caused by general supply voltage fluctuations is summarized in Figure 5.30.

The jitter estimation using Equation 5.38 takes into account the relationship between the input switching frequency and the frequency components in the supply voltage fluctuations by sampling the CTIE at each switching time, t_s. If the input switching frequency and the frequencies in the supply voltage fluctuations do not have a common divisor, the time t_s can be replaced with an arbitrary time, t. When the propagation delay skew is negligible, t_{pLH0} and t_{pHL0} can be removed from Equation 5.38. Also, for the inverter buffer, which has the same driving strength for the NMOS and PMOS, the spectra CTIE_{LH} and CTIE_{HL} are eventually the same.

The validation of the transfer functions in Equations 5.32 and 5.33 are shown in Figure 5.31.

The magnitudes of the transfer functions are compared to those simulated in HSPICE. In HSPICE, the jitter transfer functions are obtained by using the simulated jitter with single-frequency voltage fluctuations and normalizing it with the amplitude of the fluctuations. When both power and ground voltage fluctuations are present, they are assumed to have the same phase. The frequency of the buffer input switching is chosen so as not to have a common divisor with the supply voltage fluctuations. The jitter transfer functions extracted from the simulations are very close to the analytic expressions.

Consider the transfer function relating power voltage fluctuation to jitter for a low-to-high output transition. For low frequencies where the propagation delay is much smaller than the period of the supply voltage

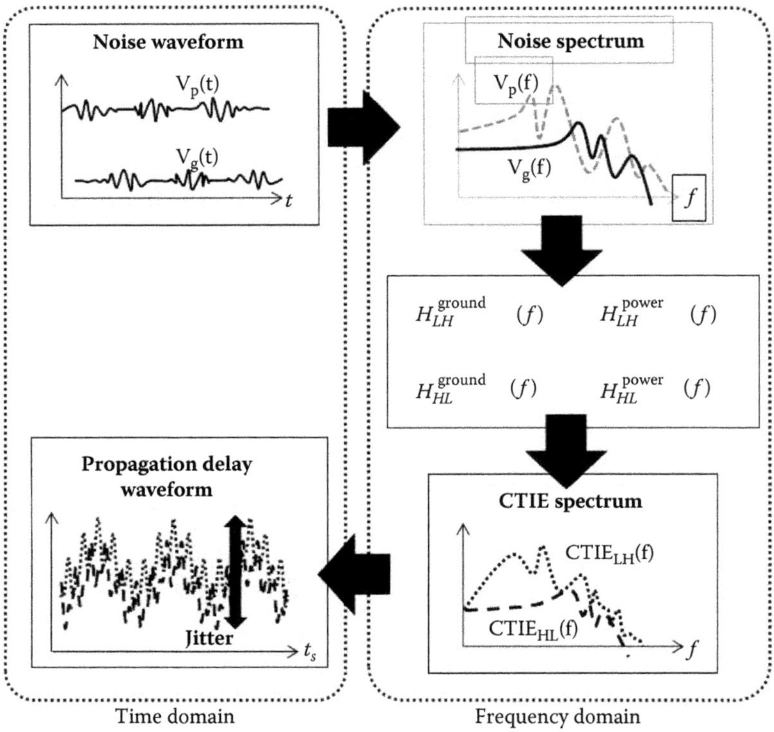

FIGURE 5.30
The overall procedure for estimating jitter caused by general power and ground noise.

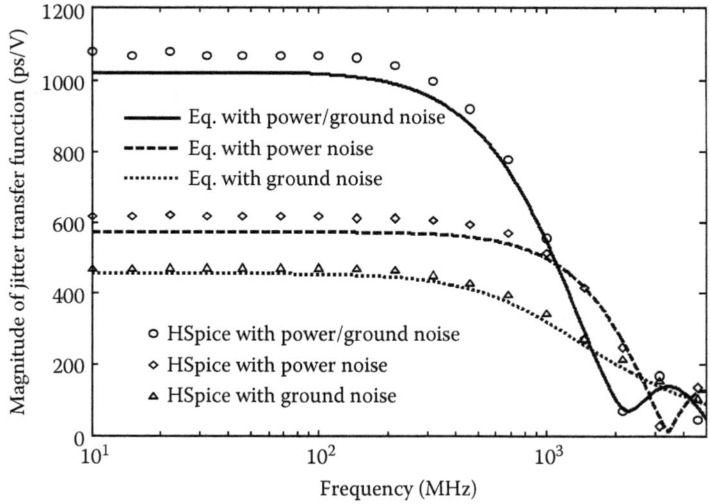

FIGURE 5.31
Transfer functions of supply voltage fluctuation to jitter obtained from HSPICE simulation and analytical expression.

fluctuations, the phase of the supply voltage fluctuations does not change much and is approximately constant during the output transition. Hence, at low frequencies supply voltage fluctuations affect jitter in the same manner as a dc offset. Thus, the transfer function is constant over frequency, because the calculated maximum time variation is determined only by the amplitude of the supply voltage fluctuations. As the frequency increases, the transfer function takes into account the ac characteristics of the supply voltage fluctuations and the resulting jitter decreases. At the frequencies where the period of the fluctuations is an integer multiple of the propagation delay, the resulting jitter approaches zero, because the overall effect of the supply fluctuations on the time variation is always constant regardless of the switching time. In the transfer function relating ground voltage fluctuation to jitter for a low-to-high output transition, as previously noted, since the magnitude of the square root of $C^2 r_{on}^2 \omega^2 + 1$ is inversely proportional to jitter, the resulting jitter is constant at low frequencies and decreases at a rate of 20 dB/Dec at high frequencies.

The jitter caused by multi-tone supply voltage fluctuations is evaluated using the jitter transfer functions. The derived analytical expressions for the jitter transfer functions have both magnitude and phase information, which can be easily applied to multi-tone voltage fluctuations. For an example, power and ground voltage fluctuations are given by

$$\begin{aligned} V_{np}(t) &= 60 \text{ mV} \cos(2\pi f_1 t) + 50 \text{ mV} \cos(2\pi f_2 t + \pi/4) + 40 \text{ mV} \cos(2\pi f_3 t + \pi/2) \\ V_{ng}(t) &= 60 \text{ mV} \cos(2\pi f_1 t + \pi) + 40 \text{ mV} \cos(2\pi f_2 t + \pi/2) + 20 \text{ mV} \cos(2\pi f_3 t + \pi) \end{aligned} \tag{5.39}$$

where f_1 = 300 MHz, f_2 = 500 MHz, and f_3 = 1 GHz.

The frequency components of the jitter are obtained as the product of the jitter transfer functions and the spectral components of the supply voltage fluctuations. The waveforms of the CTIE can be obtained by converting the frequency spectra into the time domain, and the calculated overall propagation delays, including the CTIE induced by the supply voltage fluctuations, are plotted in Figure 5.32. The jitter caused by the propagation delay skew is also included. When the frequencies in the supply voltage fluctuations are not synchronized with the buffer switching frequency, the jitter can be evaluated by observing the peak-to-peak value in the propagation delays, which is 140.3 ps, as shown in Figure 5.32. Multiple output transitions obtained from HSPICE simulations are also plotted in Figure 5.33. The simulated jitter is within a few percent of the calculated jitter.

In summary, the transfer functions can be obtained using device parameters accessible from device libraries or measurements, and so the transfer functions can be calculated in an easy and fast way without time-consuming simulations. Since the jitter at a transmitter is dominated by the output buffer, the jitter caused by supply voltage fluctuations can be estimated without knowing proprietary inner circuit information by using the jitter transfer

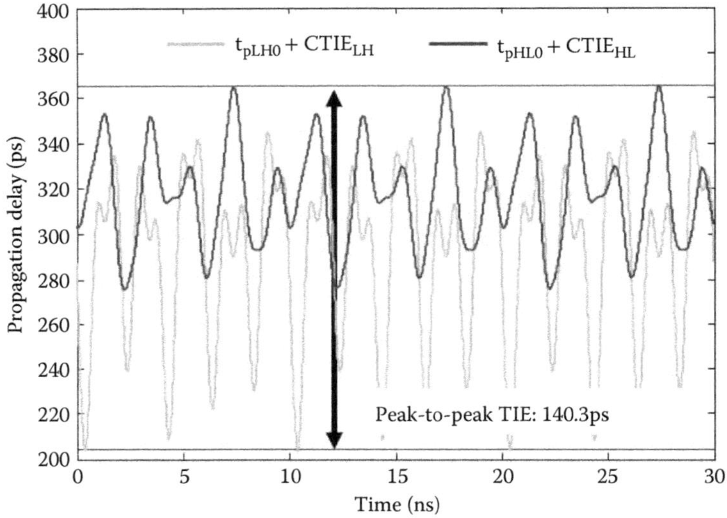

FIGURE 5.32
Calculated propagation delay variation and jitter from the peak-to-peak TIE sampled at switching time.

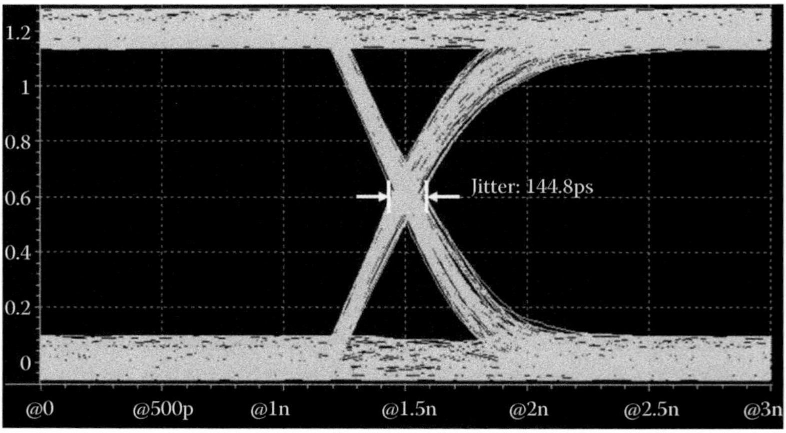

FIGURE 5.33
Multiple output transitions obtained from HSPICE simulation with the multi-tone noise described in Equation 5.39 and a buffer switching frequency of 350 MHz.

functions of the output buffer stage using the I-V curve. The analytic derivations of the transfer functions based on the CTIE concept can be applied for general circuits under a reasonable approximation. Even though only a capacitor load is assumed in the formulation, the jitter transfer function for various termination schemes can also be formulated in the same manner using appropriate termination models.

5.5.3 Extension to I/O Link BER Analysis

The previous timing specifications for double data rate (DDR) memory have provided only the total worst case jitter. The total worst jitter was simply calculated as the summation of every jitter components from I/O circuit blocks and channels. The design approach checking the set-up and hold time violations due to the total worst jitter was sufficient at lower data rates; however, it can lead to significant errors at higher data rates, resulting in over- or under-designed systems. Hence, the statistical link specification methods have been proposed to improve the predictability of maximum attainable data rate [42]. The maximum data rate of an I/O link path is determined based on the BER at a given data rate [43,44]. The BER can be simulated by considering all possible input data patterns with a long data sequence; however, it requires large computational efforts to estimate a low BER value like 10^{-16}. In the analytic statistical approaches for the I/O link analysis [45–47], the output voltage probability density function (PDF) due to inter-symbol interference (ISI) has been analytically calculated using pulse responses based on random data patterns and the linear time invariant assumptions, which are further used to effectively obtain the BER.

There are several methods to generate the ISI PDFs, such as the methods based on the single-bit responses (SBR), double edge responses (DER), and multiple edge responses (MER). The SBR method assumes the system response to any data pattern is the linear sum of the shifted version of SBR, which is not applicable for asymmetric I/O mismatches between pull-up and pull-down drivers. In the DER and MER methods, the system response is constructed in terms of step responses, instead of bit responses, to handle the asymmetric rising and falling edges.

In using the DER method, the statistical link analysis method that considers only ISI effects can be developed based on the step responses [41]. The rising and falling edge responses start from zero voltages, obtained as

$$v_r(t) = v_{outLH,0}(t) - v_{outL} \tag{5.40}$$

$$v_f(t) = v_{outHL,0}(t) - v_{outH} \tag{5.41}$$

where the $v_{outLH,0}$(t) and $v_{outHL,0}$(t) represent the responses of the rising/falling step pulses, and the v_{outL} and v_{outH} represent the DC output voltages in steady low and high states.

The rising or falling edge responses of any previous data are the sources of ISI, as depicted in Figure 5.34a. The ISI PDF can then be recursively calculated by convolving the probability of individual ISI samples. Assuming equal probability of a 1 or 0, the ISI PDFs for a 1 or 0 observed at the present sampling time, denoted as $z_0^{(1)}(v,t)$ and $z_0^{(0)}(v,t)$, are recursively calculated from $k=-\infty$ to $k=0$ as

$$z_k^{(1)}(v,t)=\frac{1}{2}\delta(v)*z_{k-1}^{(1)}(v,t)+\frac{1}{2}\delta(v-v_{r,k})*z_{k-1}^{(0)}(v,t) \tag{5.42}$$

$$z_k^{(0)}(v,t)=\frac{1}{2}\delta(v)*z_{k-1}^{(0)}(v,t)+\frac{1}{2}\delta(v-v_{f,k})*z_{k-1}^{(1)}(v,t) \tag{5.43}$$

where the $v_{r,k}$ and $v_{f,k}$ indicate the step response voltages shifted by the time of k unit-intervals (UI) at the given sampling time, whose PDFs are represented as the Kronecker delta functions. The $z_{k-1}^{(0)}(v,t)$ and $z_{k-1}^{(1)}(v,t)$ are the ISI PDFs up to the previous bits.

In Equation 5.42 for example, the following is processed: in order to obtain the present ISI PDF for the logic '1', $z_0^{(1)}(v,t)$, the previous ISI PDF of '1' is convolved with the PDF of the no-edge response, $\frac{1}{2}\delta(v)$, and the previous ISI PDF of '0' is convolved with the PDF of corresponding rising edge response, $\frac{1}{2}\delta(v-v_{r,-1})$. The recursive parameters 'k' from the initial negative value to −1 represent the ISI samples to be included. The initial conditions for the ISI PDFs of '1' or '0' at k = −∞ are given by the steady DC output voltages as

$$z_{-\infty}^{(1)}(v,t)=\frac{1}{2}\delta(v-v_{\text{out}H})\text{ and }z_{-\infty}^{(0)}(v,t)=\frac{1}{2}\delta(v-v_{\text{out}L}) \tag{5.44}$$

The statistical procedure calculating the ISI PDF is depicted in detail in Figure 5.34b. The statistical eye diagram can be obtained by sweeping the sampling time t in calculation of the ISI PDF and then combining the PDFs together.

In addition to ISI, as aforementioned, switching the numerous logic circuits and I/O buffers inside integrated circuits (ICs) can result in significant voltage drop and a ripple in the PDN, which results in power-supply induced jitter (PSIJ) and amplitude uncertainty in the output voltage of the transmitting or receiving buffers. To calculate the total BER including the effect of the supply voltage fluctuations, the output amplitude uncertainly due to the supply voltage fluctuation as well as the jitter at a logic threshold should be computed, since the BER is estimated at a receiver sampling time according to the data rate. The method to include the power-supply induced output PDF (PSI PDF) into the statistical link path analysis for random data pattern has been proposed and experimentally validated in [41,48].

The output response due to the alternate supply and ground fluctuations depends on the buffer switching time with regard to the fluctuating voltages. Assuming the supply and ground fluctuations are predominantly generated by other external circuits rather than the buffer switching itself, the PDF of the buffer switching event can be assumed to be uniform within the longest

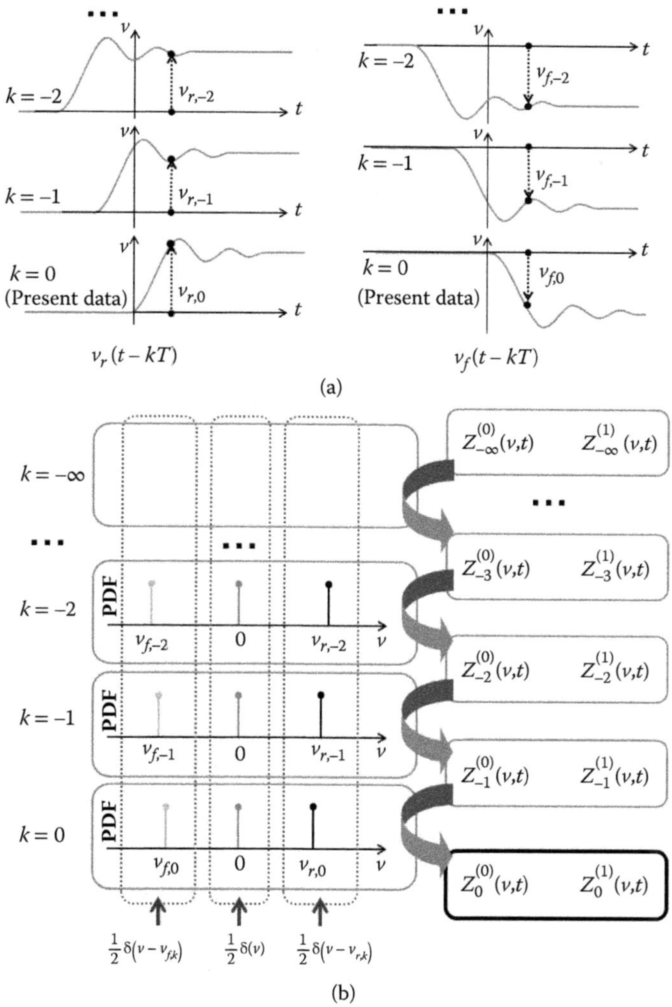

FIGURE 5.34
(a) Rising and falling step responses of the present (k = 0) and previous (k<0) data, (b) procedure for ISI PDF calculation.

period of the voltage fluctuations. Also, the output driver buffer is assumed as a linear driver, which is often used to reduce ring-back and improve the signal quality. The linear driver assumption allows all theoretical analysis to be simple and complete.

The output amplitude uncertainty due to supply voltage fluctuations, i.e., the power-supply induced output PDF (PSI PDF) can be numerically calculated from the output voltage fluctuations by using Equation 5.4. That is, the pull-up and pull-down PSI PDFs, $f_{nr}(v, t)$ and $f_{nf}(v, t)$, are computed as

$$f_{nr}(v,t)=\frac{1}{(t_{s,\text{end}}-t_{s,\text{start}})}\times\left[\frac{1}{\left|\partial v_{\text{out}LH,n}(t,t_s)/\partial t_s\right|_{t_s=t_{s,1}}}+\frac{1}{\left|\partial v_{\text{out}LH,n}(t,t_s)/\partial t_s\right|_{t_s=t_{s,2}}}+\cdots+\frac{1}{\left|\partial v_{\text{out}LH,n}(t,t_s)/\partial t_s\right|_{t_s=t_{s,k}}}\right] \tag{5.45}$$

$$f_{nf}(v,t)=\frac{1}{(t_{s,\text{end}}-t_{s,\text{start}})}\times\left[\frac{1}{\partial v_{\text{out}HL,n}(t,t_s)/\partial t_s\big|_{t_s=t_{s,1}}}+\frac{1}{\partial v_{\text{out}HL,n}(t,t_s)/\partial t_s\big|_{t_s=t_{s,2}}}+\cdots+\frac{1}{\partial v_{\text{out}HL,n}(t,t_s)/\partial t_s\big|_{t_s=t_{s,k}}}\right] \tag{5.46}$$

where the $t_{\text{s,start}}$ and $t_{\text{s,end}}$ represent the buffer switching boundary under consideration in the time frame of the supply and ground voltage fluctuations. The $v_{\text{out}LH,n}(t,t_s)$ and $v_{\text{out}HL,n}(t,t_s)$ represent the responses of the alternate supply and ground fluctuations. The $t_{\text{s},1}$, $t_{\text{s},2}$, …, and $t_{\text{s},k}$ represent all the values of t_s when the $v_{\text{out}LH,n}(t,t_s)$ or $v_{\text{out}HL,n}(t,t_s)$ has the voltage value of v.

Once the pull-up and pull-down PSI PDFs, $f_{nr}(v, t)$ and $f_{nf}(v, t)$, are found, they can be combined in the recursive convolution for ISI PDF, by extending the procedure in Figure 5.34. In addition to the rising or falling edge responses of the present and previous data, the PSI PDFs at the present and previous data transitions are also utilized, as shown in Figure 5.35a. The $f_{nr,k}$ and $f_{nf,k}$ represent the PSI PDFs shifted by the time of k-UI at the given sampling time.

The convolving Equations 5.15 through 5.17 are then extended to include each PSI PDF of every possible data transition that results in the present state. That is, the PDF of the present state should be convolved with the corresponding PSI PDF of any possible previous data transitions contributing to the present state. The total PDF including both the ISI and PSI PDFs, which are denoted as $z_{n,0}^{(1)}(v,t)$ and $z_{n,0}^{(0)}(v,t)$ for 1 and 0, respectively, are recursively calculated from k = −∞ to k = 0 as

$$z_{n,k}^{(1)}(v,t)=\frac{1}{2}\delta(v)*z_{n,k-1}^{(1)}(v,t)+\frac{1}{2}\delta(v-v_{r,k})*z_{k-1}^{(0)}(v,t)*f_{nr,k} \tag{5.47}$$

$$z_{n,k}^{(0)}(v,t)=\frac{1}{2}\delta(v)*z_{n,k-1}^{(0)}(v,t)+\frac{1}{2}\delta(v-v_{f,k})*z_{k-1}^{(1)}(v,t)*f_{nf,k} \tag{5.48}$$

where $z_{k-1}^{(1)}(v,t)$ and $z_{n,0}^{(0)}(v,t)$ represent the PDF with only the ISI effect, as given in Equations 5.15 and 5.43.

In addition to the initial conditions of Equation 5.44, the initial conditions of the total PDF with the PSI PDF should also be given as

$$z_{n,-\infty}^{(1)}(v,t) = \frac{1}{2}\delta(v - v_{\text{out}H}) * f_{nH} \text{ and } z_{n,-\infty}^{(0)}(v,t) = \frac{1}{2}\delta(v - v_{\text{out}L}) * f_{nL} \quad (5.49)$$

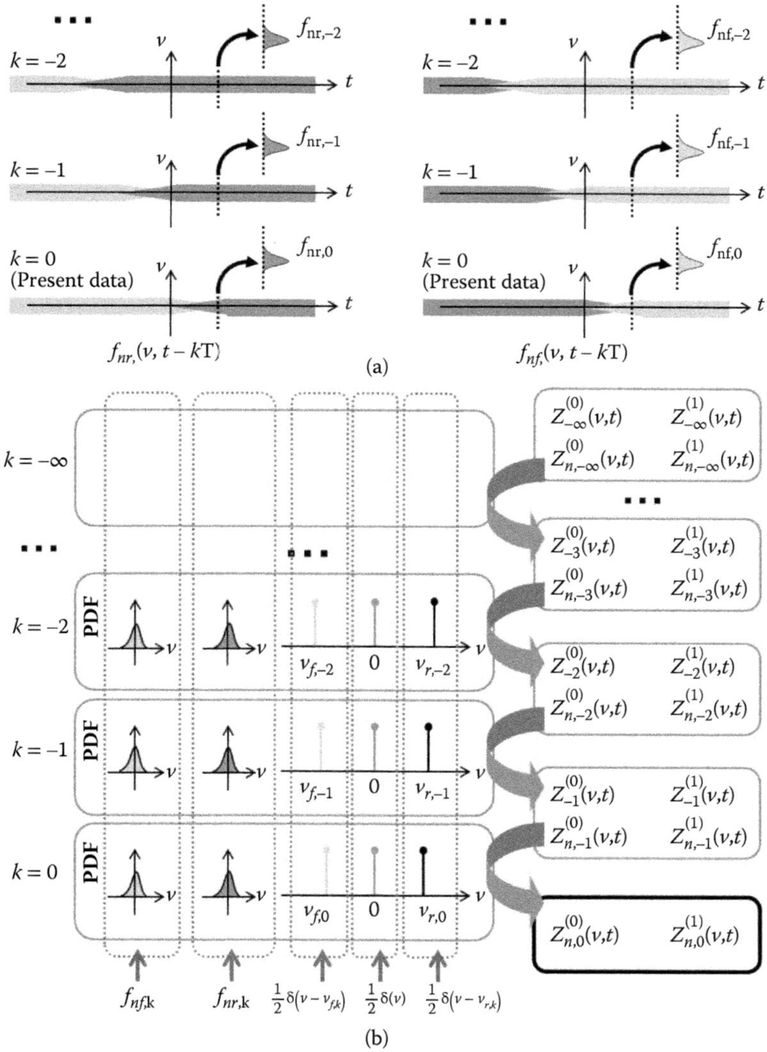

FIGURE 5.35
(a) Power supply induced (PSI) PDFs at rising and falling transitions of the present ($k = 0$) and previous ($k < 0$) data, (b) total PDF calculation combining ISI PDF and PSI PDF.

where f_{nL} and f_{nH} represent the PSI PDFs in steady, low, and high states, respectively.

The enhanced statistical procedure considering both ISI and supply voltage fluctuations is also depicted in Figure 5.35b. In order to calculate the total PDF at the present data, $z_{n,0}^{(1)}(v,t)$, $z_{n,0}^{(0)}(v,t)$, it is necessary to calculate $z_{n,k}^{(1)}(v,t)$, $z_{n,k}^{(0)}(v,t)$, $z_{k}^{(1)}(v,t)$, and $z_{k}^{(0)}(v,t)$ for every negative k.

A silicon IC, which consists of a large-sized noise-aggressing buffer and a victim linear driver of small size, has been designed and fabricated using a 0.11 µm CMOS process to validate the statistical analysis. Large switching currents are generated at every toggle of the large-sized aggressing buffer. The small-sized victim buffer shares the power and ground nets (V_{DDIO}-V_{SSIO}) with the aggressing buffer, and the output of the victim buffer is predominantly affected by the voltage fluctuations due to the aggressor. The victim buffer is excited by 2.4 Gbps pseudo random bit sequence (PRBS) data generated by a pulse pattern generator (PPG). The noise aggressing buffer is excited by a 125 MHz periodic clock pulse that is generated by a crystal oscillator and a clock buffer. The output of the victim driver is connected to a microstrip line on the PCB, and further connected to another PCB pattern with a long microstrip line to increase the effect of ISI. The eye diagram, jitter, and voltage histograms of the final output on the other end of the microstrip line was measured using a sampling oscilloscope. The V_{DDIO}-V_{SSIO} voltage fluctuations at the both chip and PCB probing pads are simultaneously measured using a real time oscilloscope, and the on-chip power and on-chip ground voltage fluctuations with regard to the PCB ground can be extracted from the measurements using the two probing pads at the IC and PCB.

As a base line, the statistical eye-diagram at 2.4 Gbps with only ISI PDF, which was calculated by the procedure in Figure 5.34 and plotted in Figure 5.36a and b, shows the measured eye diagram in the same voltage and time scales with the aggressing buffer off, where the jitter histogram crosses 235 mV DC level and the voltage PDF at a sampling time near the center of the eye are also shown together. The measured jitter and voltage histograms are also compared with the jitter and voltage PDFs calculated at the same crossing voltage and sampling time. The shape of the eye, jitter, and voltage PDFs are quite similar between the calculation and measurement.

The aggressing buffers are turned on, and Figure 5.37 shows the calculated and measured results of the eye diagram, jitter, and voltage uncertainties considering both ISI and PSI PDFs. The histograms and PDFs of the jitter and voltages are wider than those in Figure 5.36.

In the enhanced statistical I/O link analysis method shown in Figure 5.35, the supply voltage fluctuations were assumed to be external noise were generated by other noisy circuits and independent of the output buffer operation itself. In addition to the external noises, the self-generated noise induced by the switching currents of the output buffers themselves can also cause jitters or uncertainties in the output waveforms. The effects of the self-generated noise are especially serious in the single-ended simultaneous switching

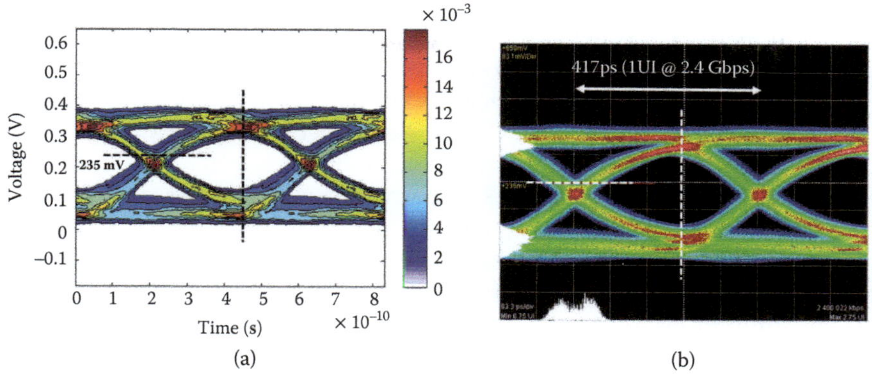

FIGURE 5.36
Eye diagrams with only ISI effect (a) calculated and (b) measured results.

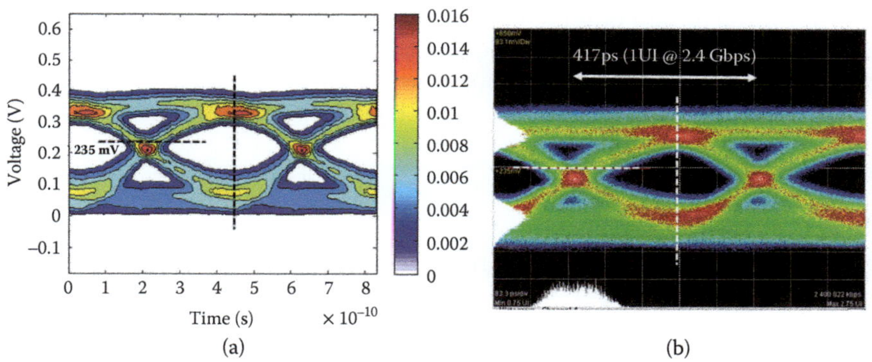

FIGURE 5.37
Eye diagrams considering both ISI and supply fluctuations (a) calculated and (b) measured results.

output (SSO) buffers, which are still widely used for wideband memory I/O interfaces and low-cost consumer devices [49]. The supply voltage fluctuations due to SSO buffers in the single-ended channels strongly depends on the SSO transition patterns at every switching, and the eye diagram at a buffer output is degraded by the supply fluctuations varying with the SSO transition patterns. The pattern-dependent SSO supply noise and resultant output jitter is one of the major performance limiters as data rate scales higher. There have been many studies to predict the PDN SSO noise and its impact on the signal transmission [50,51]. Recently, the effects of the pattern-dependent SSO supply noises have also been combined with the conventional statistical link analysis for ISI PDF extraction [52]. In the proposed statistical SSO analysis, the step response PDFs were calculated from the occurrence probability of each SSO pattern, which were then utilized in the convolution process instead of the Kronecker delta probability functions. The output voltage uncertainties due to the pattern-dependent supply-voltage fluctuations, as well as the channel

ISI, can be successfully included in the procedure. In addition, the proposed statistical method was extended and applied for the analysis of SSO buffers with the data bus inversion (DBI) –AC coding.

Although the enhanced statistical I/O link analysis methods discussed so far were based on the step pulse responses solved analytically, they can also be extracted from the simulations using a circuit solver. That is, the pull-up and pull-down step responses were individually simulated in the circuit solver and combined with the analytical probabilities. In this way, the statistical and BER eye diagrams for more complicated driver topologies could be calculated, as long as the step response waveform for the drivers is obtained by the simulations. Thus, the proposed statistical methods are not limited to the analytical solutions for just simple types of buffers, but are generally applicable to practical complicated drivers, if combined with circuit simulations. The enhanced statistical I/O link analysis methods for including the external supply noise or pattern-dependent SSO noise should be practically useful for the design of wideband memory I/O interfaces and low-cost consumer devices by drastically reducing the computational time in obtaining the output jitter and BER of various types of output buffers.

References

1. N. Wester and D. Harris, *CMOS VLSI Design*, 3rd edition, Boston, MA: Addison-Wesley, 2004.
2. I. Novak, *Power Distribution Network Design Methodologies*, International Engineering Consortium, Chicago, IL, 2008.
3. K. Shringarpure, S. Pan, J. Kim, J. Fan, B. Achkir, B. Archambeault, and J. L. Drewniak, Formulation and network model reduction for analysis of the power distribution network in a production-level multilayered printed circuit board, *IEEE Transactions on Electromagnetic Compatibility*, vol. 58, no. 3, 849–858, June 2016.
4. C. Hwang, B. Achkir, and J. Fan, Capacitance enhanced though-silicon via for power distribution networks in 3D-ICs, *IEEE Electron Device Letters*, vol. 37, no. 4, 478–481, April 2016.
5. H. Seidl, M. Gutsche, U. Schroeder, A. Birner, T. Hecht, S. Jakschik, J. Luetzen, M. Kerber, S. Kudelka, T. Popp, A. Orth, H. Reisinger, A. Saenger, K. Schupke, and B. Sell, "A Fully Integrated Al2O3 Trench Capacitor DRAM for Sub- 100 nm Technology," in *Proceedings of the IEEE International Electron Devices Meeting*, San Francisco, CA, pp. 839–842, 8–11 December 2002.
6. K. V. Rao, M. Elahy, D. M. Bordelon, S. K. Banerjee, H. L. Tsai, W. F. Richardson, and R. H. Womack, "Trench Capacitor Design Issues in VLSI DRAM Cells," in *Proceedings of the IEEE International Electron Devices Meeting*, Los Angeles, CA, pp. 140–143, 7–10 December 1986.
7. S.S. Li, *Semiconductor Physical Electronics*, New York, NY: Plenum Press, 1993.
8. P. Larsson, Parasitic resistance in an MOS transistor used as on-chip decoupling capacitance, *IEEE Journal of Solid-StateCircuits*, vol. 32, no. 4, 574–576, April 1997.

9. N. Li, et al., "Evaluation of a Multi-Line De-Embedding Technique up to 100 GHz for Millimeter-Wave CMOS Circuit Design," IEICE Transactions on Fundamentals of Electronics, Communications and Computer Sciences, vol. 93, no. 2, pp. 431–439, Feb, 2010.
10. C. Hwang, "On-Chip electromagnetic bandgap structures for suppression of simultaneous switching noise coupling in on-chip power distribution networks," PhD. Diss., KAIST, 2012.
11. J. S. Pak, J. Kim, J. Cho, K. Kim, T. Song, S. Ahn, J. Lee, H. Lee, K. Park, and J. Kim, PDN impedance modeling and analysis of 3D TSV IC by using proposed P/G TSV array model based on separated P/G TSV and chip-PDN models, *IEEE Transactions on Components, Packaging and Manufacturing Technology*, vol. 1, no. 2, 208–219, February 2011.
12. G. Katti, M. Stucchi, K. De Mayer, and W. Dehaene, Electrical modeling and characterization of through silicon via for three-dimensional ICs, *IEEE Transactions on Electron Devices*, vol. 57, no. 1, 256–262, January 2010.
13. S. C. Sun and J. D. Plummer, Electron mobility in inversion and accumulation layers on thermally oxidized silicon surfaces, *IEEE Journal of Solid-State Circuits*, vol. 15, no. 4, 562–573, August 1980.
14. Silvaco ATLAS. Available at: https://www.silvaco.com/products/tcad/device_simulation/atlas/atlas.html
15. Y. R. Kwon, V. M. Hietala, and K. S. Champlin, Quasi-TEM analysis of 'slow-wave' mode propagation on coplanar microstructure MIS transmission lines, *IEEE Transactions on Microwave Theory and Techniques*, vol. MTT-35, no. 6, 545–551, June 1987.
16. V. Milanovic, M. Ozgur, D. C. DeGroot, J. A. Jargon, M. Gaitan, and M. E. Zaghloul, Characterization of broad-band transmission for coplanar waveguides on CMOS silicon substrates, *IEEE Transactions on Microwave Theory and Techniques*, vol. 46, no. 5, 632–639, May 1998.
17. F. Stellari and A. L. Lacaita, New formulas of interconnect capacitances based on results of conformal mapping method, *IEEE Transactions on Electron Devices*, vol. 47, no. 1, 222–231, January 2000.
18. C. Hwang, K. Kim, J. S. Pak, and J. Kim, Modeling of an on-chip power/ground meshed plane using frequency dependent parameters, *Journal of Korea Electromagnetic Engineering Society*, vol. 11, no. 3, 192–200, September 2011.
19. K. Kim, J. M. Yook, J. Kim, H. Kim, J. Lee, K. Park, and J. Kim, "Interposer power distribution network (PDN) modeling using a segmentation method for 3-D ICs with TSVs," *IEEE Transactions on Components, Packaging and Manufacturing Technology*, vol. 3, no. 11, 1891–1906, November 2013.
20. A. Bansal, B. C. Paul, and K. Roy, An analytical fringe capacitance model for interconnects using conformal mapping, *IEEE Transactions on Computer-Aided Design of Integrated Circuits and Systems*, vol. 25, no. 12, 2765–2774, December 2006.
21. H. Y. Lee and T. Itoh, Phenomenological loss equivalence method for planar quasi-TEM transmission lines with a thin normal conductor or superconductor, *IEEE Transactions on Microwave Theory and Techniques*, vol. 37, no. 12, 1904–1909, December 1989.
22. W. Hilberg, From approximation to exact relations for characteristic impedances, *IEEE Transactions on Microwave Theory and Techniques*, vol. 17, no. 5, 259–265, May 1969.

23. N. V. D. Meijs and J. T. Fokkema, VLSI circuit reconstruction from mask topology, *Integration*, vol. 2, no. 2, 85–119, 1984.
24. J. Kim, W. Lee, Y. Shim, J. Shim, K. Kim, J. S. Pak, and J. Kim, Chip Package hierarchical power distribution network modeling and analysis based on a segmentation method, *IEEE Transactions on Advanced Packaging*, vol. 33, no. 3, 647–659, August 2010.
25. Y. Lo, D. Solomon, and W. Richards, Theory and experiment on microstrip antennas, *IEEE Transactions on Antennas and Propagation*, vol. 27, 137–145, March 1979.
26. J. Kim, L. Ren, and J. Fan, Physics-based inductance extraction for via arrays in parallel planes for power distribution network design, *IEEE Transactions on Microwave Theory and Techniques*, vol. 58, 2434–2447, September. 2010.
27. H. Kim, Y. Jeong, J. Park, SLee, J. Hong, Y. Hong, and J. Kim, "Accurate Measurement of Power/Ground Impedance with Embedded Film Capacitor using Two-Port Self-Impedance Measurement Technique," in *Proceeding of 5th Electronics Packaging Technology Conference*, Singapore, Singapore, 12-12 December 2003, pp. 51–54, 2003.
28. M. P. Li, *Jitter, Noise, and Signal Integrity at High-Speed*, Chap. 1, Westford, MA: Prentice Hall, 2007.
29. T. L Wu, H. H. Chuang, and T. K. Wang, Overview of power integrity solutions on package and PCB: Decoupling and EBG isolation, *IEEE Transactions on Electromagnetic Compatibility*, vol. 52, no. 2, 346–356, May 2010.
30. M. Swaminathan, J. Kim, I. Novak, and J. P. Libous, Power distribution networks for system-on-package: Status and challenges, *IEEE Transactions on Advanced Packaging*, vol. 27, no. 2, 286–300, May 2004.
31. J. Kim, J. Lee, S. Ahn, and J. Fan, Closed-form expressions for the noise voltage caused by a burst train of IC switching currents on a power distribution network, *IEEE Transactions on Electromagnetic Compatibility*, vol. 56, no. 6, 1585–1597, December 2014.
32. J. Kim, M. D. Rotaru, S. Baek, J. Park, M. K. Iyer, and J. Kim, Analysis of noise coupling from a power distribution network to signal traces in high-speed multilayer printed circuit boards, *IEEE Transactions on Electromagnetic Compatibility*, vol. 48, no. 2, 319–330, May 2006.
33. L. H. Chen, M. M-Sadowska, and F. Brewer, Buffer delay change in the presence of power and ground noise, *IEEE Transactions on Very Large Scale Integration (VLSI) Systems*, vol. 11, no. 3, 461–473, June 2003.
34. M. S-Laurent and M. Swaminathan, Impact of power-supply noise on timing in high-frequency microprocessors, *IEEE Transactions on Advanced Packaging*, vol. 27, no. 1, 135–144, February 2004.
35. K. Wong, T. Rahal-Arabi, M. Ma, and G. Taylor, Enhancing microprocessor immunity to power supply noise with clock-data compensation, *IEEE Journal of Solid-State Circuits*, vol. 41, no. 4, 749–758, April 2006.
36. R. Schmitt, H. Lan, C. Madden, and C. Yuan, "Investigating the Impact of Supply Noise on the Jitter in Gigabit I/O Interfaces," in *Proceeding of EPEP*, Atlanta, GA, October 28–31, 2007.
37. K. S. (Dan) Oh and X. (Chuck) Yuan, *High-Speed Signaling: Jitter Modeling, Analysis, and Budgeting*, Upper Saddle, NJ: Prentice Hall, 2011.

38. C. Hwang, J. Kim, B. Achkir, and J. Fan, Analytical transfer functions relating power and ground voltage fluctuations to jitter at a single-ended full-swing buffer, *IEEE Transactions on Components, Packaging and Manufacturing Technology*, vol. 3, no. 1, 113–125, January 2013.
39. J. Kim, J. Lee, S. Cho, C. Hwang, C. Yoon, and J. Fan, Analytical probability density calculation for step pulse response of a single-ended buffer with arbitrary power-supply voltage fluctuations, *IEEE Transactions on Circuits and Systems I*, vol. 61, no. 7, 2022–2033, July 2014.
40. E. Park, H. Kim, J. Shim, Y. Kim, Y. Kim, and J. Kim, Analytical calculation of jitter probability density at multi-stage output buffers due to supply voltage fluctuations, *IEEE Transactions on Electromagnetic Compatibility*, vol. 57, no. 4, 796–806, August 2015.
41. J. Kim, E. Park, J. Lee, K. Park, and J. Lee, "Improved Statistical Link Path Analysis Considering Both Channel ISI and Supply Voltage Fluctuations," In *DesignCon*, Santa Clara, CA, January 2014.
42. S. Chaudhuri, J. A. McCall, and J. H. Salmon, "Proposal for BER based specifications for DDR4," in *Proceeding of IEEE Conference on Electrical Performance of Electronic Packaging and Systems*, Austin, TX, 25–27 October, pp. 121–124, 2010.
43. J. Fan, X. Ye, J. Kim, B. Archambeault, and A. Orlandi, Signal integrity design for high-speed digital circuits: Progress and directions, *IEEE Transactions on Electromagnetic Compatibility*, vol. 52, no. 2, 392–400, May 2010
44. D. Oh, S. Chang, R. Jihong, Y. Ling, L. Hai, C. Madden, and R. Schmitt, "Statistical link analysis and in-situ characterization of high-speed memory bus in 3D package systems," in *Proceeding of 2011 IEEE International Symposium on Electromagnetic Compatibility*, Long Beach, CA, pp. 797–802, August 14–19, 2011.
45. B. K. Casper, M. Haycok, and R. Mooney, "An accurate and efficient analysis method for multi-Gbps chip-to-chip signaling schemes," in *Proceeding of IEEE Symposium on VLSI Circuits Digest of Technical Papers*, Honolulu, HI, pp. 54–57, 13–15 June 2002.
46. J. Ren, D. Oh, S. Chang, and F. Lambrecht, "Statistical Link Analysis of High-Speed Memory I/O Interfaces During Simultaneous Switching Events", in *Proceeding of IEEE conference on Electric Performance of Electronic Packaging*, San Jose, CA, 27-29 October 2008, pp. 25–28, 2008.
47. A. Sanders, Statistical simulation of physical transmission media, *IEEE Transactions on Advanced Packaging*, vol. 32, no. 2, 260–267, May 2009.
48. J. Kim, J. Lee, E. Park, and Y. Park, An enhanced statistical analysis method for I/O links considering supply voltage fluctuations and inter-symbol-interference, *IEEE Transactions on Components, Packaging and Manufacturing Technology*, vol. 5, no. 8, 1129–1141, August 2015.
49. W. T. Beyene, A. Amirkhany, and A. Abbasfar, "Statistical simulation of SSO noise in multi-gigabit systems," In *Proceeding of IEEE Conference on Electric Performance of Electronic Packaging and Systems*, Portland, OR, 19–21 October, pp. 21–24, 2009.
50. J. Ren and D. Oh, Multiple edge responses for fast and accurate system simulations, *IEEE Transaction on Advanced Packaging*, vol. 31, no. 4, 741–748, November 2008.

51. W. T. Beyene, C. Madden, J.-H. Chun, H. Lee, Y. Frans, B. Leibowitz, K. Chang, N. Kim, T. Wu, G. Yip, and R. Perego, Advanced modeling and accurate characterization of a 16 Gb/s memory interface, *IEEE Transactions on Advanced Packaging*, vol. 32, no. 2, May 2009.
52. J. Kim, Statistical analysis for pattern-dependent simultaneous switching outputs (SSO) of parallel single-ended buffers, *IEEE Transactions on Circuits and Systems* vol. 64, no. 1, 156–169, October 2016, 2016.

6

Printed Circuit Board Integration of SoC Packages and Signal Integrity Issues at Board Level

Norocel D. Codreanu and Ciprian Ionescu

CONTENTS

6.1 SoC Packages for PCB and System Integration

The main purpose of a SoC in an electronic product is to offer a complete system within a single chip [1], using no additional components or a reduced number of them (such as high precision analogue circuits, quartz crystals, high capacity memories, passive components). The advantages of SoC are package size, an easy-to-integrate package, flexibility, cost efficiency for high volume production, direct communication and full compatibility between internal blocks, low power consumption, a small printed circuit board (PCB) [2], and others. With shrinking on-chip technologies, there has been a reduction in voltage levels, lowering noise margins required to enable a correct design in terms of signal integrity (SI) [3,4] and electromagnetic compatibility (EMC) [5,6]. For this reason, SoC/chip packages are quintessential pieces of the signal integrity landscape [7] and PCB represents the backbone of the future electronic system/product [2]. Because bare silicon SoCs/chips are very sensitive, it is mandatory to encapsulate them in a proper package [8] for correct functioning in real products and for an adequate protection against environmental factors and against mechanical stress and damage [9,10]. Figure 6.1 presents the past, the present and the future of packages and packaging technologies [11].

PCB integration of SoC packages [1] is a crucial and critical issue today, representing a real challenge, due to the fact that many improvements at silicon level [12] can be dramatically limited by a non-optimal, or even wrong, integration at PCB level [13,14]. As power goes up, signals run faster and more interconnections are located on the package substrate and PCB, so the package and printed circuits design are much more difficult than they were in the past, becoming, in many cases, the bottle neck of the whole product development process. Additionally, because RoHS and WEEE directives also need to be met [15], a good practice for a successful SoC development is the chip/package/PCB co-design [8].

As Figure 6.1 shows, the trend is for a higher demand for new packages with even higher sophistication and features. On these terms, signal integrity issues between SoC packages and other components on a board become more and more important [4] and are now being studied in the early stage of pre-layout and post-layout PCB design [16].

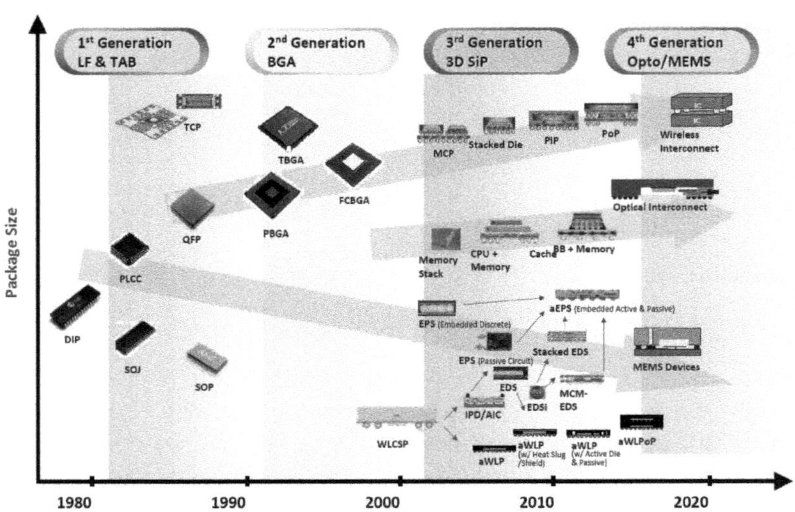

FIGURE 6.1
Packages—past, today, and future. (From A. Longford, "Trends in Advanced Packaging Technologies, An IMAPS-UK view," in *IMAPS UK–IeMRC Interconnection Event*, December 2008, and Jisso Technology Roadmap [in Japanese]).

The virtual investigation of SoC packages integration offers the advantages of understanding the functioning of electronic systems and of possible errors, creating the conditions to correct them before manufacturing [17], optimizing the development stage and decreasing the time-to-market period. The investigation methods and tools have long been used in the development of electronic products to predict performance capabilities and suitability to the given application and market. The goal is to create a design that works, after fabrication, at the specified performance on the first iteration [16]. As the complexity of electronic systems grows, and as design cycle times shortens, more attention is focused on virtual optimization of SoC package and layout for early stage maximization of features.

Highly integrated, customized and non-customized SoCs are packaged in various ways in accordance with the specific target application and the SoC based products, offering scalable performance, low costs, and flexible systems. The most used packages for SoCs are based on peripheral packages (QFP or QFN) and array packages (BGA), as follows [18,19]: QFP (Quad Flat Package), TQFP (Thin Quad Flat Package), PBGA (Plastic Ball Grid Array), MAPBGA (Molded Array Process Ball Grid Array, a smaller type of BGA package with the mold cap covering the entire topside surface and with smaller ball pitch), TEPBGA (Thermally Enhanced Plastic Ball Grid Array), SBGA (Stacked Ball Grid Array), TBGA (Tape Ball Grid Array), and FCPBGA (Flip Chip Plastic Ball Grid Array), FCPBGA being most used today [1,20]. Some of these are presented in Figure 6.2.

Additionally, the LGA (Land Grid Array) package is now an interesting alternative to the BGA package. LGA has only lands on the bottom side of the

FIGURE 6.2
Package types for SoCs and advanced chips.

FIGURE 6.3
The LGA package.

device and no solder interconnection entities (columns, balls, pins) are attached for assembling on PCB. The LGA package has the following advantages over the classical types: it can be used for either lead containing or lead-free solder pastes; it eliminates the risk of missing or damaged columns/balls/pins due to poor shipping or handling; it has a lower height than classical packages, and it has a higher durability in the case of mechanical stress. The unique disadvantage is that the LGA package is mostly used with sockets (Figure 6.3).

In the last decades, for complex chips/SoCs, the electronics industry has gradually moved from QFP/QFN (Quad Flat No-Leads) to BGA/LGA packages. Regarding the assembling, SoCs are currently connected to the package with the following assembling technologies: flip chip bumps, in-line wire bonds, and staggered wire bonds.

6.2 HDI and Embedded Components Technologies for SoC Packages and PCB

6.2.1 Introduction

The actual stage of development in electronics and microelectronics, especially in the communications and wireless applications fields, is characterized by a high level of function integration, mostly at chip level. In order to keep the balance with the innovation at the chip level, HDI (high density interconnect) [21,22] and embedded components technologies [23] were invented and introduced into the industry at the PCB level.

HDI (also called build-up board in Japan and, in the United States, either sequential build-up (SBU) or microvia technology) is an advanced PCB technology that has a higher printed circuit density than the classical technology, as well as other advanced features [24]. At the package level of SoC/chip, HDI represents a way to condense electronic circuits, minimizing the volume and weight of electronic systems for reaching a very high performance. HDI structures offer new solutions for solving SI and EMC issues [5,25,26]. Table 6.1 offers a short presentation of HDI parameters [27].

Common HDI structures are composed as follows [24,28]: (a) 1 + N + 1—PCB has one "build-up" of HDI layers; (b) $i + \mathrm{N} + i$ ($i \geq 2$)—PCB has two or more "build-up" of HDI layers (microvias on different layers can be staggered or stacked, with possible copper filling); (c) any layer HDI—PCB has only HDI layers, allowing traces on any layer to be interconnected without restrictions with copper filled stacked microvias (advanced packaging solution for highly complex large pin-count SoCs/chips).

But, even the complex packages containing SoCs/chips with very large scale integration on silicon, still require external discrete passive components (resistors, capacitors, and inductors), which are difficult or not cost effective to integrate on silicon or into the package. Studies were performed regarding effectiveness in the electronics assembling industry that revealed the possibility to decrease the cost by reducing the placement costs of SMDs (surface mounted devices). The passive components usually perform a reduced number of functions, but occupy a large PCB surface, requiring three to four time more solder joints in comparison to active SoCs/chips [23].

On these terms, embedded passives represent a different approach than using discrete passive components, a solution that can overcome some of the problems listed above. The key idea is to move the passive components from the PCB surface into the inner layers of the substrate, making them embedded components, as shown in Figure 6.4.

Embedded passives can provide solutions for system integration for modern SoC system-in-package (SiP) or system-on-package (SoP). The idea, which is not quite new (see hybrid circuits) is to include the passive components into the printed circuit board in the organic or inorganic substrate. Currently, using SMD passives mounted on the top and bottom of a PCB takes up a lot of space, and, since some discrete components must be placed away from the SoC package they serve, this can have a negative effect on system performance. By embedding these passives on the inner PCB layers, designers can

TABLE 6.1

Parameters of HDI

Parameter	Value
W	< 75 µm (3 mil)
S	< 75 µm (3 mil)
Via hole	< 150 µm (6 mil)
Via capture pad	< 400 µm (16 mil)
Pad density	> 20 pads/cm^2

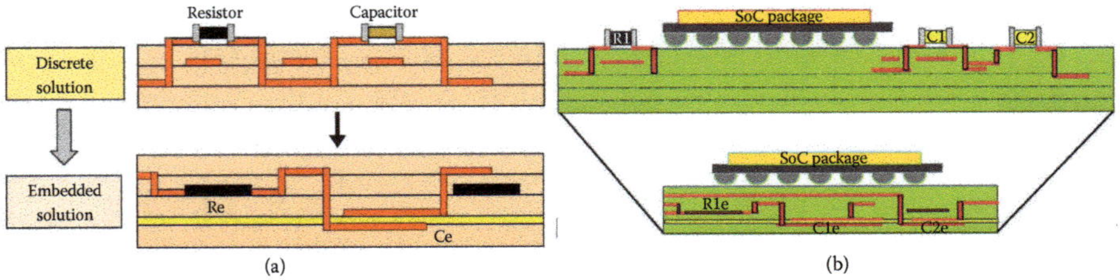

FIGURE 6.4
Transforming a classical PCB with discrete passive SMDs into an advanced PCB with embedded passives into the substrate by embedding resistive and capacitive passive components and reducing the size of SoC applications.

significantly reduce the size of the PCB (see Figure 6.4b), increase SI, and optimize the performance. The objective in implementing integrated passives is to find configurations and technologies that are cost effective and also offer improved electrical, mechanical, and thermal characteristics for these components.

The motivation in using integrated passives can be summarized by the following:

a. Reduced product mass, volume, and surface area on the board, as individual packages are eliminated and more space is made available on the PCB surface for active components.
b. Improved reliability, because there are no solder joints failures exist and embedded passives are not affected by lead-free requirements [15].
c. Improved electrical performance, because integrated passives have, in principle, lower parasitics [29,30].
d. Increased flexibility in design, because the component's values, resistance, capacitance, and inductance can be obtained at any desired value, so there are no constraints related to IEC values.
e. Reduced cost, since integrated passives can be realized simultaneously at a cost level that is practically independent of the number of components in a board, a similar approach to the case of the silicon process.

There are, of course, some difficulties in using integrated passive components on a large scale. Most of the problems are related to the technology. The technology for passive components must be available at the PCB fabricator. Also, there are some uncertainties regarding which materials to use. Cost, fabrication yield, and tolerance issues are also important. On the other side, the evolution of surface mount technology (SMT) toward very small components (such as the 01005 chip type, 0.4 × 0.2 mm) improves the packing densities at a level where integrated passives are not absolute necessary.

6.2.2 Configurations of Integrated Passive Components

The configurations of the integrated resistors mainly used in today applications are well known from the classical technologies applied to realize planar structures as thick- and thin-film technologies. The main difference in modeling the integrated passive components, as compared to "classical" components, is not their configuration, but their presence as embedded in substrate volume.

An integrated resistor is achieved by depositing a resistive film on an internal PCB layer. It results in a planar resistor with terminals; generally, it is not protected, as the protection will be ensured after PCB manufacturing. The typical rectangular structure of a resistor is shown in Figure 6.5.

As for discrete resistors with a planar resistive element, the resistance is determined by the relationship,

$$R = \rho_S \cdot k_f \tag{6.1}$$

where ρ_S is sheet specific resistance of the layer and k_f is the form factor, which also represents the number of squares which have the side W contained in the length L. It is usually used to compute resistor value. The form factor can have values, generally, from 0.5 to 100, depending on the geometry or configuration [23,18]. The most used configurations for integrated resistors are presented in Figure 6.5b.

For integrated capacitors there are three basic configurations, parallel plate, floating plate, and interdigitated configuration, as shown in Figure 6.6. For the floating plate configuration, the connections need to be made only at the top level of metal, avoiding the necessity of creating a via down to the bottom plate. The disadvantage of the floating plate capacitor is that it has four times less capacitance per unit area than the parallel configuration, because the floating plate acts as two capacitors in series, each using only one-half the available area. The interdigitated capacitors are used mainly for low capacitance values. This configuration causes low substrate

FIGURE 6.5
Configurations of integrated resistors: (a) planar resistor and (b) rectangular resistor and meander resistor. (Adapted from R. K. Ulrich and L. W. Schaper, *Integrated Passive Component Technology*, Wiley IEEE Press, 2003.)

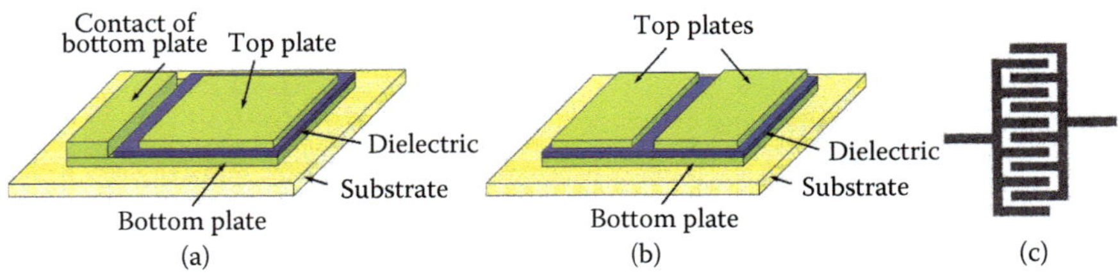

FIGURE 6.6
Configurations of integrated capacitors for (a) Parallel plate, (b) floating plate, and (c) interdigitated capacitor.

FIGURE 6.7
Configurations of integrated inductors: (a) circular spiral inductor and (b) square spiral inductor.

area utilization and the capacitance can be influenced by the copper traces of the PCB that acts as supplemental capacitance electrodes.

For integrated inductors in PCB [32,33], the most common configurations are circular spiral and square spiral, as shown in Figure 6.7.

There is no final decision regarding which inductor configuration is better, this depends on application. It is stated [23] that, for an equal area and DC resistance, the circular configuration offers a higher inductance per unit area and a higher quality factor value (Q). In some applications, hexagonal and octagonal inductors were reported as offering better results.

A further extension of the air core spiral inductors uses a deposition process compatible with PCB materials and technology to place the inductor in the vicinity of a magnetic core [34–36]. It is possible to either deposit the magnetic material on top of the inductor or to place the inductor between two layers, as can be seen in Figure 6.8.

The situation from Figure 6.8a is an open magnetic core and in Figure 6.8b there is always an air gap in the magnetic circuit, so the increase in effective permeability is much lower with the permeability of the magnetic material. For on-chip integration of inductors, a solenoid configuration was reported [37]. It uses electroplated cooper traces on two layers, connected by vias, interlayer polyimide material, and CoTaZr sputtered magnetic core. Another configuration of on-chip embedded inductor is called transmission line and is a metallic strip placed between two magnetic layers, acting as a discrete

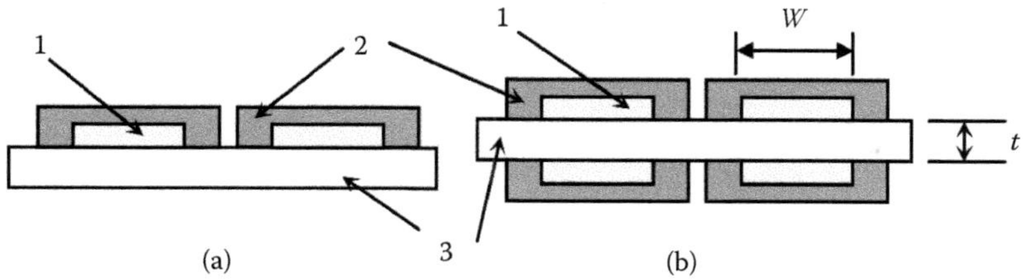

FIGURE 6.8
Planar inductors (a) with magnetic deposition on one layer; (b) with magnetic deposition on both layers; 1—copper trace, 2—magnetic layer deposition, 3—PCB substrate, *W*—inductor trace width, *t*—substrate thickness.

ferrite bead on a wire. It's used for reducing electromagnetic interference. The obtained values for integrated inductors are usually low, ranging from a few nH to a few μH.

6.2.3 Modeling and Simulation of Resistors at High Frequency

The operating frequencies of electronic circuits have rapidly increased in the last decades. The movement to higher frequencies requires the improvement of the high frequency performance of the components used in the circuit, as well as in the circuit design itself. A common approach to analyze the passive components is to use a circuit model of components that takes into account the high frequency behavior. Using SPICE-like simulations, it is possible to do impedance analysis or to study the transient response of the components [38]. In order to develop a useful model, it is important to understand the technology and the material science behind the embedded passive components. Any electronic component has different parasitic elements, which may modify the expected function of the component. For resistors the equivalent circuit, considering the parasitic effects, is shown in Figure 6.9.

In Figure 6.9 the notations were: R, the resistance of the resistive element in DC; $R\delta$, the increase in resistance due to skin effect; R_s, the resistance of the terminations; L, parasitic inductance; C, the parasitic capacitance; $\sqrt{\overline{V_n^2}}$, RMS noise voltage; V_{te}, thermoelectric voltage; R_{is}, the insulation resistance of substrate or protective element. By neglecting the skin effect and insulation losses, a resistor can be modeled by the equivalent circuit shown in Figure 6.10.

The resultant parasitic inductance and capacitance of the resistor become important parameters and they must be minimized from construction. The inductance is dependent on the structure of the resistor and can be approximated as the sum of the inductance of terminals and of the resistive element. The inductance of the PCB traces can be approximated with 10 nH/cm, and so is the inductance of terminals for THD (Through Hole Devices). SMDs present inductances in the range of 0.2 … 0.3 nH. The inductance of

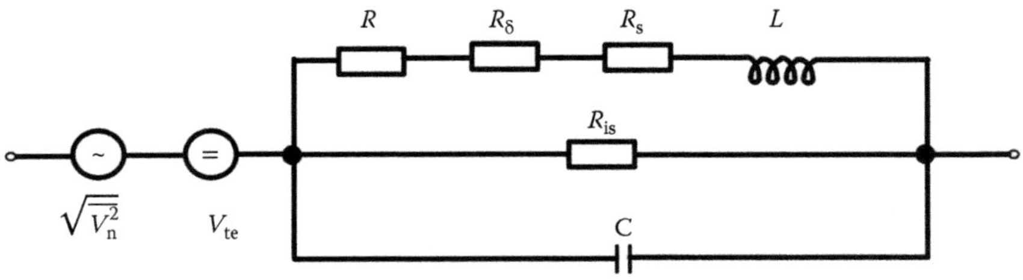

FIGURE 6.9
Full equivalent circuit for resistors. (From F. Zandman, P. Simon, and J. Szwarc, *Resistor Theory and Technology* (Revised edition), SciTech Publishing: Herndon, Virginia, 2002.)

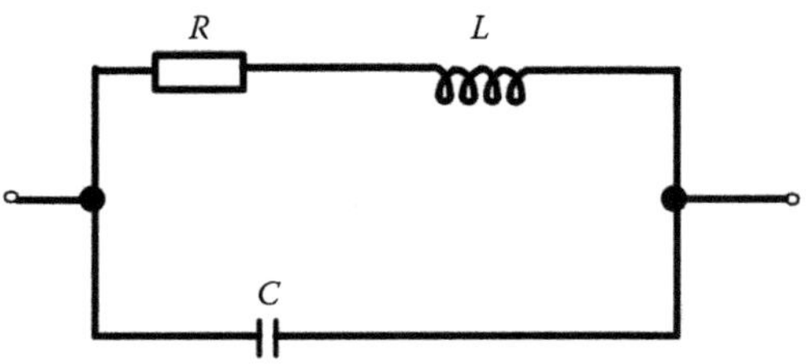

FIGURE 6.10
Simplified equivalent circuit of resistor at high frequency.

the resistive element depends on its geometry. From this point of view, the resistive elements can be either a spiral with high inductance or a plane rectangular film with low inductance. The resistor capacitance is dependent on the structure and the dielectric material. The resistor's capacitance is determined by the contribution of the dielectric substrate (FR, CEM, or other types) and the contact zone between the resistive element and terminals. Then, the capacitance is directly proportional to the substrate permittivity and the area of contact between the resistive element and terminal and inversely proportional with the distance between contacts. From the circuit in Figure 6.10, the complex admittance Y of the resistor is

$$\underline{Y} = \frac{1}{R + j\omega L} + j\omega C = \mathbf{Re}\{Y\} + j\mathbf{Im}\{Y\} \tag{6.2}$$

We try to express the real and imaginary part of the admittance. For that, we write the normalized admittance $R\underline{Y}$ as

$$R\underline{Y} = \frac{1}{1 + (\omega/\omega_r)^2 a^2} + j\frac{\omega}{\omega_r}\left(\frac{1}{a} - \frac{a}{1 + (\omega/\omega_r)^2 a^2}\right) \tag{6.3}$$

where a is a new defined parameter $a = \sqrt{L/C}/R$ and $\omega_r = 1/\sqrt{LC}$. The expression of real and imaginary parts is thus straightforward.

At high frequency, the impedance of a resistor may be capacitive or inductive, depending on R, L, or C values. Figure 6.11 shows the normalized admittance in complex plane **Im**{RY}–**Re**{RY}.

From this figure, where the frequency (f) was used instead of angular frequency (ω) as parameter, we can comment as follows:

a. For $\sqrt{L/C} < R$ (high values of resistance), the impedance is capacitive at high frequency and the maximum working frequency decreases with the increasing of the resistance value.

b. For $\sqrt{L/C} = R$ at high frequency, the impedance is still capacitive, but the resistor behaves better, keeping the imaginary part **Im**{RY} lower for high frequencies. This case ($a = 1$) corresponds to a situation where the capacitive time constant of the resistor equals the inductive time constant ($L / R = RC$), which means that there is a compensation for capacitive and inductive effects. This situation can be exploited in resistor design, to tailor the value of resistance to obtain an operation to high frequencies with minimum deviation from the DC value.

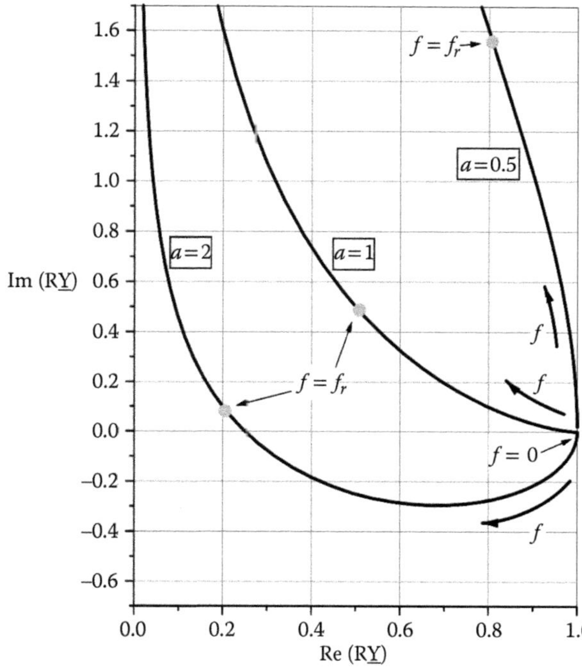

FIGURE 6.11
Normalized admittance in complex plane **Im**{RY}- **Re**{RY} for resistors. (Adapted from O. Zinke and H. Seither, *Widerstände, Kondensatoren, Spulen und ihre Werkstoffe*, Springer Verlag, Berlin/Heidelberg, 1982.)

c. For $\sqrt{L/C} > R$ (small value of R), with increasing frequency the impedance is inductive and beyond a certain frequency, it becomes capacitive.

Setting the condition $\mathbf{Im}\{R\underline{Y}\} = 0$ for Equation 6.3 results in the expression of the real axis crossing, point ω_0, similar to a resonance frequency:

$$\frac{1}{a} - \frac{a}{1+\left(\frac{\omega_0}{\omega_r}\right)^2 a^2} = 0 \tag{6.4}$$

The result is

$$\omega_0 = \omega_r \sqrt{\frac{a^2-1}{a^2}} \tag{6.5}$$

The crossing point ω_0 with a real axis exists only for values of a greater than 1, as it was supposed initially in this case. Another variant of the plot used to estimate the frequency behavior of resistors, as an alternative to the graph from Figure 6.11, is the graph of impedance modulus versus frequency (Figures 6.12 and 6.13, for different values of resistance and parasitics). The graphs can be used to estimate the maximum working frequency of a resistor and to estimate the reactive character, capacitive or inductive at even higher frequencies. As long the modulus of impedance remains close to R, we can use the resistor with no problems at these frequencies. If there is an increase in modulus, this corresponds to an inductive character.

In the graph from Figure 6.13, the resistors in range 1 MΩ to 1 kΩ show a capacitive character at high frequencies and the resistors with 10 Ω and 100 Ω show an inductive behavior for frequencies about 1 GHz, followed by a capacitive behavior at higher frequencies [39]. Figure 6.14 presents the impedance modulus graph for a given resistance and capacitance values and have the inductance as a parameter. All graphs show an inductive behavior. Similarly, Figure 6.15 presents the impedance modulus plotted for a given resistance and inductance values and have the parasitic capacitance as parameter. Here the impedance shows capacitive character for high capacitances and inductive behavior for low capacitances.

Now we will consider the response of the resistor to digital signals [40], more precisely a current step of value I_o applied to the circuit at the moment $t = 0$, when the K switch is closed (Figure 6.16).

The voltage transient response to a current step pulse of magnitude I_o can be estimated by solving the differential equation:

$$\frac{d^2u}{dt^2} + \frac{R}{L}\frac{du}{dt} + \frac{1}{LC}u = \frac{RI_0}{LC} \tag{6.6}$$

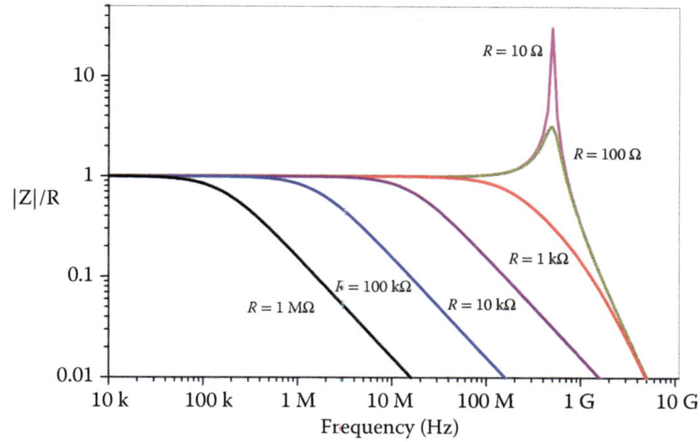

FIGURE 6.12
Normalized impedance of a resistor for C = 1pF, L = 100 nH, ($\sqrt{L/C} = 316\Omega$), R =(10 Ω; 100 Ω; 1 kΩ; 10 kΩ; 100 kΩ; 1 MΩ).

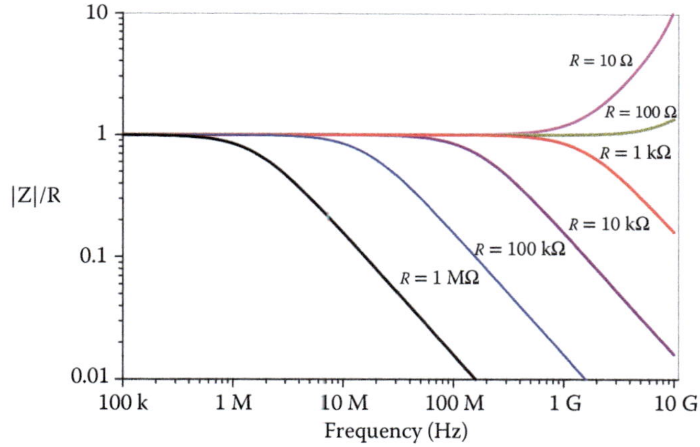

FIGURE 6.13
Normalized impedance of resistor for C = 0.1 pF, L = 1 nH, $\sqrt{L/C} = 100\ \Omega$, R = (10 Ω; 100 Ω; 1 kΩ; 10 kΩ; 100 kΩ; 1 MΩ).

The solution of the equation depends on the nature of the roots of characteristic equation. These roots have the form

$$p_{1,2} = -\alpha \pm \beta \tag{6.7}$$

where $\alpha = \dfrac{R}{2L}$, $\beta = \sqrt{\dfrac{R^2}{4L^2} - \dfrac{1}{LC}} = \sqrt{\alpha^2 - \omega_0^2}$ are some working parameter, and $\omega_0 = 1/\sqrt{LC}$.

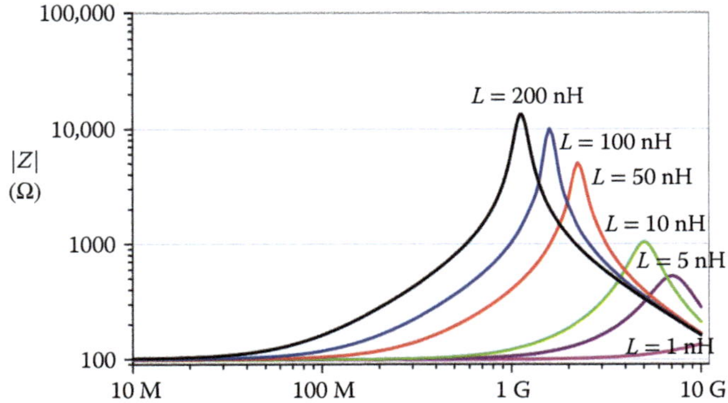

FIGURE 6.14
Impedance of resistor for R = 100 Ω, C = 0.1 pF, L = (1; 5; 10; 50; 100; 200) nH.

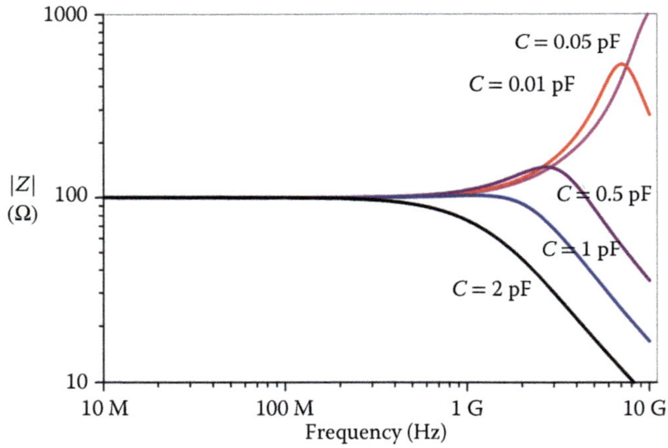

FIGURE 6.15
Impedance of resistor for R = 100 Ω, L = 5 nH, C = (0.05; 0.1; 0.5; 1; 2) pF.

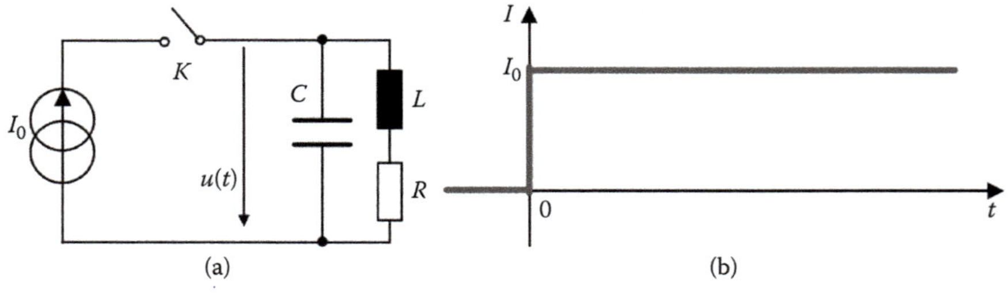

FIGURE 6.16
Transient response of the circuit to a current step: (a) electric circuit and (b) current waveform.

We distinguish three cases for β: real values, imaginary values, or v.

1. β is real, or $R > 2\sqrt{\frac{L}{C}}$

   In this case the characteristic equation has two different real roots and the solution is

$$u(t) = R \cdot I_0 \left[1 - \frac{e^{-\alpha t}}{\beta \cdot R \cdot C} \sinh(\beta t + k)\right] \tag{6.8}$$

   with k a constant given by $\tanh k = \frac{\beta}{\alpha - \frac{1}{RC}}$. This is called the over-damped regime.

2. $\beta = 0$ or $R = 2\sqrt{\frac{L}{C}}$. This is called the critically damped regime.

   In this case the characteristic equation has one double root and the solution can be written as

$$u(t) = R \cdot I_0 \left[1 - \left[1 - \left(\alpha - \frac{1}{RC}\right) \cdot t\right] e^{-\alpha t}\right] \tag{6.9}$$

3. β is complex, or $R < 2\sqrt{\frac{L}{C}}$. This is called the oscillating regime and $\beta = j \cdot \omega_p$.

$$u(t) = R \cdot I_0 \left[1 + \frac{e^{-\alpha t}}{\omega_p \cdot R \cdot C} \sin\left(\omega_p t + k\right)\right] \tag{6.10}$$

with $$\tan k = \frac{\omega_p}{\alpha - \frac{1}{RC}}$$

   So, the pulse behavior of the resistor is somehow similar to behavior in the AC regime.

In the first case of above $\left(R > 2\sqrt{\frac{L}{C}}\right)$, the response of the resistor increases practically exponential to the final value. The rise time increases with the resistance value. In the second case $\left(R = 2\sqrt{\frac{L}{C}}\right)$, the response of the resistor is similar to case 1 and the rise time is lower than in the first case. In a similar way, the rise time increases with the value of C. In case 3, $\left(R < 2\sqrt{\frac{L}{C}}\right)$, the response of the resistor has oscillations. The frequency of these oscillations is ω_p, which is very close to the resonance frequency of the resistor ω_0. The amplitude of oscillations increases when R values decrease and the transition time increases with the decreasing of resistance values. So, the behavior of a resistor at current pulses depends on the parasitic elements value in a similar way as in alternating current.

One can use Equations 6.8 through 6.10 to estimate the transient response or use a SPICE type simulator [38,40]. Figures 6.17 through 6.20 present the results of the transient voltage *u(t)* across the resistor, normalized to steady state value, as a response to a current step transition I_0 of 1 mA.

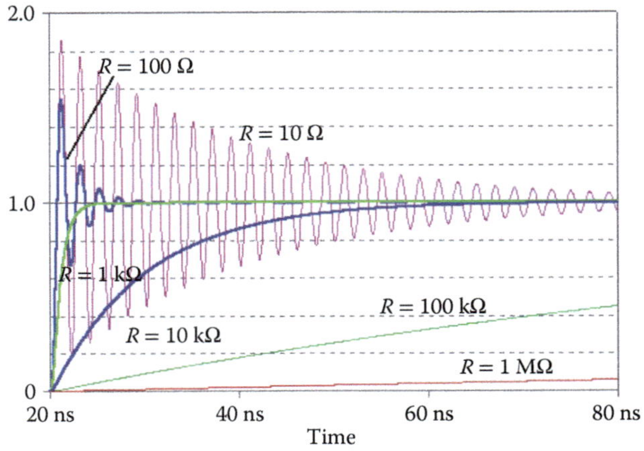

FIGURE 6.17
Normalized transient response of resistor for C = 1 pF, L = 100 nH, R = (10 Ω; 100 Ω; 1 kΩ; 10 kΩ; 100 kΩ; 1 MΩ).

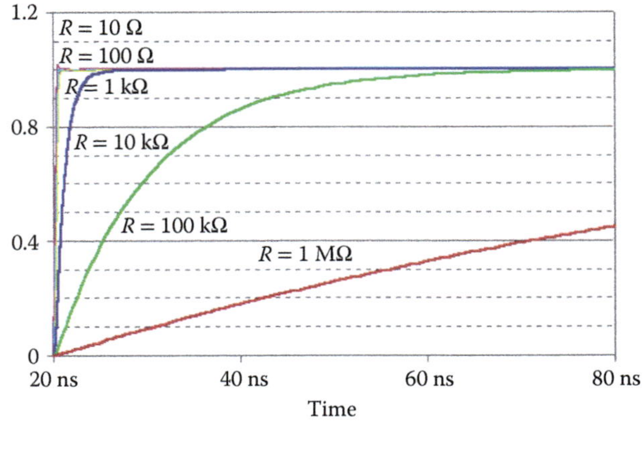

FIGURE 6.18
Normalized transient response of resistor for C = 0.1pF, L = 1nH, R = (10 Ω; 100 Ω; 1 kΩ; 10 kΩ; 100 kΩ; 1 MΩ).

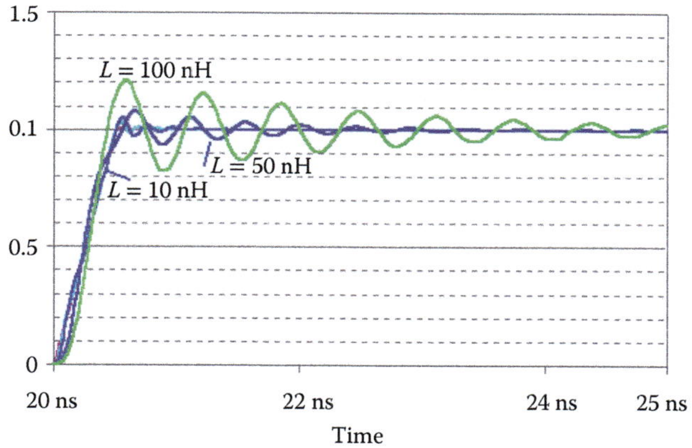

FIGURE 6.19
Normalized transient response of resistor for R = 100 Ω, C = 0.1pF, L = (1; 5; 10; 50; 100) nH.

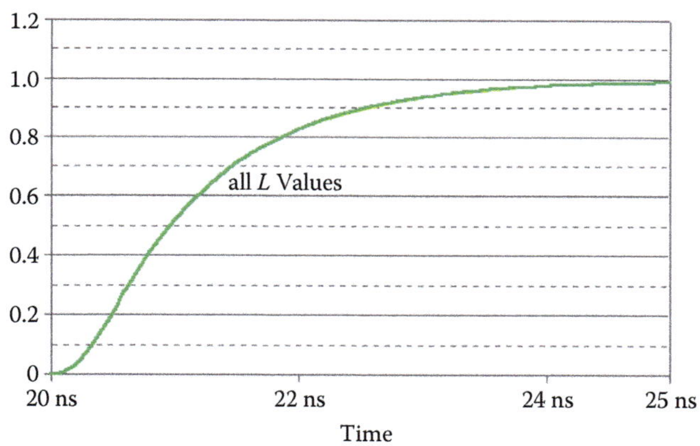

FIGURE 6.20
Normalized transient response of resistor for R = 10 kΩ, C = 0.1pF, L = (1; 5; 10; 50; 100) nH.

The conclusion is that the inductive impedance is associated with oscillations and the capacitive impedance determines the damped response. For higher resistance values, the influence of inductance can be neglected. With decreasing R values, depending on the reactive character, the amplitude of oscillations becomes higher or the transition rise time becomes smaller.

6.2.4 Modeling Passive Components Using the Finite Element Methods (FEM)

Although, for resistors, capacitors, or inductors, there are many closed form expressions that help in determining the values of characteristic parameter

(R, C, or L), the relations that are used for determining the parasitics of these structures are usually semi-empirical.

The irregular shape and complex geometries of some configurations make it helpful to use dedicated software tools or electromagnetic simulators in order to determine the parasitics of passive components. The simultaneous presence of resistive, capacitive, and inductive effects makes the analysis a real "full wave" analysis. This means that the electric and magnetic fields are treated as coupled, as in the case of wave propagation. These simulators offer the scattering parameters matrix (S parameters) as results, so the component is always seen as a two port network [29]. Designers usually wish to analyze the integrated passive components as dipole components, integrated in organic materials for Radio Frequency (RF) or Low Frequency (LF) applications, where they cannot apply the distributed model, but where the parasitics of the components can alter their functionality. In some cases, the lumped parameters obtained from a full-wave electromagnetic analysis [41] do not quite exactly model the components' behavior in the domain of less than 1 GHz. This loss of accuracy at low frequency is caused by a relation that must always be assured between wavelength and cell (discrete element) dimension. Another problem with using this type of simulator (such as ANSYS HFSS [42], IE3D [43], and Momentum), is the fact that these are destined mainly for microwave circuits and the application of excitation sources is conceived accordingly, using the concept of ports and de-embedding arms, and so on. It is difficult, in these programs, for instance, to apply just a voltage of 1 Volt to a circuit and to see the resulting current in order to compute the impedance.

So, in order to analyze the integrated passive components as dipole components, a low frequency analysis, or, more exactly, a quasi-static analysis is recommended. In this analysis type, the electric and magnetic fields are separated and treated individually. This is equivalent to saying that the displacement current $J_D = \partial D / \partial t$ is neglected in magnetic analysis. The simulation must be consequently divided in magnetic and electric conduction analysis and these must be done separately. This is because the phenomena are not coupled, as in the case of time dependent "full wave" electromagnetic field. The quasi-static approach is a very common technique in modeling and analysis of electronic structures. It starts from the assumptions that the fields are quasi-static, so their value does not change considerably with incrementing frequency. This assumption is valid if the circuit is small compared to the wavelength of the electromagnetic field [44]. A rule of thumb to appreciate the quasi-static behavior of a circuit, structure or component is

$$l_{max} < \lambda_{min} / 30 \div \lambda_{min} / 10 \quad (6.11)$$

where l_{max} is the largest dimension of the component and λ_{min} is the shortest wavelength used in the application. The circuits for which the above formula can be applied are also called lumped or are made of lumped elements. So, this is equivalent to saying quasi-static fields or modeling using lumped components.

Using the above assertions for a frequency of 1 GHz, the maximum dimension for a component that can be modeled with the lumped model lies between 10 mm and 30 mm. most of integrated passive components used in today applications are in this situation. For example, take a resistor; considering the parasitic effects relevant for high frequency domain, the equivalent circuit is the one shown in Figure 6.10, where the resistor is considered "isolated," meaning that only the intrinsic effects are taken into account. In a real circuit, the resistor operates in an interconnection structure and the capacitances with respect to ground, or the mutual capacitances, must be taken into account.

As mentioned above, the modeling of passive components using analytical methods has some limitations. There are precise closed form expressions for the characteristic parameter (*R*, *C*, or *L*) but many "semi-empirical" expressions for estimating the parasitic elements. In order to characterize the passive dipole in a more realistic way, a low frequency or quasi-static Finite Element Analysis (FEA) was proposed [29]. As presented, due to the principle of operation of FEM/FEA simulators, there is a separation between the electrostatic field and the magnetic field, involving the necessity of solve two separate problems. In the first case, the elements are defined using, as material parameters, the relative permittivity and electrical conductivity and there is either no possibility of defining magnetic permeability, or it is possible, but is not taken into account. In the other analysis, the reverse is true and the magnetic permeability can be defined. In this way, the possible analyses are "conduction + electrostatic" and "conduction + magnetic." Both of these analyses can be realized as harmonic or transient. For our purposes, in most cases, the AC, or harmonic analysis, is suitable [29].

For the simulation of integrated resistors, the user needs to perform an electric and a magnetic analysis, which will give the capacitance and inductance, respectively (Figure 6.21) [45].

For our harmonic analysis, a constant voltage *V*, which is considered to have a real value (phase is null), will be applied to the circuit. The admittance of the circuit is then

$$Y = \frac{1}{R} + j\omega C = \frac{I}{V} \tag{6.12}$$

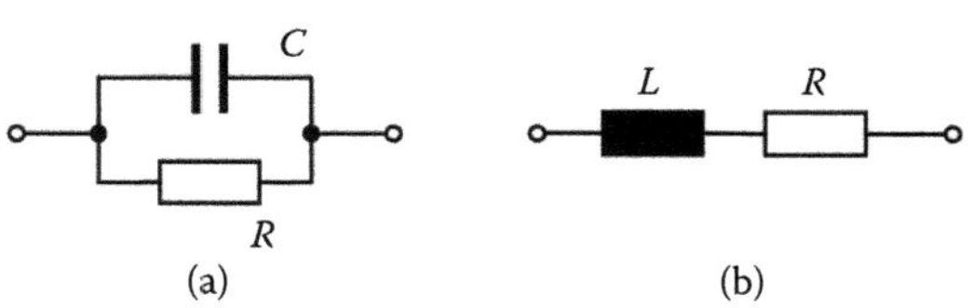

FIGURE 6.21
Circuit for (a) electric analysis, and (b) magnetic analysis.

The electric current obtained will be a complex quantity, of magnitude $|I|$ and phase angle φ. It results from the capacitance and the resistance of the circuit (Equation 6.12):

$$C = \frac{|I|}{V} \frac{\sin\phi}{2\cdot\pi\cdot f}, R = \frac{V}{|I|\cos\phi} \tag{6.13}$$

In a similar way, the magnetic analysis will result in the complex current I and, from the impedance formula, we can compute the inductor elements L and R: ($\underline{Z} = R + j\omega L$).

$$L = \frac{V\sin\phi}{2\cdot\pi\cdot f\cdot|I|}, R = \frac{V\cos\phi}{|I|} \tag{6.14}$$

In both simulations, the solid model is the same, only the elements are switched from electric elements to magnetic elements. The magnetic analysis takes into account the eddy currents, so it is expected to give more accurate results for the AC resistance. Normally, if the eddy currents are neglected, the analysis must give the same result for the resistance R. The investigations required for inductors are the same as in Figure 6.21. The difference is that the main parameter is now the inductance obtained from magnetic analysis. The first electrostatic analysis will give the parasitic capacitance of the inductor.

A more complex problem appears for capacitors. In the electric analysis, as above, the capacitance and the loss resistance are obtained in a similar manner. For the determination of parasitic inductance, a little problem appears. In this case the analysis requires finite elements of a magnetic type, but the source is always the current density. In this situation, there is no path for current flow, not even in AC, because there are no electrostatic elements (no permittivity) that can contribute to current flow (Figure 6.22).

The solution in this situation is to declare the elements used for the dielectric in the first analysis to be conductive elements in order to have a current path through the capacitor. The structures from Figure 6.23 [46] demonstrate the modeling and simulation techniques presented above.

Authors have used the ANSYS software system [42], which includes the modeling and simulation flow, to build the solid model, define and assign material properties and finite elements, mesh the model, apply loads and

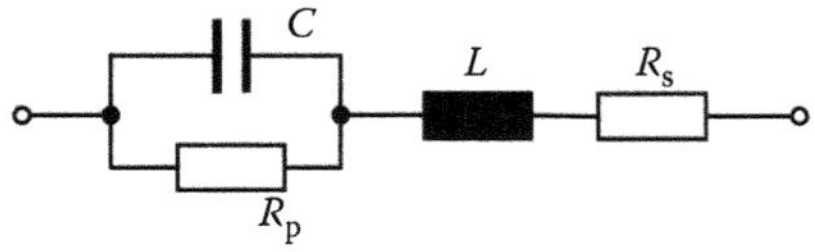

FIGURE 6.22
Equivalent model of a capacitor.

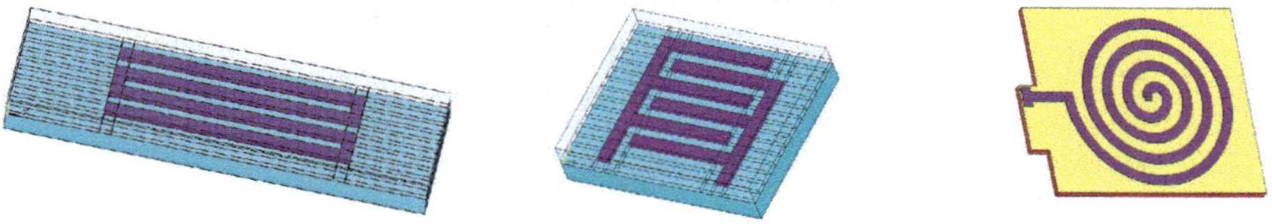

(a) (b) (c)

FIGURE 6.23
Finite element models with surrounding air not shown: (a) meander resistor with 4 segments, (b) interdigitated capacitor, and (c) spiral circular inductor. (From C. Ionescu et al., in *SIITME 2006, International Symposium for Design and Technology of Electronic Packages,* 12th edition, Iasi, Romania, pp. 39–43, 2006.)

boundary conditions, solve the problem, and post-process the results. A constant potential of 1 V was applied and the other terminal was considered to be grounded. ANSYS has three methods for modeling the effect of the open field decay [42]. The one used here was to extend the free space region. As a rule of thumb, the free space needs to be extended approximately 2–4 times than the structure size. Another characteristic is that the full 3D structure must be modeled, no symmetry reduction being possible. A major problem in modeling planar structures is the large number of elements that can be generated by the very thin layers that model the resistive, conductive, or dielectric depositions. In mechanical and thermal analysis, special elements were developed, the so-called "shells" that model very thin structures, such as metal or plastic sheets. Unfortunately, there are no electromagnetic shell elements available. Authors have used a special modeling technique, which involves building solid model by extruding areas along the OZ direction, which dramatically reduces the number of finite elements. For all the models presented here, there were up to 15000 elements, a number which is reasonable low. In order to reduce the number of elements, an assumption was made. Since copper thickness has no major contribution to parasitic capacitance and inductance, it was increased at 50 μm. The inductors were manufactured on a classical FR-4 substrate and were measured using a HP4396B impedance analyzer in the 100 kHz to 1.8 GHz frequency range. The analyzer can compute the elements according to the series-parallel RLC model, as in Figure 6.10. The inductance values from measurements and from simulations are presented in Table 6.2 [34,36].

6.2.5 Modeling Embedded Inductors Using Closed Form Expressions

For practical considerations, in the design process where FEM is difficult to apply, one alternative is to use engineering formulae based on closed form expressions. The dependency between inductance and its parameters (number of turns, distance between windings, trace width, and PCB thickness) is achieved directly from electromagnetic laws. The model's geometry

TABLE 6.2

Spiral Inductor Parameters from Measurements and from FE Simulation

Inductor Layout	Number of Turns	Data Sources	L(nH)	C(pF)	R(Ω)
Round spiral—single layer	2	Measurement	55.9	2.06	1.4
		FE Simulation	46.2	2.42	1.6
	4	Measurement	257.7	3.14	2.44
		FE Simulation	240	3.25	2.63
	8	Measurement	1679.4	5.04	2.78
		FE Simulation	1552.3	6.03	2.92

is considered to be a one-dimensional wire. The distance between the traces of the model will be the sum of the distance between the traces of the real inductor and the trace width. This is the only approximation. Other parameters, such as parasitic capacitance, are strongly influenced by the track width, but are not subject to this modeling. We shall make the following notations: L is inductance; Φ is magnetic flux; $\vec{B}$ is magnetic induction; $\vec{A}$ is vector potential; d$\vec{l}$ is the length unit; N is the number of turns; $\hat{n}$ is the normal to the surface; i is the current through the inductor; μ is the magnetic permeability; dS is the surface unit.

The definition of inductance is $L = \Phi / i$, with $\Phi = \iint_{\Sigma} \vec{B}\hat{n}dS$. From Maxwell laws, a vector potential $\vec{A}$ exists, so that $\vec{B} = rot\vec{A}$. Additionally, from Gauss theorem $\iint_{\Sigma} rot\vec{A}\hat{n}dS = \int_{\Gamma_{\Sigma}} \vec{A}d\vec{l}$; so,

$$\Phi = \iint_{\Sigma} rot\vec{A}\hat{n}dA = \int_{\Gamma_{\Sigma}} \vec{A}d\vec{l} \tag{6.15}$$

and

$$\vec{A} = \frac{\mu i}{4\pi}\int_{0}^{2\pi} \frac{d\vec{l}}{r} \tag{6.16}$$

The magnetic permeability is $\mu = \mu_0 \cdot \mu_{ef}$, where the last term represents the effective permeability, a characteristic of the entire non-homogenous environment. The first case is for the spiral inductor on one layer. For determining the geometrical inductance polar coordinates are used, as shown in Figures 6.24 and 6.25 [33,34,47].

In Figure 6.24, $x = R\cos\theta; y = R\sin\theta; x_1 = b\cos\phi; y_1 = b\sin\phi$, $d\vec{l} = Rd\vec{\theta}$; $R = \frac{\theta}{2\pi}d$ and $r^2 = R^2 + b^2 - 2Rb\cos(\theta - \varphi)$. So, according to the formulas in Equations 6.15 and 6.16:

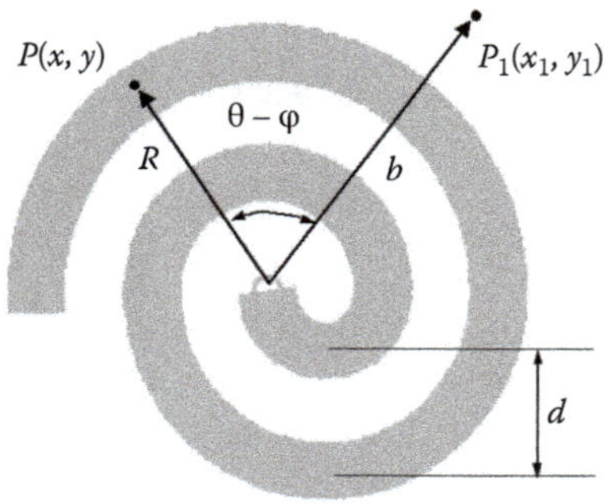

FIGURE 6.24
Relevant definitions of parameters for a spiral inductor.

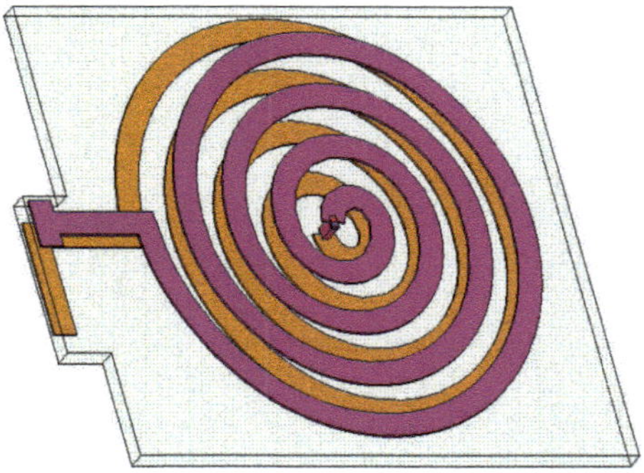

FIGURE 6.25
Structure of a spiral inductor on two layers.

$$\vec{A} = \frac{\mu i d}{8\pi^2} \int_0^{2N\pi} \frac{\theta}{\sqrt{\dfrac{\theta^2 d^2}{4\pi^2} + b^2 + \dfrac{\theta d}{\pi} b\cos(\theta - \varphi)}} d\vec{\theta} = \vec{A}(b,d,\phi) \tag{6.17}$$

This result gives us the vector potential (*A*) of the geometry of the inductor (given by its parameters "*d*" and "*N*") in every point of the plane (given, in polar coordinates, by "*b*" and "φ"). To obtain the magnetic flux, it is necessary to integrate A over the same geometry that generated it:

$$L = \int_0^{2N\pi} \vec{A}(R,d,\theta) R d\vec{\theta} = \frac{d}{2\pi} \int_0^{2N\pi} \bar{A}\left(\frac{\theta}{2\pi} d, d, \theta\right) d\theta = L(N,d) \tag{6.18}$$

The inductance is now written as a function of the number of turns and the distance between them, which is closer to practical situations.

The spiral inductor on two layers behaves like 2 inductors on a single layer in series and magnetically coupled. The mutual inductance represents the effect of the magnetic flux produced by the inductor from one layer to the inductor from the other layer, $M = \frac{\Phi_{12}}{i_2}$.

It is necessary, when representing the potential vector, to take into account a third dimension "z," which perpendicular to the surface of the inductors, so that we can use the effect of the potential vector of the coil, for instance, of the "bottom" layer, over the one from the "top" layer. In the case of the single layer inductor, the same plane, "z," was 0. Instead of polar coordinates, cylindrical coordinates will be used and A becomes

$$\vec{A} = \frac{\mu i d}{8\pi^2} \int_0^{2N\pi} \frac{\theta}{\sqrt{\frac{\theta^2 d^2}{4\pi^2} + b^2 + z^2 + \frac{\theta d}{\pi} b \cos(\theta - \varphi)}} d\vec{\theta} = \vec{A}(b, d, \phi, z) \tag{6.19}$$

The quantity "z" represents the distance between the two layers. The mutual inductance is

$$M(N, d, h) = \frac{d}{2\pi} \int_0^{2N\pi} \vec{A}\left(\frac{\theta}{2\pi} d, d, \theta, h\right) \theta d\theta \tag{6.20}$$

Finally, the total inductance has the following formula:

$$L_2 = 2(L_1 + M) \tag{6.21}$$

where L_1 is the inductance of one layer, given by Equation 6.18.

Sometimes, rectangular geometries are used. A rectangular spiral inductor on one layer is similar to a circular one, but the computational method uses Cartesian coordinates instead of polar, as follows: $\mathrm{d}\vec{l} = \mathrm{d}\vec{x} + \mathrm{d}\vec{y}$; $r^2 = (x - x_0)^2 + (y - y_0)^2$. The vector potential and the inductance can be found using Equations 6.22 and 6.23:

$$\vec{A} = \frac{\mu i}{4 \cdot \pi} \sum_{i=1}^{n} \left[\int_{\Gamma} \frac{1}{\sqrt{(x - x_0)^2 + (y - y_0)^2}} (\mathrm{d}\vec{x} + \mathrm{d}\vec{y}) \right] = \vec{A}(n, d, x_0, y_0) \tag{6.22}$$

$$L = \sum_{i=1}^{n} \left[\int_0^{nd} \vec{A}(n, d, x, y) \mathrm{d}\vec{x} + \int_0^{nd} \vec{A}(n, d, x, y) \mathrm{d}\vec{y} \right] \tag{6.23}$$

Analog with the spiral inductor on two layers, the rectangular inductor can be investigated on two layers. The inductance of this geometry is formed of the inductance of the two layers and the mutual inductances between them.

$$\vec{A} = \frac{\mu i}{4 \cdot \pi} \sum_{i=1}^{n} \left[\int_{\Gamma} \frac{1}{\sqrt{(x - x_0)^2 + (y - y_0)^2 + z^2}} (d\vec{x} + d\vec{y}) \right] \tag{6.24}$$

$$\vec{A} = \vec{A}(n, d, x_0, y_0, z)$$

$$M = \sum_{i=1}^{n} \left[\int_{0}^{nd} \vec{A}(n, d, x, y, h) dx + \int_{0}^{nd} \vec{A}(n, d, x, y, h) dy \right] = M(n, d, h) \tag{6.25}$$

The values of inductance from the equations above can be obtained using software tools as MathCAD or MATLAB, which can easily compute the integrals and plot the data.

For the one-layer round spiral geometry, are the plots for the inductance and mutual inductance over the spacing between the tracks, respectively the distance between layers and the number of turns, can be found in Figures 6.26 and 6.27.

One can observe from these graphs that, for an inductor having, for instance, 2 mm trace width and spacing, the mutual inductance does not depend much on the distance between layers, but the contribution to overall inductance given by Equation 6.21 is very strong. A double layer inductor has a much higher inductance than two one-layer inductors connected in series.

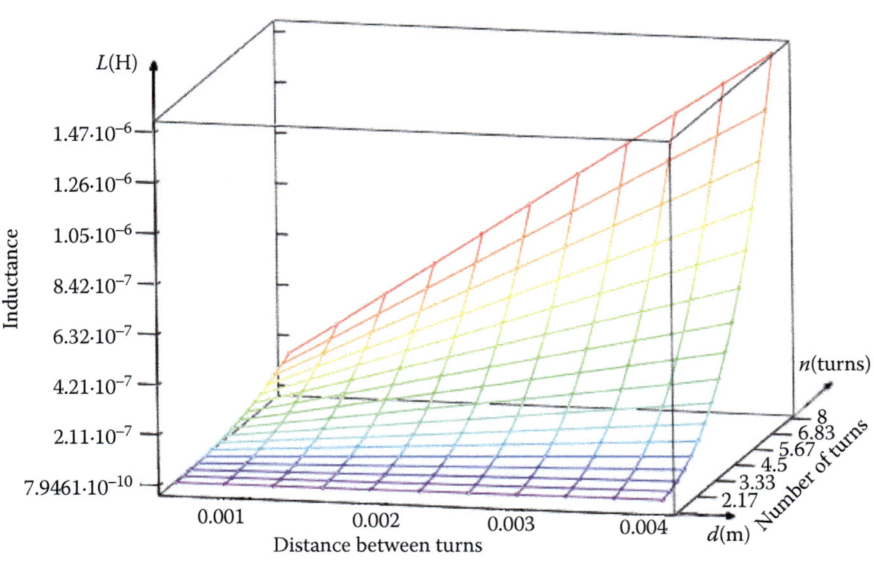

FIGURE 6.26
Dependencies between L, d, and N for one-layer spiral inductor.

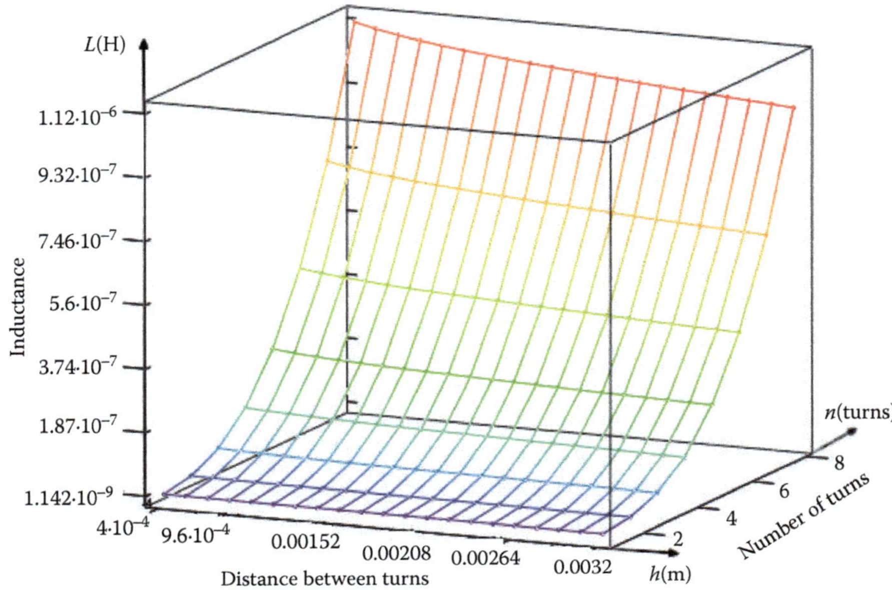

FIGURE 6.27
Dependencies between mutual inductance M and distance between the layers h and N for a spacing between rounds and a trace width of 2 mm.

6.3 Virtual Investigation Setup for Evaluation of SoC Based PCB Modules

The PCB design of SoC based electronic modules/systems need electrical validation before manufacturing, so it is necessary to perform virtual investigations in the pre-production stage. Today's SoCs require very strict design strategies for package. For PCB and controlled impedances, characteristic impedance matching techniques, optimal return paths, low power supply impedances, minimum discontinuities, and a.s.o. have all become critical for reaching the target quality of the modules/systems under development. Electrical problems relate to both signal propagation between SoC packages and power distribution required to operate the module/system.

In order to start from one point (on a SoC/chip) and to end at another point (on another SoC/chip), the signal must travel from the chip through its internal interconnections, through substrate balls and wire bonds to the package, through package balls/terminals, PCB traces, PTH vias/microvias placed between various electrical layers, finally reaching the end point of another chip, and then back through the balls/wire bonds of the end package and through the internal interconnections of the end chip. Based on virtual investigations, high-quality designs will minimize the electromagnetic interference in SoC based systems, offering an appropriate methodology for the project under development and for future projects. Therefore, the usage of electromagnetic methods and software tools is a major challenge for the development of the SoC based systems. Figure 6.28 presents an example of a packaging structure modeled for virtual investigations, and which emphasizes the presence of wire bonds from the chip to the lead frame of the package, the internal lead frame, the PCB structure, vias, and SMDs [48,49].

Planar electromagnetic simulators, also referred to as 2.5D EM simulators, rely on the method of moments (MoM) to discretize and solve Maxwell's equations for planar circuits embedded in a multilayered environment. These simulators were traditionally used to model and simulate microwave integrated circuits, planar antennas, and radio-frequency structures, but today they are more and more involved in advanced investigations and design of complex PCBs and related packaging elements.

The planar electromagnetic simulators use an appropriate mesh to discretize the currents on the planar structures, and the current distribution are modeled for allowing a piecewise linear current distribution. The corresponding charge distribution is piecewise constant and satisfies the local current continuity relation. The size and number of cells in the mesh (Figure 6.29) is mainly determined by the electrical wavelength of the signals travelling in the PCB planar structure. Due to the accurate modeling of the surface current distributions, planar electromagnetic simulators allow generating accurate electrical models for arbitrary power/ground and signal configurations up to the higher RF and microwave frequencies and are able to offer solutions for

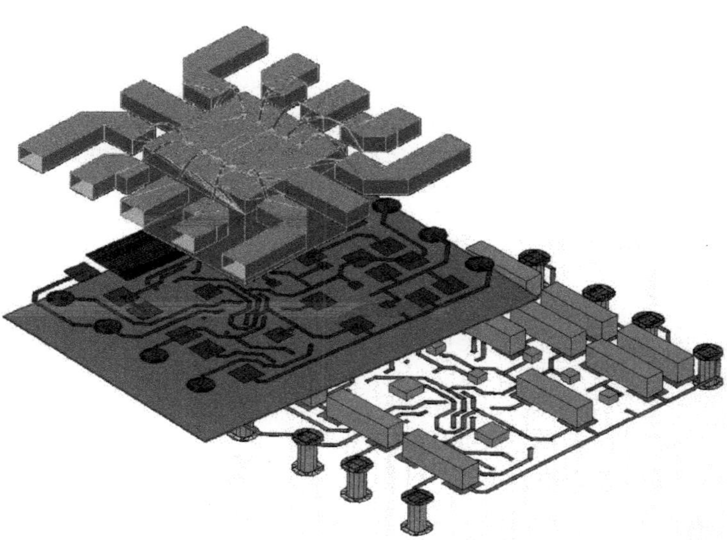

FIGURE 6.28
Example of a packaging structure modeled for virtual investigations [48,49].

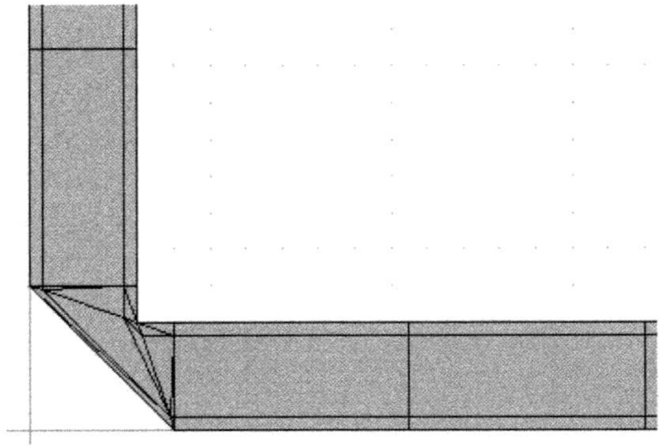

FIGURE 6.29
Part of a PCB structure and associated cells (From *IE3D User's Manual Release 6*, Zeland Software, 2000).

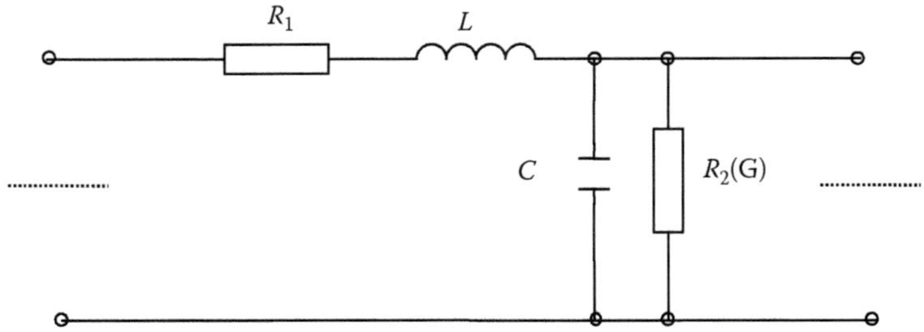

FIGURE 6.30
RLCG equivalent circuit model of a PCB interconnection.

modern SoC based modules/systems. The cells of the structure (Figure 6.29) can limit, in some cases, the simulation, increasing the processing time and memory resources and reducing the simulator efficiency (in the case of complex SoC based PCB structures with very high number of cells).

To start, it is necessary to offer an overview of the MoM technique behind the simulator and discuss the derivation of an equivalent RLCG circuit model. This allows giving the MoM interaction matrix equation a physical interpretation by constructing a RLCG equivalent model (Figure 6.30).

The surface currents on the planar structure are modeled with basis functions defined over cells in the mesh. The boundary conditions are imposed by applying the Galerkin method (a projection method in which a set of approximated operator equations is solved exactly). For obtaining the final results, a matrix equation (Equation 6.26) is generated.

$$\text{For } i = 1 \ldots \text{N}, \quad \sum_{j=1}^{N} Z_{i,j} I_j = V_i \text{ or } [\text{Z}] \cdot [\text{I}] = [\text{V}] \tag{6.26}$$

where impedances and voltages are

$$Z_{i,j} = \iint_S B_i(r) \cdot dS \cdot \iint_S \bar{\bar{G}}(r,r') \cdot B_j(r) \cdot dS' \tag{6.27}$$

$$V_i = \iint_S B_i(r) \cdot E(r) \cdot dS \tag{6.28}$$

Each impedance element ($Z_{i,j}$) represents the electromagnetic interaction between two basis functions, $B_i(r)$ and $B_j(r)$, placed in the substrate layer stack. Each element V_i in the excitation vector represents the discretized contribution of the sources applied at the ports of the PCB structure. The solution of the matrix equation yields the expansion coefficients for the unknown surface currents (Equation 6.29), which contribute to the electromagnetic field in the interconnection structure by means of the Green function of the substrate layer stack.

$$J(r) = \sum_{j=1}^{N} I_j B_j(r) \tag{6.29}$$

The Green function is decomposed into a sum of contributions from the vector potential A(r), the scalar potential V(r), and a term related to DC losses and AC losses (skin effect).

$$\bar{\bar{G}}(r,r') = j\omega G^A(r,r')\bar{\bar{I}} + \frac{1}{j\omega}\nabla[G^V(r,r')\nabla'] + Z_s\delta(r-r')\bar{\bar{I}} \tag{6.30}$$

The scalar potential originates from the dynamic surface charge distribution derived from the surface currents related to the vector potential through the Lorentz gauge. The surface impedance, Z_s, depends on the conductivity and thickness of the PCB conductive structure (usually manufactured in copper) and models the frequency dependent skin effect losses. Substituting the expression Equation 6.30 into the expression of impedance elements $Z_{i,j}$ (Equation 6.27), produces the following form:

$$Z_{i,j} = R_{i,j} + j\omega L_{i,j} + \frac{1}{j\omega C_{i,j}} \tag{6.31}$$

with

$$Z_{ij}^L = j\omega L_{ij}(\omega) = \iint_{S_i} dS \iint_{S_j} G_m(\omega, r-r') B_i(r) \cdot B_j(r') \cdot dS'$$

$$Z_{ij}^C = \frac{1}{j\omega C_{ij}(\omega)} = \iint_{S_i} dS \iint_{S_j} G_e(\omega, r-r') \nabla \cdot B_i(r) \nabla \cdot B_j(r') \cdot dS'$$

$$Z_{ij}^{R} = R_{ij}(\omega) = Z_s(\omega)\iint_{S_i} dS \iint_{S_j} \delta(r - r')B_i(r) \cdot B_j(r') \cdot dS'$$

This allows giving the MoM matrix from the first formula a physical interpretation by generating the RLCG equivalent circuit model (Figure 6.30). Figure 6.31 presents the association between the meshed planar structure and the equivalent circuit, though only the capacitive and inductive elements are depicted [17,50,51].

The MoM matrix equation follows from applying Kirchhoff voltage laws to the equivalent circuit. The inductor branch currents in the circuit represent the coefficients of the basis functions in the expansion of the surface currents. The use of basis functions allows giving a physical interpretation of currents in the circuit, with the currents corresponding to the total normal current flowing across the edges in the structure:

$$I_j = \int_{edge[j]} J(r) \cdot u_n \cdot ds \tag{6.32}$$

The physical interpretation for the voltages in the circuit follows from power conservation considerations. The complex power delivered to the surface current $J(r) = I_j B_j(r)$ is given by

$$P_j = \frac{1}{2}\iint_S E(r) \cdot J^*(r) \cdot dS = \frac{1}{2}\left[\iint_S E(r) \cdot B_j(r)\right] \cdot I_j^* \cdot dS \tag{6.33}$$

Based on a full-wave approach, in IE3D the magnetic and electric Green's functions follow from Maxwell electromagnetic equations, which include all coupling, radiation, and dispersion effects in the substrate. These Green's

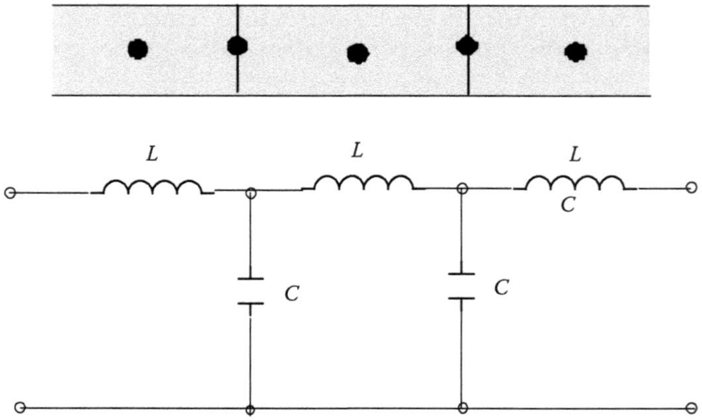

FIGURE 6.31
Segment with capacitive and inductive elements and the LC equivalent circuit.

functions and, as a result, the inductances and capacitances in the RLCG equivalent circuit, are complex and frequency dependent. Equation 6.34 presents the case of free space:

$$G_m(\omega, r-r') = \frac{j\omega\mu_0}{4\Pi|r-r'|} e^{-jk_0|r-r'|}$$

$$G_e(\omega, r-r') = \frac{1}{j\omega\varepsilon_0 4\Pi|r-r'|} e^{-jk_0|r-r'|} \tag{6.34}$$

where $$k_0 = \omega\sqrt{\varepsilon_0\mu_0}$$

The elements of the equivalent RLCG circuit can be calculated using either a full-wave approach or a quasi-static approach. These two approaches differ in the type of Green's functions used to calculate inductances and capacitances. Due to the use of the electrostatic and magnetostatic Green's functions, the resulting inductances and capacitances are real, frequency independent, and do not include the radiation effects radiation. For low frequencies or electrically small PCB structures, the static terms dominate and the quasi-static approach will generate similar results to the full-wave approach [41]. As the frequency increases and the electrical length of the PCB structure becomes significant, the quasi-static results gradually deviate from the accurate full-wave results.

For general purpose electromagnetic simulators, a meshed structure (with triangular, rectangular, or polygonal elements) is used. The efficiency of the meshing can be greatly enhanced by removing the restriction imposed by the use of rectangular and triangular cells in the mesh. This is realized by adopting the concept of mesh reduction. Starting from an initial mesh of rectangular and triangular cells, a new mesh is constructed by merging two or more adjacent cells. This results in a "reduced" mesh with a lower number of "generalized" polygonal shaped cells. In the resulting equivalent circuit, the capacitors model the charges built up in the generalized cells and the inductors model the current flowing from one generalized cell to another. Mesh reduction provides a way to overcome the limitations of traditional MoM meshes that use only rectangular and triangular cells. An important problem is whether the mesh has to be uniform or non-uniform. For uniform grid based simulators, the PCB structure is divided based on this uniform grid. Then, the user creates the circuit based upon the existing grid. The process is basically the procedure of fitting the particular circuit into the grid (Figure 6.32). If the circuit cannot be fitted into the uniform grid, there are two solutions. One is to cut the portion cannot be fitted and the other is to make the uniform grid finer in order to get a better approximation. In the case of MoM, making the grid twice as fine means multiplying the number of cells by 4, leading to a much longer processing time and other problems, as presented previously.

The uniform grid offers the most accuracy and efficiency, but creates a large number of cells. For example, Figure 6.32a presents a zone of a PCB trace, with the structure placed in a uniform grid, but the mitered corner is not matched with the grid. To solve this problem, the corner is changed in order to fit into the uniform grid (Figure 6.32b). The cells inside the fitted structure are the cells used for calculation (83 cells). A non-uniform grid approach solves the problem in a better way. The user creates the circuit inside a layout editor, having no concerns about the number and shape of cells. Then, the simulator tries to fit a non-uniform triangular and rectangular mesh into the circuit (instead of trying to fit the circuit into a pre-defined mesh, as was done for uniform grid). Figure 6.33 shows the same PCB trace structure with two particular discretization schemes and, by comparison between Figures 6.32 and 6.33, the non-uniform grid meshing schemes are shown to be more flexible, efficient, and accurate than the uniform meshing scheme. The circuit based on the non-uniform grid and non-uniform cells can approximate more accurately the real planar interconnection structure, especially in case of HDI (where dimensions are smaller than in classical PCBs) because in the zones of highest interest (corners, vias, wire bonds, etc.) the number of cells is much higher and very well fitted with the specific geometry.

The non-uniform grid is flexible, efficient, and, also, accurate. Figure 6.33a presents a coarse non-uniform meshing (with 8 cells) of the structure from Figure 6.32a. Figure 6.33b presents small edge cells which are automatically created on the edge of the PCB trace in order to enhance accuracy (29 cells).

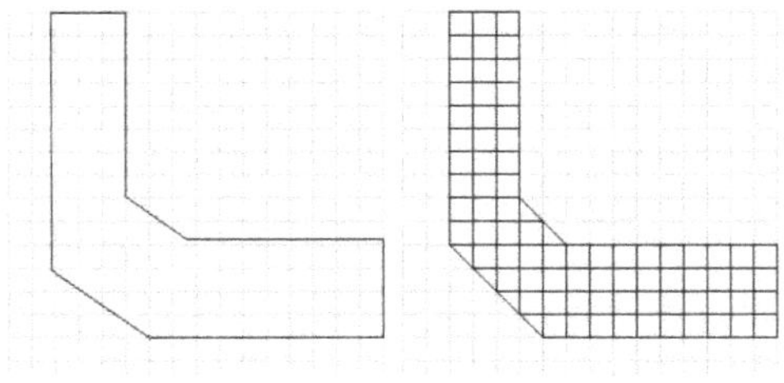

(a) (b)

FIGURE 6.32
PCB structure configured with a uniform grid. (From *IE3D User's Manual Release 6*, Zeland Software, 2000.)

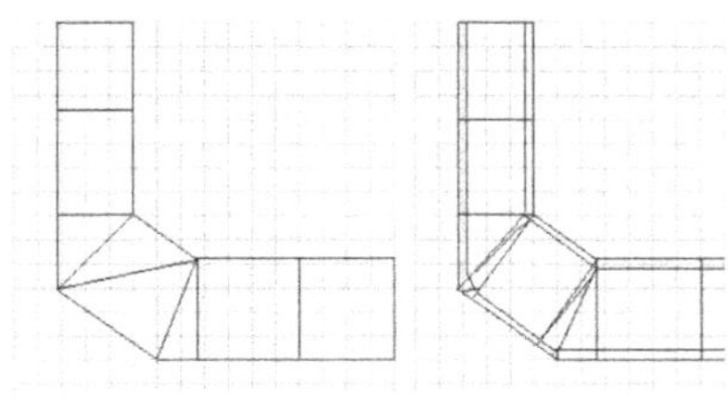

(a) (b)

FIGURE 6.33
PCB structure configured with a non-uniform grid. (From *IE3D User's Manual Release 6*, Zeland Software, 2000.)

The small cells along edges of the structure are used to model the edge singularity of the current distribution (Figure 6.34). While a circuit based on a uniform grid creates more freedom in the calculation, it does not mean that it is more accurate than one based on a non-uniform grid, since uniform grid based simulators do not use all their cells correctly, even though they have more cells. For current density simulation in the transverse direction, 1–5 cells can be created (Figure 6.34), offering sufficiently high accuracy for current density at the edges of the planar circuit.

Regarding the layer stack-up and thickness, the graphical editor of the electromagnetic simulator uses a two layers system for planar interconnection structures description: dielectric layers and metallic layers. The dielectric layer system is defined in the "Substrate" zone, in the "Param" menu. By default, the no.0 dielectric is always at z = 0. The default no.0 dielectric is always the ground plane. Mathematically, a ground plane of finite thickness and a ground plane of infinite thickness produce the same result when they are large in the horizontal directions. The metallic layer system is created based on the dielectric system. Table 6.3 [52] shows two examples of the layer stack-up in the cases of microstrip [53] and stripline geometries.

Regarding the thickness of interconnection structures, there are two methods to model it. The default and simplest way is to use an infinitely thin strip to model the structure (Figure 6.35a, for microstrip geometry). In addition, IE3D uses analytical formula for the surface impedance (Zs) in order to model the loss effect of the strip. The skin effect is included in the analytical formula for the Zs. The infinitely thin strip model is good when the trace width (W_{trace}) is much bigger than the trace thickness (t):

$$W_{trace} >> t \tag{6.35}$$

The second method is a more accurate model for thick traces. Instead of approximating the traces as infinitely thin traces, the software tool models

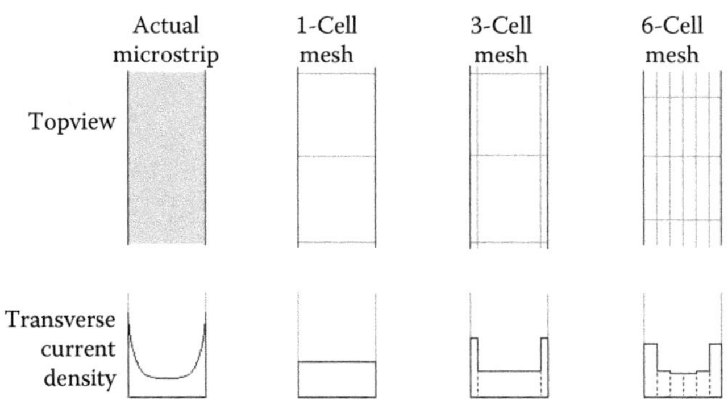

FIGURE 6.34
Current density modeling (1–5 cells) in the transverse direction. (From *IE3D User's Manual Release 6*, Zeland Software, 2000.)

TABLE 6.3

Layer Stack-Up for EM Simulation

Cases	Layer Stack-Up	Parameters
Microstrip H = 1 mm $E_r = 4.7$ $\tan\delta = 0.02$	Air Substrate $E_r = 4.7$ GND plane 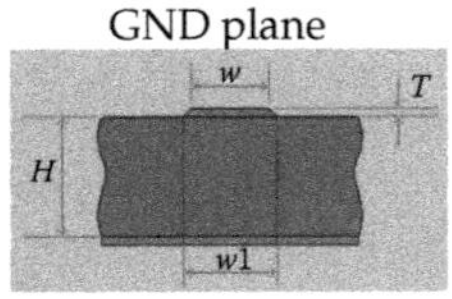	No.0, $Z_{top} = 0$, $\varepsilon_r = 1$, $\sigma = (5.8\ e7)$ S/m No.1, $Z_{top} = 1$ mm, $\varepsilon_r = 1$, $\sigma = (0,0)$ S/m
Stripline Metallic track is in the middle. H = 1.5 mm $\varepsilon_r = 4.7$ $\tan\delta = 0.02$	GND plane Substrate $\varepsilon_r = 4.7$ Substrate $\varepsilon_r = 4.7$ GND plane 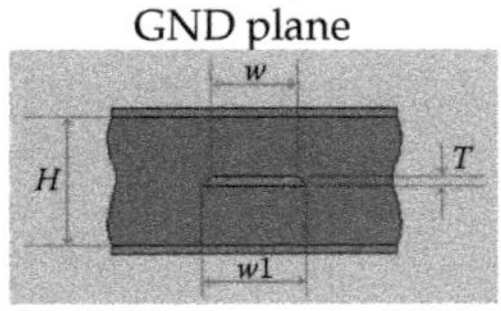	No.0, $Z_{top} = 0$, $\varepsilon_r = 1$, $\sigma = (5.8\ e7)$ S/m No.1, $Z_{top} = 1.5$ mm, $\varepsilon_r = 4.7$, $\sigma = (0,0)$ No.2, $Z_{top} = 1.0\ e15$, $\sigma = (5.8\ e7)$ S/m

Source: From *IE3D User's Manual Release 6*, Zeland Software, 2000; www.polarinstruments.com/.

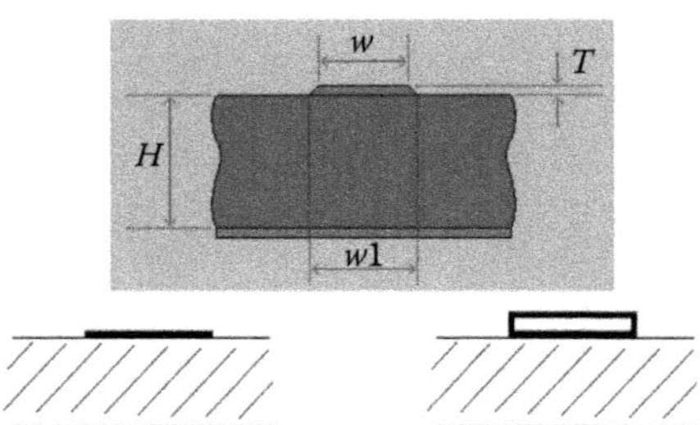

FIGURE 6.35
Modeling of PCB trace thickness. (From *IE3D User's Manual Release 6*, Zeland Software, 2000; www.polarinstruments.com/.)

the current on the 4 sides of the track (Figure 6.35b). The obtained model generates a better L and C calculation. The trace information will be used in finding the Z_s for loss calculation.

6.4 Investigations of PCB and Interconnection Elements

For SoC based complex PCB layouts [2,7], Z, Y, and S are very important parameters, because they are directly related to the total current on the transverse direction of a port. For solving the problem, the PCB structure under evaluation is decomposed into a stack-up with various conductive and dielectric layers and into PCB/interconnection elements, which are meshed using different cells [50]. Although, in the literature, numerous closed form formulas can be found for the calculation of different parameters, they are usually semi-empirical and cover only a narrow frequency bandwidth. The irregular shape and complex geometry of modern PCBs make useful the dedicated software tools or electromagnetic simulators for finding the parasitics of PCB/interconnection structures and for evaluating the discontinuities which affect the signal integrity. The simultaneous presence of conductive, capacitive and inductive effects makes the analysis a "full-wave" investigation. This means that the electric and magnetic fields are treated as coupled, as in the case of wave propagation. When using electromagnetic simulators, it is standard to combine field analysis with circuit analysis, in order to perform a high quality signal integrity evaluation (see the simplified flowchart from Figure 6.36). Finally, when the electrical parameters of interconnection structures are obtained [54], the signal integrity analysis is a simple electrical circuit simulation.

In the combined field-circuit analysis, it is necessary to use SPICE macromodels [38], which are approximations from the electromagnetic point of view of passive linear networks by equivalent electric circuits. Usually, the Z parameters are synthesized from S parameters that are obtained from full-wave electromagnetic investigation. Even without SPICE circuit extraction, a conversion between S matrix and Z matrix is always possible. In order to perform the

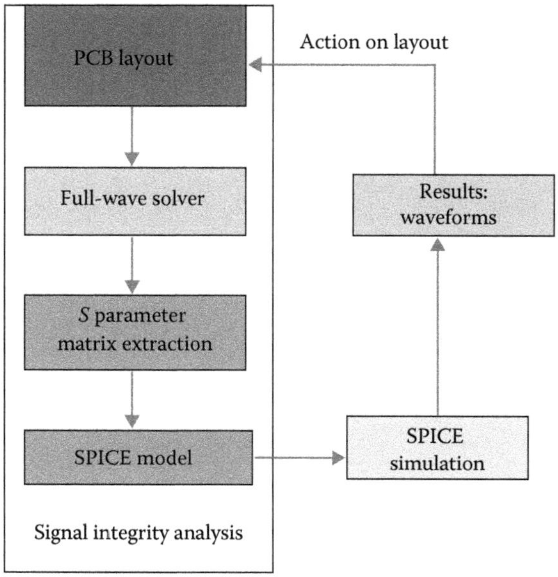

FIGURE 6.36
Signal Integrity Analysis (SIA) based on full-wave simulation, with SPICE parameters extraction.

full-wave MoM electromagnetic simulation, 2.5D modeling was used to create the elements and structures of interest. Figure 6.37 presents the investigation flow. The work was done with IE3D and Fidelity simulation engines. IE3D is suitable for 2D or 3D layered configurations and Fidelity for full 3D structures [55]. The simulators can emphasize parasitic effects, reflections, crosstalk, radiation patterns, and others, taking into account the distributed character of circuits and the non-ideal behavior of PCB interconnection structures.

To perform an electromagnetic simulation, the user needs to first create the geometry. After finishing constructing the element/structure, it is necessary to define the signal ports. It is possible also to develop the structure in classical PCB design tools and to import the structure in the simulator. The transfer is done using the CADNEX editor, which generates CIF files, as seen in Figure 6.38 [55,56].

The IE3D environment includes five applications (Figure 6.39): MGRID, the layout editor for creating the geometry; IE3D, the electromagnetic simulator for numerical analysis; MODUA, the schematic editor, graphical display, and

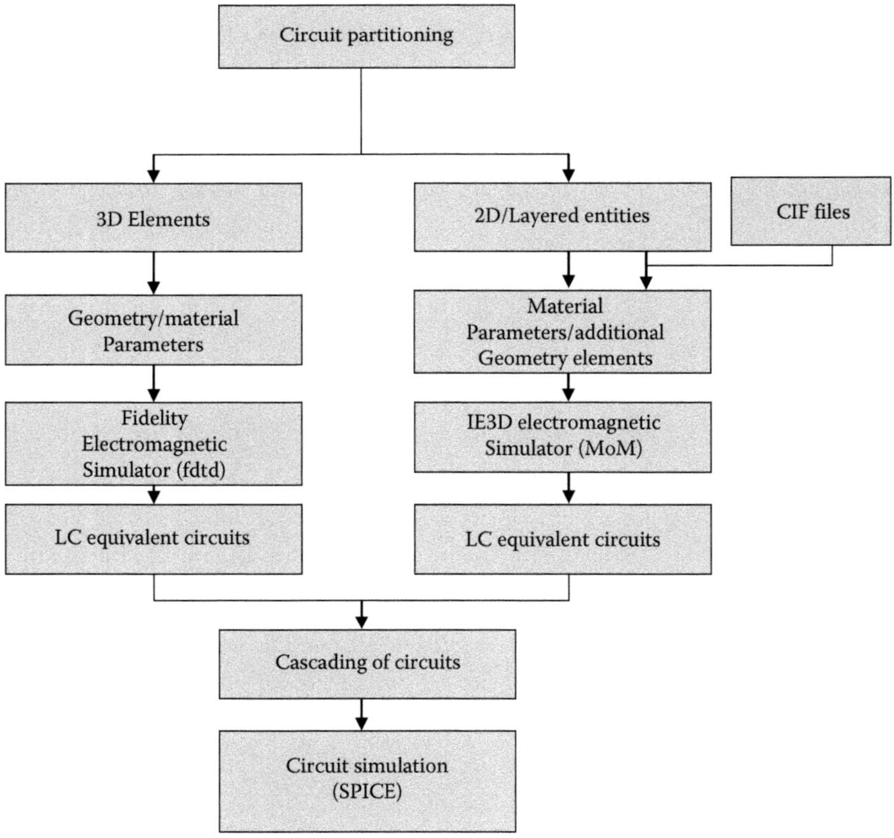

FIGURE 6.37
Full-wave electromagnetic investigation flow.

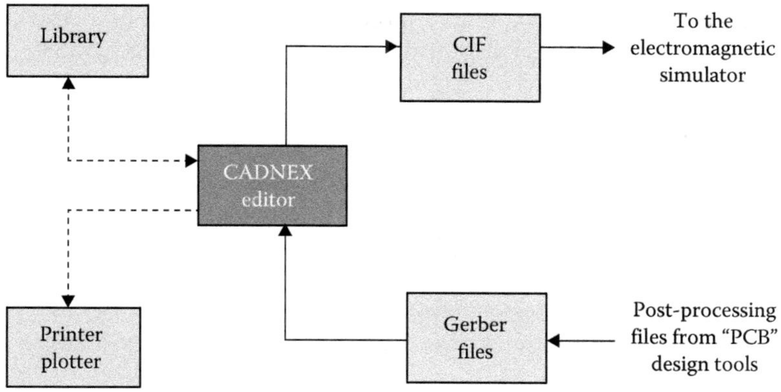

FIGURE 6.38
Gerber—CIF conversion. (From *Fidelity User's Manual Release 1*, Zeland Software, Fremont, CA, 1998; Bay Technology, *Cadnex User and Reference Manual*, Release 6, 1998.).

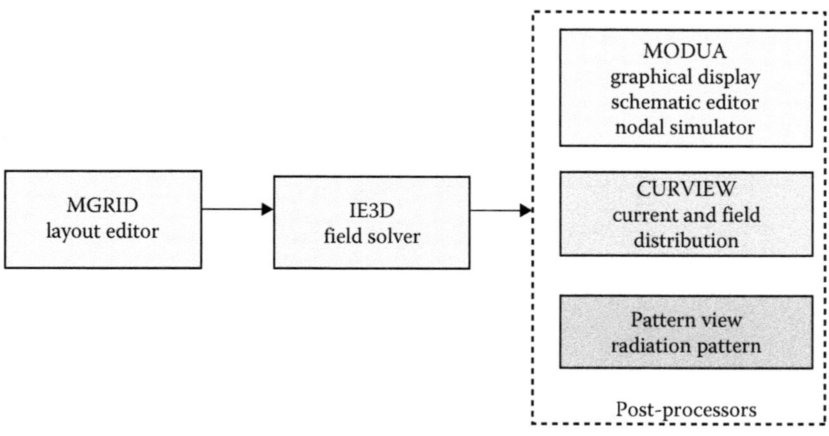

FIGURE 6.39
Main modules of the electromagnetic simulation environment. (From *IE3D User's Manual Release 6*, Zeland Software, 2000.)

nodal circuit simulator; CURVIEW, the post-processor for displaying the current distribution, respectively the field distribution; and PATTERNVIEW, the post-processor for displaying the radiation pattern [43].

The simulation results can be imported into classical nodal circuit simulators and can also be displayed and processed by MODUA schematic editor. MODUA takes the simulation result files from IE3D and pre-simulated geometry files from MGRID. A powerful feature of MODUA is that it can combine the results of electromagnetic simulation and perform a nodal simulation. In this way, complex structures can be split into elementary elements/entities, which can be investigated and then interconnected as equivalent circuits. One major advantage of MODUA is that field and current distributions of a simulated structure can easily be displayed [57,58].

In this chapter, the MoM electromagnetic investigation was focused on a few elements, as presented in the Figure 6.40. Each element was transformed into a layered finite element structure which was investigated in the frequency range, at different frequencies, and with different number of cells [54]. Furthermore, the elements can be physically interconnected in order to define a transmission line for the signals travelling in the package/module. Dividing the whole structure into elements was neccessary due to hardware limitations and processing time optimization. After modeling and simulation, a link to classical circuit simulators was created by extracting RLCG

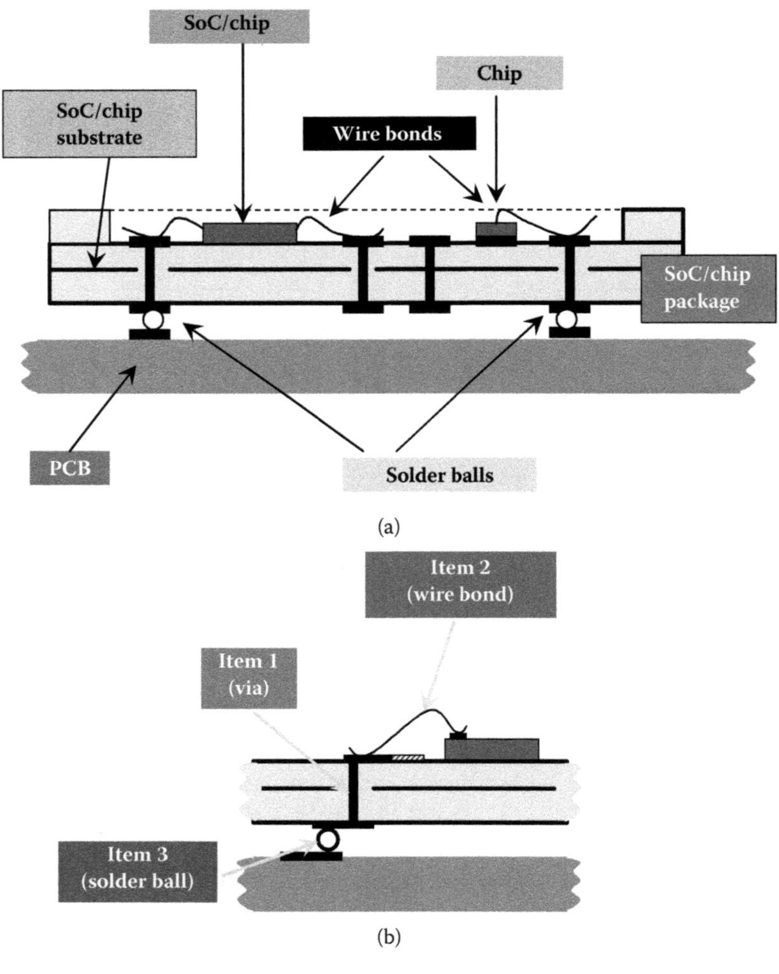

FIGURE 6.40
(a) SoC based package placed on a PCB; (b) elements under interest.

parameters, which are directly related to the HDI structures of SoC based packages and PCBs.

Figure 6.41 presents the case of a multiple stacked dice/chips in the frame of a HDI BGA package [59–61].

Regarding the PCB and interconnection elements, Figure 6.42 shows virtual 3D views of some important elements/entities (PCB trace, wire bond, via, solder ball) [62].

Next will be presented virtual investigations on some of the above elements, with a focus on via [63–65], which represents an important PCB object and a small discontinuity in the signal propagation path (the plated through hole/via invention was one of the major inventions in electronic packaging and, generally, in electronics since the system could be used on double-layer and multi-layer PCBs, dramatically increasing the interconnection density and, finally, the complexity of the product) [49,66–68].

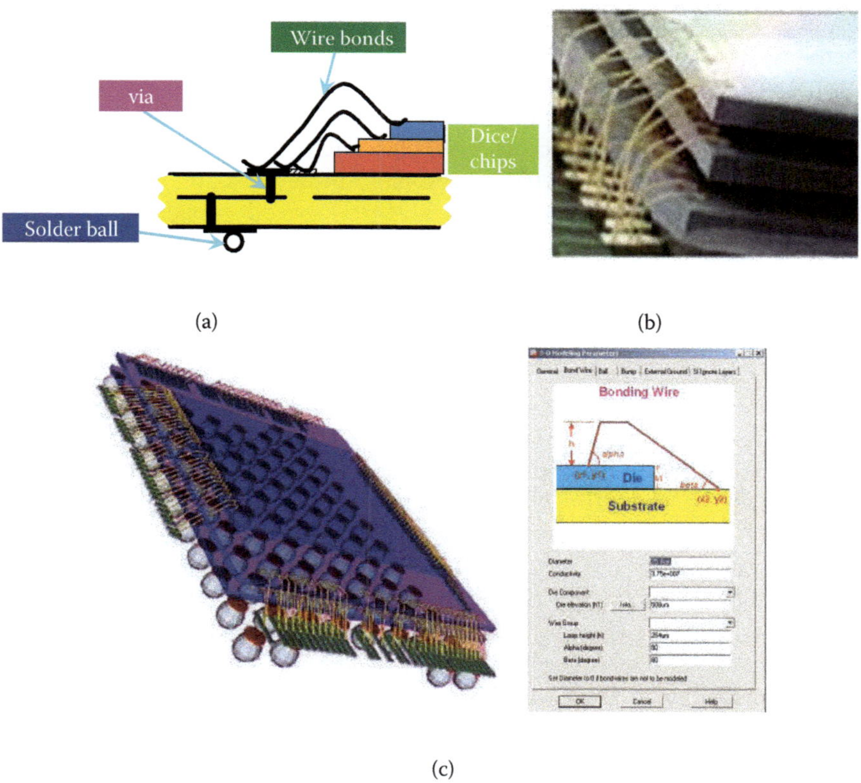

FIGURE 6.41
(a) Multiple stacked dice/chips in the frame of a HDI BGA package (25); (b) photo of the multiple stacked dice/chips structure; (c) virtual 3D view and setup of parameters using Cadence's Allegro Package Designer tool. (From www.cadence.com/community/allegro/Resources/resources_pkg/tech/tpartwork_3DModeling.pdf; *Allegro Package Designer, 610/620 Datasheet*, Available at: www.cadence.com.)

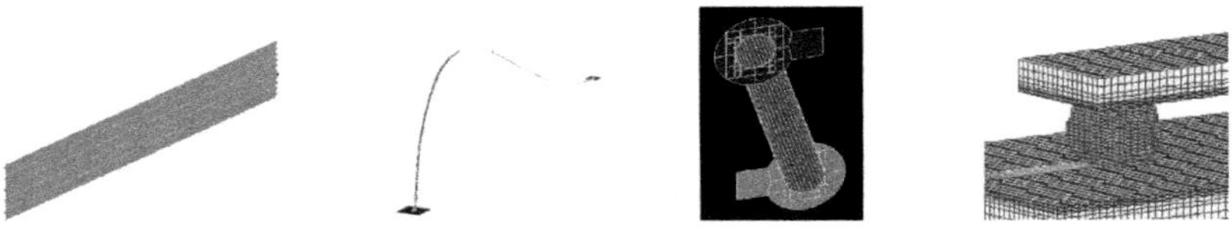

FIGURE 6.42
Overview of various interconnection elements (trace, wire bond, via, solder ball).

Because today via density is an important electrical issue, various vias were studied. The more vias on a board, the more discontinuities in the PCB interconnection structure are present. For this reason, the effect of different vias on the characteristics of high speed/high frequency interconnection lines was evaluated.

Figure 6.43 presents various types of vias and a comparison between conventional vias and microvias [22,69,72]. It is good to mention that values for them are currently 0.2–0.25 mm (8–10 mil) hole/0.5–0.6 mm (20–24 mil) capture pad for conventional vias, based on Through Hole Technology (THT), and 0.025–0.1 mm (1–4 mil) hole/0.2–0.3 mm (8–12 mil) capture pad for microvias. Shortly, a few techniques for via manufacturing are available: mechanical (punching and drilling), laser (CO_2, YAG and excimer lasers), photo-chemical (liquid or dry film resists), and etching (wet or dry etching).

The via represents a discontinuity with an inherent capacitive effect, which cause a step-function change in the characteristic impedance of the signal line. The accumulation of excess charge in the spatial zone of via creates the parasitic capacitance. Also an inductive effect is present in the zone due to the corners introduced by via. The effects are larger with increasing of frequency, since the parasitics are considered important in the microwave range. For analyzing of vias and microvias, the first step is to do a comparison between the hole diameter of via ("d") and the height/length of the via (the distance "h" between the two layers connected by it; Figure 6.44) [49].

If $h \gg d$, via can be approximated with a conductive barrel placed between "i" and "j" layers, the discontinuity will appear as a reunion of two corners. If $h \cong d$ or $h < d$, the approximation from above is no longer valid and the influence of the via barrel and via capture pads must be taken into account using electromagnetic simulators.

In Figure 6.45a through c, a complete via model, ready for full-wave analysis, is presented. It is possible to observe the two reference planes, specifically for 4 layer and multi-layer PCBs, and the "anti-pads" in the planes. Authors have checked the simulation results with various measurement results and have reached the conclusion that it is a high precision model that, simulated with enough hardware resources and accurate software tools, offers very good results, in accordance with measurements. For this reason, in many cases, virtual full-wave investigations are used as input data in other software systems [73,74].

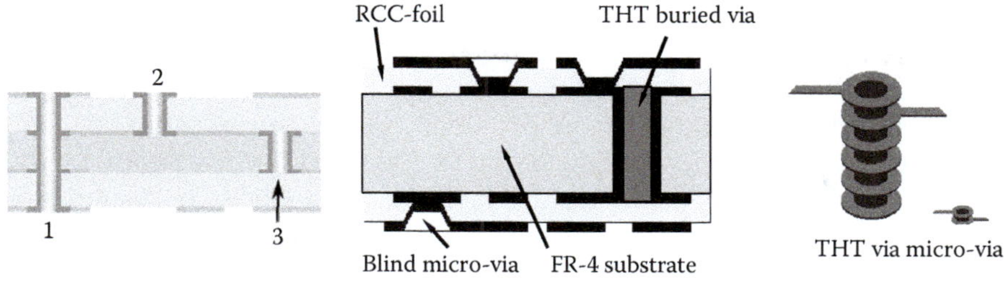

FIGURE 6.43
Types of vias: (1) through hole via, (2) blind via, and (3) buried via and microvias in complex PCB structures. (From IPC-2226, *Sectional Design Standard for High Density Interconnect (HDI) Printed Boards*, April 2003; C. Choi and A. Dasgupta, *Proceedings of ASME International Mechanical Engineering Congress and Exposition*, 5, 15–22, 2010; Y. Ning, M. H. Azarian, and M. G. Pecht, "Simulation of the Influence of Manufacturing Quality on Thermomechanical Stress of Microvias," in *IPC APEX 2014 Technical Conference*, March 25–27, 2014; IPC/JPCA-2315, *Design Guide for High Density Interconnects (HDI) and Microvias*, June 2000; IPC-2226, *Sectional Design Standard for High Density Interconnect (HDI) Printed Boards*, April 2003.)

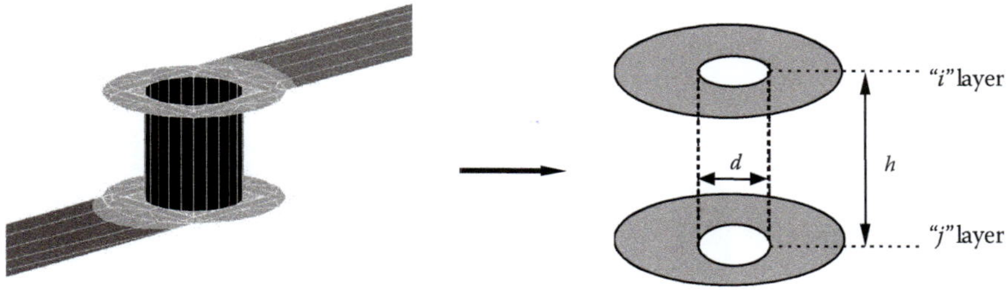

FIGURE 6.44
Geometrical sizes for vias evaluation ("d" and "h").

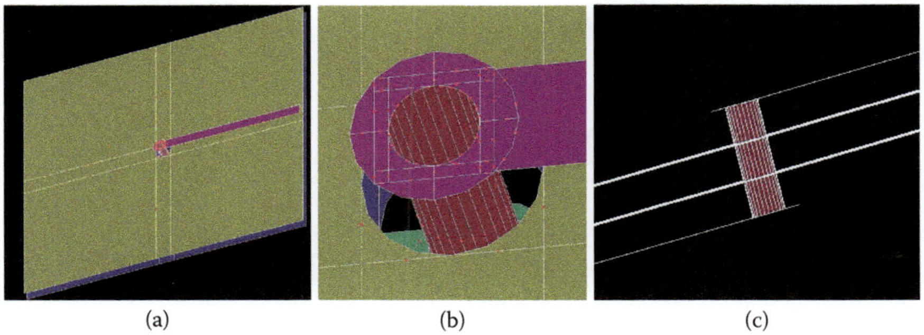

FIGURE 6.45
Model of a via in the frame of a multi-layer structure: (a) overview, (b) zoom in on the via area, and (c) zoom and rotation for showing the layer stack-up.

The correspondence between real vias and finite elements models is shown in Figure 6.46 [75]. For a proper management of hardware resources and for minimizing the processing time, the cylinder of the via and the capture pads were created using triangular, quadrilateral, and octagonal cells.

In the case of THT via, the hole diameters selected for investigations were (mm): 0.3, 0.4, 0.5, 0.6, 0.7, 0.8, 1, and 2. As an example, the "Grid Generation" process has created 286 cells for "via 0.3" and 366 cells for "via 2." Signal traces were designed for reaching a characteristic impedance of 50δ. The PCB parameters were: FR-4 dielectric type, *h* (dielectric thickness) = 1.5 mm, *t* (copper thickness) = 17.5 μm, and εr = 4.7. During the electromagnetic simulation process S-, Y-, and Z-parameters of vias were calculated. The link with circuit simulators [38] was performed by extracting RLCG parameters in

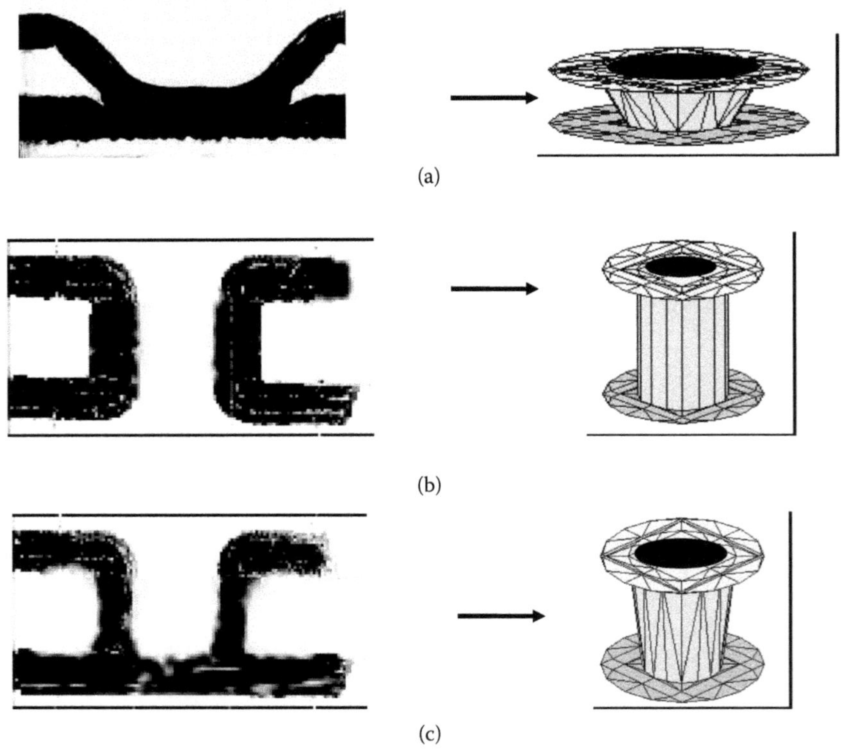

FIGURE 6.46
Correspondence between real vias and finite elements models (not to scale): (a) plasma ablation via, (b) mechanically drilled via, and (c) laser ablation via. (From J. Paulus and M. Petti, Microvia materials: Enablers for high density interconnects, *IPC Printed Circuit Expo 1997*, S17 1–6, March 9–13, 1997; P. Svasta et al., "Simulation of Multichip Module Structures, EC-MCM '99," in *5th European Conference on MultiChip Modules*, London, UK, pp. 48–54, February 1–2, 1999; P. Svasta et al., "New Aspects In Modeling And Simulation Of PCB/MCM Vias," *The 12th IMAPS Europe Conference*, Harrogate, UK, pp. 205–213, June 7–9, 1999.)

SPICE format. Additionally, a comparison with values given by formulas was done in order to establish their limits and to decide in which frequency range designers can trust simple geometrical formulas and where they will need the help of powerful full-wave electromagnetic/signal integrity tools. With "Curview" it was possible to display 3D views of discretized structures, as well as field/current distributions on them. Listed below are some of the results obtained for the 0.3 mm via. Because the via is a "short" entity, the lossless model is considered (as in Figure 6.31), R1 (series resistance), and R2 (shunt resistance) being neglected (see Figure 6.30) (Figure 6.47) (Tables 6.4 and 6.5).

Furthermore, an evaluation of plasma and laser microvias was performed. A special laser/plasma ablatable dielectric material, manufactured as resin coated copper, was selected (Figure 6.48b). Figure 6.48a offers an image of a two coatings material (C-Staged: cured and B-Staged: semi-cured).

The analyzed structure (Figure 6.48b) consists in a single resin layer, coated with a copper foil, in a semi-cured state. The resin has the properties both to fill and encapsulate the underlying circuitry and to offer the dielectric function. Usual values for this material are: B-Staged epoxy resin: 80 μm, copper: 17.5 μm, εr = 3.4–3.6, tanδ = 0.025–0.028. This material was placed on a conventional single sided FR-4 substrate with a ground plane onto the bottom side. Below are listed some results obtained for a plasma microvia (with the parameters: capture pads—0.3 mm, via diameter on top—0.175 mm, via

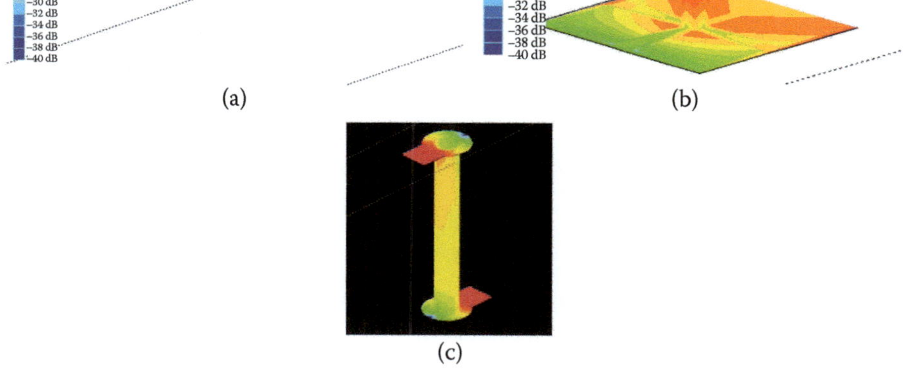

FIGURE 6.47
Current distribution in the spatial region of the "via 0.3:" (a) overview, (b) zoom in of the via region, and (c) zoom in of "via 0.3," which has a greater dielectric thickness.

TABLE 6.4
LC SPICE Circuit/Model for "Via 0.3"

Parameters → Frequency (MHz) ↓	L [nH]	C [pF]
50	1.469	0.5186
100	1.468	0.5187
500	1.46	0.5205
800	1.452	0.5230
1000	1.444	0.5252

FR-4 dielectric type, h = 1.5 mm, t = 17.5 μm, and $\varepsilon_r = 4.7$.

TABLE 6.5
LC SPICE Circuit/Model for "Via 2"

Parameters → Frequency (MHz) ↓	L [nH]	C [pF]
50	1.261	0.4715
100	1.26	0.4715
500	1.248	0.4727
800	1.243	0.4745
1000	1.24	0.4759

FR-4 dielectric type, h = 1.5 mm, t = 17.5 μm, and $\varepsilon_r = 4.7$.

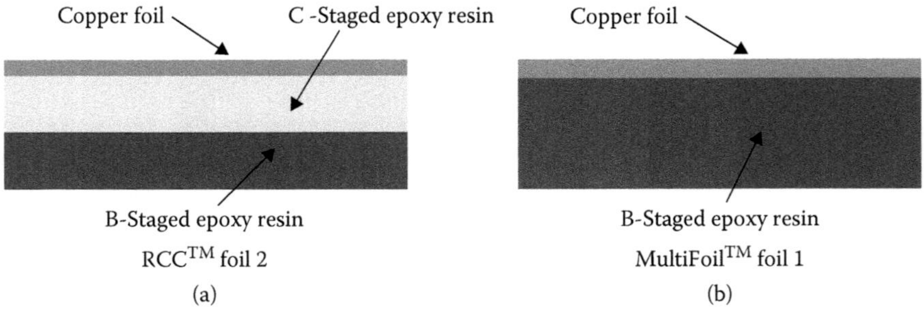

FIGURE 6.48
Resin coated copper materials for multi-layer circuits.

diameter on bottom—0.1 mm) and a laser microvia (with the parameters: capture pads—0.2 mm, via diameter on top - 0.1 mm, via diameter on bottom—0.075 mm). Using the same procedure as for conventional vias, current distribution in the region of microvias and post-processed files for signal integrity evaluation of circuitry containing microvias were prepared, too. Below are offered only the LC parameters tables (Table 6.6).

An important advantage of microvias is that it is not necessary to have a capture pad larger than the trace, the pad diameter being equal with the trace width and, thus, practically eliminated due to the presence of the trace.

Investigations of other two interconnection elements, wire-bond and solder ball, are presented shortly in the following. The wire bond element has a non-symmetrical configuration, as presented in Figure 6.49. The wire for bonding is made of Gold (Au), having 25 µm in diameter and 2.5 mm in length. The pads parameters are: 0.2 × 0.2 mm on PCB and 0.1 × 0.1 mm on SoC/chip (Figure 6.49).

As in the case of the via, because the wire bond is also a "short" element, the lossless model is used and the R1 (series resistance) and the R2 (shunt resistance) are neglected. LC parameters were obtained and presented in Table 6.7. Figure 6.50 presents an example of the 3D average current density in the wire bond (Table 6.8).

In the case of the solder ball, the element of the BGA package/substrate which contains the SoC/chip, the configuration from Figure 6.51 was created. It presents a solder ball structure with short interconnection lines. In this case, the Fidelity simulator performs a better modeling and simulation of the 3D object than a software tool based on the Finite Difference Time

TABLE 6.6

Lumped LC SPICE Model for Plasma Microvia

Parameters→ Frequency (MHz) ↓	L [nH]	C [pF] × 10^{-2}
50	0.8387	2.090
100	0.8374	2.090
500	0.8317	2.092
800	0.8299	2.096
1000	0.8289	2.099

B-Staged epoxy resin: 80 µm, copper: 17.5 µm, ε_r = 3.4–3.6, tanδ = 0.025–0.028 placed on a standard FR-4 structure, h = 1.5 mm, t = 17.5 µm, ε_r = 4.7.

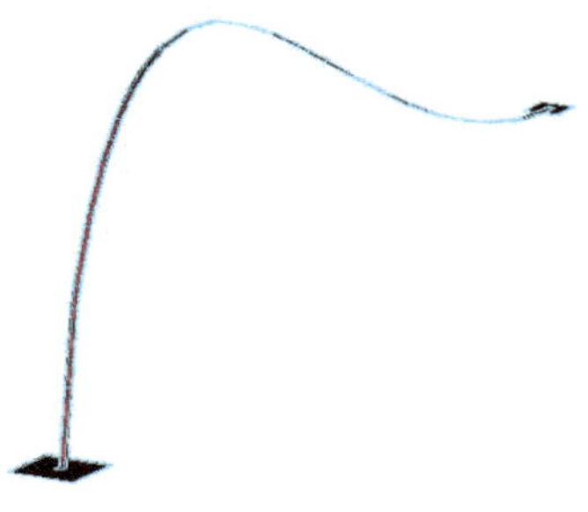

FIGURE 6.49
3D view of the wire bond interconnection element.

TABLE 6.7

Lumped LC SPICE Model for Laser Microvia

Parameters → Frequency (MHz) ↓	L [nH]	C [pF] × 10^{-2}
50	0.9587	1.564
100	0.9569	1.564
500	0.9485	1.565
800	0.9459	1.568
1000	0.9460	1.570

B-Staged epoxy resin: 80 µm, copper: 17.5 µm, ε_r = 3.4 – 3.6, tanδ = 0.025–0.028 placed on a standard FR-4 structure, h = 1.5 mm, t = 17.5 µm, ε_r = 4.7.

TABLE 6.8

Lumped LC SPICE Model for the Wire Bond

Parameters → Frequency (MHz) ↓	L [nH]	C [pF]
50	4.41	0.044
100	4.46	0.044
500	4.62	0.045
800	4.74	0.046
1000	4.82	0.047

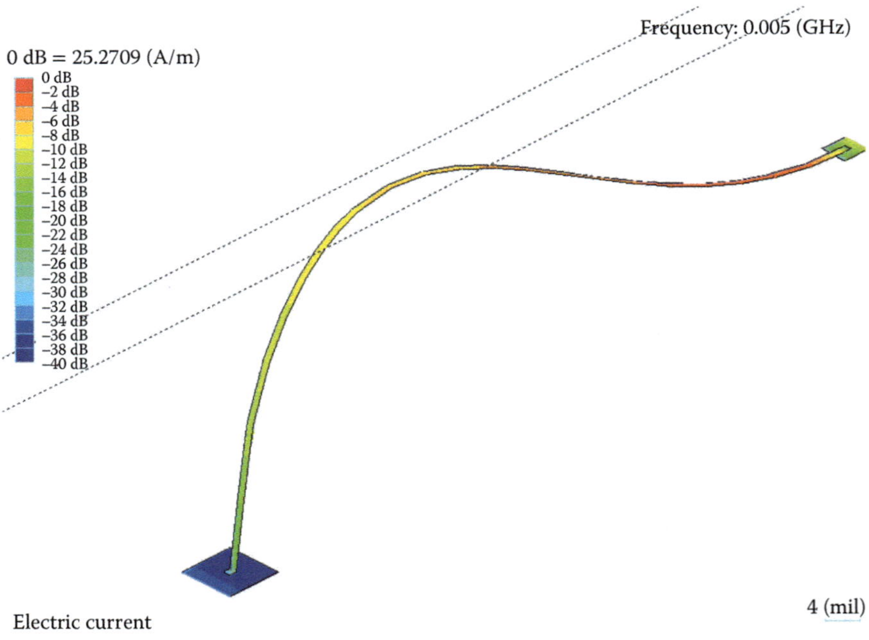

FIGURE 6.50
Average current density in the wire bond.

Domain (FDTD) method. Fidelity creates a non-uniform rectangular meshing using a pre-defined 3D grid.

The structure from Figure 6.51a comprises two balls on a PCB, a connecting trace on the bottom side of the BGA package/substrate and a small zone of the BGA. This structure is a symmetrical construction, composed of two balls. This spatial arrangement was required by the simulator, which requires the field excitation to be made by means of transmission lines ports, two microstrip lines being placed at the input and output of the structure. Because the main purpose was the computation of the LC parameters of the structure, the equivalent circuit was obtained using the symmetrical description of the LC equivalent circuit. The non-symmetrical equivalent circuit from Figure 6.30 can be rearranged in a symmetrical configuration, as shown in Figure 6.52a and b.

Based on the symmetry, one can simply divide the values of L and C, obtained after simulation, in order to obtain the ones valid for one ball. The solder balls used in this investigation were of 1 mm diameter, placed with 2.54 mm spacing. The width of the copper traces and connecting bridge

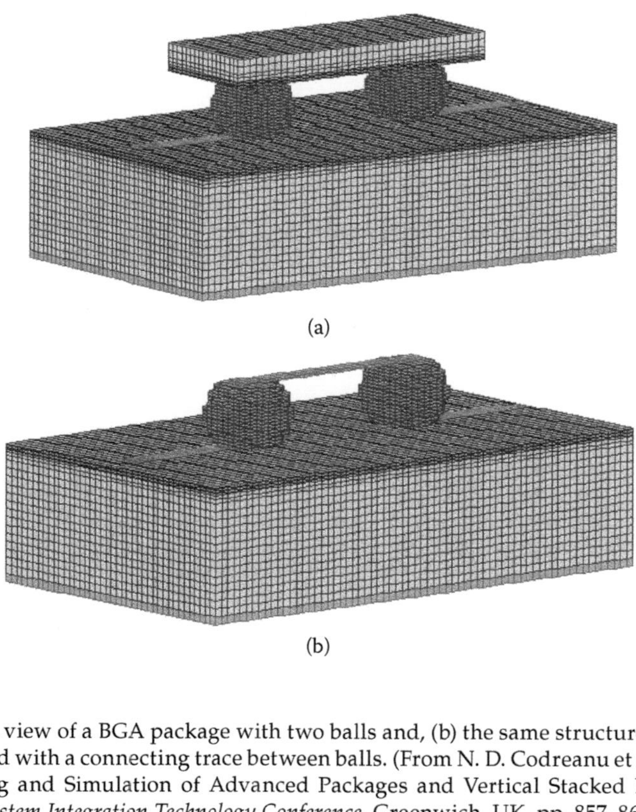

(a)

(b)

FIGURE 6.51
(a) 3D partial view of a BGA package with two balls and, (b) the same structure without the BGA zone and with a connecting trace between balls. (From N. D. Codreanu et al., "Accurate 3D Modelling and Simulation of Advanced Packages and Vertical Stacked Dice," in *2nd Electronics System Integration Technology Conference*, Greenwich, UK, pp. 857–861, September 1–4, 2008.)

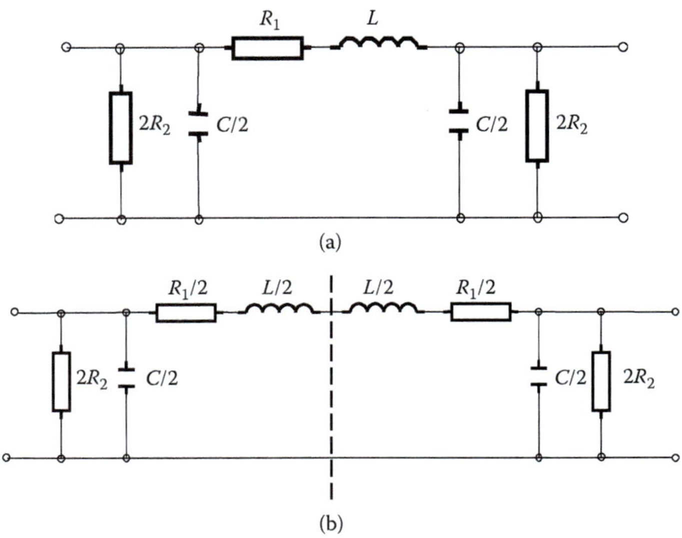

FIGURE 6.52
(a) and (b) The symmetrical circuit of a PCB interconnection element.

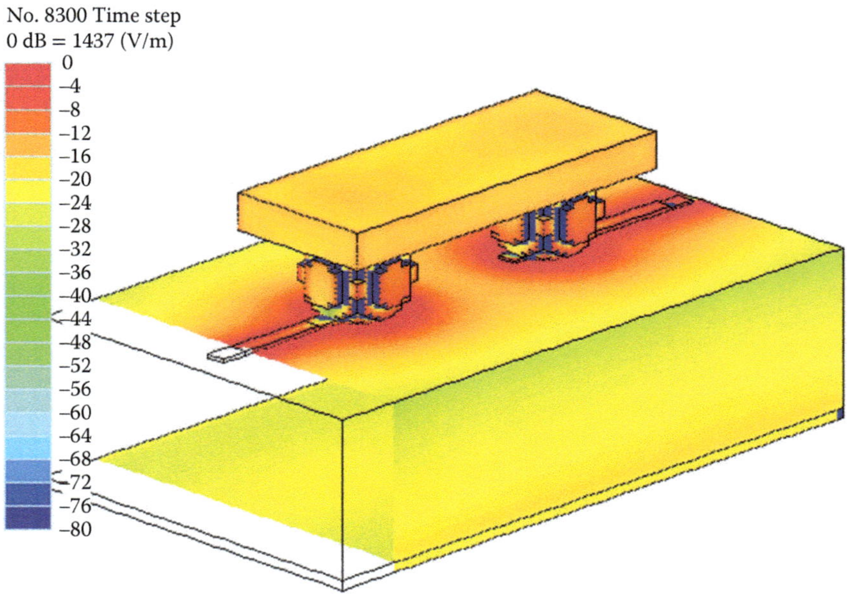

FIGURE 6.53
Near field display (electrical strength) for the solder ball structure.

TABLE 6.9
LC SPICE Parameters of the Solder Ball

Parameters → Frequency (MHz) ↓	L [nH]	C [pF]
50	7.1	0.139
100	7.4	0.14
500	7.7	0.144
800	7.9	0.147
1000	8.1	0.149

between the balls was 0.2 mm. The excitation ports were placed at 1 mm distance from the center of the balls or, in other words, the length of the copper traces was of 1 mm. The substrate thickness was 1.5 mm for the PCB and 0.5 mm for the BGA package substrate. The grid was chosen to be finer in the area of interest and coarser in rest, in order to achieve a reasonable compromise between accuracy and processing time.

As in the case of via and wire bond, because the solder ball is a "small" element, the losses model was used (see Figure 6.31). LC parameters were obtained and presented in Table 6.9. For a qualitative evaluation, Figure 6.53 presents the near field display for the structure, a feature that is useful for observing the possible field coupling between the elements of the structure and other entities placed in the vicinity.

6.5 Conclusions

Printed Circuit Board (PCB) integration of SoC packages represents a real challenge today, since many improvements at Silicon level (first packaging level [9]) can be dramatically limited by a non-optimal integration and signal integrity degradation at PCB level (second packaging level [9]).

In the last decades, for complex SoCs/chips, the electronics industry has gradually moved from QFP/QFN (Quad Flat No-leads) to BGA/LGA packages, the packaging solution currently used for modern and complex SOCs.

In order to take full advantage of the higher level of integration realized in a SoC, there is a high demand to minimize the area and volume of the final module/system. Next to HDI structures, this is possible by using embedded passive components. There are many benefits that came from using embedded passives, such as the release of free surface on PCB, low mass and volume, less components to be assembled, and less solder joints, leading improved reliability of the system. Additionally, embedded passives show, in general better, high frequency behavior due to smaller dimensions and, hence, low parasitics. The decoupling of power supply based on embedded capacitors in the PCB dielectric is also a benefit. Last, but not least, in a large scale production, the overall costs can be reduced by using integrated passive components instead of corresponding discrete components.

Even though parasitic elements of embedded passives are low in comparison to discrete components, they must be taken into consideration where signal integrity issues are present.

For resistors in a sine wave regime and at higher frequencies, a general rule is that high value resistors behave in a capacitive mode and low value resistors shows, first, an inductive behavior, followed by a capacitive one. For every resistor, there is a specific configuration for which the capacitive and inductive effects can be compensated. In this case, the working frequency up to which the parasitics produce no significant change in the impedance of resistor can be increased. The parasitic elements L and C influence also the pulse behavior of resistors. The oscillating and damped character of response to an electric pulse depends on the parasitics mentioned above with respect to resistance value R, as presented in Section 6.2.3.

Regarding the calculation of parasitics for a passive component, the problem is still complicated, even using modern investigation tools, such as FEM simulators. Only a "full wave" electromagnetic analysis, where the coupling between electric and magnetic field is addressed, can offer accurate results. For the designing and dimensioning of embedded capacitors and inductors, FEM simulators offer very good results. In some cases, computations based on closed form integrals can offer faster and quite accurate results.

The authors have also presented a full-wave virtual investigation method of SoC based PCB structures, method which helps designers to better understand the PCB and interconnection structures and their influence on signal/power integrity. The electromagnetic investigation based on MoM (Method of Moments) planar simulators (named 2.5D simulators) represent a high-quality method for design and development of modern SoC based PCBs, a method that discretizes and solves Maxwell equations for planar circuits. The simulation was performed with IE3D and Fidelity engines. IE3D is suitable for 2D or 3D layered structures and Fidelity handles 3D structures. The simulators can emphasize important effects, taking into account the distributed character of circuits and the non-ideal behavior of PCB interconnection structures. The simulation tools used for investigation combine an accurate modeling of HDI structures with an accurate description of interconnections and elements using distributed, frequency-dependent parameters. This allows monitoring and keeps under control the combined influence of various parameters related to signal integrity, the output information available for the design stage. Regarding the grids used, the non-uniform grid was considered to be more efficient than the uniform grid.

A SoC based BGA package and the associated PCB structure were investigated, from an electromagnetic point of view, by partitioning in 3D elements (solder balls) and 2D layered elements (vias, PCB traces, wire bonds). The full-wave electromagnetic investigation has offered useful results related to these elements, leading to the conclusion that, for high speed/high frequency applications, the quality of the design is increased with the support of a powerful full-wave electromagnetic/signal integrity tool. Among other outputs, LC lumped parameters were extracted, which are useful for nodal circuit simulations to investigate the influence of PCB and interconnection structures on signal integrity and, generally, on the development of electronic modules/systems.

References

1. C. A. Harper, *Electronic Packaging and Interconnection Handbook,* New York, NY: McGraw-Hill, 2000.
2. C. F. Coombs, Jr., *Printed Circuits Handbook,* 6th edition, New York, NY: McGraw-Hill, 2008.
3. M. I. Montrose, *Printed Circuit Board Design Techniques for EMC Compliance,* 2nd Edition, Piscataway, NJ: IEEE Press, 2000.
4. E. Bogatin, *Signal and Power Integrity—Simplified,* Boston, MA: Prentice Hall, 2010.
5. A. I. Schwab, *Electromagnetische Verträglichkeit,* Berlin, New York, NY: Springer-Verlag, 1993.
6. M. Mardiguian, *Electromagnetic Control in Components and Devices,* Vol. 5, Gainesville, VA: Interference Control Technologies, 1988.
7. D. Behera, S. Varshney, S. Srivastava, and S. Tiwari, A brief primer on embedded SoC packaging options, *Freescale Semiconductor,* November 20, 2011.
8. eSilicon, "eSilicon Package Types and Assembly Flow," in *eSilicon Packaging Department,* April, 2015. Available at www.esilicon.com.
9. R. Tummala, *Fundamentals of Microsystems Packaging,* 1st edition, New York, NY: McGraw-Hill, 2001.
10. Y. Jin, Z. Wang, and J. Chen, *Introduction to Microsystem Packaging Technology,* Boca Raton, FL: CRC Press, 2011.
11. A. Longford, "Trends in Advanced Packaging Technologies, An IMAPS-UK view," in *IMAPS UK–IeMRC Interconnection Event,* TWI, Cambridge, UK, December 2008.
12. G. E. Moore, Progress in digital integrated electronics, *IEDM Technical Digest,* 11–13, 1975.
13. M. I. Montrose, *EMC and the Printed Circuit Board–Design, Theory, and Layout Made Simple,* Piscataway, NJ: IEEE Press, 1999.
14. H.W. Johnson and M. Graham, *High-Speed Digital Design: A Handbook of Black Magic,* Englewood Cliffs, NJ: PTR Prentice Hall, 1993.
15. S. Ganesan and M. Pecht, *Lead-Free Electronics,* Hoboken, NJ: John Wiley & Sons, New Jersey, 2006.

16. P. Svasta, N. D. Codreanu, *Computer Aided Design of Electronic Modules*, Bucharest: Technical Publishing House, 1998 (in Romanian).
17. N. D. Codreanu, *Advanced Investigation Methods of PCB Structures*, Bucharest: Cavallioti Publishing House, 2009.
18. R. K. Ulrich and W. D. Brown, *Advanced Electronic Packaging*, 2nd edition, Hoboken, NJ: Wiley IEEE Press, 2006.
19. www.amkor.com/go/packaging/packages-by-family
20. J. Lau, C. P. Wong, J. L. Prince, and W. Nakayama, *Electronic Packaging–Design, Materials, Process and Reliability*, New York, NY: McGraw-Hill, 1998.
21. H. Holden, D. Neer, J. Andresakis, E. Bogatin, M. Carano, K. A. Carpenter, K. H. Dietz, M. Laing, C. Vaucher, P. Viklund, and M. Wuensch, *The HDI Handbook*, 1st edition, Seaside, OR: BR Publishing Inc. Available at: www.hdihandbook.com/.
22. IPC-2226, *Sectional Design Standard for High Density Interconnect (HDI) Printed Boards*, April 2003.
23. R. K. Ulrich and L. W. Schaper, *Integrated Passive Component Technology*, Hoboken, NJ: Wiley IEEE Press, 2003
24. *HDI (High Density Interconnect PCBs)*, Available at: www.ttmtech.com/products/prod_hdi.aspx.
25. N. D. Codreanu, C. Ionescu, P Svasta, and I. Plotog, "Accurate 3D Modelling and Simulation of Advanced Packages and Vertical Stacked Dice," in *2nd Electronics System Integration Technology Conference*, Greenwich, UK, pp. 857–861, September 1–4, 2008.
26. D. R. J. White, *EMI Control in the Design of Printed Circuit Boards and Backplanes*, Gainsville, VA: Don White Consultants, 1982.
27. What is HDI?, Available at: www.acreo.se.
28. IPC-6016, *Qualification and Performance Specification for High-density Interconnect (HDI) Structures*, May 1999.
29. C. Ionescu, N. D. Codreanu, V. Golumbeanu, and P. Svasta, "Modeling and Simulation of Integrated Passive Components," in *SIITME 2006, International Symposium for Design and Technology of Electronic Packages*, 12th edition, Iasi, Romania, pp. 39–43, September 21–24, 2006.
30. C. Ionescu, N. D. Codreanu, V. Golumbeanu, and P. Svasta, "Thermal Capabilities of Integrated Resistors in Organic Substrates," in *16th European Microelectronics and Packaging Conference & Exhibition, EMPC 2007*, Oulu, Finland, pp. 95–99, June 17–20, 2007.
31. F. Zandman, P. Simon, and J. Szwarc, *Resistor Theory and Technology* (Revised edition), Herndon, Virginia: SciTech Publishing, 2002.
32. O. Zinke and H. Seither, *Widerstände, Kondensatoren, Spulen und ihre Werkstoffe*, Berlin/Heidelberg: Springer Verlag, 1982.
33. R. Rodriguez, J. M. Dishman, F. T. Dickens, and E. W. Whelan, Modeling of two-dimensional spiral inductors, *IEEE Transactions on Components, Hybrids and Manufacturing Technology*, vol. 3, no. 4, 535–541, 1980.
34. P. Svasta, C. Ionescu, V. Golumbeanu, N. D. Codreanu, A. Romanescu, A. Pietrikova, and J. Neamtu, "Electromagnetic Behavior of Inductive Planar Structures with Non-homogenous Magnetic Environment," in *1st Electronics Systemintegration Technology Conference, ESTC 2006*, Dresden, Germany, pp. 966–973, vol. 2, Septemer 5–7, 2006.
35. C. Ionescu, V. Golumbeanu, and P. Svasta, "Modeling and Simulation of Inductive Planar Structures with Non-homogenous Magnetic Medium," in *29th*

International Spring Seminar on Electronics Technology, ISSE 2006, St. Marienthal, Germany, pp. 448–453, May 10–14, 2006.
36. P. Svasta, C. Ionescu, A. Romanescu, and A. Pietrikova, "Integrate Inductive Structure in Cost Efficient System in Package Application," in *Proceedings of the 4th European Microelectronics and Packaging Symposium (EMPS 2006)*, Terme Catež, Slovenia, pp. 91–95, May 22–24, 2006.
37. D. Lu and C. P. Wong, *Materials for Advanced Packaging*, New York, NY: Springer, 2009.
38. D. Fitzpatrick, *Analog Design and Simulation using OrCAD Capture and PSpice*, Oxford: Newnes/Elsevier, 2012, ISBN 978-0-08-097095-0.
39. C. Ionescu, V. Golumbeanu, and N. D. Codreanu, "Frequency Response of Resistors," *The 8th International Symposium for Design And Technology of Electronic Modules, SIITME2002*, Cluj Napoca, Romania, pp. 20–24, September 19–22, 2002.
40. G. Becke, P. Forstner, E. Haseloff, and J. Huchzermeier, *Digital Design Seminar*, Texas Instruments, 1996.
41. L. Williams, S. Rousselle, and B. Boots, "Cray Supercomputer 3.2 Gb/s Serial Interconnect Simulation Using Full-wave Electromagnetics," in *DesignCon 2004, Conference Proceedings*, Santa Clara, CA, February 2–5, 2004.
42. *ANSYS 6.1, Theory Reference Manual*, 2002. Available at: www.ansys.com
43. *IE3D User's Manual Release 6*, Zeland Software, 2000.
44. M. I. Montrose, "Analysis on the Effectiveness of Clock Trace Termination Methods and Trace Lengths on a Printed Circuit Board," in *Proceeding of the IEEE 1996 International Symposium on Electromagnetic Compatibility*, Santa Clara, CA: pp. 453–458, 1996.
45. C. Ionescu, V. Golumbeanu, and J. Sitek, "Finite Element Modeling of Resistor Parasitics," *The 9th International Symposium for Design And Technology of Electronic Packages, SIITME 2003*, Timisoara, Romania, pp. 238–242, September 18–21, 2003.
46. L. Zu, Y. Lu, R. C. Frye, M. Y. Lau, S.-C. S. Chen, D. P. Kossives, J. Lin, and K. L. Tai, high Q-factor inductors integrated on MCM si substrates, *IEEE Transactions on Components, Packaging and Manufacturing Technology, Part B*, vol. 19, no. 3, 635–641, August 1996.
47. J. D. Jackson, *Electrodinamica Clasica* ("Classical Electrodynamics," in Romanian, trans. from English), Bucharest: Technical Publishing House, pp. 154–176, 1991.
48. P. Svasta, N. D. Codreanu, C. Ionescu, and V. Golumbeanu, "Simulation of Multichip Module Structures, EC-MCM '99," in *5th European Conference on MultiChip Modules*, London, UK, pp. 48–54, February 1–2, 1999.
49. P. Svasta, N. D. Codreanu, Zs. Illyefalvi-Vitez, and V. Golumbeanu, "New Aspects In Modeling And Simulation Of PCB/MCM Vias," *The 12th IMAPS Europe Conference*, Harrogate, UK, pp. 205–213, June 7–9, 1999.
50. N. D. Codreanu, C. Ionescu, P. Svasta, and T. Cucu, "HDI Printed Circuits Electromagnetic Modeling and Simulation," in *4th European Microelectronics and Packaging Symposium*, Terme Catež, Slovenia, pp. 259–264, May 22–24, 2006.
51. B. Young, *Digital Signal Integrity: Modeling and Simulation with Interconnects and Packages*, Upper Saddle River, NJ: Prentice Hall, 2001.
52. www.polarinstruments.com/
53. F. Gardiol, *Microstrip Circuits*, New York, NY: John Wiley & Sons, 1994.
54. K. J. Scott, *Practical Simulation on Printed Circuit Boards and Related Structures*, New York, NY: Wiley-Blackwell, 1994.

55. *Fidelity User's Manual Release 1*, Fremont, CA: Zeland Software, 1998.
56. Bay Technology, *Cadnex User and Reference Manual*, Release 6, 1998.
57. P. Meuris, W. Schoenmaker, and W. Magnus, Strategy for electromagnetic interconnect modeling, *IEEE Transactions on CAD*, vol. 20, 753–763, 2001.
58. A. E. Ruehli, Equivalent circuit models for three dimensional multi-conductor systems, *Micro-wave Theory*, vol. MTT-22, 216–221, 1974.
59. C. Ionescu, N. D. Codreanu, P. Svasta, and D. Leonescu, "Electromagnetic Simulation of BGA Structures in PCB/MCM Configurations," in *ISSE 1999 Conference Proceedings*, Dresden, Germany, pp. 182–187, May 18–22, 1999.
60. www.cadence.com/community/allegro/Resources/resources_pkg/tech/tpartwork_3DModeling.pdf.
61. *Allegro Package Designer, 610/620 Datasheet*. Available at: www.cadence.com.
62. N.D. Codreanu, P. Svasta, V. Golumbeanu, and C. Ionescu, *Full-Wave Electromagnetic Simulation of Printed Circuit Board Structures*, IPC Printed Circuits EXPO 2002, S06-2-1-S06-2-7, Long Beach, CA, March 24–28, 2002.
63. A. O. Ogunjimi, S. Macgregor, and M. G. Pecht, The effect of manufacturing and design process variabilities on the fatigue file of the high density interconnect vias, *Journal of Electronics Manufacturing*, vol. 5, no. 2, 111–119, July 1995.
64. A. S. Prabhu, D. B. Barker, M. G. Pecht, J. W. Evans, W. Grieg, E. S. Bernard, and E. Smith, A thermo-mechanical fatigue analysis of high density interconnect vias, *Advances in Electronic Packaging*, vol. 10, no. 1, 1995.
65. F. Liu, J. Lu, V. Sundaram, D. Sutter, G. White, D. F. Baldwin, and R. Rao, Reliability assessment of microvias in HDI printed circuit board, *IEEE Transactions on Components and Packaging Technologies*, vol. 25, no. 2, 254–259, June 2000.
66. G. Ramakrishna, F. Liu, and S. K. Sitaramana, "Experimental and Numerical Investigation of Microvia Reliability," *The 8th Intersociety Conference on Thermal and Thermomechanical Phenomena in Electronic Systems*, San Diego, CA, USA, pp. 932–939, 2002.
67. T. Wang and Y. Lai, Stress analysis for fracture potential of blind via in a build-up substrate, *Circuit World*, vol. 32, no. 2, 39–44, 2006.
68. N. Sherwani, Q. Yu, and S. Badida, *Introduction on Multichip Modules*, New York, NY: John Wiley & Sons, pp. 36–38, 1995.
69. www.ipc.org.
70. C. Choi and A. Dasgupta, Microvia non-destructive inspection method, *Proceedings of ASME International Mechanical Engineering Congress and Exposition*, vol. 5, 15–22, 2010.
71. Y. Ning, M. H. Azarian, and M. G. Pecht, "Simulation of the Influence of Manufacturing Quality on Thermomechanical Stress of Microvias," in *IPC APEX 2014 Technical Conference*, Las Vegas, U.S.A., March 25–27, 2014.
72. IPC/JPCA-2315, *Design Guide for High Density Interconnects (HDI) and Microvias*, June 2000.
73. N. D. Codreanu, P. Svasta, C. Ionescu, and V. Golumbeanu, "New Investigations of High Performance "PCB" Structures for Signal Integrity Compliance", in *15th European Microelectronics and Packaging Conference*, Brugge, Belgium, pp. 534–539, June 12–15, 2005.
74. N. D. Codreanu, C. Ionescu, P. Svasta, and H. Schlenker, "Electromagnetic Investigations of New PWB Structures," in *Proceedings of ISSE 2003 Conference*, High Tatras, Slovak Republik, pp. 44–48, May 8–11, 2003.

75. J. Paulus and M. Petti, Microvia materials: Enablers for high density interconnects, *IPC Printed Circuit Expo 1997*, S17 1–6, March 9–13, 1997.
76. Z. Felendzer, B. Janiczak, and T. Gorecki, "Computer-Aided Design of Multilayer Printed Circuit Boards with Respect to Reflection and Crosstalk Effects," in *Proceeding of the 20th Conference of the International Society for Hybrid Microelectronics*, Poland Chapter, Jurata, Poland, pp. 23–30, September, 1996.

7

Modeling and Characterization of TSV-Induced Noise Coupling

Xiao Sun, Martin Rack, Geert Van der Plas, Jean-Pierre Raskin, and Eric Beyne

CONTENTS

7.1 Introduction: Background and Driving Forces

Planar scaling of integrated circuits (ICs) that follow Moore's law is approaching its limits. As an alternative, three-dimensional (3D) technologies follow a more-than-Moore strategy, which has the potential to significantly boost the performance and capabilities of state-of-art ICs while supporting the integration of heterogeneous technologies [1]. Among such 3D technologies, the through-silicon via (TSV) process has become a key breakthrough to shorten global interconnects and enable the heterogeneous integration of different technologies. However, TSVs are also an important source of noise coupling in high speed and high frequency applications and can interfere with the operation of neighboring devices and circuits through the TSV oxide liner and the conductive silicon substrate. This can cause circuit malfunctioning and signal integrity problems.

Moreover, a TSV is a metal-insulator-semiconductor (MIS) structure and its capacitance depends on both frequency and bias. The presence of fixed charges at the backside passivation layer of the TSV after wafer thinning can induce inversion at that interface, and cause an anomalous TSV capacitance behavior, which, in turn, can significantly increase the capacitive coupling between TSVs and neighboring devices. An inversion layer at the back side can act as a conductive channel between neighboring TSVs and affects the RF behavior of TSVs in terms of noise coupling and signal integrity in neighboring devices. Therefore, the understanding and evaluation of the electromagnetic (EM) coupling mechanisms to and from TSVs have become crucial to the design of three-dimensional integrated circuits (3D-ICs).

In the first part of this chapter, we will address different scenarios of electromagnetic (EM) noise coupling in 3D-ICs. In the next step, we will discuss different mitigation techniques to reduce the EM coupling. Finally, we will propose different modeling and simulation methodologies to accurately evaluate and predict the noise coupling with or without mitigation techniques.

3D full-wave EM solvers require an enormous amount of computing resources and are not appropriate for the simulation at the circuit and system level. Thus, a fast and accurate circuit-based TSV model that can be easily integrated into standard circuit simulators is highly desired for 3D-IC design optimization. However, classic TSV circuit models, for example, reported in [2,3], are not capable of capturing complex field distributions in lossy Si substrates. Such models are, thus, limited to single TSV pairs without close substrate ground contacts, such as guard rings. To overcome these limitations, the 3D transmission-line matrix (TLM) method proposed in [4] divides the TSV and the substrate into a network of unit cells to properly model the substrate effects (such as crosstalk and losses). Still, the 3D-TLM does not include anomalous TSV capacitance behavior [5], which can be expected to affect the RF behavior of TSVs. To overcome these shortcomings, calibrated equivalent circuit models of the TSV in contact with a back side (BS) inversion layer are proposed for the first time in the context of 3D integration. Such models can incorporate both the RF field distribution in bulk Si and the anomalous TSV capacitance behavior. They are fast and accurate and can be used to study the noise coupling between adjacent devices by using standard circuit-based simulators with or without the influence of the anomalous TSV capacitance behavior. Such models will be detailed in Section 7.4.2.

Contrary to most publications that address the noise coupling through electromagnetic simulations and modeling without experimental verification [6–8], we also focus on an experimental coupling study based on RF S-parameter measurements up to 40 GHz. This is complemented by calibrated TCAD models. Such an approach allows for a complete analysis of the coupling mechanisms between TSVs and different active devices, both planar and FinFET based [9].

7.2 TSV Introduced EM Noise Coupling in 3D-ICs

A TSV can be considered as a vertical conductor connecting two dies of a 3D stack. As with any conductor, the main electrical parameters of a TSV are resistance, capacitance, and inductance. Other parameters relevant to the definition of the electrical characteristics of a TSV are leakage current to the substrate, breakdown voltage of the liner, maximum current density to avoid electromigration, and time-dependent dielectric breakdown (TDDB). Most of these parameters can be simulated or directly measured on suitable test structures, *i.e.* capacitance, resistance, leakage current, breakdown voltage, and TDDB lifetime. Table 7.1 reports typical electrical parameters that were measured for a TSV with 5 μm-diameter, 50 μm-depth, and a tetraethylorthosilicate (TEOS) liner thickness of 200 nm. All measurements were performed under static bias condition (DC) except the capacitance that was measured at low frequency (100 kHz).

The TSV capacitance is determined by the cylindrical metal insulator semiconductor (MIS) capacitor formed by the Cu conductor, the insulator, and the substrate around the TSV, as shown in Figure 7.1. The metal line

TABLE 7.1

Typical Electrical Parameters of a TSV with 5 μm-diameter, 50 μm-length, and 200 nm TEOS Liner Thickness

Resistance	Capacitance (accumulation) @ 100 kHz	Capacitance (inversion, LF) @ 100 kHz	Leakage current @+/− 20 V	Break down voltage
50 mΩ	130 fF	60 fF	1 pA	> 100 V

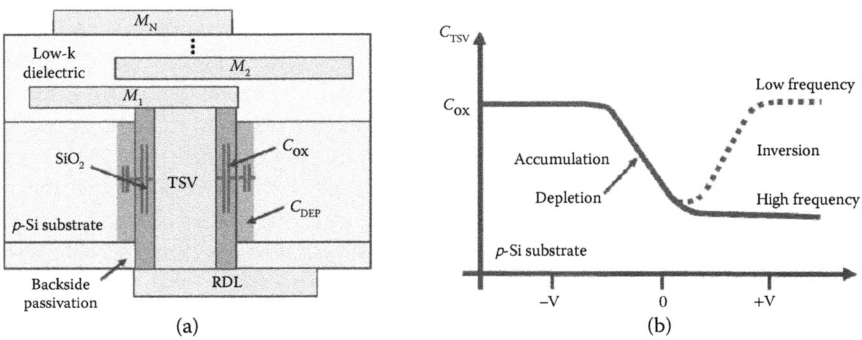

FIGURE 7.1
(a) Typical cross-section of a via-middle TSV. (b) Typical C-V curve showing the voltage-dependent behavior of a TSV in a p-type substrate.

connections to the TSV are not considered in this case. MIS capacitors are known for being voltage-dependent: typical capacitance versus voltage characteristics (C-V curve) measured at low frequency (50–100 Hz) and high frequency (10–100 kHz) are shown in Figure 7.1 (right) where the accumulation, depletion, and inversion voltage ranges are also indicated.

However, the 3D thinning processing may cause anomalies in the C-V curves of TSVs. For example, the low frequency inversion region may still appear at higher frequencies. The capacitance value at positive bias (for a p-type substrate) would be, in this case, identical to the value in accumulation, not in deep depletion. This anomalous C-V behavior will be discussed in Section 7.2.4.

7.2.1 Equivalent Circuit Model of a TSV and Two Major Paths of Noise Coupling Introduced by TSVs

The TSV behavior at low frequencies can be represented by a lumped element model, which is composed of the series resistance and the inductance of the TSV as well as the TSV capacitance. In most cases, the inductance is less significant than the resistance at low frequencies and can be neglected. Neglecting the inductance leads to the simplified RC model.

However, this low frequency model is not valid at microwave frequencies, where the Si substrate loss will become dominant. In order to capture this

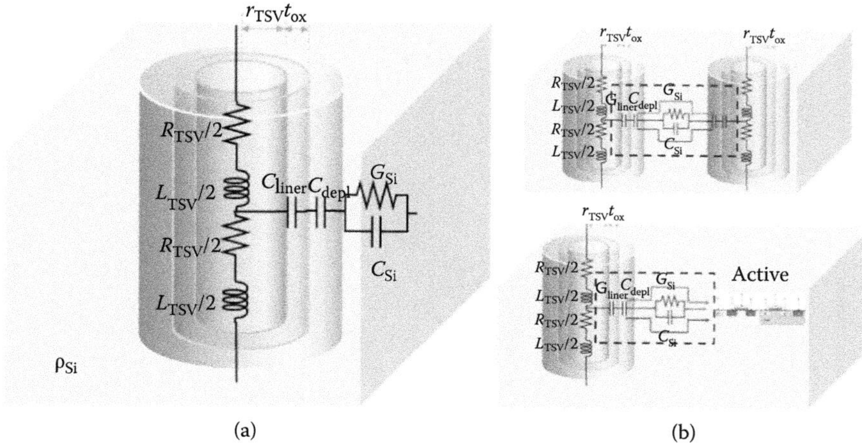

FIGURE 7.2
(a) TSV high frequency model. (b) TSV coupling paths to TSV and active device, respectively.

effect, a more complex model of the TSV is required, as shown in Figure 7.2. Here, the silicon substrate is modeled by a parallel conductance (G_{si}) and a capacitance (C_{si}). This model also includes the series inductance (L_{TSV}) and resistance (R_{TSV}) the oxide liner capacitance (C_{liner}) between the TSV and the Si substrate, the depletion capacitance (C_{depl}) and the Si substrate capacitance shunted by the substrate resistance.

At low frequencies, the conductive term (G_{Si}) dominates and the substrate has a purely resistive behavior. The electric (E) fields are primarily concentrated in the TSV oxide liner, and the mode of propagation is called the slow-wave mode. At frequencies above the majority carrier relaxation frequency (f_{maj}) as shown in Equation 7.1, the Si capacitance dominates because the majority carrier's finite inertia in the substrate prevents it from following the AC signal. Then, the substrate acts as a dielectric and the mode of propagation is called the quasi-transverse electromagnetic (TEM) mode, allowing complete penetration of both the electric and magnetic fields.

$$f_{\text{maj}} = 1/(2\pi\varepsilon_{si}\rho_{si}) \tag{7.1}$$

There are two major paths of noise coupling introduced by a TSV, as indicated by the dashed lines in Figure 7.2: TSV to TSV noise coupling and TSV to active device coupling. Both cases have the same coupling mechanism and, thus, lead to similar frequency dependence for the coupling. The general TSV noise coupling behavior is shown in Figure 7.3. At low frequencies, the TSV oxide capacitance dominates the noise coupling, which is frequency dependent with a slope of 20 dB per decade. By contrast, at intermediate frequencies, the silicon substrate resistance is the

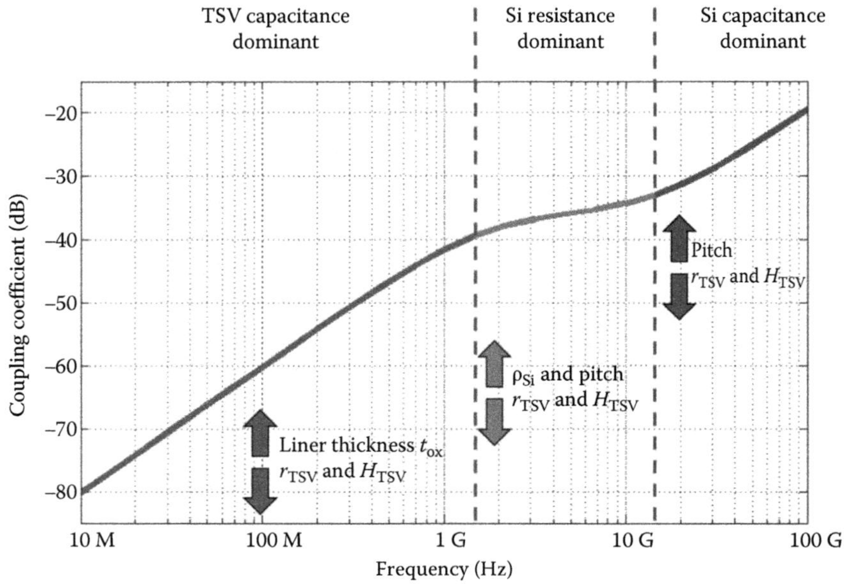

FIGURE 7.3
General frequency behavior of TSV noise coupling.

dominant contributor to noise coupling, which leads to a region where noise coupling is independent on frequency. At high frequencies, for example, above the substrate's dielectric relaxation frequency f_{maj} given by Equation 7.1, the coupling path is dominated by the silicon substrate capacitance and becomes frequency dependent once more with a 20 dB per decade slope in this region. The transition frequencies from low to intermediate and to high frequency depend on various factors, such as the TSV dimensions, the liner thickness, the TSV pitch, and the Si substrate resistivity.

7.2.2 Passive EM Coupling: TSV to TSV Noise Coupling

Based on the noise coupling mechanism and the general shape of the TSV coupling, the impact of the TSV dimensions on electromagnetic (EM) coupling can be deduced. Results are shown in Figure 7.4 for a standard diameter of 5 μm and a TSV height of 50 μm with an oxide liner thickness of 200 nm. In general, increasing the oxide liner thickness is only efficient at low frequency. By contrast, increasing the Si resistivity can improve the EM noise coupling immunity in the intermediate frequency range. Increasing the TSV pitch can reduce the noise coupling at both low and intermediate frequency, whereas decreasing the TSV radius and height leads to reduced noise coupling over the entire frequency range. This gives a general guideline for the selection of different TSV parameters based on the frequency range of interest.

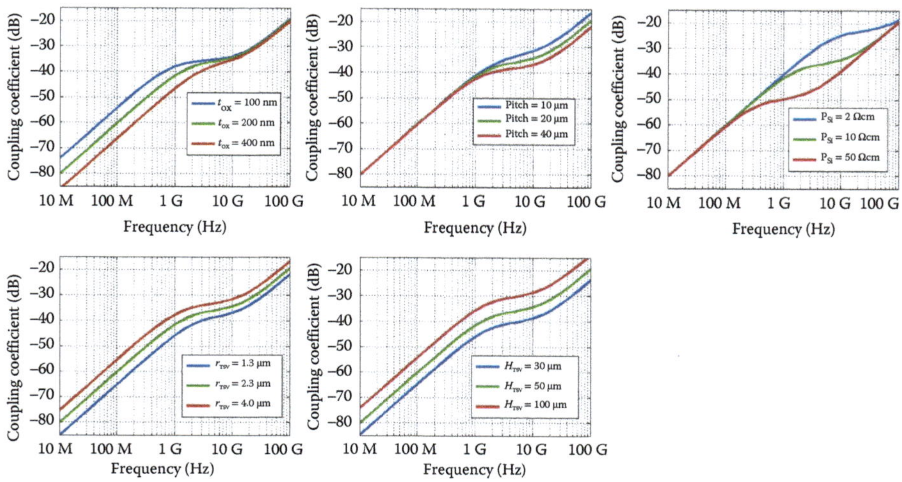

FIGURE 7.4
Impact of TSV dimensions and Si substrate resistivity on the noise coupling.

7.2.3 Active EM Coupling: TSV to Planar and FinFET Transistor

Although TSVs can introduce noise coupling to transistors through the Si substrate, this coupling occurs via the same mechanism as the noise coupling between adjacent TSVs. Recently, alternative transistor architectures, such FinFET devices, which have replaced "conventional" planar structures, have been introduced to enable the continuation of transistor scaling and Moore's law. Such alternative architectures can, however, lead to different noise coupling mechanism inside the transistor itself. In this section, we will focus on the noise coupling from TSVs to active devices, including both FinFETs and planar transistors, based on two-port S-parameter measurements up to 40 GHz [9]. In the first part of this section, the TSVs to active devices test structures and the RF measurement setups are presented. DC and two-port (drain-to-TSV) S-parameter measurements were performed under different biasing conditions of the TSV as well as of the gate and drain of the transistors. Subsequently, the source is grounded, the measured coupling coefficients (S_{12}) are analyzed for both types of active devices in both their ON and OFF states, and the dominant coupling mechanisms are discussed.

7.2.3.1 Test Structure Description and RF Measurements

The RF noise coupling test structures were built using state of the art 5 μm/50 μm TSV technology and two types of active devices: 65 nm planar nMOSFETs and 32 nm nFinFETs. The specifications of these two types of transistors are given in Table 7.2.

TABLE 7.2
FinFET and Planar Transistor Specifications Based on RF Test Structures

Parameter	Planar	FinFET
TSV to active spacing	10 μm	10 μm
TSV diameter/height/liner	5/50/0.2 μm	5/50/0.2 μm
W_{total}	15 μm	6.4 μm
L	50 nm	40 nm
#	5 fingers	80 fins

7.2.3.2 TSV–Planar nMOSFETs Noise Coupling

In a first step, DC measurements for planar nMOSFETs under different TSV biasing conditions were performed. The initial DC measurements showed that biasing the TSVs (-15 V to 10 V) had no impact on the DC behavior of planar nMOSFETS and nFinFET transistors.

Two-port RF measurements of the noise coupling coefficients (S_{12}) between the TSVs and the drains of planar device were performed and are shown in Figure 7.5. This coupling coefficient represents the noise that can be coupled from a RF signal into the TSV to the active devices through the TSV oxide liner and the bulk silicon substrate. Two coupling mechanisms exist between TSVs and active devices: noise amplification through the bulk transconductance (g_{mb}) as well as capacitive coupling to the drain/source through the substrate. For planar nMOSFETs, the first mechanism dominates in the ON state, leading to large noise coupling with respect to the OFF state. As shown in Figure 7.5, differences up to 25 dB are observed for nMOS planar transistors in the ON and OFF states.

Noise coupling was also investigated as a function of V_{gate} in both linear and saturation regions (Figure 7.6). Above 0.4 V, the increase of V_G decreases the coupling in the linear region because it turns on the inversion channel in the transistor, creating a resistive path from drain to source (R_{on}). The TSV signal that couples to the drain can thus be sunk through this resistive channel to the source, which is connected to ground. This reduces the noise coupling coefficient.

7.2.3.3 TSV–nFinFETs Noise Coupling

For nFinFETs, the noise coupling level was similar in the ON and OFF states of the transistors and even slightly smaller in the ON state than in the OFF state (Figure 7.7). This phenomenon can be explained by the small bulk transconductance (g_{mb}) of nFinFETs. Thus, the dominant coupling mechanism between TSVs and nFinFETs is the capacitive coupling to the drain in both the ON and OFF states. When the nFinFET is ON, the depletion region increases and the capacitance decreases which reduces the capacitive coupling to the drain.

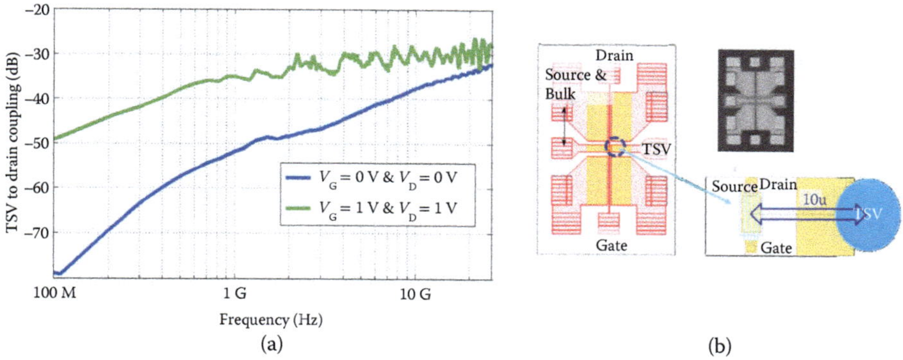

FIGURE 7.5
TSV to planar nMOSFET noise coupling measurement with transistors in ON/OFF state (TSV bias = 0 V).

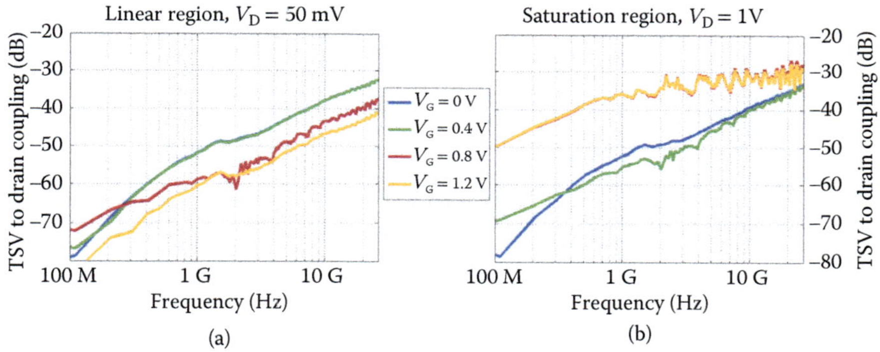

FIGURE 7.6
(a) Measured TSV-planar nMOSFET coupling versus $V_{G\,bias}$ in the linear region. (b) Measured TSV-planar nMOSFETS coupling versus V_G bias in the saturation region.

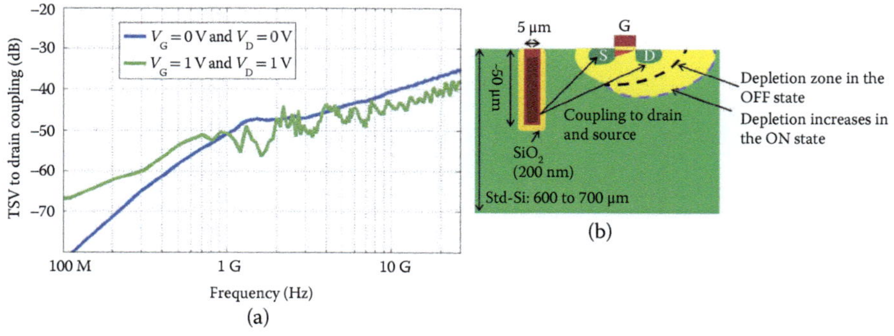

FIGURE 7.7
(a) TSV–nFinFETs noise coupling measurement with transistors in ON/OFF state; (b) TSV-nFinFET device EM coupling path.

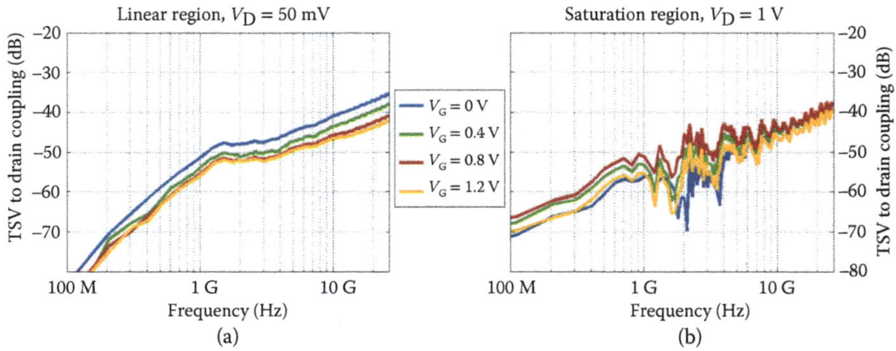

FIGURE 7.8
(a) TSV-nFinFET measurements: noise coupling versus V_G bias in linear region. (b) Noise coupling versus V_G bias in the saturation region.

Moreover, noise coupling was also investigated as a function of V_{gate} both in linear and saturation regions (Figure 7.8). Increasing V_G decreases the coupling in the linear region because it turns on the inversion channel in the transistor, as in the case of planar nMOSFET. Again, this connects the drain through this resistive channel to the grounded source and thus reduces the noise coupling coefficient. Increasing V_G in the saturation regime increases the coupling because of the bulk transconductance (g_{mb}). However, the increase is insignificant because of the small g_{mb} of FinFETs.

7.2.3.4 TSV–Active Device Coupling Mechanisms: Planar nMOSFETs versus nFinFETs

By analyzing and comparing the impact of TSV noise on FinFET and planar device performance, the dominant coupling mechanisms were identified in Table 7.3 for both types of active devices. For planar nMOSFETs in the ON state, because of their substantial bulk transconductance (g_{mb}) noise amplification through the bulk transconductance (g_{mb}) dominates, leading to large noise

TABLE 7.3
TSV to Active Device Coupling Mechanism Summary

Parameter	Planar nMOSEFT		nFinFET	
	OFF $V_D = 0$ V $V_G = 0$ V	ON $V_D = 1$ V $V_G = 1$ V	OFF $V_D = 0$ V $V_G = 0$ V	ON $V_D = 1$ V $V_G = 1$ V
g_{mb} values	Not applicable	Big	Not applicable	Small
Dominant coupling mechanism	Capacitive coupling	g_{mb}	Capacitive coupling	Capacitive coupling (g_{mb} is too small)
Coupling level	Small	Big	Small	Smaller

coupling. For FinFETs, this mechanism, however, is negligible due to its much smaller transconductance (g_{mb}). In this case, the main coupling mechanism was capacitive even in the ON state. Moreover, the increased bulk depletion around the FinFET in the ON state led to a reduced capacitive noise coupling. In this case, the coupling was even lower in the ON state than in OFF state, in contrast to the planar device. The comparison in Figures 7.5 and 7.7 clearly demonstrates that nFinFETs have better noise coupling immunity than planar nNMOSFETs. TSV noise coupling to FinFET devices was typically 20 dB lower than it was to planar devices at 1 GHz and 4–8 dB smaller at 40 GHz.

7.2.4 Parasitic Surface Conduction

Above, we have studied the TSV-introduced EM coupling considering the semiconductor substrate as a material with homogeneous electrical properties. However, the TSV capacitance is both frequency and bias dependent since a TSV is a metal-insulator-semiconductor (MIS) structure. The semiconductor volume directly around a TSV can be in a state of inversion, depletion, or accumulation, as shown in Figure 7.1, depending on the amount of fixed charges [10] in the oxide and the DC bias applied to the via. Furthermore, the top and bottom interfaces of the passivation layers and bulk Si are also unlikely to be in a state of flatband, that is, showing a homogeneous carrier distribution, since fixed charges in these passivation layers are likely to induce strong parasitic conduction channels. This is more of an issue at the back side interface, because, at the top-side interface, well doping structures are implemented to help suppress the parasitic conduction. Close to such interfaces, the electrical resistivity can differ significantly from the rest of the bulk material (which is in a flatband state) and can vary by up to 6 orders of magnitude depending on the applied (by a voltage source) or induced (by fixed charges in dielectrics) potential at the Si/passivation interfaces.

In most cases, the amount and nature of the fixed charges at the Si/passivation interfaces depends strongly on the process technology used. However, in most cases for SiO_2 or Si_3N_4 passivation layers, the fixed charges are positive and are usually present in a density above 10^{11} cm^{-2}. Such an interfacial charge density is high enough to induce strong inversion in usual Si substrates with nominal resistivities higher than 1 Ω.cm.

7.2.4.1 Extension of the Depletion Capacitance along the Back-Side Passivation Layer

The existence of a parasitic surface conduction channel significantly influences the EM coupling path in 3D-ICs. Positive charges at the backside passivation layer have been reported to be responsible for inducing an inversion channel at the bottom of the Si bulk. At low frequencies, the conductive term (G_{Si}) dominates and the substrate shows a purely resistive behavior.

At frequencies above the majority carrier relaxation frequency, the Si capacitance dominates and the substrate has majority carriers in the bulk that cannot follow the AC signal. Then, the substrate acts as a dielectric with complete penetration of both electric and magnetic fields.

In this case, an AC signal coupled from the TSV through the oxide can propagate along the inversion layer around the TSV (due to bias or fixed charges in the liner) into the inversion layer at the back side (caused by fixed charges in the passivation layer) and disperse along the entire back of the wafer. The AC current is then distributed over an extended area and, therefore, has a much larger depletion surface through which it can couple to the conductive bulk. As a result, the depletion capacitance is much larger than the one solely surrounding the TSV. Figure 7.9 illustrates this effect by plotting the electron concentration in a TSV structure with an inverted TSV surface as well as an inverted back side interface. Superimposed on this profile, a small-signal equivalent circuit model is depicted. Conductive regions rich in free carriers are modeled as resistive elements and depleted regions low in carriers are modeled as capacitive elements. In essence, the signal from the TSV sees the impedance of the oxide capacitance C_{ox-TSV} in series with an *extended* depletion capacitance $C_{dep-TSV} + C_{dep-BS}$. A very large depletion capacitance C_{dep-BS} has been added in parallel to the original one surrounding the TSV: $C_{\text{dep-TSV}}$. This makes the total effective depletion capacitance much larger than the oxide capacitance.

$$C_{\text{dep-eff}} = \left(C_{\text{dep-TSV}} + C_{\text{dep-BS}}\right) \gg C_{ox} \tag{7.2}$$

As a consequence, the capacitance seen in inversion is no longer $\left(C_{ox} // C_{\text{dep-TSV}}\right)$ but is instead

$$\left(C_{ox} // C_{dep-eff}\right) = \left(C_{ox} // \left(C_{dep-TSV} + C_{dep-BS}\right)\right) \approx C_{ox} \tag{7.3}$$

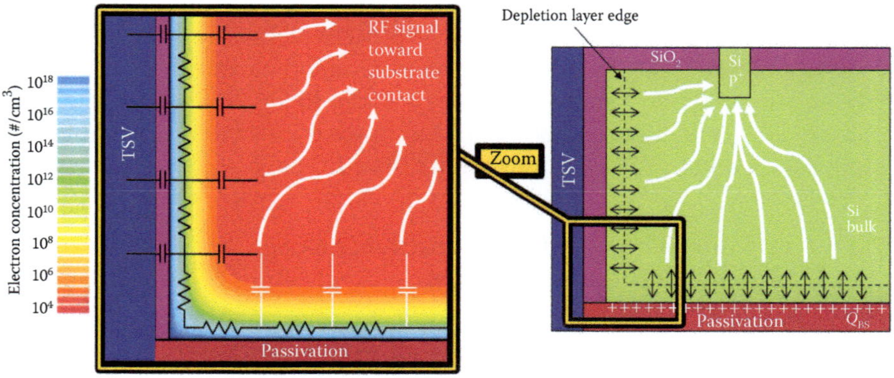

FIGURE 7.9
Cross section view of the electron concentration in a cylindrical TSV structure that has been revealed and passivated with a charged BS layer.

The right hand cross section of Figure 7.9 shows the physical movement of the majority carriers in response to an AC signal on the TSV that flows to and from the depletion layer edge. The edge oscillates horizontally around the TSV, but also vertically where it has been extended parallel to the back side.

7.2.4.2 BS Bridge Effect between Neighboring TSVs

When considering coupling between neighboring TSVs, the presence of a backside inversion layer effectively shorts out the depletion capacitances around the TSVs, thereby increasing the coupling. Figure 7.10 plots the electron concentration in such a structure that has inverted TSV surfaces as well as an inverted back side interface. The equivalent circuit model that has been superimposed shows that the back side resistance R_{BS} electrically connects the two inverted TSV oxide liners, diverting the signal from passing through the depleted regions.

Once the transition frequency $f_t = 1/\left(\pi R_{BS} C_{dep\text{-}TSV}\right)$ is reached, which is equal to 20 MHz in the case of Figure 7.10, the impedance of the coupling

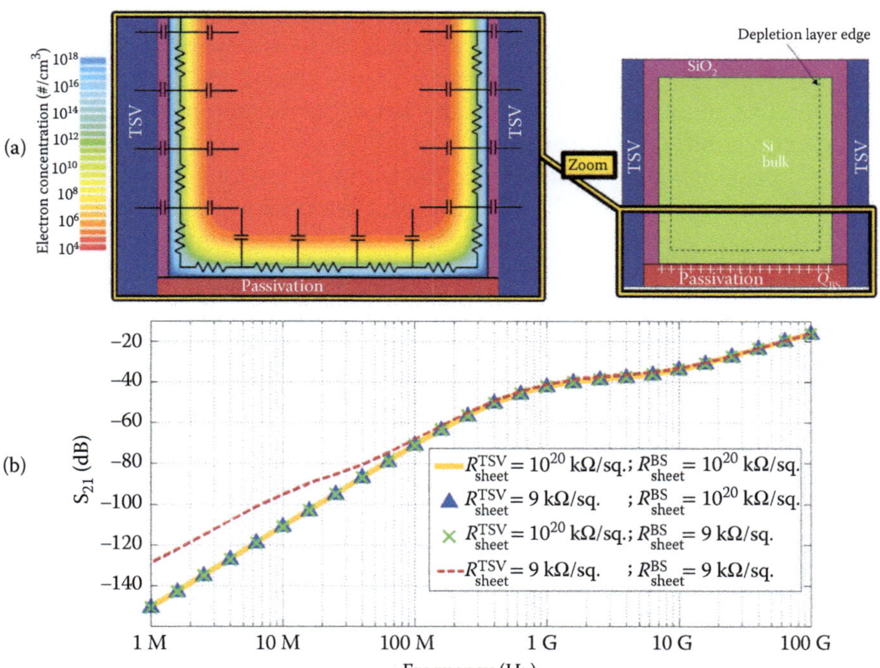

FIGURE 7.10
(a) Cross section view of the electron concentration in a two-TSV structure in contact with a charged BS passivation layer. (b) TSV-to-TSV coupling with and without BS inversion and with and without TSV inversion.

path through $C_{dep-TSV}$ becomes lower than the impedance of the propagation path along the back side channel R_{BS}. The simulated TSVs were implemented using 5/50 technology (see Table 7.2) with a pitch distance of 10 µm. In depletion, the sheet resistances are very large and the S_{21} coupling was not affected, as shown in Figure 7.10. The low-frequency capacitive coupling is then determined by $1/C_{tot} = 1/C_{ox} + 1/C_{depTSV}$. However, when inversion is present at both the TSV and the BS interfaces, the TSV depletion capacitance is shunted by the BS inversion channel at low frequencies. The low-frequency capacitive region is then determined by $C_{tot} = C_{ox}$, which increases the noise coupling between the TSVs by about 15 dB at 10 MHz.

7.2.4.3 Discussion of Possible Solutions

The solution of grounding the BS inversion layer was considered to alleviate the coupling. This could be accomplished by adding a grounded n+ contact at the BS. This enables an ohmic contact with the BS inversion layer and could efficiently reduce the coupling at low frequencies. The TSV-substrate capacitance is actually rendered lower than $\left(C_{ox}//C_{dep-TSV}\right)$, demonstrating that this technique enables the compensation of the parasitic BS conduction effects. It is important to note that these benefits only occur as long as the coupling fields are concentrated in the BS inversion layer. Once the frequency is high enough, the E field no longer penetrates significantly into this layer and the grounded n+ plug at the BS has no alleviation power anymore.

A second solution is to use an improved passivation method. Simulation results (not shown here) demonstrate that the introduction of a trap-rich layer (inspired by [11]) between the substrate and the BS passivation layer makes the anomalous electrical behavior vanish and, thus, reduces the parasitic coupling effect. The traps introduced effectively freeze the inversion layer carriers at the BS and, consequently, the conduction of that layer becomes close to zero. Trap-rich layers, such as amorphous or polycrystalline silicon, rich in defects have been proven to be the most effective while being compatible with industrial silicon-on-insulator (SOI) fabrication and standard CMOS processes.

7.3 Coupling Mitigation Techniques and Signal Integrity

Noise coupling is a real issue in TSV-based 3D-ICs, as vertical interconnects are often densely packed to achieve high density I/O throughputs. TSVs are MIS structures that possess high capacitance to the silicon bulk, meaning that the TSV signal is easily coupled to and from the Si substrate at high frequency. Since TSVs are large and noisy structures, the evaluation of electromagnetic coupling to and from TSVs has become crucial to the design of 3D integrated circuits. It is required that the noise coupling levels be properly assessed to ensure that the tight noise tolerance budget of a current high-performance 3D-IC is satisfied. If necessary, coupling mitigation and reduction techniques need to be introduced.

In this section we investigate the two most common methods used to alleviate coupling to and from TSVs, namely, grounded substrate guard rings (GR) and grounded shielding TSVs. All coupling alleviation schemes involve adding a quiet ground network in close proximity to a noisy TSV and/or around a sensitive circuit node, so that the injected noise will find its way to the ground network instead of to the sensitive node, which can be anything from another TSV to active circuit elements. In essence, the idea is to divert the power injected into the substrate by the TSV away from sensitive nodes of the system.

These schemes inherently rely on increasing the total conductance to ground, which will inevitably lead to a higher insertion loss for the TSV itself. This decrease in TSV transmission performance must be taken into account in parallel with the gain in coupling performance for a valid design.

7.3.1 Substrate Contact and Guard Ring

To suppress noise coupling between TSVs, a grounded guard ring (GR) can be implemented around it. It is physically implemented by means of a substrate contact connected to a quiet ground network. Guard rings are commonly used in 2D-IC systems to capture the substrate noise, in particular in mixed-mode systems in which the analog and RF modules require shielding from noisy digital circuits on the same chip.

In 2D-ICs, the circuitry and interconnects are all coplanar at the top surface of the silicon substrate. For these reasons, guard-rings always show high noise alleviation in 2D-ICs, as they can be placed optimally across the noise coupling path. In 3D-ICs, however, the coupling between TSVs occurs along the entire via length and deep into the substrate, so a noise sink defined only at the surface may not be as optimal as in 2D-ICs. For this reason, TSV noise alleviation using guard rings is, in some cases, limited for densely packed TSVs characterized by low pitch to height ratios.

7.3.1.1 Coupling Reduction of GR and Importance of TSV Pitch to Height Ratio

Figure 7.11 shows the coupling curves between two TSVs, implemented in 5/50 μm technology (see Table 7.2) with a varying pitch distance between their centers, both with and without a guard ring (GR) alleviation scheme. The crosstalk coefficient between two TSVs with a pitch p between them is denoted by $S_{xt}^{p}(f)$, where the index p is referenced in micrometers. The crosstalk coefficient between two TSVs in an identical set-up with an added guard

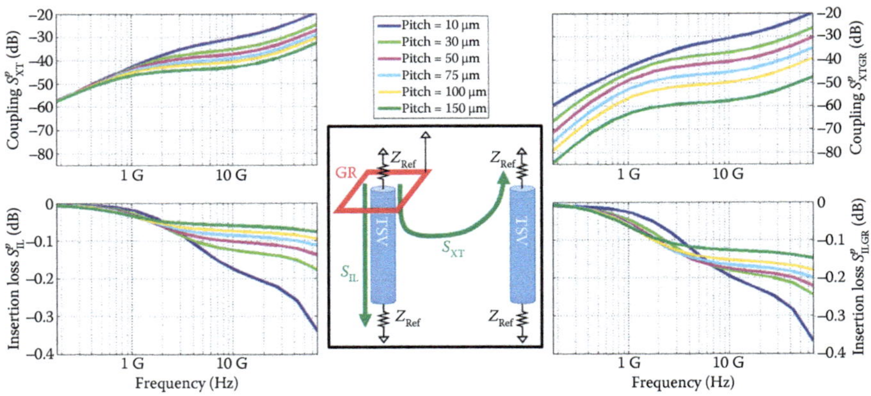

FIGURE 7.11
Crosstalk and insertion loss coefficients between two TSVs with and without GR shielding for a varying pitch.

ring is denoted by $S^p_{xtGR}(f)$. The guard ring is square and surrounds a single TSV. It is 2 μm away from the TSV wall and is 2 μm-wide. The corresponding results in Figure 7.11 are denoted with the *GR* subscript. The notations $S^p_{xt}(f)$ and $S^p_{xtGR}(f)$ are considered to be the magnitude of the coupling parameters expressed in dB.

Comparing the two sets of curves in Figure 7.11, we notice that there is almost no improvement at low pitch to height ratio. Indeed, we have that $S^{10}(f) \approx S^{10}_{GR}(f)$. On the other hand, we notice that the guard ring alleviates the coupling to a much higher extent when the pitch to height ratio is increased: $S^{150}(f) \gg S^{150}_{GR}(f)$.

To better appreciate the coupling alleviation due to a guard ring, the isolation improvement factor is defined by $I^p_{xt}(f) = S^p_{xt}(f) - S^p_{xtGR}(f)$ and is expressed in dB. The curves $I^p(f)$ in Figure 7.12a highlight the importance of the pitch to height ratio between TSVs for the efficiency of a guard ring for alleviation. The isolation improvement when adding a guard ring increases from 0.3 dB for a pitch of 10 μm to 15 dB for a pitch of 150 μm (for ports with 50 Ω reference impedances).

The isolation is better at low frequencies when the dominant coupling term is the TSV sidewall capacitance (region A). This is due to a change from a 20 dB/decade coupling mechanism to a 40 dB/decade one when the bulk is biased with a ground contact. The coupling gain in this region will strongly depend on the position of the closest substrate ground contact.

The isolation gain factor I^p is be plotted as a function of the pitch at 4 GHz in Figure 7.12b. The coupling isolation improvement shows an exponential growth with a rising pitch (almost linear dependency in a dB scale) and corroborates the importance of the pitch to height ratio on the effectiveness of a guard ring alleviation scheme (with standard substrates and 50 Ω termination impedances).

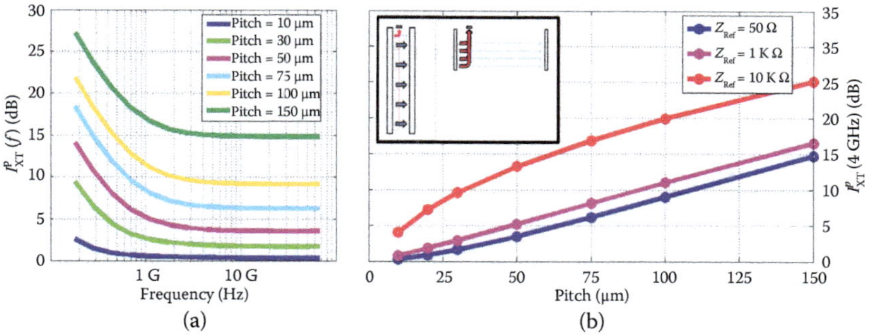

FIGURE 7.12
(a) Isolation improvement factor for the added GR versus frequency for a varying pitch. (b) Isolation improvement factor at 4 GHz versus pitch and load impedance. Inset: Schematic representation of the electric field distribution in the TSV pair structure for different aspect ratios. Left: a low pitch to height ratio. Right: a high pitch to height ratio.

As discussed above, the performance of a guard ring in 3D-ICs is strongly dependent on the pitch to height ratio parameter. The inset in Figure 7.12b depicts the electric field distribution in the TSV pair structure for two different configurations. In the left hand side configuration, the pitch to height ratio is low; whereas in the right hand side configuration, it is high. In both depictions, the red arrows represent the electric fields coupling from a TSV to the guard ring and the blue arrows represent the electric fields coupling from one TSV to the other.

When the pitch to height ratio is high, the fields arrange themselves according to the right side depiction of the inset in Figure 7.12b. The impedance paths from the aggressor TSV to the guard ring and other TSV are dominated, essentially, by the bulk Si. Simply due to the fact that the guard ring is much closer to the aggressor TSV than the victim TSV, the majority of the power is captured by the grounded guard structure and the alleviation is effective.

However, by reducing the TSV pitch, the bottom section of the aggressor TSV moves closer to the bottom section of the victim TSV while staying at the same distance from the guard ring. The fields then distribute according to the left side depiction of the inset of Figure 7.12: the electromagnetic noise injected deep into the substrate sees a lower impedance path to the very close victim TSV than to the guard ring far away up on the surface and the noise power distribution changes as a function of the pitch to height ratio. The noise injected by the aggressor TSV close to the surface is still captured by the guard ring, but, for certain aspect ratios, it can be totally negligible, especially in the GHz frequency range. This is the case for a pair of TSVs in 5/50 technology when the pitch is 10 μm. Because of the low pitch to height ratio conferred by these dimensions, the effect of adding a guard ring has little influence on the coupling coefficient; the added isolation is inferior to 1 dB. This isolation improvement is increased to 15 dB when the inter-TSV pitch is increased to 150 μm. However, low pitches are in demand for highly integrated systems and it is difficult to thin the handle substrate to much less than 50 μm for mechanical reasons associated with the wafer thinning process. Therefore, guard rings are limited solutions for highly integrated 3D-ICs in which the pitch to height ratios are low.

Nonetheless, the results so far have been assessed under the assumption of 50 Ω impedances connected to the TSV driver and load terminals. This assumption means that the impedance path between two TSVs is dominated by the oxide capacitance or silicon substrate. For example, the frequency at which the impedance path between two 5/50 TSVs with a pitch of 10 μm is equal to 50 Ω around 250 GHz, so a TSV load impedance of 50 Ω is never the dominant term of the coupling path in the usual frequency band. However, raising the value of the reference impedance of each TSV port results in it dominating this coupling path at lower frequencies, in the GHz region, but of course its value has no effect on the coupling impedance toward a grounded guard ring. We therefore expect the guard ring's effectiveness to be reliant upon the TSV port impedance in a positive manner.

7.3.1.2 TSV Driver and Load Impedances

Figure 7.13 shows the coupling curves between two TSVs implemented in 5/50 technology (see Table 7.2) with a varying pitch distance between their center, both with and without a guard ring (GR) alleviation scheme. The S-parameters are no longer represented with reference to 50 Ω source and load impedances at the TSV terminals, but are instead represented with reference to 10 kΩ impedances.

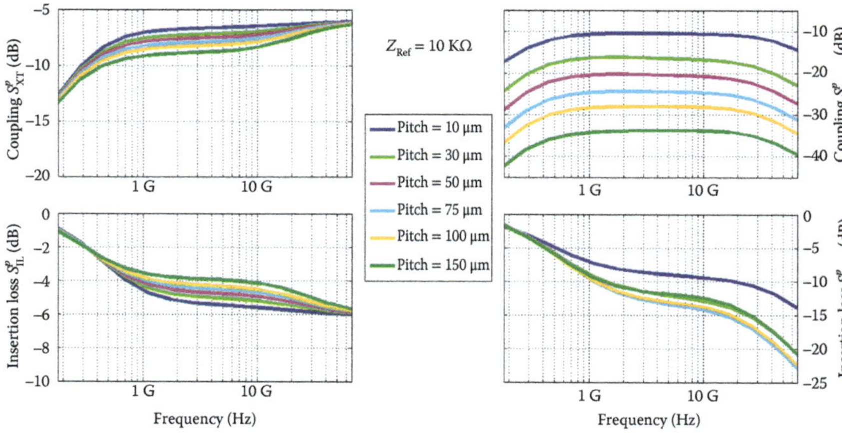

FIGURE 7.13
Isolation crosstalk and insertion loss coefficients between two TSVs, with and without GR shielding, for a varying pitch for 10 kΩ port impedances.

The addition of the guard ring alleviates coupling by 4 dB or so even at low pitch to height ratios. Furthermore, there is a general improvement in isolation gain achieved by adding a grounded guard ring around the TSV when it's loaded with a higher impedance. Figure 7.12b plots the isolation improvement factor $I_{xt}^{p}(f)$ for several driver/load conditions at the TSV ports.

A guard ring, then, seems like an effective scheme to reduce crosstalk between TSVs. It is simple to implement and provides decent crosstalk shielding with no notable degradation to the TSV's transfer characteristics. However, this scheme's effectiveness is very much layout dependent and is of limited use in highly dense TSV arrays.

7.3.2 Ground Shielding TSVs

As explained in the previous section, guard ring technology is no longer an efficient coupling alleviation scheme for TSVs at low pitch to height ratios because the fields are only captured by the guard ring on the top surface of the structure. Alleviation using grounded TSVs is a solution, providing shielding deep inside the silicon bulk.

In order to investigate the effect of a grounded TSV on coupling alleviation, the three simulation set-ups of Figure 7.14a are proposed.

The first set-up is the single TSV pair, without any coupling alleviation system in place, using 5/50 µm technology with a pitch of 20 µm. These simulations are compared with the results of the second and third set-ups, in which a TSV is inserted between the aggressor and victim TSVs in a collinear manner to alleviate coupling. This TSV is either connected to 50 Ω load impedance, and, therefore, acts as a third signal TSV, or is directly connected to ground and acts purely as an EM shield. The simulation results are plotted

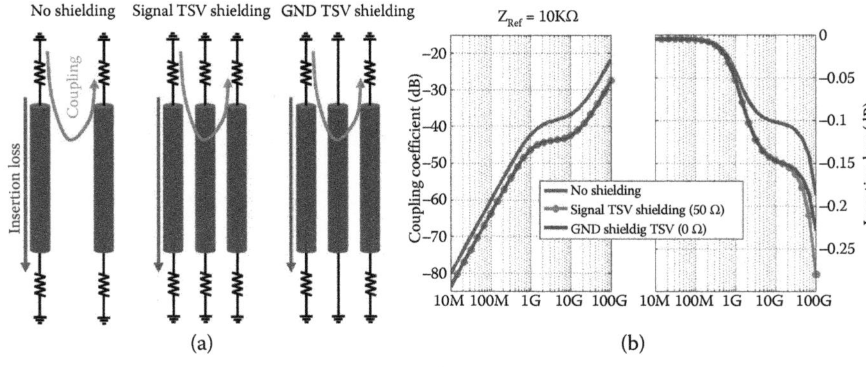

FIGURE 7.14
(a) Implemented simulation set-ups to investigate the effect of adding a grounded TSV between two signal TSVs as an alleviation technique. The reference impedances are set to 50 Ω. (b) Comparison of the coupling alleviation and insertion loss performance of adding a shielding TSV between two signal TSVs.

in (b) where the near end crosstalk S_{21} and insertion losses S_{31} are compared for each set-up.

The results show that the isolation is improved when a grounded TSV is placed between the two signal TSVs. The results also show that the insertion of a 50 Ω terminated TSV acts as well as a shield with a grounded TSV. This is due to the fact that the 50 Ω reference impedance of the ports is negligible in the coupling paths below 200 GHz. The silicon resistance between two 5/50 μm TSVs with a pitch of 10 μm has a value of approximately 700 Ω and is the dominant term of the impedance path between 2 and 8 GHz. This explains why there is very little difference in the coupling alleviation using a grounded or a signal TSV.

We can say once again that we expect grounded TSVs to have a higher shielding effect for higher TSV load impedances, because the load may dominate the coupling path between TSVs in the frequency band of interest, increasing the impedance of such a path relative to a path toward a grounded TSV. In such scenarios, the shielding from grounded TSV is significantly better than the shielding from signal TSVs terminated with larger port impedances. This will be discussed below in the following section in the context of large TSV arrays.

7.3.2.1 Coupling in Large TSV Arrays

The drastic increase in simulation power brought to substrate coupling modeling by the 3D circuit model (see Section 7.4.2) in the context of 3D integration allows for the investigation of the full-coupling matrix in very large TSV arrays. A square array of 11 by 11 TSVs is depicted in Figure 7.15a and

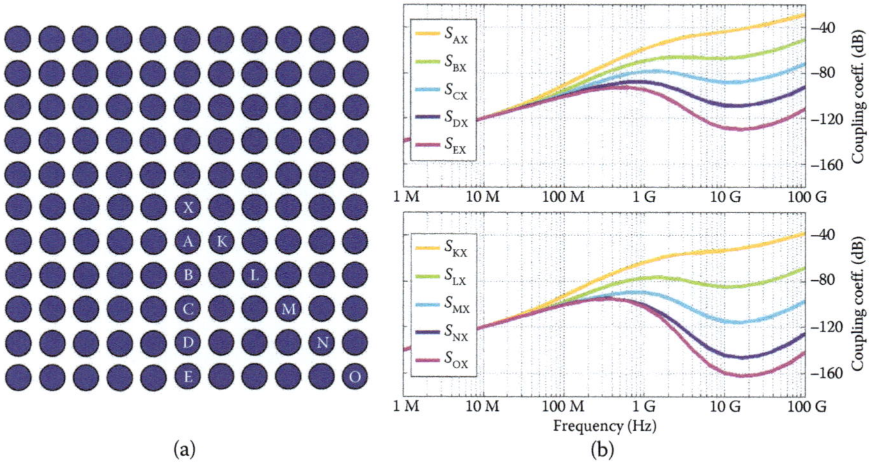

FIGURE 7.15
(a) TSV top port lettering in the 11 × 11 TSVs array simulation. (b) Coupling results of the 11 × 11 TSVs array simulation.

introduces the port letterings used for the ports on the top side of the wafer. The TSVs are implemented in 5/50 μm technology (see Table 7.2) with a pitch of 15 μm between them.

The RF port, denoted as X, is connected to the central TSV of the 11 × 11 array on the top side. The entire 242 × 242 S-matrix of the 11 × 11 TSVs array (two 50 Ω ports per TSV, top and bottom) was simulated and a few of the coupling results are presented in Figure 7.15b, referring to Figure 7.15a for the port naming. Many of these coupling curves $S_{iX}(f)$ present a negative slope just after the point where the frequency is high enough for the electric fields to start to penetrate into the silicon bulk. Indeed, at low frequency, all of the coupling curves become almost equal. This is because the oxide capacitances are dominating the impedance paths between the nodes. Between any two TSVs of the array, the impedance path consists of two times the oxide capacitance in series with the silicon bulk resistance. Below a few hundred MHz, this second term is negligible compared with the first, making the impedance paths between any two TSVs essentially equal, which is the reason why all the curves intersect at low frequency. At these frequencies, the bulk is essentially acting as an equipotential volume. At higher frequencies however, the electric fields start to penetrate into the silicon bulk and are captured predominantly by the surrounding TSVs. The noise power is no longer divided evenly between all the victim ports of the array. This is because the bulk impedance becomes the dominant term in the coupling paths. Therefore, the noise power predominantly couples to the neighboring TSVs (TSVs A and K). This has the effect of drastically reducing the power coupled to the TSVs that are further away (D, E, N, and O, for example) to such an extent that the slopes of the coupling coefficients as a function of frequency become negative for these nodes. Although the impedance path to these far away, TSVs reduce, as a function of frequency, the impedance paths to the neighboring TSVs, which are competing impedance paths and reduce much faster. For this reason, such coupling curves experience a negative slope once the potential starts to distribute itself inside the bulk. Curves of such shapes have been presented in [12,13] although no comments were given about the negative slope regions. In [12] the introduction of a shielding TSV between two signal TSVs had the effect of alleviating coupling and introduced a negative slope region in the crosstalk curve at a few GHz. In [13], a 5 × 5 array of TSVs was simulated and negative slope regions in the coupling curves were obtained.

7.3.2.2 TSV Driver and Load Impedances

The coupling toward a victim node, as expressed in S-parameters, is dependent on the load impedance to ground at that node. It is common to represent the coupling S-parameters with reference to standard 50 Ω terminations and/or evaluate the crosstalk voltage between an aggressor TSV and a victim TSV loaded with 50 Ω [14–16]. Such low loading conditions are the standard for measurement equipment; however, they are not realistic in the context of 3D integration, as device input impedances are closer to the kΩ range at RF frequencies.

Figure 7.16a plots the coupling coefficient between two TSVs of the 11 × 11 array for different loading conditions. The coupling is evaluated between the two TSVs shown in yellow in Figure 7.16b. The TSV denoted as "A" is considered as the aggressor TSV. Its top port is excited by an RF power source with driver impedance of Z_{Ref} and is loaded with an impedance Z_{Ref} at its bottom end. The TSV denoted as "V" is the victim TSV. Its top port is considered as the victim node, and is loaded with an impedance Z_{Ref}. Its bottom port is also loaded with an impedance of Z_{Ref}. All other TSVs of the array are loaded with impedance values of Z_{Ref} at their top and bottom ports. Throughout the simulations, the value of Z_{Ref} is varied between 50 Ω (red curves) 1 kΩ (green curves) and 10 kΩ (blue curves). Several coupling alleviation configurations are investigated by grounding TSVs in the vicinity of the victim TSV, that is, by replacing the Z_{Ref} loads of these TSVs by ground terminals. First, a single grounded shielding TSV between the two signal TSVs of interest is considered (dashed lines) and then the eight TSVs immediately surrounding the victim are grounded (dot curves).

The results show that the coupling coefficient increases with the value of the reference impedance. This is due to the fact that the signal TSVs that are excited from RF ports at the top of the wafer are loaded by impedances Z_{Ref} at the bottom side of the wafer. The insertion loss through the via (not shown) is degraded and the coupling into the bulk becomes higher. It is indeed these two coupling paths that are in competition to sink the RF current toward a ground node: through the via then into the load Z_{Ref}, or into the bulk and then toward other loaded TSV nodes. For low loading conditions, the path

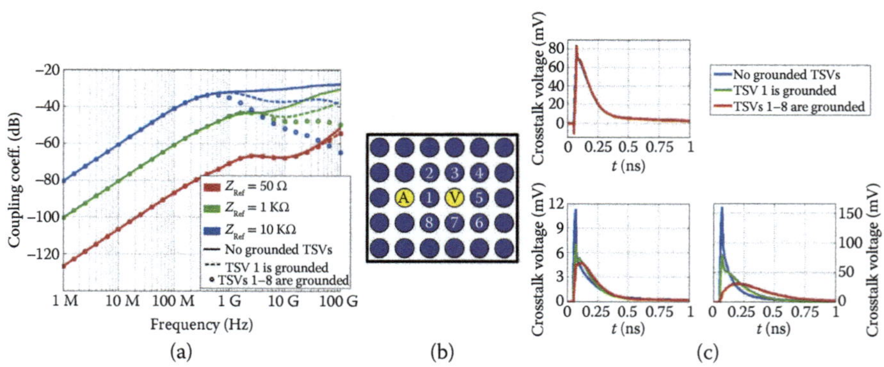

FIGURE 7.16

(a) Coupling coefficient simulation results of the 11 × 11 TSV array simulation under three general loading conditions: 50 Ω, 1 kΩ, and 10 kΩ with different levels of EM shielding using grounded TSVs. (b) Depiction of the aggressor (A) and victim (V) TSVs in the array. (c) Voltage crosstalk results of the 11 × 11 TSVs array simulation under three general loading conditions: 50 Ω, 1 kΩ, and 10 kΩ with different levels of EM shielding using grounded TSVs.

through the via is much lower impedance than the path through the oxide liner, into the silicon bulk and toward some load or ground load elsewhere in the system (neighboring loaded TSV for example). However, for higher loading conditions, in the kΩ range, the coupling dramatically increases, because the ratio between the impedance path through the TSV and into the bottom load and between the impedance path through the oxide liner and toward a neighboring load is much higher. It is the ratio of these two impedance paths that defines where the RF power will be sunk, and therefore the insertion loss and coupling levels.

For similar reasons the results of Figure 7.16a also highlight that in the case of 50 Ω reference impedances (red curves), substituting the 50 Ω loads of TSVs 1–8 for ground nodes (zero impedance loads) does not affect much at all the coupling coefficient curve. This is because the impedance path between two neighboring TSVs is above 700 Ω in our simulations for the considered dimensions up to 20 GHz. This means that 50 Ω loaded TSVs are basically already acting as virtually grounded, and so shielding using a 50 Ω signal-TSV has the same effect as shielding using a grounded (0 Ω) one. However, as the TSV load impedances are increased, an appreciable difference can be seen when substituting a signal TSV for a grounded TSV in the overall coupling coefficient, as can be seen from the green and blue curves in Figure 7.16a. The effect of the coupling alleviation of the shielding TSVs starts at the frequency at which the impedance path between TSVs becomes equal to the loading impedances. Before this frequency is reached, the TSV sidewall capacitances or the bulk resistance dominate the impedance paths between TSVs. However, once the aforementioned frequency has been reached, the electric fields between TSVs couple preferentially to the grounded vias that now have an appreciably lower impedance path to ground than the loaded signal-TSVs. The shift in this frequency point can be observed for different loading conditions, and is indeed seen to be inversely proportional to the load's value.

Figure 7.16c shows the results of the time-domain voltage crosstalk simulations performed in the same context as the previous frequency domain results of Figure 7.16a. In these simulations, a 1 V amplitude voltage step with a ramp-time of 20 ps is applied to the aggressor TSV of the array and the crosstalk voltage coupled to the victim TSV is obtained and plotted for the same set of loading and shielding configurations. The voltage driver's impedance was set to 200 Ω in all time-domain simulations.

First, we notice again the increase in the coupling for higher loading conditions: the crosstalk voltage amplitude shows a strong dependence on the TSV load's value. Once again, as expected, the 50 Ω loaded TSVs act as being virtually grounded, and no improvement in the crosstalk voltage is to be found by replacing signal TSVs with shielding TSVs. However, when the loading conditions are increased, improvements are observed when introducing the shielding TSVs. This reduction of the peak crosstalk voltage is around a factor of 2 and 4, respectively, for the 1 kΩ and 10 kΩ loading cases.

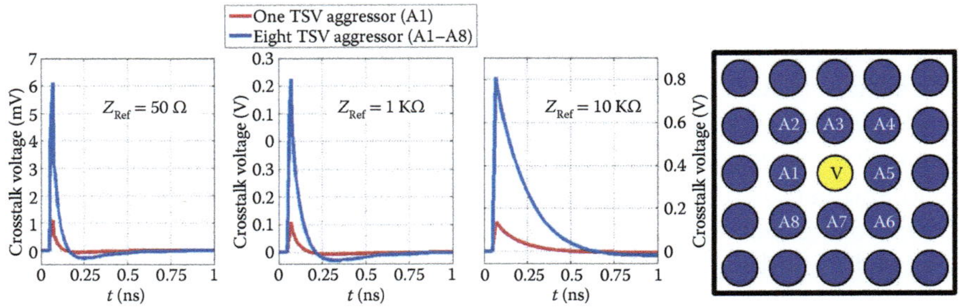

FIGURE 7.17
Voltage crosstalk results of the 11 × 11 TSVs array simulation under three general loading conditions: 50 Ω, 1 kΩ and 10 kΩ, with different amounts of in-phase aggressor TSVs. Inset: depiction of the aggressor TSVs (A1–A8) and victim TSV (V) in the array.

Figure 7.17 plots the crosstalk voltage levels on a victim TSV in the case where one of its directly neighboring TSVs is acting as an aggressor (red plots: TSV A1 shown to the right is the only aggressor), and also in the case where its eight closest neighbors are all acting as aggressors (blue plots; TSVs A1–A8 are all acting as aggressors). In this second case, all voltage steps on the eight aggressing TSV nodes are synchronized in-phase, which enables us to extract the worst-case total crosstalk voltage at the victim node. This is also the most realistic switching scenario in a digital IC system because flip-flops will have a high chance of toggling together on the same rising or falling edge of the clock. Figure 7.17 shows that higher impedance loads are more sensitive to voltage crosstalk. This crosstalk is however more easily alleviated for higher loaded TSVs, as indicated by the results of Figure 7.16c. For the 50 Ω terminated TSVs, the total electrical crosstalk obtained on the victim TSV through the contributions of all eight aggressor TSVs amounts to a peak value of 6 mV. This result is coherent as the victim TSV is highly tied to ground through its 50 Ω load, which is much smaller than the other impedances of the TSV system. Once again, by increasing the TSV load values, the crosstalk voltage is raised. We can see from the 1 kΩ and 10 kΩ multi-aggressor results in Figure 7.17 that a total voltage crosstalk can easily reach a significant fraction of a volt in practical systems, and that the voltage crosstalk results obtained using 50 Ω loads can be highly misleading.

7.4 Modeling and Simulation Methodologies

To address the noise coupling by TSVs, different modeling and simulation methodologies can be used, from commercial software such as ANSYS HFSS, Silvaco Atlas, and Synopsys TCAD to the "in house" circuit-based

3D extended circuit model. All these methods have their own strengths and limitations and should be properly used based on the specific cases.

7.4.1 HFSS (High Frequency Structure Simulators)

Ansys HFSS (high frequency structure simulators) utilizes a 3D, full-wave frequency domain electromagnetic field solver based on the finite element method (FEM). With HFSS, reliable extraction of SYZ parameters can be performed; the electromagnetic fields can be visualized in 3D; and it can generate component models to evaluate signal quality, transmission path loss, impedance mismatch, parasitic coupling, and far-field radiation. However, while HFSS solves Maxwell's equations, it does not include semiconductor equations.

To illustrate the usage of HFSS, three TSV coupling test structures were fabricated and measured (see Figure 7.18 for a micrograph). The first test structure in Figure 7.18a realizes the coupling between two TSVs connected in parallel to two neighboring TSVs, also connected in parallel. The second type of structure tests, the single TSV to single TSV coupling, and the third coupling structure includes a single TSV shielded by a guard-ring at the surface (a grounded p^+ doped region that defines an Ohmic contact to the bulk) and an adjacent single TSV, also shielded by such a guard-ring. All measurements were performed using GSG on-wafer probes and a vector network analyzer with a wideband frequency sweep from 10 MHz to 67 GHz. The TSV coupling structures are all accessed through 50 Ω CPW lines isolated from the silicon bulk by a 400 nm oxide. All three coupling structures were fabricated using the TSV technology parameters in Table 7.4.

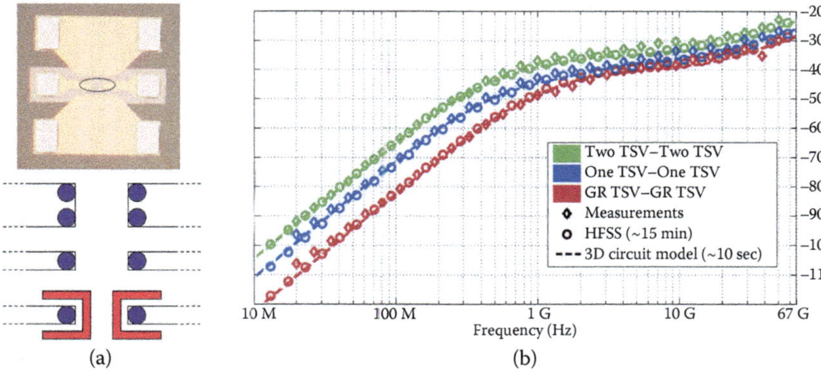

FIGURE 7.18
(a) Details of the three two-port TSV coupling test structures that were fabricated, measured, and simulated. A photograph of one of the two port devices, including the feed CPW lines and the GSG probe landing pads, is shown. Beneath it representations of the three coupling structures can be found: the two-TSV to two-TSV coupling structure, the one-TSV to one-TSV coupling structure, and the guard-ring-TSV to guard-ring-TSV coupling structure. (b) Simulated and measured coupling curves relative to the three different test structures.

TABLE 7.4
TSV Specifications

Symbol	Description	Value
r_{TSV}	TSV metal conductor radius	2.5 μm
H_{TSV}	TSV height	50 μm
t_{ox}	Oxide liner thickness	200 nm
p	Center to center TSV pitch	35 μm
ρ_{Si}	Silicon substrate resistivity (measured)	5 Ω.cm

The good agreement between HFSS full-wave simulations and measured S-parameters in Figure 7.18b for all three test structures demonstrates the accuracy of the HFSS simulations. However, such simulations are limited to TSVs with a relatively small volume and relatively few ports. Also, such simulations are not suitable for large-scale analysis and circuit-level design optimization including active devices. To overcome this, a new "in house" extended 3D circuit model has been developed and compared with the S-parameter measurements using the test structures in Figure 7.18. The equally good agreement between the 3D circuit model simulations and measurements demonstrates that the complex RF field distributions in the bulk Si between TSVs can be accurately represented by the 3D circuit model. Moreover, the 3D circuit model is also capable of handling multiple TSVs as well as incorporating the presence of grounded substrate nodes or guard rings. This will be discussed in more detail below in Section 7.4.2.

7.4.2 3D Circuit Model

In the context of 3D-IC design, a fast and accurate model capable of simulating substrate volumes with multiple nodes and ports is highly desirable. Classic compact lumped elements equivalent circuit simulations are very often limited to a single TSV pair, such as the works of [5,6]. Such simulations are then difficult to extend and apply in different contexts. In addition, they cannot capture the complex field distributions in the silicon substrate in the presence of additional TSVs or circuit nodes (such as transistors or guard rings).

7.4.2.1 3D Circuit Model Approach

The 3D circuit model is based on a very large distributed network of discrete lumped elements [17]. The lumped elements form up to 100,000 unit cells and the unit cells are used to represent the 3D volume and its properties. Simulations can then be performed in several ways, either in the frequency domain to yield an S-parameter matrix over a wide frequency band, or in the time domain to obtain voltage and current waveforms at key circuit nodes in response to a time-varying signal.

The entire space is discretized into many unit cell elements, which are electrically interconnected to represent the desired volume. Several unit cell elements with fundamental differences are implemented for the model, and are shown in Figure 7.19. The physical dimensions of all unit cells are described by three parameters, h_{unit}, l_{unit}, and w_{unit}, which represent the height, the length, and the width of the unit cell, respectively. The TSV unit cell shown in Fig. 7.19b is constructed using self-resistance and self-inductance elements as well as shunt capacitances. The properties of the TSV unit cell elements are described by Equations 7.4 through 7.10. The total TSV resistance per unit cell is given by the square root of the squared AC and DC resistance terms [18,19]. The frequency dependence mainly stems from the skin effect that occurs at high frequencies.

$$R_{\mathrm{TSV}} = \sqrt{(R_{\mathrm{TSV}-DC})^2 + (R_{\mathrm{TSV}-AC})^2} \tag{7.4}$$

$$R_{\mathrm{TSV}-DC} = \frac{\rho_{\mathrm{TSV}} h_{\mathrm{unit}}/2}{\pi r_{\mathrm{TSV}}^2} \tag{7.5}$$

$$R_{\mathrm{TSV}-AC} = \frac{\rho_{\mathrm{TSV}} h_{\mathrm{unit}}/2}{\pi\left(2 r_{\mathrm{TSV}}\delta - \delta^2\right)} \tag{7.6}$$

$$\delta = \sqrt{\frac{\rho_{\mathrm{TSV}}}{\pi \mu f}} \tag{7.7}$$

Here, r_{TSV} is the TSV's radius, ρ_{TSV} and μ are the resistivity and magnetic permeability of the TSV conductor, respectively, and δ is the skin depth inside the conductor at frequency f.

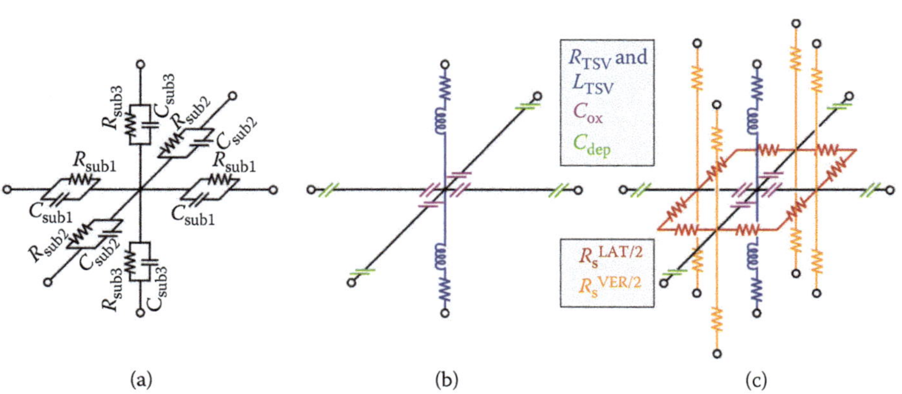

FIGURE 7.19
Equivalent circuit model for the 3D unit cells: (a) Silicon substrate unit cell, (b) TSV unit cell, and (c) TSV unit cell including parasitic surface conduction.

The self-inductance [19,20] of a single TSV cell is given by

$$L_{\mathrm{TSV}} = \frac{\mu h_{\mathrm{unit}}/2}{2\pi}\left[\ln\left(\frac{h_{\mathrm{TSV}}}{r_{\mathrm{TSV}}}+\sqrt{\left(\frac{h_{\mathrm{TSV}}}{r_{\mathrm{TSV}}}\right)^2+1}\right)+\frac{r_{\mathrm{TSV}}}{h_{\mathrm{TSV}}}-\sqrt{\frac{r_{\mathrm{TSV}}}{h_{\mathrm{TSV}}}+1}\right] \tag{7.8}$$

The capacitive elements of a TSV unit cell are modeled as cylindrical capacitors, as per Equations 7.9 and 7.10. The properties of the TSV oxide liner define the oxide capacitance, C_{ox}. If the interface between the TSV and the Si bulk is in a state of depletion, an additional depletion capacitance, C_{dep}, needs to be taken into account.

$$C_{\mathrm{ox}} = \frac{1}{4}\cdot\frac{2\pi\varepsilon_{\mathrm{ox}}h_{\mathrm{unit}}}{\ln\left(\dfrac{r_{\mathrm{TSV}}+t_{\mathrm{ox}}}{r_{\mathrm{TSV}}}\right)} \tag{7.9}$$

$$C_{\mathrm{dep}} = \frac{1}{4}\cdot\frac{2\pi\varepsilon_{Si}h_{\mathrm{unit}}}{\ln\left(\dfrac{r_{\mathrm{TSV}}+t_{\mathrm{ox}}+t_{\mathrm{dep}}}{r_{\mathrm{TSV}}+t_{\mathrm{ox}}}\right)} \tag{7.10}$$

It is common for the TSV/Si interface to be in depletion or inversion due to the presence of fixed positive oxide charges [21] in the TSV liner for standard *p*-type substrates with resistivities of 5–20 Ω.cm. The depletion widths can either be estimated by solving Poisson's equation in cylindrical coordinates or through the use of a semiconductor simulation tool, such as Atlas TCAD from Silvaco, for a known nominal substrate resistivity. If the TSV is in inversion or accumulation, the TSV unit cell needs to account for the highly conductive regions around the oxide liner, particularly if they are in contact with an inverted top or bottom surface, as discussed in Section 7.2.4. Such an extended TSV unit cell is depicted in Figure 7.19c and will be discussed further in Section 7.4.2.3.

The substrate unit cell, shown in Figure 7.19a, is modeled using parallel resistances and capacitances whose expressions are given by Equations 7.11 through 7.16.

$$R_{\mathrm{sub1}} = \rho_{Si}\frac{l_{\mathrm{unit}}/2}{w_{\mathrm{unit}}h_{\mathrm{unit}}} \tag{7.11}$$

$$C_{\mathrm{sub1}} = \varepsilon_{Si}\frac{w_{\mathrm{unit}}h_{\mathrm{unit}}}{l_{\mathrm{unit}}/2} \tag{7.12}$$

$$R_{\mathrm{sub2}} = \rho_{Si}\frac{w_{\mathrm{unit}}/2}{l_{\mathrm{unit}}h_{\mathrm{unit}}} \tag{7.13}$$

$$C_{\text{sub2}} = \varepsilon_{Si} \frac{l_{\text{unit}} h_{\text{unit}}}{w_{\text{unit}}/2} \tag{7.14}$$

$$R_{\text{sub3}} = \rho_{Si} \frac{h_{\text{unit}}/2}{w_{\text{unit}} l_{\text{unit}}} \tag{7.15}$$

$$C_{\text{sub3}} = \varepsilon_{Si} \frac{w_{\text{unit}} l_{\text{unit}}}{h_{\text{unit}}/2} \tag{7.16}$$

7.4.2.2 Model Validation and Power

The comparison of the 3D circuit model to on-wafer measurements in Figure 7.18 shows excellent agreement for all data. This validates the model in the case of multiple TSVs and also in the presence of grounded substrate nodes or guard rings. This approach is capable of accurately evaluating the coupling in 3D-IC structures between two nodes up to 200 μm (1 mm) apart for frequencies up to 50 GHz (10 GHz). Indeed, at these frequencies the wavelength of the electromagnetic waves in the materials considered are more than 20 times larger than the largest distance in the considered structures.

Furthermore, Figure 7.18 shows that the 3D-circuit model is also in excellent agreement with the HFSS full-wave EM simulations while being approximately 100 times faster. Such a lumped element modeling approach simplifies Maxwell's equations to linear Kirchhoff relations, which are much faster and easier to solve, leading to the substantial increase observed in the 3D-circuit model's simulation throughput. This increase in simulation speed opens the door to a wide range of large and complex simulations, including a very large number of TSVs and other circuit nodes that would be computationally too heavy and complicated for full-wave simulators such as HFSS. The simulation of an 11 × 11 TSVs array, presented in Section 7.3.2.1, which evaluated the coupling to and from 242 nodes, is a good demonstrator of the 3D-circuit model's simulation power. Unlike the other simulation tools discussed so far this model approaches a complete system-level simulation, as it is capable of simulating very large structures, taking into account the coupling paths between a large number of ports and circuit nodes.

7.4.2.3 3D TSV Extended Circuit Model and Validation

As described in Section 7.2.4, parasitic surface conduction is likely to occur in Si-based substrates at their interfaces with passivation layers. As discussed in that section, such highly conductive regions can have an important impact on the coupling mechanisms and paths and should be taken into account for the accurate evaluation of the substrate crosstalk levels between devices.

In order to incorporate these parasitic conduction channels into the 3D circuit model and assess their influence on the coupling to and from adjacent devices, it is necessary to insert additional elements into the TSV unit cell and to modify the substrate unit cell at the bottom and/or top interfaces [22]. The TSV unit cell is therefore adapted from the one depicted in Figure 7.19b to the one depicted in Figure 7.19c. In the latter, new resistive elements are introduced in the vertical and horizontal directions that account for the inversion channel between the TSV oxide and the depletion capacitance elements. The R_{sTSV} added in red and orange colors models the resistive paths around the TSV at the oxide/silicon interface in the vertical and horizontal directions and can be calculated by Equations 7.17 and 7.18, respectively. This interface may be highly conductive due to the presence of an inversion layer or highly resistive if depleted. The sheet resistance of the inversion layer is calculated using the semiconductor simulation software Silvaco Atlas and is found to be 9 kΩ/Square for a 20 Ω cm silicon substrate with an interfacial charge density of $Q_{ox} = 10^{12}$ cm^{-2}. This sheet resistance is in good agreement with DC measurements of an inversion layer induced around a TSV.

$$R_{sTSV}^{VER} = R_{sheetTSV}^{inv} \frac{H_{cell}}{\pi(r_{TSV} + t_{liner})} \tag{7.17}$$

$$R_{sTSV}^{LAT} = R_{sheetTSV}^{inv} \frac{\pi(r_{TSV} + t_{liner})}{2H_{cell}} \tag{7.18}$$

In addition to the TSV and the substrate unit cells, a backside unit cell is also developed to model the interface with the backside passivation layer. This interface may be either in inversion or in depletion. An example of connecting a network of unit cells in a 3D volume is shown in Figure 7.20b. The element C_{dBS} in the backside unit cell represents the depletion capacitance that may be present at the back side; whereas, R_{sBS} models the resistive path at the back side passivation layer/silicon interface. Equations 7.19 and 7.20 are used to calculate the resistances in the x and z directions:

$$R_{sBS}^{X} = R_{sheetBS}^{inv} \frac{l_{unit}/2}{W_{unit}} \tag{7.19}$$

$$R_{sBS}^{Z} = R_{sheetBS}^{inv} \frac{W_{unit}/2}{l_{unit}} \tag{7.20}$$

C-V plots of TSVs can be used to calibrate the results of the adapted model using Equations 7.17 through 7.20.

The blue and red curves in the C-V plot in Figure 7.21a were measured at 10 kHz on a 64 TSV array before and after wafer thinning and backside passivation. While the blue curve shows the expected depletion behavior for positive bias at this frequency, the red curve (after thinning and passivation) shows inversion at the same frequency. Figure 7.21b shows the TSV capacitance versus frequency at positive bias. The inversion capacitance persists up to much higher frequencies after thinning and backside passivation

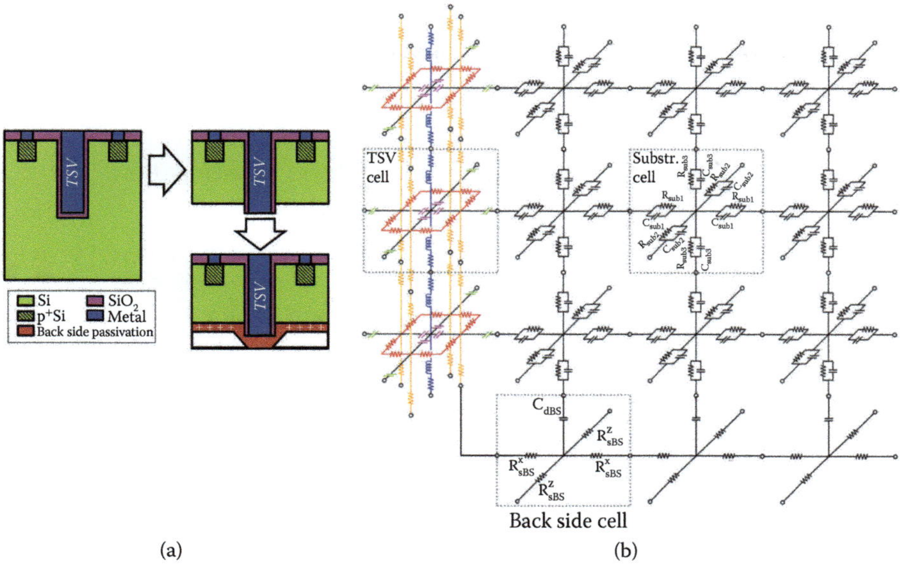

FIGURE 7.20
(a) Key steps in the TSV reveal and passivation process: (1) TSV before wafer thinning, (2) Wafer thinning and TSV reveal, and (3) Backside passivation. (b) Example of a 3D interconnect including parasitic surface conduction around the TSV and at the back side, constructed using three types of unit cells: TSV unit cell, substrate unit cell, and back side unit cell.

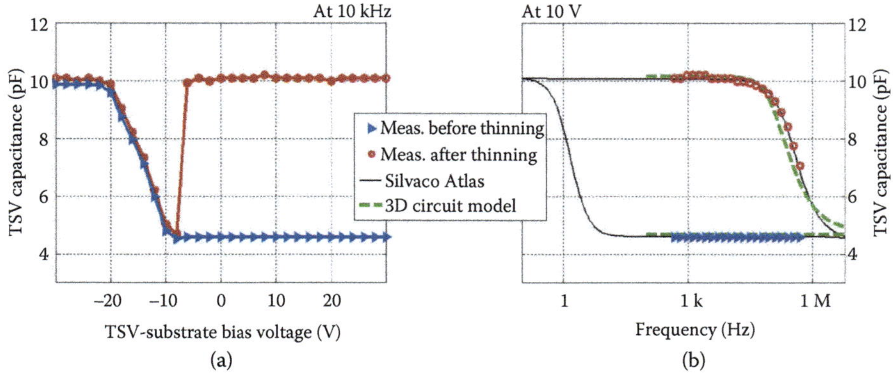

FIGURE 7.21
(a) C-V measurements. (b) C-f measurements (at positive bias) between p-substrate and a TSV array consisting of 64 parallel connected TSVs.

because of positive charges in the backside passivation layer, which cause substrate inversion at backside and provides minority carriers flowing into the TSV and causing the inversion of the TSV capacitance. The excellent agreement between the 3D extended circuit model (green dashed line) and the measurements validates the model and the correct understanding of the physical coupling paths and mechanisms. The black lines in Figure 7.21b were obtained from the 2D models using Silvaco Atlas (as further detailed in Section 7.4.3) and also agree well with experiments and the 3D circuit model.

7.4.3 Finite-Element Semiconductor Simulators: Silvaco Atlas

Full-wave electromagnetic (EM) simulation tools, such as HFSS from Ansys, are often used as the standard for coupling evaluation in 3D-ICs. However, one may doubt the validity of using pure EM simulations in analyzing semiconductor-based problems. Indeed, pure EM simulations do not include semiconductor physics, such as drift, diffusion, generation, recombination, *etc.* For such full-wave methods, the materials considered have linear properties that cannot depend on the local electric field. Semiconductors, however, are, by nature, nonlinear materials because of the local carrier concentration, and, hence, the electrical resistivity depends on the applied voltage, the oxide thickness, the location of substrate ground contact, the presence of interfacial charges or defects, and so on. A TSV is, in fact, a metal-insulator-semiconductor (MIS) structure and thus the silicon-oxide interface can be in different regimes: accumulation, depletion, or inversion. This leads to a frequency- and voltage-dependent TSV capacitance, which cannot be captured by pure EM simulations.

However, such simulations can be modified to account for these effects under small-signal operation so that they may still be used to analyze the high frequency characteristics of the structure. For example, a depletion region around the TSV can be modeled as a region of low conductivity in EM simulations or in the 3D circuit model. The boundary of this layer is obtained by solving semiconductor equations (Poisson and boundary conditions) or by using a semiconductor simulation tool. The Atlas software from Silvaco is such a tool and it has been used extensively throughout this work to complement HFSS and 3D circuit model simulations in aspects where semiconductor physics need to be taken into account. In particular, this concerns

- The extraction of depletion layer widths
- The simulation of fixed charges in oxide layers
- The extraction of carrier density and carrier mobility in inversion layers

Silvaco Atlas is a 2D and 3D device simulator that performs electrical DC, AC, and transient analysis for Si or other semiconductor material-based devices, and enables the characterization and optimization of semiconductor devices for a wide range of technologies. In the context of coupling evaluation between TSVs in Si substrates, it allows you to take into account parasitic surface conduction (PSC) phenomena that can radically influence the physical coupling path and mechanisms. Its results are mainly used to calibrate

HFSS and 3D circuit model simulations with appropriate substrate resistivity profiles, in particular near interfaces with passivation layers.

7.4.4 SYNOPSYS TCAD Modeling

In order to have a complete overview of the coupling mechanisms between TSVs and different active devices, Synopsys TCAD models were developed. Synopsys Sentaurus Device is a semiconductor device simulator that can accurately describe capacitive coupling, albeit being time consuming. However, it is capable of incorporating the coupling of transistors with different nodes and technologies into the model.

First, TCAD models were built and calibrated by RF measurements of TSV-to-active noise coupling test structures. The specifications of the test structures are listed in Table 7.2. Secondly, the calibrated TCAD models are used to further investigate the impact of the substrate contact location and different TSV architectures on the noise coupling level to active devices.

7.4.4.1 TCAD Model Calibration

The Synopsys TCAD Sentaurus device simulator can assess substrate coupling effects between TSVs and active devices by performing transient analyses of the transistor behavior. A time-domain analysis was performed by applying high frequency sine signals to the TSV and extracting the active device saturation current variations. These variations are converted to S-parameter and compared with 2-port (TSV to drain) S-parameter measurements of the reference RF test structures described in Section 7.2.3.1.

The good agreement between the TCAD simulations and the RF measurements in Figure 7.22 verifies our TCAD model. It also verifies our

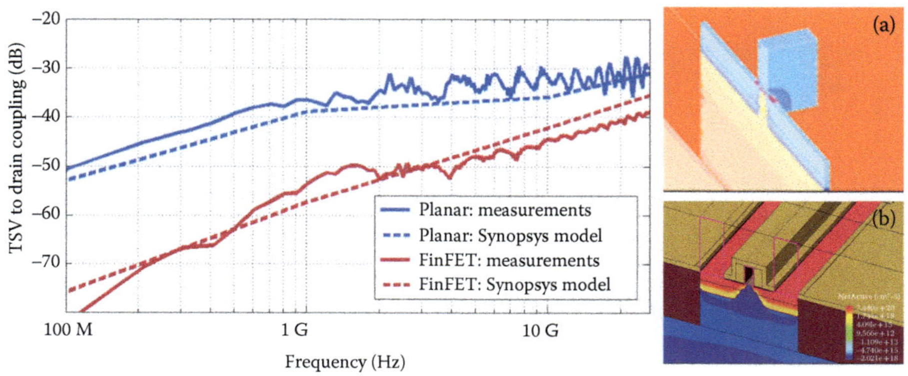

FIGURE 7.22
TCAD simulation versus measurements for FinFET and (a) planar FET, (b) device: $V_g = 1$ V, $V_d = 1$ V, $V_{TSV} = 0$ V.

understanding of the mechanisms responsible for the TSV noise coupling. In general, FinFET devices show a better TSV noise immunity compared with equivalent planar device, as discussed above.

7.4.4.2 Impact of TSV Diameter and Liner Thickness on Noise Coupling

Based on the calibrated TCAD model, four different TSV architectures were investigated further to predict their impacts on the noise coupling to active devices: (1) 5/50 µm via-middle TSVs with 200 nm oxide liners; (2) scaled 3/50 µm via-middle TSVs with 200 nm oxide liners; (3) 5/50 µm via-middle TSVs with 400 nm oxide liners; and (4) via-last TSV architecture with thick (3 µm) polymer liners ("donut" TSV).

As shown in Figure 7.23, scaling the TSV diameter using thicker liner materials or polymer based liner materials with smaller dielectric constant reduces the coupling level mainly at low frequencies where TSV liner capacitance is the dominant factor for noise coupling. At higher frequency, all noise coupling curves for the different TSV architectures tend to become similar. At 100 MHz, for the 3/50 µm and "Donut" TSVs, the noise coupling reduces by about 4 dB and 20 dB, respectively.

7.4.4.3 Impact of TSV Height on Noise Coupling

With recent advances in TSV and wafer thinning technology, the TSV diameter/height can be scaled down to 1/5 µm, which may lead to a strong reduction of the EM coupling between the TSV and active devices, as shown in Figure 7.24. 1/5 µm TSVs (with 100 nm liners) show better noise immunity than the 5/50 µm TSV (with 200 nm liners) over the entire frequency range.

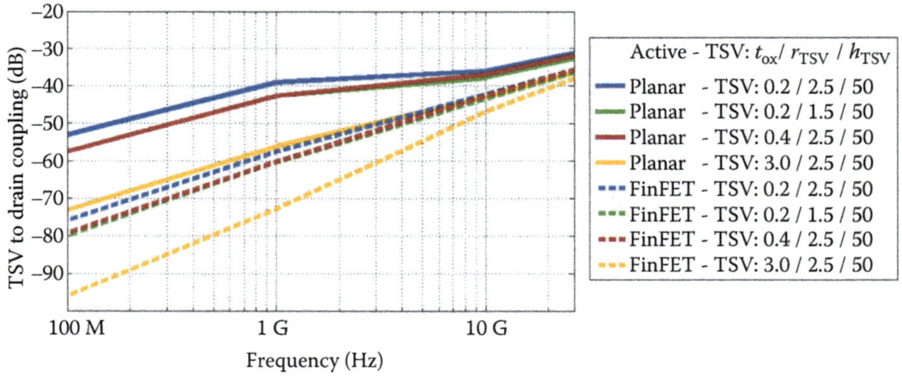

FIGURE 7.23
TCAD simulation: impact of TSV architectures on noise coupling.

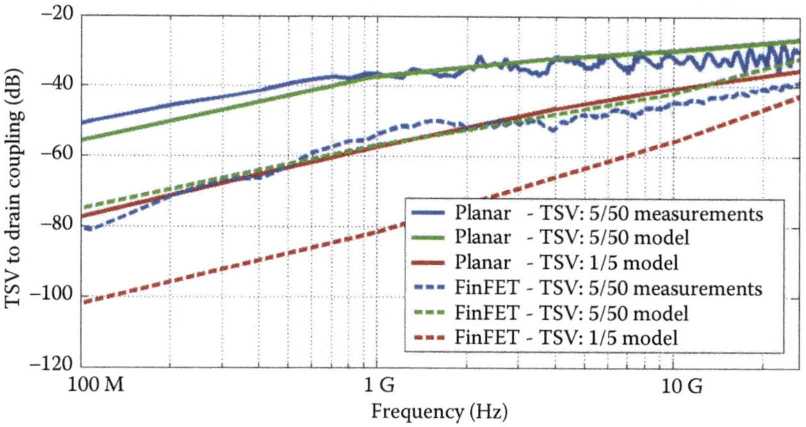

FIGURE 7.24
Noise coupling with 1/5 μm or 5/50 μm TSV to our referred 65 nm planar nMOS (solid line) and 32 nm nFinFET (dash line) device, respectively.

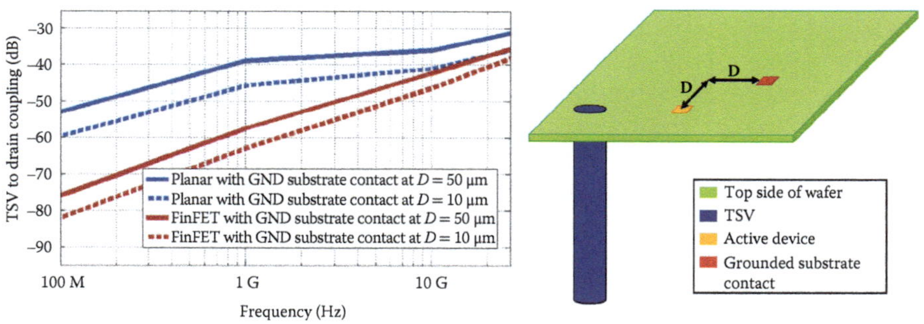

FIGURE 7.25
TCAD simulation: impact of substrate contact location on noise coupling.

7.4.4.4 Impact of Substrate Contact Location on Noise Coupling

The grounded substrate contact offers a resistive path for the substrate noise to be sunk into the ground, reducing the noise coupling level. The location of the substrate contact with respect to the active device determines the impedance of this resistive path: the closer the substrate contact is to active device, the smaller the impedance of the resistive path is, which reduces noise coupling. Figure 7.25 shows the noise coupling of 5/50 μm TSVs to planar/FinFET devices with a substrate contact either 10 μm or 50 μm away from the active devices. As expected, Figure 7.25 indicated a 4–6 dB noise reduction by placing the substrate contact closer to the active device.

7.5 Summary and Conclusions

In this chapter, we have addressed different scenarios of TSV-induced noise coupling in 3D-ICs and their impact on the signal integrity of neighboring devices, namely TSVs and active devices. The impact of TSV noise coupling on both FinFET and planar transistors has been investigated up to 40 GHz based on S-parameter measurements. The two key coupling mechanisms were identified for both types of active devices: noise amplification through the bulk transconductance (g_{mb}) as well as capacitive coupling. For planar nMOSFETs, noise amplification through the bulk transconductance (g_{mb}) mechanism dominates in the ON state, leading to large noise coupling. For FinFETs, the dominant coupling mechanism is capacitive even in the ON state. This demonstrates that FinFETs have better noise coupling immunity than planar nNMOSFETs, which makes FinFETs a strong candidate for future system integration with TSVs.

Moreover, the effect of an inversion layer after wafer thinning at the backside of the substrate was studied for the first time in the context of 3D integration and is successfully explained and accounted for using equivalent electrical circuit models. Furthermore, these parasitic surface conduction paths are implemented well within the extended 3D circuit model, enabling accurate coupling evaluation in such scenarios. It is important to be aware of this anomalous electrical behavior because it is common practice to tailor the oxide charges within the oxide liner such that an inverted state of the TSV is insured in order to guarantee a depletion layer and reduce EM coupling. Both the extension of the depletion region and the bridge effect are at the origin of bypassing the impedance of the depletion capacitance, substantially increasing the EM coupling. It is therefore of great importance for 3D-IC designers to take the reported effects into account or to implement one of the proposed solutions to counterbalance them.

In addition, different modeling methodologies based on commercial software were developed to perform a complete analysis of the coupling mechanisms between TSVs and adjacent devices. These approaches are either based on EM simulations to solve Maxwell's equations or based on semiconductor device simulations to solve semiconductor equations. Both the impact of the TSV architectures and the substrate contact location on the noise coupling have been simulated based on calibrated TCAD or HFSS models.

Furthermore, an accurate "in-house" 3D TSV extended circuit model including the anomalous TSV inversion, capacitance, and parasitic surface conduction effects has been proposed and calibrated against electrical measurements. It has been integrated into standard circuit simulators to evaluate the noise coupling between TSVs and adjacent devices. In contrast to 3D full-wave EM solvers, this model is able to easily incorporate different devices or transistor models while capturing the complex field distributions that arise in the Si substrate. Noise mitigation techniques can also be easily implemented in the calibrated 3D circuit model. The model proves to be very powerful, allowing for large-scale analysis and design optimization in both the frequency and the time domains. A large 11 × 11 TSVs array was fully simulated in a short time frame (5 seconds).

References

1. K. Banerjee, S. J. Souri, P. Kapur, and K. C. Saraswat, "3-D ICs: a novel chip design for improving deep-submicrometer interconnect performance and system-on-chip integration," *Proceedings of the IEEE* , vol. 89, no. 5, 602–633, May 2001.
2. C. Ryu, J. Lee, H. Lee, K. Lee, T. Oh, and J. Kim, "High Frequency Electrical Model of Through Wafer Via for 3-D Stacked Chip Packaging," pp. 215–220, *2006 1st Electronic System Integration Technology Conference*, Dresden, September 2006.
3. K. Yoon, G. Kim, W. Lee, T. Song, J. Lee, H. Lee, K. Park, and J. Kim, "Modeling and Analysis of Coupling between TSVs, Metal, and RDL interconnects in TSV based 3D IC with Silicon Interposer," *Proceedings of the Electronics Packaging Technology Conference*, pp. 702–706, December 2009.
4. J. Cho, J. Shim, E. Song, J. S. Pak, J. Lee, H. Lee, K. Park, and J. Kim, "Active circuit to through silicon via (TSV) noise coupling," *2009 IEEE 18th Conference on Electrical Performance of Electronic Packaging and Systems*, pp. 97–100, Portland, OR, October 2009.
5. M. Rack, M. Stucchi, X. Sun, C.R. Neve, G. Van der Plas, E. Beyne, P.A. Absil, and J.-P. Raskin, "Modeling the effect of charges in the back side passivation layer on through silicon via (TSV) capacitance after wafer thinning," *2015 IEEE MTT-S International Microwave Symposium (IMS)*, 17–22, May 2015.
6. C. Xu, R. Suaya, and K. Banerjee, "Compact modeling and analysis of coupling noise induced by through-Si-vias in 3-D ICs," *2010 International Electron Devices Meeting*, pp. 8.1.1–8.1.4, San Francisco, CA, December 2010.
7. M. Lee, J. Cho, and J. Kim, "Noise coupling analysis between TSV and active circuit," *2012 IEEE Electrical Design of Advanced Packaging and Systems Symposium (EDAPS)*, pp. 45–48, Taipei, December 2012.
8. H. Wang, M. H. Asgari and E. Salman, "Efficient characterization of TSV-to-transistor noise coupling in 3D ICs," *2013 ACM international conference on Great lakes symposium on VLSI (GLSVLSI)*, pp. 71–76, Paris, France, May 2013.
9. X. Sun, A. Rouhi Najaf Abadi, W. Guo, K. Ben Ali, M. Rack, C. Roda Neve, M. Choi, V. Moroz, I. De Wolf, J.-P. Raskin, G. Van der Plas, E. Beyne, and P. Absil, Noise coupling between TSVs and active devices: Planar nMOSFETs vs. nFinFETs, *2015 IEEE 65th Electronic Components and Technology Conference (ECTC)*, pp. 260–265, San Diego, CA, July 2015.
10. E. Nicollian and J. Brews, *MOS Physics and Technology*, New York, NY: Wiley, 1982.
11. D. Lederer and J.-P. Raskin, RF performance of a commercial SOI technology transferred onto a passivated HR silicon substrate, *IEEE Transaction on Electron Devices*, vol. 55, no. 7, 1664–1671, July 2008.

12. C. Liu, T. Song, J. Cho, J. Kim, J. Kim, and S. K. Lim, "Full-chip TSV-to-TSV coupling analysis and optimization in 3D IC," *2011 48th ACM/EDAC/IEEE Design Automation Conference (DAC)*, pp. 783–788, New York, NY, June 2011.
13. B. Xie, M. Swaminathan, K. J. Han, and J. Xie, "Coupling analysis of through-silicon via (TSV) arrays in silicon interposers for 3D systems," *2011 IEEE International Symposium on Electromagnetic Compatibility (EMC)*, pp. 16–21, Long Beach, CA, USA, August 2011.
14. X. He, W. Wang, and Q. Cao, "Crosstalk modeling and analysis of Through-Silicon-Via connection in 3D integration," *PIERS Proceedings*, pp. 857–861, Taipei, March 25–28, 2013.
15. W. Yao, S. Pan, B. Achkir, J. Fan, and L. He, Modeling and Application of Multi-Port TSV Networks in 3-D IC, *IEEE Transactions on Computer-Aided Design of Integrated Circuits and Systems*, pp. 32, 487, April 2013.
16. S. Piersanti, F. De Paulis, A. Orlandi, M. Swaminathan, and V. Ricchiuti, Transient Analysis of TSV Equivalent Circuit Considering Nonlinear MOS Capacitance Effects, *IEEE Transactions on Electromagnetic Compatibility*, vol. 57, no. 5, 1216–1225, October 2015.
17. M. Rack, J. P. Raskin, X. Sun, G. Van der Plas, P. Absil, and E. Beyne, "Fast and Accurate Modelling of Large TSV Arrays in 3D-ICs Using a 3D Circuit Model Validated Against Full-Wave FEM Simulations and RF Measurements," *2016 IEEE 66th Electronic Components and Technology Conference (ECTC)*, pp. 966–971, Las Vegas, NV, May–June 2016.
18. M. Brocard, P. Le Maître, C. Bermond, P. Bar, R. Anciant, A. Farcy, T. Lacrevaz, P. Leduc, P. Coudrain, N. Hotellier, H. Ben Jamaa, S. Chéramy, N. Sillon, J-C. Marin, and B. Fléchet, "Characterization and modelling of Si-substrate noise induced by RF signal propagating in TSV of 3D-IC stack," *2012 IEEE 62nd Electronic Components and Technology Conference (ECTC)*, pp. 665–672, San Diego, CA, May–June 2012.
19. J. Kim, J. Cho, and J. Kim, "TSV modeling and noise coupling in 3D IC," *Electronic System-Integration Technology Conference (ESTC)*, pp. 1, 6, 13–16, September 2010.
20. J. Cho and J. Kim, "Signal integrity design of TSV and interposer in 3D-IC," *2013 IEEE Fourth Latin American Symposium on Circuits and Systems (LASCAS)*, pp. 1–4, Cusco, February–March 2013.
21. G. Katti, M. Stucchi, J. Van Olmen, K. de Meyer, and W. Dehaene, Through-silicon-via capacitance reduction technique to benefit 3D IC performance, *IEEE Electron Device Letters*, vol. 31, no. 6, 549, 551, June 2010.
22. X. Sun, M. Rack, G. Van der Plas, M. Stucchi, J. De Vos, P. Absil, J.-P. Raskin, and E. Beyne, "Investigation of TSV noise coupling in 3D-ICs using an experimental validated 3D TSV circuit model including Si substrate effects and TSV capacitance inversion behavior after wafer thinning," *2016 IEEE MTT-S International Microwave Symposium (IMS)*," pp. 1–4, San Francisco, CA, May 2016.

8

Layout Strategies for Substrate Crosstalk Reduction in Low Cost CMOS Processes

Pedro Mendonça dos Santos, Luís Mendes, João Caldinhas Vaz, and Henrique Quaresma

CONTENTS

8.1 Introduction

As complementary metal–oxide-semiconductor (CMOS) technology continues scaling down to the nanometer range, processing it is becoming more and more complex, with an increasing number of masks and related processing steps. The high integration level achieved in current integration technologies led to the development of system-on-chip (SoC) solutions, which allow the integration of several distinct systems into the same chip, with obvious advantages in terms of size, power consumption, and cost. Modern integrated circuits (ICs) include a diversity of digital, analog, mixed-signal, and radio-frequency (RF) functional blocks, with different power, bias levels, and signals of different frequencies all sharing the same CMOS die. Also, very high-frequency digital and power management parts are responsible for the increase in the noise level in the IC, which can seriously affect the performance of sensitive analog and RF parts, like band-gap references, low-noise amplifiers, and oscillator circuits.

The interfering signals in the ICs, commonly referred to as noise, although essentially deterministic, appear in two ways. One is through the power and ground lines, if care is not taken by the designer. The use of strategically placed decoupling capacitors and differential circuit topologies, rather single-ended, and separating different power and ground domains (e.g., analog from digital blocks) as much as possible are examples of good design practices. The other way noise is injected into ICs is through coupling between metal interconnects, substrate, and wells.

Figure 8.1 shows a simplified cross section of a typical CMOS integrated circuit. In this case, an active component (or circuit) is placed close to a passive component (or circuit). Each component has its own guard-ring (GR) to bias the local substrate below, in this case a p-well. The active element placed on top of the p-well should be an NMOS transistor. For its correct operation, a depletion region should exist between the active physical structure and the p-well. In this way, the device is isolated from the p-well and the chip substrate. The level of isolation is very high for direct-current (DC) operation, but as the operation frequency increases, the isolation decreases because of the junction (source-body and drain-body) and channel-body capacitances. For the passives case, since they are fabricated inside the dielectric region and above the shallow trench isolation (STI) layer, further isolation from the p-well is achieved. Just like the actives, the dielectric isolating material's nature creates a capacitive effect, which also reduces isolation levels at high frequencies. Thus, due to the imperfect isolation between the components (or

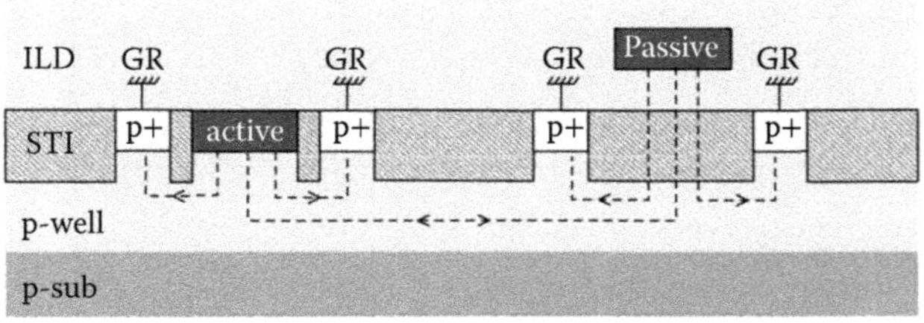

FIGURE 8.1
CMOS typical substrate coupling mechanisms.

circuits), parasitic currents are generated in the IC. For the example above, the parasitic currents paths (dashed lines) through the p-well to the GR for both components (circuits) and the coupling current path between both components/circuits through the p-well region are illustrated in Figure 8.1.

To circumvent these problems, CMOS technology has evolved, not only above (back-end-of-line) silicon (e.g., thinner dielectrics layers, low and high dielectric constant isolation materials, such as k dielectrics; an increased number of metal levels; and accessibility to thick copper layers), but also at silicon level (front-end-of-line), with the availability of trench isolation structures, multiple wells processes and, in some specific cases, buried layers (e.g., isolation and deep n-well). Although some of these improvements allow the development of high performance isolation structures, the cost of these technologies can be prohibitive for some applications. In cases where the usage of lower nodes does not bring significant circuit performance gains and where layout area is not the major concern, the "older" technology nodes (above 180 nm) can be a solution. However, since deep n-well formation is not available in those CMOS processes, the best performance isolation structures cannot be implemented. To overcome this problem, mask manipulation in "older" CMOS processes can be employed to implement pseudo deep n-wells and, in this way, make possible the implementation of high isolation structures in low cost CMOS technologies.

This chapter presents several layout strategies for substrate crosstalk reduction in low cost CMOS processes. These strategies are all based on mask manipulations without additional masks or modifications of the CMOS process flow. The purpose of the techniques presented here is to increase the resistance of any possible path among neighboring components (or circuits) at silicon level. The chapter is divided in 3 sections. In Section 8.2, an overview of the state-of-art of CMOS isolation techniques and structures is presented and, in Section 8.3, the layout and performance results of several isolation structures, some using pseudo n-wells, developed in a low cost single-well CMOS technology are presented and discussed.

8.2 Isolation Techniques Overview

Isolation techniques created by simple mask design without adding extra fabrication steps are compatible with almost all silicon integrated technologies. In this case, there is no cost increase and inexpensive nodes can be used. On the other hand, in modified standard CMOS technologies, where post-processing or extra fabrication steps are employed, the isolation structures can achieve better results than the ones implemented in standard CMOS technologies, but the fabrication process frequently becomes more expensive, because, in order to protect the circuit, it is surrounded by an isolation structure. A brief review of the major isolation methods will be presented in this section.

The usual approach when studying substrate isolation techniques is to start with a structure with no isolation, which is used as the reference thru standard in the performance assessment of each isolation structure, i.e., the different isolation structures are measured and the results are compared with the ones of the reference thru. Typically, noise is injected at a node (NE, noise emitter) and collected at another node (NR, noise receiver). The first node represents the point where an "aggressor" circuit that produces an electrical perturbation is placed and the second point is where the "victim" circuit, affected by the perturbation, is located. In the following description, it was assumed that the substrate (p-sub) is lightly-doped.

8.2.1 Isolation Structures for Standard CMOS

The guard-ring is one of the simplest structures used to isolate a single component, a group of components, or even a larger circuit. It is compatible with all technologies and does not require any additional process steps. To simplify the following sections, it was assumed and represented in all figures that all p-type and n-type GR are connected to the alternating-current (AC) ground, even though p-type GRs are actually connected to the lower supplied voltage (usually the reference or ground), while the n-type GRs are connected to the higher supplied voltage (VDD).

8.2.1.1 P-Type and N-Type Guard-Rings

P-type and n-type GRs are commonly used to connect the body of active components to VSS (lower voltage supply) or ground (GND) and to VDD, respectively, but they also contribute to isolating them from neighboring circuit generated noise [1–3]. P-type and n-type GR structures are presented in Figure 8.2a and b, respectively, and the noise current flow paths from nodes NE to NR are represented by dashed lines. In the standard CMOS technology, path 1 is less resistant than path 2 and, therefore, most of the noise will go through path 1. Isolation occurs because part of the current reaching the

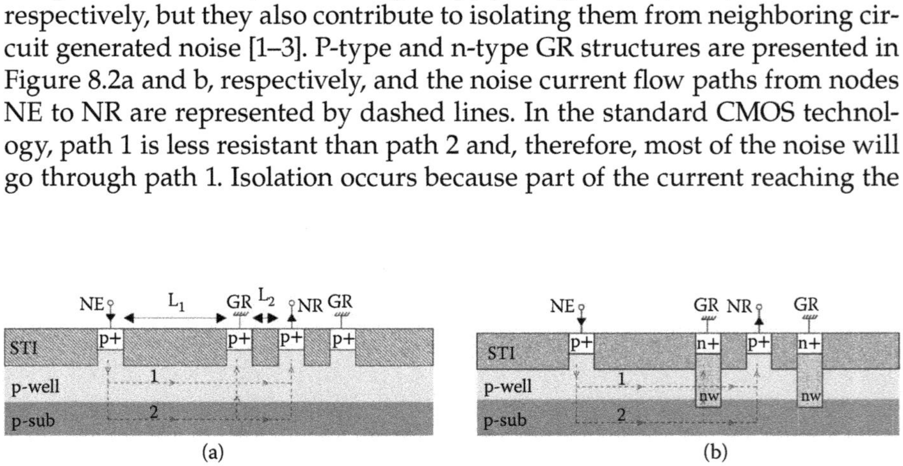

FIGURE 8.2
Classical guard-rings: (a) p^+ and (b) n^+.

region below p^+ GR is deviated to GND through the grounded GR node. However, a remaining current will inevitably reach NR node.

Higher isolation can be obtained by mitigating the main noise current in path 1. For that purpose, an n-type GR can be placed around the component, as shown in Figure 8.2b. This n-well GR trench has a depth slightly larger than the p-well, so it blocks path 1 current due to the two lateral p-n junctions surfaces, which are reverse biased. Figure 8.2a also denotes distances L_1 and L_2, which are the aggressor to isolation structure boundary and isolation structure to victim distances, respectively. Of course, isolation is increased if p-sub and p-well regions resistivity is increased, but that is not always possible in standard CMOS processes. In the case of a p-type GR, a typical isolation improvement of up to 30 dB can be obtained for L_1 close to 100 µm, and an extra improvement of up to 10 dB if the GR is n-type. Of course, these values are very dependent on geometry and the electrical properties of the materials used in the fabrication process.

In both cases the isolation can be improved by increasing L_1 and decreasing L_2 [3]. Increasing the GR width by placing more contacts rows in parallel is beneficial, up to a certain point. In all cases, the GRs must be connected to an on-chip efficient AC ground providing a very low impedance path to off-chip AC ground [3].

8.2.1.2 P-Sub Guard-Ring and Deep-Trench Isolation

Another option that can be used to increase the path 1 resistance is creating a region without a *p*-well around the component, as depicted in Figure 8.3a. This is called the *p*-sub guard-ring [4,5]. One of its characteristics is that it has no node connection, so it has a simple design.

The inclusion of the p-sub GR can lead to a typical isolation increase, ranging from 5 to 15 dB, larger for a wider guard-ring. The p-sub GR is more effective in CMOS processes where p-well has a higher conductivity. Many times, the p-sub GR is used in combination with a p-type GR, providing extra isolation.

For a process that includes a deep-trench isolation (DTI) flavor, further isolation will be achieved, since DTI is deeper than STI (Figure 8.3b). This DTI is fabricated with the same dielectric material as STI.

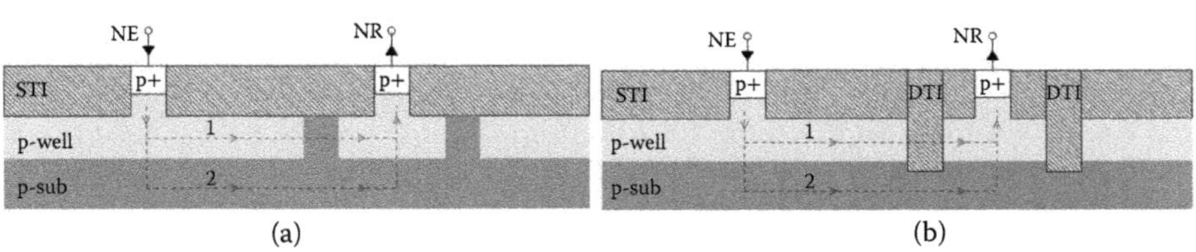

FIGURE 8.3
Isolation with (a) *p*-sub GR and (b) deep-trench isolation.

8.2.1.3 N-Well and Deep n-Well

Wells or tubs are commonly used, in combination with guard-rings, to provide active devices with body biasing. Moreover, isolation structures with both elements are more effective than the ones presented so far. Figure 8.4a shows the classical n-well isolation. Both path 1 and 2 currents must cross a reverse biased p-n junction surface. If a deep n-well option is available, as presented in Figure 8.4b, the same currents suffer an additional attenuation due to the second internal reverse biased p-n junction surface [6,7].

Usually these two structures that place reversed biased p-n surfaces in the current path create a high isolation at very low frequencies, both being preferable to the n-type GR. But, the advantages of the structures represented in Figure 8.4 become equivalent to the n-type GR with increasing frequency due to the capacitive nature of the junction.

8.2.1.4 Active Suppression

The principle behind this technique is to sense the noise in the neighborhood of the circuit that requires protection and reinjecting the noise out-of-phase in the same region, cancelling the noise. In this isolation method, active circuits are required, which leads to a power consumption increase. This technique is more effective in terms of wideband cancelation than passive structures. A graphical representation of this method can be seen in Figure 8.5, where the sample GR picks a sample of the noise generated by NE,

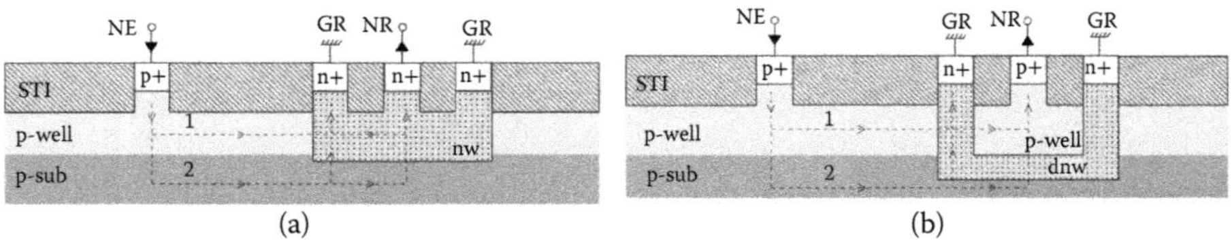

FIGURE 8.4
Isolation with (a) n-well and (b) deep n-well.

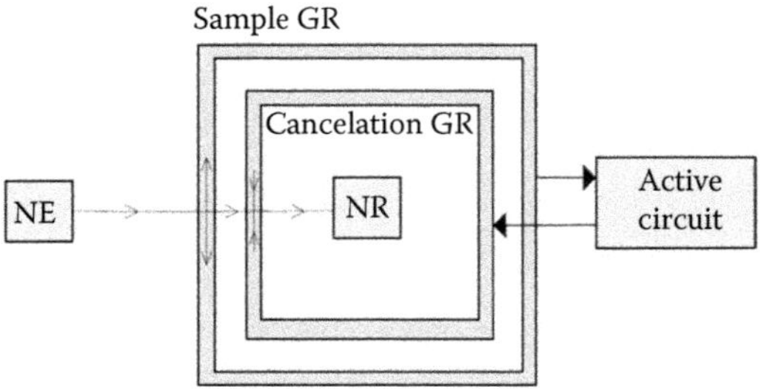

FIGURE 8.5
Active noise suppression.

which is signal inverted and amplitude adjusted by an external circuit, and adds it to the noise path through the cancelation GR. The objective is to cancel the noise that reaches NR by choosing the correct amplitude and phase adjustment. This technique allows a wideband noise cancelation but has the power consumption drawback [8].

Using a simple active circuit in [8] the authors were able to extend the noise isolation from DC to 250 MHz with 9 dB, and 3 dB up to 1.2 GHz, with a power consumption of 2.5 mW. The cancelation design requires a careful guard-rings and substrate characterization.

8.2.2 Isolation Structures in Non-Standard CMOS

The following techniques have been used to improve isolation, but they require process steps usually not available in standard CMOS technology.

8.2.2.1 Buried Dielectric Layer Isolation

Silicon on Insulator (SOI) is a CMOS variant where an insulating layer is grown on top of the silicon p-sub. The active devices are then grown on top of this insulating layer. The insulating layer can be sapphire or a buried oxide (box). The second option is represented in Figure 8.6. The active devices grown on top of the oxide layer achieve a higher isolation from each other and from the substrate than the ones implemented in standard CMOS, providing faster devices and low leakage devices [9].

With absence of wells, other advantages arise. For example, devices can be designed closer to each other, since latch-up is not an issue. A drawback is the nonexistence of ESD-protecting substrate diodes. Although this approach successfully improves the substrate noise isolation, the higher cost of CMOS SOI technology makes this option impracticable in many situations.

8.2.2.2 Through-silicon Via Isolation

Through-silicon Via (TSV) is an effective way to isolate a region from an aggressor noise, but the cost is also high, since it requires non-standard fabrication steps. After fabrication, the wafer is turned upside down and a trench

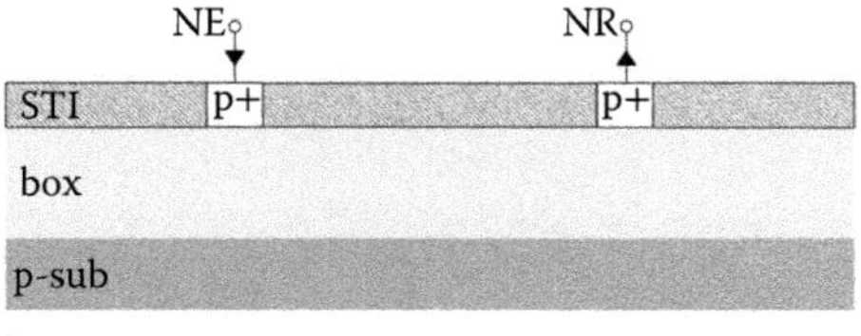

FIGURE 8.6
Silicon-on-Insulator (SOI) structure, with a buried-oxide (box) grown on the top of the substrate.

is excavated through the silicon and oxide up to metal 1 (see Figure 8.7), which is connected to ground. In this technique, all noise currents that flow from NE to NR are blocked if the NR is completely surrounded by a continuous TSV GR. In practice, this is not possible because of wafer stress. Thus, some noise will always reach NR [10].

In [10], a 30 dB and 40 dB isolation improvement were achieved up to 100 MHz and 1 GHz frequencies, respectively, by protecting the victim from the aggressor with an H-shaped TSV. According to the authors, these results are better than conventional structures such as GRs, deep n-well, or deep trench isolation.

8.2.2.3 Particles Irradiation Isolation

The idea behind the particles irradiation isolation technique is to create a high resistive region between the aggressor and the victim. In this case, a post-processing particles radiation or bombardment is made in selected regions. In the example presented in Figure 8.8, the irradiated region blocks the p-well path and partially blocks the p-sub path. Once again this is not a usual procedure or compatible with the standard nodes [11].

For example, in [11], Helium-3 was irradiated to create a semi-insulated region in silicon, which improved its resistivity to 1 kΩ·cm. With this method, noise isolation was improved 10 dB at 2 GHz.

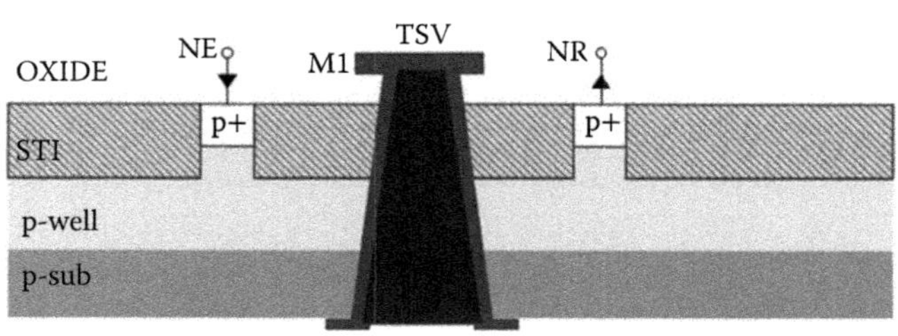

FIGURE 8.7
TSV isolation.

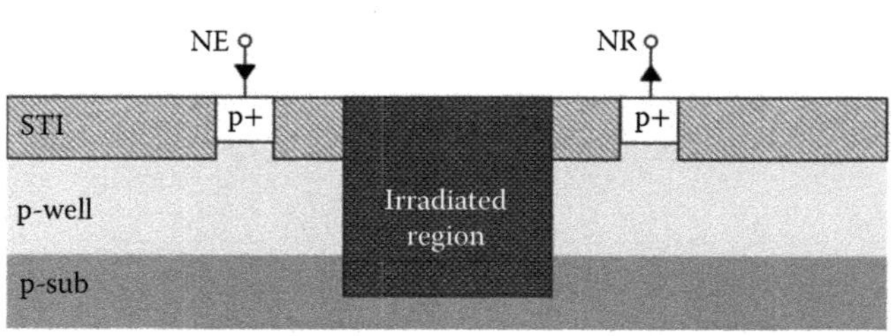

FIGURE 8.8
Particles irradiation isolation.

8.2.3 Noise Propagation through Substrate

The isolation techniques and structures that were presented in the above sections have a common purpose, to avoid that noise propagating through the doped silicon part of the integrated circuit (p-sub and wells) from reaching the point where the victim is placed. Thus, a good way to understand the substrate noise propagation is to study the electrical behavior of doped silicon. This section presents this study supported by electrical measurement results.

8.2.3.1 Substrate Characterization Structures

To carry out a wide frequency range analysis, several integrated circuits with test structures suitable for DC and AC on-wafer measurements were fabricated. The basic test structure used for the substrate characterization is presented in Figure 8.9a and comprises two p^+ substrate contacts at different distances (L). The substrate contacts are used to inject (NE) into the p-type substrate and sense (NR) the noise from the p-type substrate.

This structure assures a good ohmic contact between the metal and the p^+ implant and a fairly good resistive contact from the p^+ implant to the p-well substrate. In this way, moderate low frequency attenuation can be accurately measured between the NE and NR points. The opposite is shown in Figure 8.9b, where the n^+/p-well junction capacitance that is used to inject and sense the noise in the p-well completely masks the low frequency behavior of the substrate [12].

8.2.3.2 Substrate DC Characterization

Experimental results were obtained from DC measurements of several test structures identical to the one in Figure 8.9a, using a packaged chip, a test board, an ohmmeter, and a controlled temperature chamber, for p^+ contact distances (L) ranging between 20 μm to 170 μm in steps of 10 μm, for two temperatures (278 K and 301 K). The obtained resistance versus distance results are presented and compared in Figure 8.10. The resistance behavior for the two temperatures follows a non-linear trend with the contact distance, which confirms the most recent results presented by [13,14]. In addition, it can

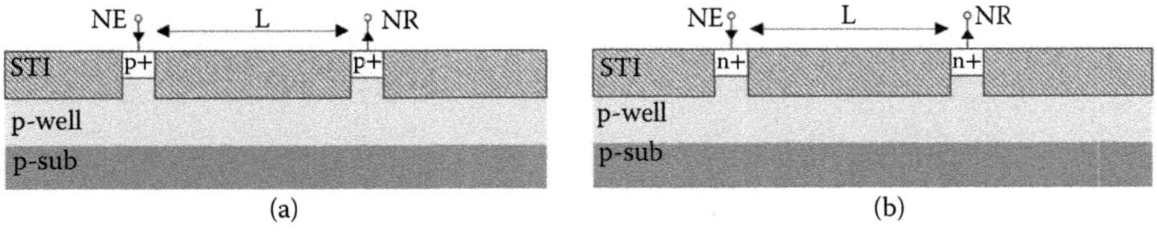

FIGURE 8.9
Typical substrate characterization test structure with (a) p^+ and (b) n^+ implant contacts.

also be seen in Figure 8.10 that the natural logarithmic curve fitting closely describes the resistance trend with the contact separation, agreeing with the synthesized compact models (SCM) recently proposed in [13].

The isolation between neighbor circuits is improved if the substrate resistance between those circuits increases, at least at low frequencies. Then, circuit parts that have different functions and domains (e.g., analog and digital) should be placed as far away from each other as possible. The isolation improvement obtained by doing that is more noticeable in circuits that were initially placed near each other, since the resistance derivative is higher for lower NE to NR distances. Thus, the isolation improvement that can be achieved by further separating the circuits that are already at distances higher than 100 μm is low. Although the isolation between different circuit parts improves as distance increases, the required area also increases, leading to higher costs. It is also interesting to note that, as the substrate temperature increases, the resistance between the aggressor and victim regions increases. This indicates that, as the operation temperatures increases, the isolation between different circuit parts tends to increase.

8.2.3.3 Substrate AC Characterization

To evaluate the effect of the distance L between the NE and NR on the substrate noise coupling in the frequency domain, S-parameters measurements were performed for the distances of 20, 40, 60, 80, 120 and 160 μm. The test structures were once again identical to the one illustrated in Figure 8.9a but, in this case, probe pads were included to allow on-wafer measurements.

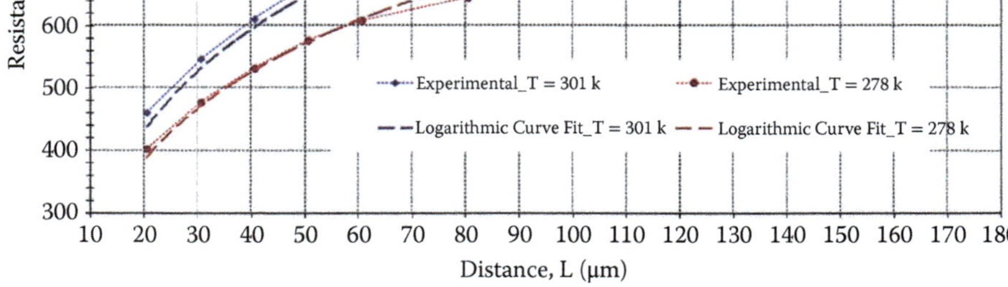

FIGURE 8.10
Comparison of experimental results for the substrate resistance as a function of p^+ regions (NE and NR) distance.

The experimental results are presented in Figure 8.11. The measurements were made with chip-on-board assembly and coaxial connectors, because they were performed inside a controlled temperature chamber. As can be seen, the forward transmission data presents a flat level up to around 400 MHz. This frequency behavior is in accordance with the results presented by [14], which used similar test structures.

For a given operating frequency, Figure 8.11 also shows a decrease of the amount of signal arriving to the sensing contact (NR) with the increase of contacts distance, agreeing with the results in [15] and with the trend expected from the DC analysis and the conclusions presented in Section 8.2.3.2. The flat response up to a given frequency suggests that the substrate propagation path can be modeled by a series resistance that can be obtained from the DC measurements illustrated in Figure 8.10. This low frequency model of the substrate is presented in Figure 8.12.

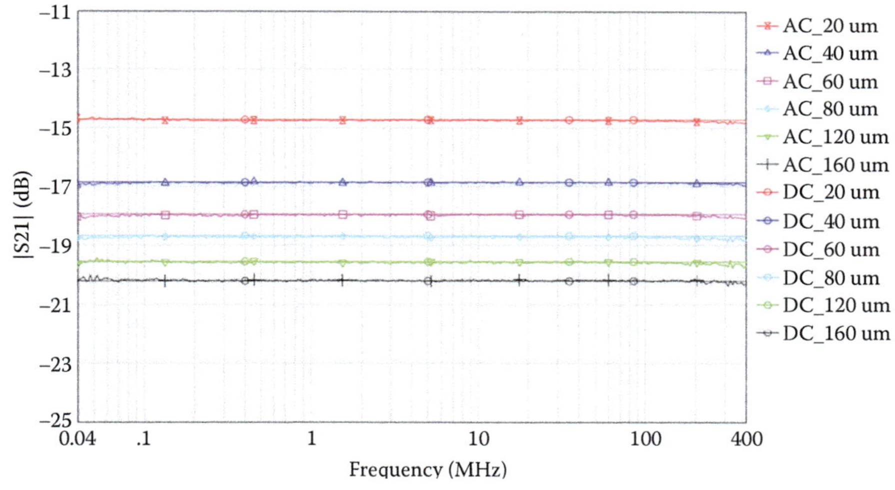

FIGURE 8.11
Experimental results and DC extrapolated values (obtained from Figure 8.10) of |S21| for the p^+ substrate contacts distance L of 20, 40, 60, 80, 120, and 160 μm for a substrate temperature of 301 K.

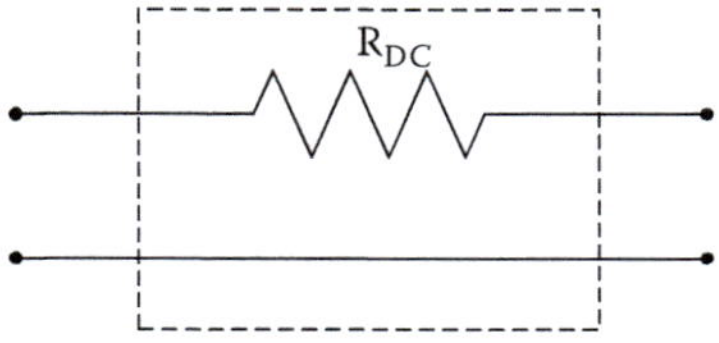

FIGURE 8.12
Two-port network substrate model valid for low frequency.

In Figure 8.11, additional $|S_{21}|$ horizontal curves, calculated from Figure 8.10 resistances, were included. These extrapolated $|S_{21}|$ values were determined from

$$S_{21} = \frac{1}{1 + R_{DC}/2R_0} \tag{8.1}$$

where R_0 is the 50 Ω reference impedance. As can be seen, good matching is attained in the flat region with the de-embedded results. From the results presented in Figure 8.11, similar conclusions to the ones drawn in Section 8.2.3.2 can be made.

As the frequency noise surpasses a given frequency, the reactive effects of the substrate begin to be noticeable. Thus, with a larger bandwidth measurement, it is expected that the $|S_{21}|$ will be almost constant, until a certain frequency is reached, and then start to increase, that is, the isolation will start to decrease. This is a consequence of the dielectric behavior of the substrate that should be considered and can no longer be described by the simple resistive model of Figure 8.12. A substrate model that accounts for this effect has been proposed in [16–18]. The model consists of replacing the pure resistor in the model of Figure 8.12 with a parallel resistor and capacitor, as shown in Figure 8.13. In this way, the absolute value of the parallel impedance decreases with frequency, causing an isolation decrease.

In fact, there is a cut-off frequency, f_c, given by Equation 8.2, for which the substrate resistive and capacitive effects are equal [3],

$$f_c = \frac{1}{2\pi\, \varepsilon_{sub}\, \rho_{sub}} \tag{8.2}$$

where ε_{sub} and ρ_{sub} are the substrate permittivity and resistivity, respectively. As can be seen by Equation 8.2, the cut-off frequency is only dependent on the substrate electrical properties. Knowing R_{sub}, a first guess for C_{sub} can be obtained using the measured cut-off frequency value. For the substrate used here, f_c is around 7.5 GHz.

Figure 8.14 shows the on-wafer measurements up to 6 GHz for the same structures presented in Figure 8.11, but build on a different chip where high

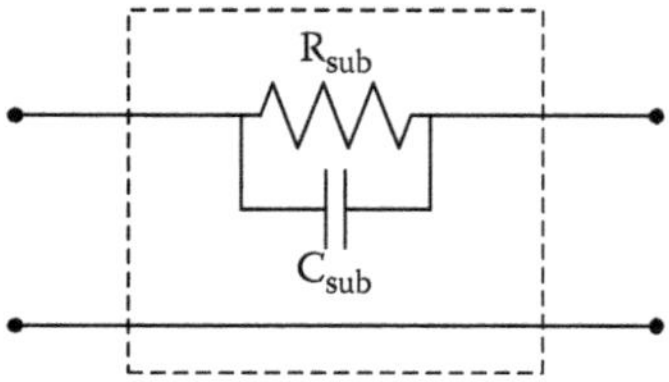

FIGURE 8.13
Equivalent two-port network model assumed for the substrate impedance.

frequency accurate de-embedding standards were included. Moreover, additional test structures with opens, shorts, and thru standards were used in the de-embedding phase to remove the text-fixture influence. The $|S_{21}|$ curves presented in Figure 8.14 have an almost flat behavior up to 6 GHz, which is a frequency value not high enough to denote the substrate capacitive influence. As in DC and sub-GHz frequency range, the substrate isolation at several GHz increases with the distance between the NE and NR.

In general, published work focusing on the AC characterization of the substrate commonly use test structures such as the one shown in Figure 8.9b, [12,19,18], or variations of this structure that replace one of the n^+ implants with a transistor gate [20,15] or with a p^+ implant [21,22]. The reason pointed out in [23] for this is that it is a configuration that can also be applied to other circuit components that use n-doped layers, such as the n-wells of PMOS transistor, the deep isolation wells for PNP transistors, and the n^+-buried layers used for triple-well isolation [3] or used as the lower electrode of integrated capacitors. However, those structures do not provide any insight into the small-signal behavior of the substrate, at least up to a few GHz. This is because there is a capacitance involved in the noise injection and reception mechanisms due to the p-n junction from the n^+ to the p-well. In fact, the capacitive substrate noise injection and/or reception mechanism is not being separated from the noise propagation within the substrate. Yet, no comments regarding this issue in the substrate characterization have been found, even in more recent works [4,14,24] where there is some tendency to use simple p^+ implant substrate contacts (as in Figure 8.9a), which eliminates

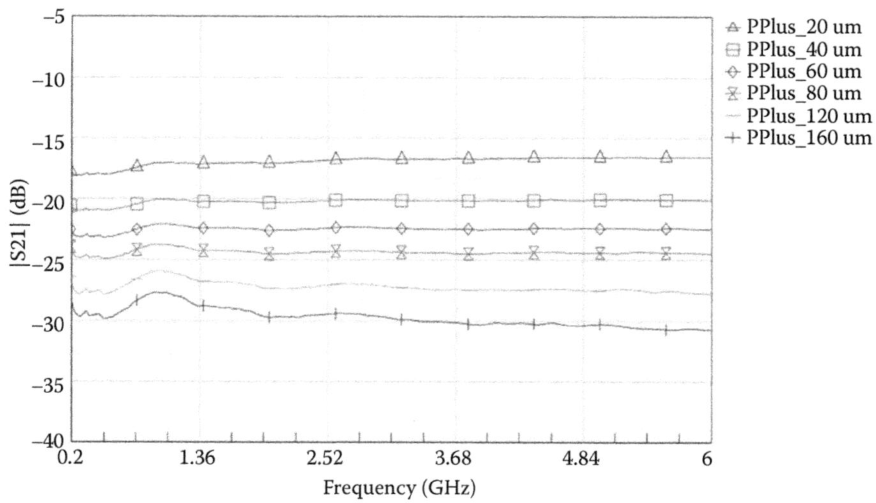

FIGURE 8.14
Experimental results of |S21| for 20, 40, 60, 80, 120, and 160 µm distances between the two p^+ substrate contacts.

this capacitance and provides a low resistive noise signal path to and from the substrate.

The impact of these capacitive coupling to the substrate is shown in Figure 8.15 by presenting the measured forward transmission ($|S_{21}|$) of the test structure of the Figure 8.9b for different contacts distances (L). In Figure 8.15, this effect is visible above 1 MHz due to the 20 dB/dec increase of the $|S_{21}|$ measured curves.

Around 200 MHz, the contact impedance Z_c becomes comparable with the substrate impedance and starts to unveil some of the substrate's influence, which is visible through the branching of the different curves for each contact distance. Then, above a certain frequency, depending on the substrate characteristics, the n^+ impurity concentrations, the size of the n^+ region, and on the bias condition, the Z_c impedance becomes so small that the results obtained for the structures similar to the one in Figure 8.9b became flat, like the isolation results for the structures that only use p^+ regions (Figure 8.9a), although a minor additional isolation is visible in Figure 8.15, when compared with the one presented in Figure 8.14.

8.2.4 Modeling Isolation Structures

Isolation technique effectiveness depends on many factors. For example, the dimensions of the isolation structures and the electrical properties of the materials have great influence on the obtained isolation level. Usually, CAD/CAE tools are oriented to electric circuit simulation where components are described by electric models provided by the technology manufacturers. These models are scalable with component dimensions and valid up to a

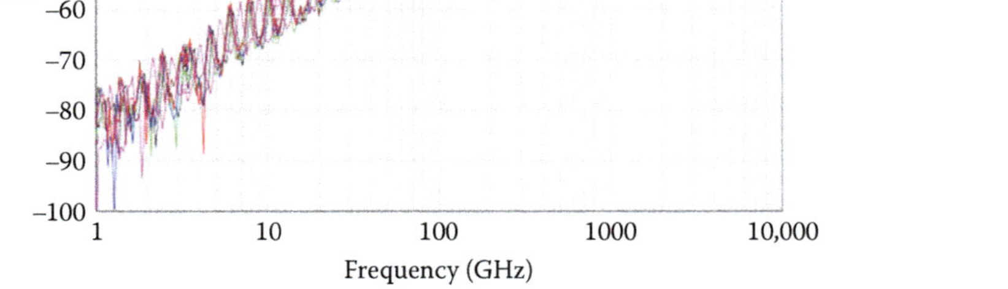

FIGURE 8.15
Experimental results of the test structure shown in Figure 8.9b for different contact distances, namely, 20, 40, 60, 80, 120, and 160 μm.

certain frequency. For radio frequency applications, the models present the influence of the substrate (or well), known as the component body, and usually are only valid when considering a predefined guard-ring around it. This is because, at high frequencies, the substrate influence is no longer negligible and depends on frequency and guard-ring-to-component distance. After designing the layout (post-layout), it is possible to perform parasitic components extraction. The extractor software tool can be used and a final netlist will include, besides the original components in the active-area, the parasitic capacitors, resistances, and inductors essentially related to the regions outside the active-areas. Although these tools are highly developed and capable of extracting substrate parasitics influence, they cannot model p-n junctions or semiconductor behavior accurately. Furthermore, the parasitic current flow between devices and substrate is a three-dimensional (3D) phenomenon, difficult to predict and model. This is what happend in many of the isolation structures and is the reason why design engineers try to model the isolation structures they use, so that they can predict their behavior for different dimensions than the ones they experimentally characterize.

Next, for some of the structures presented in Section 8.2.1, an electric model will be suggested. These models also consider the high frequency behavior of the isolation structures that arise because of the dielectric nature of silicon and p-n junctions' capacitive behavior.

In Figure 8.16a and b, it is possible to observe equivalent models for both isolation structures shown in Figure 8.2. The models are similar except for the GR region. The contacts p^+ to p-well and n^+ to n-well are modeled by R_{pc} and R_{nc} resistances, respectively. A parallel RC circuit models the longer p-sub and p-well paths, since the resistance value is high and the capacitive effect cannot be neglected, as stated in Section 8.2.3.3. For shorter paths, only the resistive part is considered. The n-type GR region interface with neighboring p-type regions is modeled by capacitances. The resistive part is now neglected because the junctions are reversed biased and only leakage current exists. It was also assumed that the n-type GR depth is small, thus the resistive effect is negligible. For low frequencies, the superior isolation of the n-type GR is evident, but with frequency increase both structures became equivalent.

Figure 8.3 p-sub GR can be modeled by the circuit shown in Figure 8.17. Since the GR has no connection to a metal contact, the model is composed of

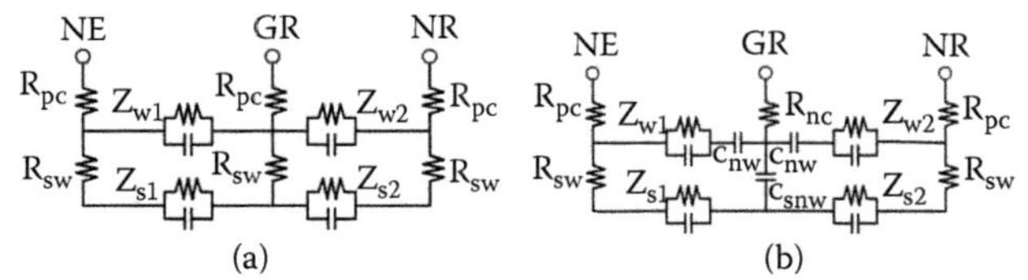

FIGURE 8.16
Models for the (a) p^+ and (b) n^+ guard-rings of Figure 8.2.

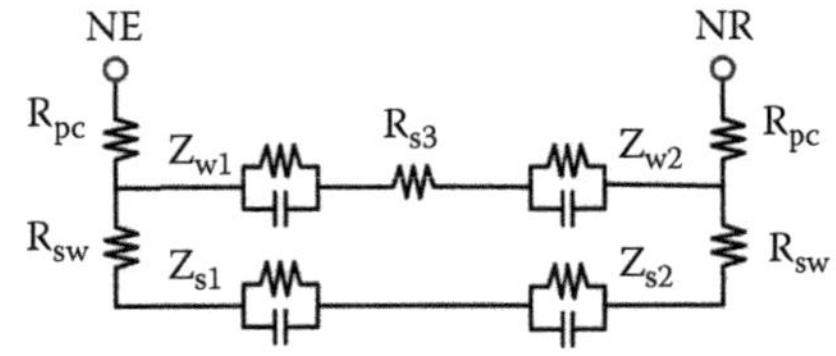

FIGURE 8.17
Model for substrate guard-ring for Figure 8.3a.

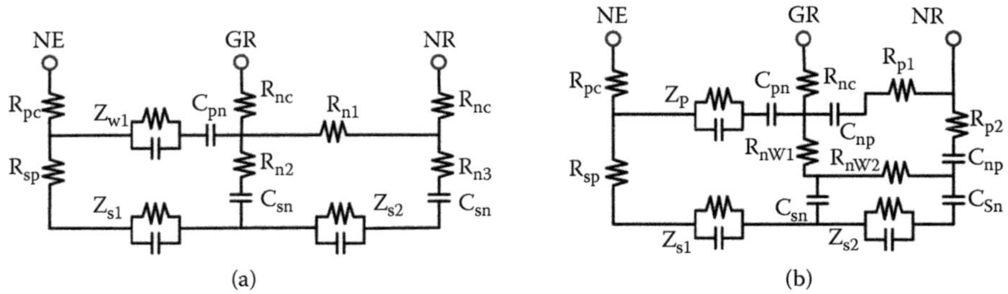

FIGURE 8.18
Models for the (a) n-well and (b) deep n-well isolation structures presented in Figure 8.4.

a moderate resistance R_{s3} that models the moderate resistivity path of p-sub. Of course, the effectiveness of p-sub GR is improved by increasing the GR thickness or by using high resistivity substrates.

Figure 8.18a and b presents two possible circuit models for n-well and deep n-well guard-rings, respectively. The n-well model presents the C_{pn} and C_{sn} capacitors for the n-well surface junction with p-well and p-sub regions, respectively. The deep n-well creates another p-n surface junction for both currents paths, modeled by the additional capacitors C_{np}.

8.3 Isolation Techniques and Structures in Low-Cost CMOS Nodes

This section describes a methodology that can be used to design a pseudo deep n-well in low-cost standard CMOS nodes, which makes available an isolated lightly doped p-type region inside the single n-well. The methodology can also be used to remove field implant at regions where the naked p-type substrate is needed. After describing the design methodology of the pseudo deep n-well, several isolation structures are presented. Some of the structures are equal to the ones already presented in previous sections and others are based on those structures. The isolation structures are then characterized through on-wafer measurements up to several GHz.

Usually, when a device is measured, it is necessary to discount the influence of the structures needed to connect the device ports to the test equipment. This operation is called de-embedding. For high frequency measurements, this influence can be very important, because it becomes the same order of magnitude as the device measured values. Many techniques have been presented to perform de-embedding for on-wafer measurements. The de-embedding difficulty depends on what desired the accuracy is for device measurement results. For the presented results, the de-embedding was performed for the input and output pads and respective interconnect lines structures. Also, the vector network analyzer (VNA) used in AC measurements was calibrated at the probes' tips by using standards built in an Impedance Standard Substrate (ISS).

8.3.1 The Design of Pseudo-Deep N-Wells

The implant of a lightly doped p-type volume inside an n-well region is an option in triple well CMOS processes, due to the availability of the deep n-well. In these cases, the isolated region will be formed by a p-well implant over the pre-processed deep n-well.

Single-well processes are based on a p-type substrate (p-sub), which is the bulk of the NMOS transistors, and an n-well to form the body of the complementary PMOS transistors. Figure 8.19a shows an example of part of the mask arrangement design of a complementary pair. Usually, a shallow p-type region, denoted as field implant or P-block region, is formed in all regions where the n-well is not implanted. Depending on CMOS layout design rules, this p-type region is usually defined by an automatic mask generation procedure, which draws a field implant layout layer (FIMP) coincident or complementary with the n-well layout layer (NW), depending on if this FIMP layer is a "dark" or "light" field mask. For the example shown, FIMP is dark field and field implantation will be processed outside n-well regions.

Taking into account that the field implant is shallower than the n-well, it is possible to consider its complete involvement by the n-well, as long as some *ad hoc* mask layout manipulation occurs. Figure 8.19b shows an extra field implant mask at the left, which has an opening window, all covered by the n-well mask. According to this mask information, the opening window in the field implant mask will allow acceptor ion penetration over the already formed n-well, establishing an isolated lightly doped p-type region. This p-type volume is completely immersed in the n-well volume, which can then be considered as a pseudo deep n-well.

Like in CMOS processes with an available deep n-well, this fully CMOS-compatible technique allows the isolation of analog, digital, or mixed mode circuits inside this pseudo-deep n-well.

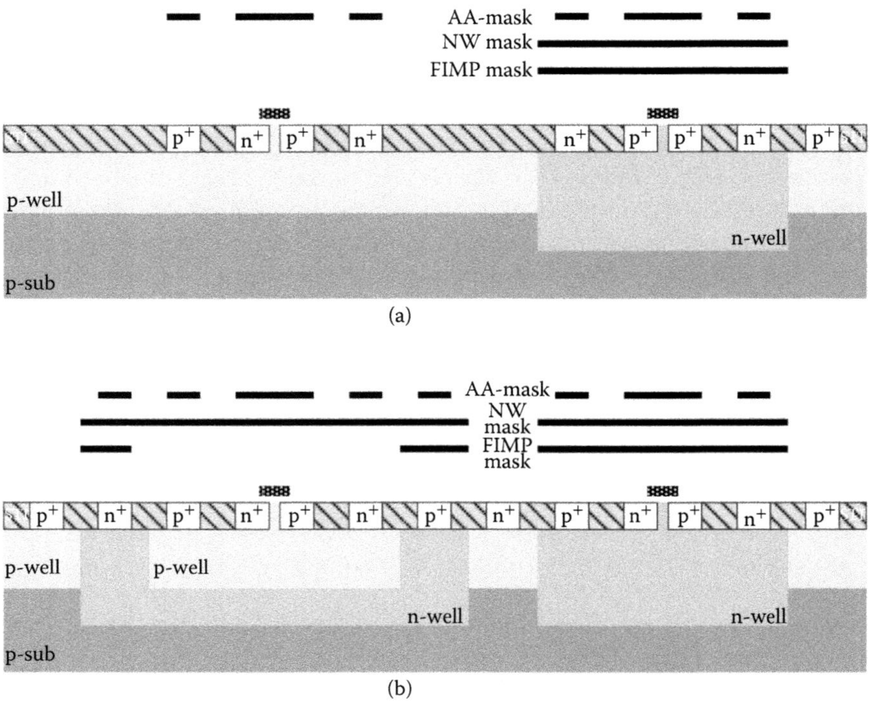

FIGURE 8.19
Cross sections of the structures under study, showing the proposed masks arrangements: (a) the conventional NMOS/PMOS complementary pair and (b) the "opening" in the FIMP mask layer (left) for the formation of a p-well inside the pre-processed n-well (pseudo deep n-well).

8.3.2 High Frequency Characterization of Noise Isolation Structures

This section shows the feasibility and practical performance of several isolation structures, some of which use the pseudo deep n-well, developed in a low-cost CMOS process where the deep n-well option is not available. As mentioned above, this is possible because the typical n-well depth is higher than that of the p-well and so, with mask manipulation, it is possible to grow a p-well inside a n-well creating a pseudo deep n-well.

The implemented structures are based on the one illustrated in Figure 8.9a, where the noise isolation structures, classified as A, use p^+ contacts on both NE and NR; whereas, in the structures of type B, the NR contact is replaced by a n^+ implant. The type A structures are illustrated in Figure 8.20a and the B structures are depicted in Figure 8.21a. For both types, 4 different structures were implemented. The structure 1 (1A and 1B) is used as a reference, i.e., is used to assess the isolation improvement of each of the other noise isolation structures.

The type A structures in the test are the following ones: structure REFA without isolation (reference structure), structure 1A with a p-type GR, 2A showing a pseudo deep n-well and, finally, in structure 3A, the isolation is improved by placing a p-type GR inside the p-well structure, as illustrated in Figure 8.20a. Figure 8.20b presents the measured $|S_{21}|$ (de-embedded) for each structure in Figure 8.20a. Analyzing the results, it can be concluded that the pseudo deep n-well gives greater isolation improvement compared to the classical p-type GR for up to 6 GHz. Additional isolation is obtained if an extra p-type GR is placed around NR inside the p-well.

Figure 8.21a presents isolations structures similar to the ones of Figure 8.20a but with the p^+ of the NR replaced by a n^+ diffusion/implant. As previously explained, this creates a low frequency additional isolation due to the reversed biased p-n junction n+/p-well. As can be seen, 4 structures were implemented: a reference structure (REFB), an isolation structure with only a p-type GR (1B), a pseudo deep n-well isolation structure without (2B), and an isolation structure with a p-type GR (3B). These isolations structures were developed using the same technique referred above to form the pseudo deep n-well. In Figure 8.21b, the measured (de-embedded) $|S_{21}|$ of the isolation structures is shown. In this case the effectiveness of the pseudo deep n-well isolation versus the p-type GR is less evident because the extra isolation of the n+/p-well junction is dominant, at least until 3 GHz. Above this frequency, the results tend toward those of the structures in Figure 8.20a, since the capacitive effect of the reversed biased p-n junction n+/p-well becomes negligible.

8.3.3 P-Sub Guard Ring

As explained in Section 8.2.1.2, the p-sub guard-ring is a simple way to improve isolation of a certain region. In this section, some measurements are

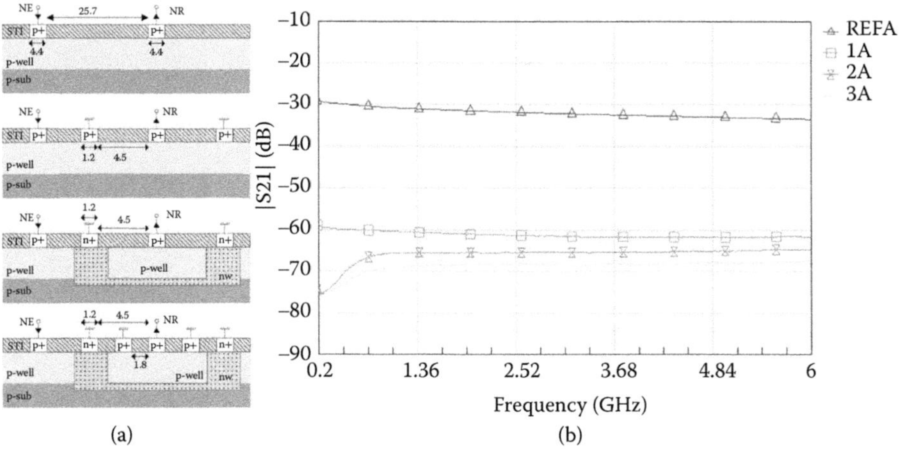

FIGURE 8.20
Pseudo deep n-well with p^+ NE and NR: (a) reference and isolation structures and (b) experimental results. The distances are all in microns (μm).

shown for structures where the p-type GR is complemented with regions where the p-well is blocked, leading to p-sub GRs. The structures developed and measured are presented in Figure 8.22 with different p-sub guard-ring placements and designs in relation to the classical p-type GR.

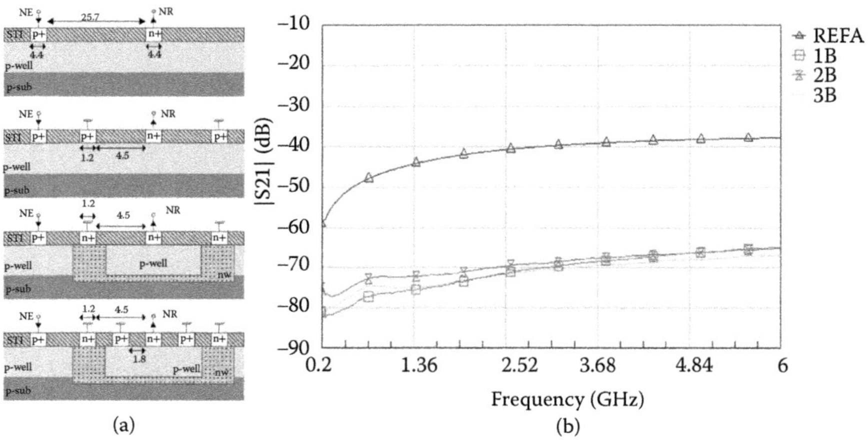

FIGURE 8.21
Pseudo deep n-well with n⁺ receiver contact: (a) isolation structures and (b) experimental results. The distances are all in microns (μm).

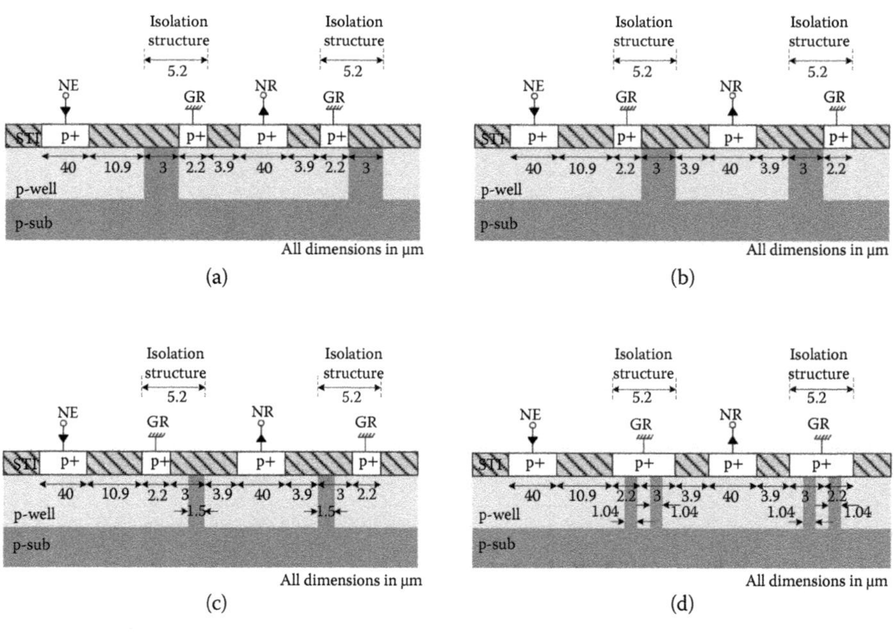

FIGURE 8.22
Details of the p-sub guard-ring isolation structures named (a) p-sub1, (b) p-sub2, (c) p-sub3 and, (d) p-sub4. The distances are given in microns (μm).

The experimental results are exhibited in Figure 8.23, where it is possible to conclude that a single p-sub guard-ring is more effective if placed inside the p-type GR. The double p-sub guard-ring has even better isolation performance, if placed exactly below the p+ contacts.

8.3.4 Effect of the Guard-Ring Width

Among the guard-ring geometrical dimensions depicted in Figure 8.24, its width (GRw) is known, from literature, to affect the substrate noise isolation provided by the guard-ring structure. In an early study [15], the isolation obtained in the function for the frequency of a typical p+ guard-ring is presented, showing the improvement of the noise isolation as the guard-ring gets wider. More recently, the effect of the guard-ring width on the isolation between different circuit parts has again received some attention [14,20,25,26],

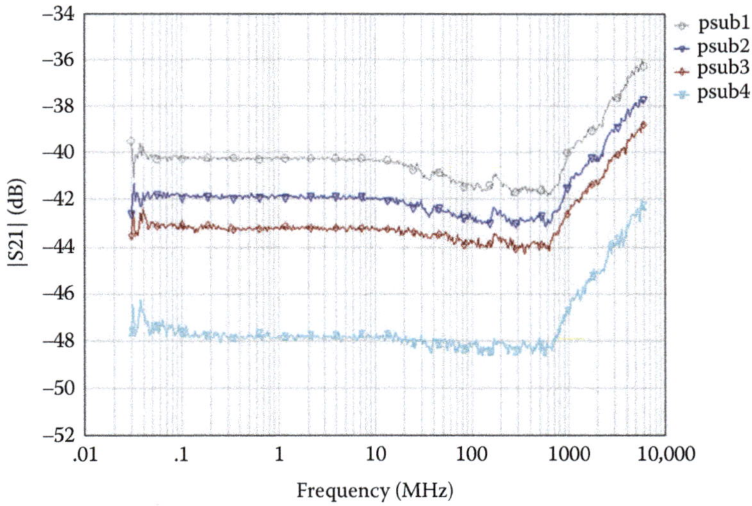

FIGURE 8.23
Experimental results for the p-sub guard-ring isolation structures.

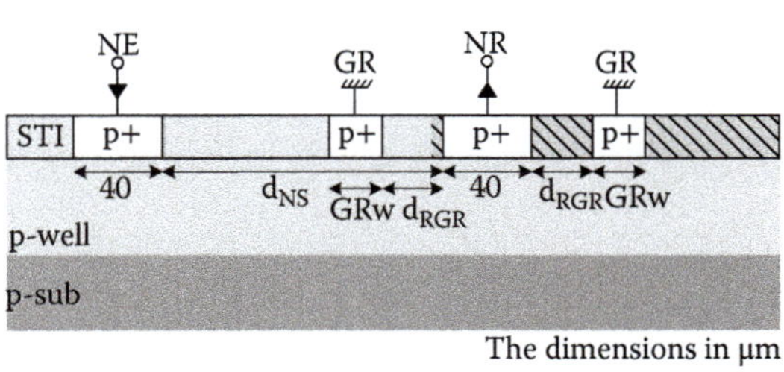

FIGURE 8.24
Geometrical dimensions of the guard-ring test structures.

where the reported results confirm the improvement of the isolation provided by the p^+ guard-ring by increasing GRw.

However, it is important to verify how the isolation evolves directly as a function of GRw, to provide the designer with the information that enables him to foresee the area occupancy of the guard-ring for a desired isolation level. Moreover, this will also let the designer know if increasing the guard-ring width will always bring improvement in isolation and at which rate. Except for [25], this information is not clear from the previous reported results. Therefore, to confirm this behavior, six test structures, similar to the one presented in Figure 8.24, but with different GRw distances, were fabricated and measured. Dimensions d_{RGR} and d_{NS} are fixed with 3.9 μm and 20 μm, respectively. Guard-ring widths are varied from the minimum value of 0.7 μm (allowed by the layout rules) up to 15.2 μm.

The experimental results are illustrated in Figures 8.25 and 8.26. In Figure 8.25, the experimental de-embedded $|S_{21}|$ in the function for the frequency of several guard-ring widths is shown, while Figure 8.26 illustrates the trend of the isolation with the guard-ring width. From both figures, it is possible to see that, qualitatively, the results agree with published ones, confirming the improvement of the isolation provided by the guard-ring as the GRw dimension increases. Besides that, and from the results presented in Figure 8.26, it can be concluded that the isolation has a logarithmic behavior with GRw, indicating that the isolation improvement is more pronounced when narrow guard-rings are extended. When GR widths are already large, the isolation improvement is lower. However, the use of wider guard-rings leads to a large silicon footprint.

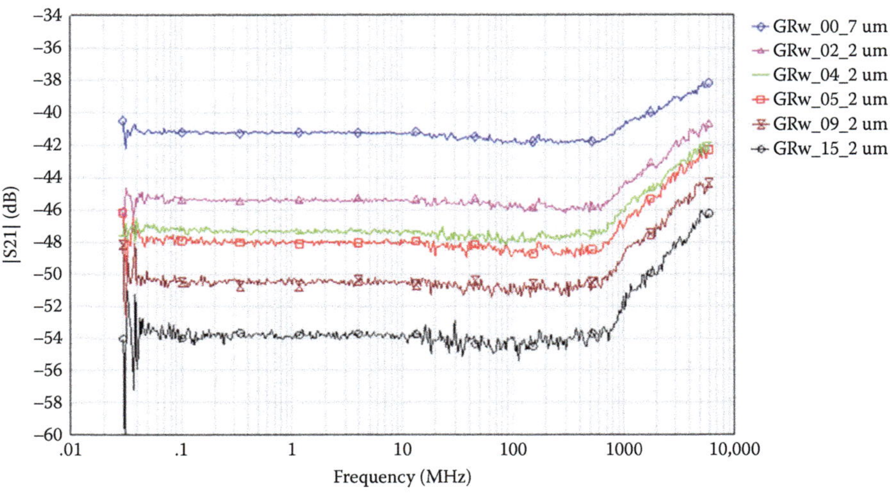

FIGURE 8.25
Experimental results of the substrate noise isolation provided by a typical p^+ guard-ring for guard-ring widths ranging from 0.7 μm to 15.2 μm.

The isolation degradation above 600 MHz is illustrated in Figure 8.25 is related to the parasitic inductance of the guard-ring connection to GND. At high frequencies, the guard-ring is no longer a good AC ground point and, because of that, less noise is sunk to ground through it.

8.3.5 Effect of the Guard-Ring Distance to the Noise Receiver

Another geometrical parameter that influences the isolation efficiency of the guard-ring is the distance from the noise receiver to the guard-ring, represented by the design distance d_{RGR} in Figure 8.24. Its effect on the noise isolation provided by the guard-ring was reported in [4,14,15,27,28], showing the improvement of the noise isolation as the guard-ring comes closer to the noise receiver.

Despite its importance, a detailed analysis of the isolation efficiency trend as a direct function of d_{RGR} has not been found in literature. This analysis provides further insight into the effects of d_{RGR} in the efficiency of the guard-ring; specifically, if reducing the guard-ring d_{RGR} always brings isolation improvement and, also important, at which rate. This provides the designer with useful information that enables him to foresee the area occupancy of the guard-ring for a desired isolation level and how to optimise the guard-ring implementation. Therefore, to confirm this behavior, seven test structures like the one presented in Figure 8.24, but with different d_{RGR} distances, were fabricated and measured. Dimensions GRw and d_{NS} are fixed with 2.2 μm and 20 μm, respectively. The distance d_{RGR} was varied from 0.6 μm to 13.9 μm.

Looking at Figure 8.27, it is possible to see that, qualitatively, the results agree with published ones, confirming the improvement of the isolation provided by the guard-ring as the design distance d_{RGR} decreases. The isolation

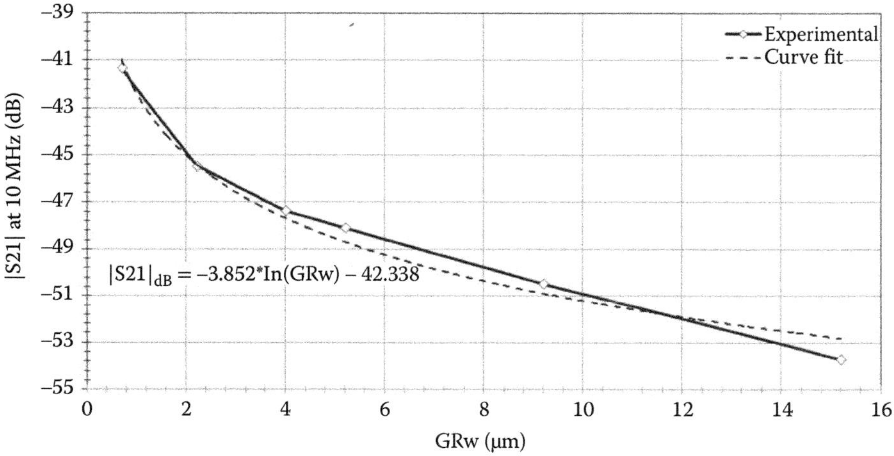

FIGURE 8.26
Measured de-embedded $|S_{21}|$ at 10 MHz versus the guard-ring width GRw.

degradation above 600 MHz, clearly seen in Figure 8.27, is again related with the parasitic inductance of the guard-ring connection to ground. The isolation trend with d_{RGR} can be analyzed in more detail using the results presented in Figure 8.28, where the $|S_{21}|$ at 10 MHz is shown as a function of the noise receiver to the guard-ring distance. As can be seen, the behavior follows a logarithmic trend. The rate of the isolation improvement is increased when the boundary of the guard-ring is very close to the NR node.

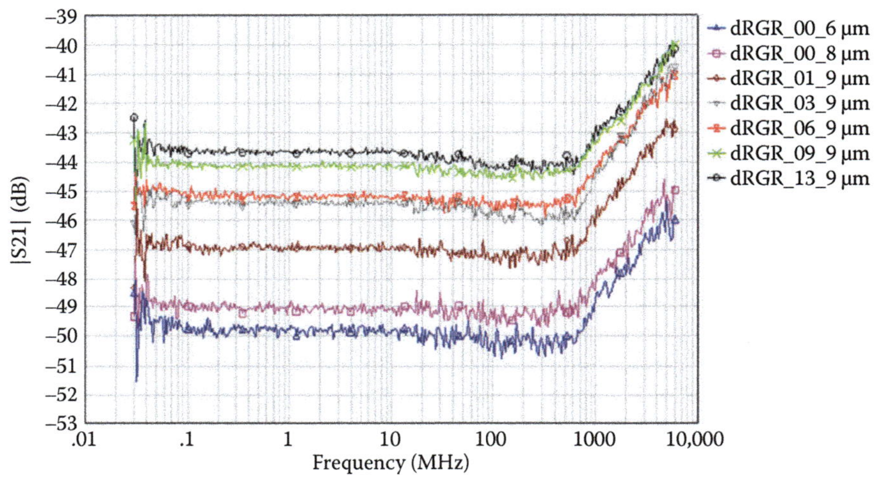

FIGURE 8.27
Experimental results of the substrate noise isolation provided by a typical p⁺ guard-ring for d_{RGR} ranging from 0.6 µm to 13.9 µm.

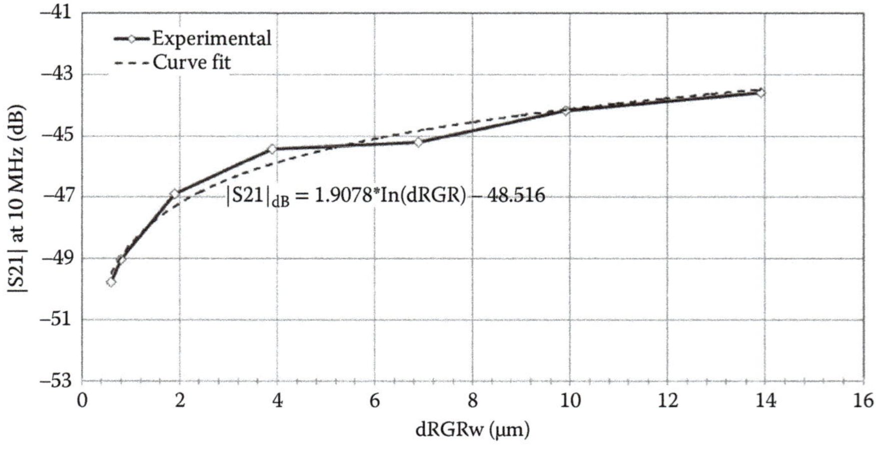

FIGURE 8.28
Measured de-embedded $|S_{21}|$ at 10 MHz versus d_{RGR}.

As a final remark, it can be concluded that by simultaneously maximizing guard-ring width and minimizing the distance to the circuit or device under protection, further isolation can be achieved for an isolation structure.

References

1. S. Bronckers, G. Vandersteen, G. van der Plas and Y. Rolain, On the P+ guard ring sizing strategy to shield against substrate noise, in *2007 IEEE Radio Frequency Integrated Circuits (RFIC) Symposium*, Honolulu, HI, 2007.
2. W. C. Pflanzl and E. Seebacher, Investigation of Substrate Noise Coupling and Isolation Characteristics for a 0.35UM HV CMOS Technology, in *2007 14th International Conference on Mixed Design of Integrated Circuits and Systems*, Ciechocinek, 2007.
3. A. Helmy and M. Ismail, The CHIP—A Design guide for reducing substrate noise coupling in RF applications, *IEEE Circuits and Devices Magazine*, vol. 22, pp. 7–21, September-October 2006.
4. Tsun-Lai Hsu, Yu-Chia Chen, Hua-Chou Tseng, V. Liang and J. S. Jan, P-sub guard ring design and modeling for the purpose of substrate noise isolation in the SOC era, *IEEE Electron Device Letters*, vol. 26, pp. 693–695, September 2005.
5. E. P. Li, L. Zhang and X. P. Yu, P-minus substrate guard ring modeling for the purpose of noise isolation in CMOS substrates, in *2015 IEEE Electrical Design of Advanced Packaging and Systems Symposium (EDAPS)*, Seoul, 2015.
6. M. Molina, X. Aragones and J. L. Gonzalez, Experimental analysis of substrate isolation techniques for RF-SOC integration, in *2009 IEEE International SOC Conference (SOCC)*, Belfast, 2009.
7. H. Lin, J. Kuo, R. Sobot and M. Syrzycki, Investigation of Substrate Noise Isolation Solutions in Deep Submicron (DSM) CMOS Technology, in *2007 Canadian Conference on Electrical and Computer Engineering*, Vancouver, BC, 2007.
8. H. M. Chao, W. S. Wuen and K. A. Wen, An active guarding circuit design for wideband substrate noise suppression, *IEEE Transactions on Microwave Theory and Techniques*, vol. 56, pp. 2609–2619, November 2008.
9. M. A. Karami, N. Masoumi and B. Forouzandeh, New high frequency model for substrate crosstalk noise injected by SOI devices, in *Proceedings of the International Conference Mixed Design of Integrated Circuits and System, 2006. MIXDES 2006*, Gdynia, 2006.
10. S. Uemura, Y. Hiraoka, T. Kai and S. Dosho, Isolation techniques against substrate noise coupling utilizing through silicon via (TSV) process for RF/mixed-signal SoCs, *IEEE Journal of Solid-State Circuits*, vol. 47, pp. 810–816, April 2012.
11. Li, N. et al., Substrate noise isolation improvement by helium-3 ion irradiation technique in a triple-well CMOS process, in *45th European Solid State Device Research Conference (ESSDERC)*, Graz, 2015.

12. K. A. Jenkins, Substrate coupling noise issues in silicon technology, in *Digest of Papers. 2004 Topical Meeting on Silicon Monolithic Integrated Circuits in RF Systems, 2004*, 2004.
13. H. Lan, T. W. Chen, C. O. Chui, P. Nikaeen, J. W. Kim and R. W. Dutton, Synthesized compact models and experimental verifications for substrate noise coupling in mixed-signal ICs, *IEEE Journal of Solid-State Circuits*, vol. 41, pp. 1817–1829, 2006.
14. K. Rajendran, J. B. Johnson, S. Furkay, M. Kumar and S. Fischer, Measurement and 3D simulations of substrate noise isolation and resistance for mixed signal applications, in *Digest of Papers. 2004 Topical Meeting on Silicon Monolithic Integrated Circuits in RF Systems*, 2004.
15. K. Joardar, A simple approach to modeling cross-talk in integrated circuits, *IEEE Journal of Solid-State Circuits*, vol. 29, pp. 1212–1219, October 1994.
16. N. K. Verghese and D. J. Allstot, Computer-aided design considerations for mixed-signal coupling in RF integrated circuits, *IEEE Journal of Solid-State Circuits*, vol. 33, pp. 314–323, March 1998.
17. B. R. Stanisic, N. K. Verghese, R. A. Rutenbar, L. R. Carley and D. J. Allstot, Addressing substrate coupling in mixed-mode ICs: simulation and power distribution synthesis, *IEEE Journal of Solid-State Circuits*, vol. 29, pp. 226–238, March 1994.
18. M. Pfost, H. M. Rein and T. Holzwarth, Modeling substrate effects in the design of high-speed Si-bipolar ICs, *IEEE Journal of Solid-State Circuits*, vol. 31, pp. 1493–1501, October 1996.
19. J. P. Raskin, A. Viviani, D. Flandre and J. P. Colinge, Substrate crosstalk reduction using SOI technology, *IEEE Transactions on Electron Devices*, vol. 44, pp. 2252–2261, December 1997.
20. S. Magierowski, K. Iniewskir and C. Siu, Substrate noise coupling effect characterization for RF CMOS LC VCOs, in *The 3rd International IEEE-NEWCAS Conference, 2005*, 2005.
21. W. Steiner, H. M. Rein and J. Berntgen, Substrate coupling in a high-gain 30-Gb/s SiGe amplifier-modeling, suppression, and measurement, *IEEE Journal of Solid-State Circuits*, vol. 40, pp. 2035–2045, October 2005.
22. Wen-Kuan Yeh, Shuo-Mao Chen and Yean-Kuen Fang, Substrate noise-coupling characterization and efficient suppression in CMOS technology, *IEEE Transactions on Electron Devices*, vol. 51, pp. 817–819, May 2004.
23. M. Pfost, P. Brenner, T. Huttner and A. Romanyuk, An experimental study on substrate coupling in bipolar/BiCMOS technologies, *IEEE Journal of Solid-State Circuits*, vol. 39, pp. 1755–1763, October 2004.
24. D. Kosaka et al., Isolation strategy against substrate coupling in CMOS mixed-signal/RF circuits, in *2005 Symposium on VLSI Circuits*, 2005.
25. S. Ardalan and M. Sachdev, An overview of substrate noise reduction techniques, in *International Symposium on Signals, Circuits and Systems. Proceedings, SCS 2003*, 2003.
26. G. A. Rezvani and Jon Tao, Substrate isolation in 0.18 μm CMOS technology, in *Proceedings of the 2005 International Conference on Microelectronic Test Structures, 2005*, 2005.
27. X. Aragones, J. L. Gonzalez, A. Rubio, *Analysis and Solutions for Switching Noise Coupling in Mixed-Signal ICs*, Kluwer Academic Publishers, 1999.

28. D. K. Su, M. J. Loinaz, S. Masui and B. A. Wooley, Experimental results and modeling techniques for substrate noise in mixed-signal integrated circuits, *IEEE Journal of Solid-State Circuits,* vol. 28, pp. 420–430, April 1993.
29. J. Su, H. Hsu, S. Wong, C. Chang, T. Huang and J. Y. Sun, "Improving the RF performance of 0.18 μm CMOS with deep n-well implantation," 10 ed., vol. 22, *IEEE Electron Device Letters,* 2001, pp. 481–483.

9

*Wireless Communications System-on-Chip Substrate Noise Real Time Sensing**

Thomas Noulis, Stefanos Stefanou, Errikos Lourandakis, Panayotis Merakos, and Yiannis Moisiadis

CONTENTS

9.1 Introduction

There has been an increase in the development of mixed-signal system-on-chip (SoC) that have analog, RF, and digital blocks implemented on the same die for many applications, especially in wireless communications and wireless sensor node interfaces. However, integrating these blocks onto the same silicon causes important performance degradation, especially in applications requiring fast digital circuits and high performance analog/RF blocks. Fast switching, in digital blocks, generates noise, which is propagated through the common substrate to the analog/RF circuits (demonstrated in Figure 9.1). Coupling occurs between a noise transmitter, which is usually a fast switching digital block or blocks, and a noise receiver, which, in most cases, is a sensitive analog or RF block. Coupling takes place due to the capacitive/resistive nature of the substrate and the devices interface (Figure 9.1) [1,2].

* The data presented in this chapter are Reprinted from Noulis, P. Merakos, E. Lourandakis, S. Stefanou, and Y. Moisiadis, Wide-band substrate crosstalk sensor for wireless SoC application, *Sensors and Actuators A: Physical*, vol. 239, 144–152, March 1, 2016, reproduced by permission from Elsevier and from T. Noulis, E. Lourandakis, S. Stefanou, Pe. Merakos., "CMOS 65 nm 'on chip' broadband real time substrate noise measurement," Issue: 21, pp. 1710–1711, October 2015, reproduced by permission of the Institution of Engineering & Technology.

The noise impact of the substrate crosstalk is both linear and nonlinear, since signals from the digital blocks are added to the analog signals and are, also, introduced by mixing in the desired carrier signals spectrum, which imposes detrimental effects on the performance of mixed-signal circuits, especially in systems where digital blocks are integrated in close proximity to sensitive analog and radio frequency (RF) circuits. Therefore, simulating and identifying the critical substrate crosstalk paths can lead to solutions for reducing/optimizing substrate crosstalk and improving mixed-signal circuit's performance [1,2].

On the application side, currently the general trend in CMOS technology is to shrink technology nodes below 28 nm; the extremely dense integration minimizes the product area/cost. The new design concept makes analog/RF functions using digital architectures and renders substrate crosstalk as the most crucial problem, especially in the case of technology multi-sourcing. This becomes even more severe at the SoC level, due to complete Radio Spectrum usage from 3 kHz to 30 GHz and the wireless application specifications where signal/power has to be below background noise. This will become even tighter going forward into millimeter wave IC design and from 4G/LTE mobile communication to the 5G mobile communication design era [1,2].

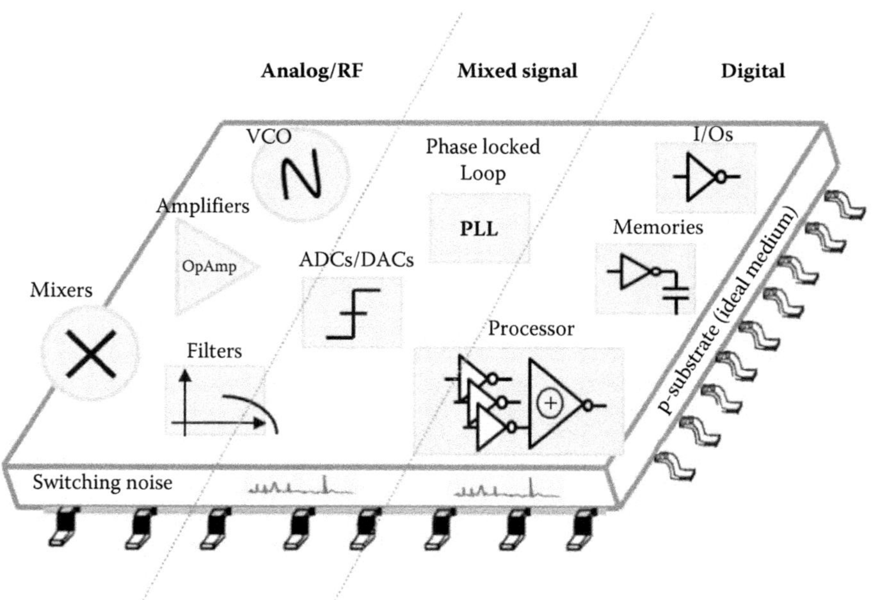

FIGURE 9.1
Representation of a full SoC substrate crosstalk with the related analog, RF, digital, and mixed signal blocks integrated onto a common silicon substrate. *Source:* T. Noulis et al., Wide-band substrate crosstalk sensor for wireless SoC application, *Sensors and Actuators A: Physical*, vol. 239, 144–152, March 1, 2016.

Traditionally, interactions between the analog and digital parts through the common silicon substrate were not part of the design flow for several reasons. Firstly, the design process of a digital block is totally different due to its need for large, complex systems consisting of billions of transistors. In addition, digital design is also highly automated. On the other hand, the analog design is a full custom process that involves blocks of smaller complexity occupying more silicon area. Several methodologies and modeling approaches [3–14] to simulate substrate crosstalk were so far suggested, but currently none is thoroughly validated due to signal measurement difficulties and are inadequate to address this phenomenon in its full extent. In particular, among the related methodologies, only Badaroglou et al. [13] and Nagata [14] addressed the substrate coupling problem at the full SoC level and provided a methodology that was addressed the problem in its full extent. Other more recent analyses and methodologies were also presented [15–19] so as to simulate substrate crosstalk both in terms of accuracy and physical phenomenon behavioral trends. On the Electronic Design Automation (EDA) industry status and the respective software tools capabilities, only two solutions are commercially available [19,20] but both have questionable accuracy [1,2].

While the need for a substrate coupling aware design flow or a related EDA tool are of extreme importance, even more challenging and crucial is to be able to validate whether the proposed substrate crosstalk noise simulation methodologies/flows are accurate, correlating simulations with direct measured substrate noise signals and to identify critical substrate coupling paths. The measurement of substrate and power supply noise enables the monitoring and the analysis of digital switching noise propagation, and it is greatly important for understanding substrate crosstalk in several processes, for validating the accuracy of respective EDA tools and simulation techniques and for identifying critical coupling paths across the chip and analyzing the efficiency of various noise suppression methodologies [1,2].

In this chapter, a design approach is presented for designing and implementing an extremely broadband noise sensor architecture in a 65 nm CMOS process. A direct and broadband real time sensing topology is provided and its operation is confirmed with silicon measurements on a CMOS 65 nm wireless communication pin-grid array (PGA) SoC, capturing substrate noise injected from a 120 kGate programmable Input–Output (IO) and core digital logic in a frequency range from few MHz to 4.5 GHz [1,2].

9.2 State of the Art and Related Work

In terms of substrate noise sensing, several requirements make the definition of such a design, both in schematic (architecture) and in mask design level (layout), as an extremely challenging task. In particular, the sensor should not affect the noise propagation or inject additional noise. Its operation, *BW*, should be high enough to reach 4 to 5 Ghz, to capture the high frequency components, and low enough to measure the low frequency components. In addition, the sensor should occupy as little active chip area as possible, so as to enable probing in multiple locations on the chip. Finally, it is also desirable to be able to multiplex the signals from multiple sensors (since the noise sensing is commonly done in various locations across the chip) to reduce the number of the required pins [1,2].

Several substrate noise sensing techniques were presented so far in [20–29]. A simple measurement technique involved the use of a voltage threshold modulation of a single MOSFET [19,20]. In this technique, the substrate noise sensor consists of a single NMOS transistor that senses the substrate noise through the body effect and capacitive coupling to its gate, drain, and source nodes. The specific method is an indirect measurement technique since it isn't the substrate noise that is measured, but the influence of the substrate voltage on the MOS current. In addition, it is extremely limited in terms of the related *BW*, since, with a bulk driven MOSFET operation, the signals that can be sensed could have a frequency below a few MHz [22–31] because it uses voltage comparators as substrate crosstalk noise sensors. A differential latch comparator was used for the substrate noise detector circuits, because a high temporal resolution can be expected due to the large gain in positive feedback. An unavoidable drawback of the specific topology is the limitation on probe points, mainly due to the large size of these detectors. In addition, this is also an indirect measurement technique, since the influence of the substrate voltage on a comparator state is detected and *BW* is limited to hundreds of MHz, making it not as suitable for wireless communications crosstalk sensing. A more practical topology was also proposed by Nagata et al. [14,24,25]. This topology consists of a source follower (SF) that senses noise voltage and a transistor's transconductance (Gm) and converts the SF's output voltage to current signal. Because of the small device count in this simple front-end structure, it is comparable to a standard flip-flop cell. While this technique enables direct power supply/ground noise to be probed at sufficient points within a digital circuit for mapping noise distributions, it is also limited by a related low operation *BW* that cannot reach the GHz region and it is not suitable for wireless communications SoCs. M. van Heijningen et al. [26] implemented direct substrate noise measurement circuitries with differential amplifiers to probe substrate and power supply noise with reference to an arbitrary voltage. However, the operation *BW* obtained was in the range of half a GHz. Two extra topologies were also proposed [27,28] based on digital operations, but both were indirect techniques and limited in *BW* for wireless communications crosstalk paths identification.

The main challenge in wireless SoCs, in terms of substrate noise sensing, are the really broadband gain behavior needed in order to amplify the sensed substrate signals and capture all the intermodulation's products in the wireless communication GHz region of interest and the direct substrate noise sensing feature needed to identify sensitive coupling paths. Both of these challenges are addressed with the noise sensor architecture presented in this chapter.

9.3 Substrate Noise Sensor

The noise sensor implemented in a 65 nm LP CMOS process by TSMC is provided in Figure 9.2. It is a differential amplifier with 1.2 V RF transistors M1 and M2, for the differential pair, and with one input connected to a dedicated ground and the other connected onto the substrate, so as to implement common mode rejection. The main objectives during the design have been a broad band-pass *BW* response from few MHz over 4 GHz, a satisfactory gain performance, and the ability to deliver output single ended signal to the 50 ohm load of a spectrum analyzer [1,2].

The differential pair M1, M2 is biased through the resistor dividers into the subthreshold region. The DC drain current for an n-channel MOSFET in the subthreshold region is given by [29]

$$I_D = \frac{W}{L} I_0 e^{\frac{V_{GS}-V_{th}}{m\,V_T}\left(1-e^{-\frac{V_{DS}}{V_T}}\right)} = \frac{W}{L} I_0 \bullet e^{\frac{V_{GS}-V_{th}}{m\,V_T}} \tag{9.1}$$

where *W/L* is the aspect ratio of the transistor, V_{GS} and V_{DS} are the gate-to-source and drain-to-source voltages, V_{th} is the transistor threshold voltage,

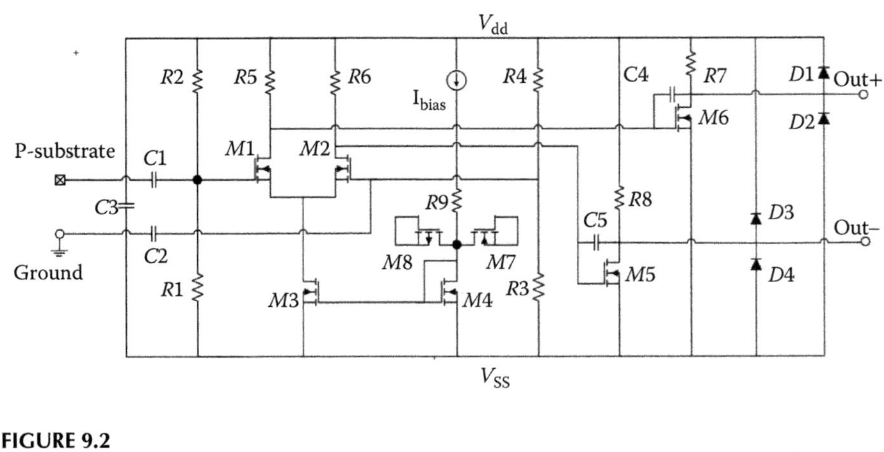

FIGURE 9.2
Substrate noise sensor schematic. *Source:* T. Noulis et al., Wide-band substrate crosstalk sensor for wireless SoC application, *Sensors and Actuators A: Physical*, vol. 239, 144–152, March 1, 2016.

and $V_T = k_B T/q$ is the thermal voltage (k_B is the Boltzmann constant, T the absolute temperature, and q the elementary charge). The subthreshold slope parameter m is a technology-dependent constant, usually with a value of $1 < m < 2$, and I_0 is process parameter that also depends on temperature.

The respective transconductance and the drain–source resistance are given as

$$g_m = \frac{\theta I_D}{\theta v_{GS}} = \frac{I_D}{mV_T} \tag{9.2}$$

$$r_D = \left[\frac{\theta I_D}{\theta v_{GS}}\right]^{-1} = \frac{mV_T}{\lambda_D\, I_D} \tag{9.3}$$

where λ_D is the *DIBL* effect coefficient. The DC gain provided from the differential gain of the proposed topology is provided as

$$A_{VDC} = \frac{g_{m2}}{\left(\frac{1}{R6}\right) + g_{ds2}} \tag{9.4}$$

The amplified (by the differential pair stage) substrate and the quite ground reference signals are driven to M5 and M6. These output stages are open drain topologies and were designed with thick gate 2.5 V devices, so as to achieve the 50 Ω driving capability lowering the output impedance and providing the necessary current to drive the 50 Ω instrument port. R7 and R8 act as pull up resistors for the open drain outputs. The respective gain of these stages cannot exceed 0 dB.

The back end of line metal capacitors C1 and C2 together with the resistor dividers R1, R2 and R3, R4 form high pass filters for the differential input. This high pass *CR* filter defines the lower cut off frequency of the band pass behavior. The resistance dividers also define the DC operating point of the differential pair. Resistors R5, R6 act as loads. Low value capacitors *C4, C5* are used to match the outputs *OUT+* and *OUT−* through the inductances of the SoC PGA package bond wires. M7, M8, and R6 form an ESD protection for the input bias current and D1, D2 and D3, D4 for the differential RF output signals. C3 Metal-insulator-Metal (MiM) decoupling capacitance is placed between the supply power routing rails for supply ripple minimization. A bias current of 750 μA and power supplies of $V_{dd} = 1.2\,\text{V}$ and $V_{SS} = 0\,\text{V}$ were used. A respective current mirror topology comprised by M3 and M4 provides the bias current for the differential pair.

In relation to the mask design level, the noise sensor layout was fully symmetrical and special attention was paid to isolating sensitive devices using biased guard rings. In particular, all of the RF MOSFETs were isolated using an inversely biased deep n-well. The differential pair and the output stages had a fully symmetrical layout. All the resistances were designed with separate guard rings. The full sensor footprint lies on the reversely biased deep n-well (*dnw*) and has an active area of 332 μm × 224 μm. The respective layout is given in Figure 9.3.

Special attention was paid to each signal and each pin in relation to the isolation of the sensor. The device types used and the respective bias settings are provided in Table 9.1.

In relation to the measurement setup, an off-chip *5320B* broadband transformer (balun), commercially available from Picosecond, was used (and simulated using an s-parameter model) for differential to single ended conversion. On chip routings parasitics, PGA package, and PCB board parasitics were extracted and simulated, all in *RLCk* mode, using *VeloceRaptor Electromagnetic (EM)* modeling engine [29] in order to capture all the effects of high frequency on performance and ensure silicon functionality. This way, complex electromagnetic effects, such as self and mutual inductances, frequency-dependent resistance, and capacitive couplings were captured [1,2].

Figure 9.4 provides post-layout simulation results (in nominal conditions and in 25°C temperature) of the gain versus the *BW*, including full RLCk parasitic extraction [30] and adds into the simulation test bench the bond wire package and the PCB board RLCk parasitics model (netlist). The substrate crosstalk sensor simulated gain is from a 6.6 dB in an operation band pass *BW* from 1 Mz to 4.6 GHz.

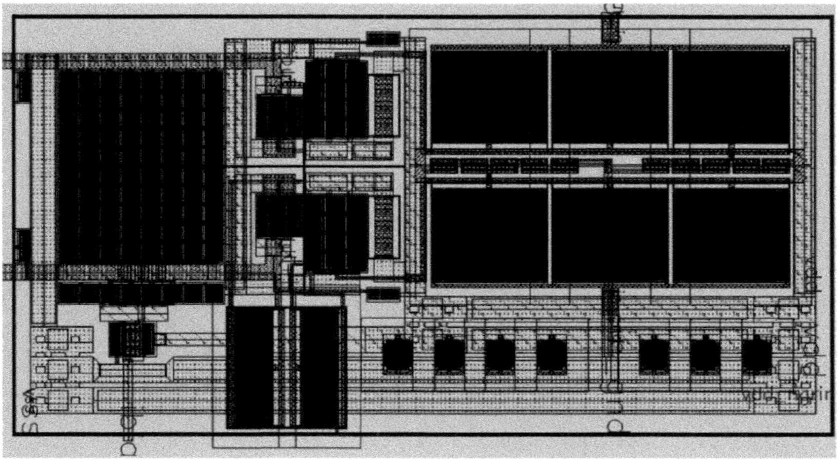

FIGURE 9.3
CMOS 65 nm noise sensor physical design. *Source:* T. Noulis et al., Wide-band substrate crosstalk sensor for wireless SoC application, *Sensors and Actuators A: Physical*, vol. 239, 144–152, March 1, 2016.

TABLE 9.1

Noise Sensor Bias Settings, Pins, and Devices Types

PIN List		Description
V_{dd}	Positive power supply 1.2V	"clean" V_{dd}—no connection to nwell
V_{SS}	Negative power supply 0V	"clean" V_{SS}—all the NMOS devices into deep nwells—each pwell biased to its source dc level (no body effect and no connection to substrate)
vdd_gr	Positive bias of 1.2V for the deep nwell (DNW) ring surrounding the topology	Separate deep nwell noise isolating ring biasing routed separately to a 1.2 V (no visibility on the s hematic)
		DC biasing current 0.75 mA
Ibias	External current source	
Ground	Quite ground reference	Separate ground pin available
p-substrate	Substrate sensing input signal	Substrate noise RF signal captured using a substrate tap
OUT+/-	Positive and negative voltage outputs	Two outputs are delivered through an external balun to the 50 ohm load of a spectrum analyzer
Device Type List		**Description**
MOSFET	1.2 V NMOS RF / M1–M4, M7–M8	Used in the dif. Pair / nwell ring availability—biased to V_{dd} for pn junction inverse biasing and isolation from substrate
	2.5 V IO NMOS RF / M5–M6	IO 2.5 V std RF NMOS in Deep n-well (DNW) are used at the open drain output stage for electro-migration (EM) reasons
Resistors	RF p+ polysilicon resistors without silicide / R1–R4	p+ Poly resistors in n-well without silicide are used for low noise performance/ substrate noise coupling minimization
	RF p+ polysilicon resistors / R5–R8	p+ Poly resistor in n-well with silicide—high current density capability
Capacitors	RF Mim-Capacitor / C3	Placed onto the parallel V_{dd} and V_{ss} routings for area minimization and decoupling
	Metal back end of line (BEOL) capacitors / C1–C2, C4–C5	M2 to M5 metal caps for high capacitance density

Source: T. Noulis et al., Wide-band substrate crosstalk sensor for wireless SoC application, *Sensors and Actuators A: Physical*, vol. 239, 144–152, March 1, 2016 and T. Noulis et al., "CMOS 65 nm 'on chip' broadband real time substrate noise measurement," *Electronics Letters*, vol, 51, no. 21, 1710–1711, October 2015.

The respective through s-parameter simulation and stability factor *Kf* plot is provided in Figure 9.5. It shows that the circuit is unconditionally stable, since the *Kf* value is above 1 for the whole range of the operating *BW*.

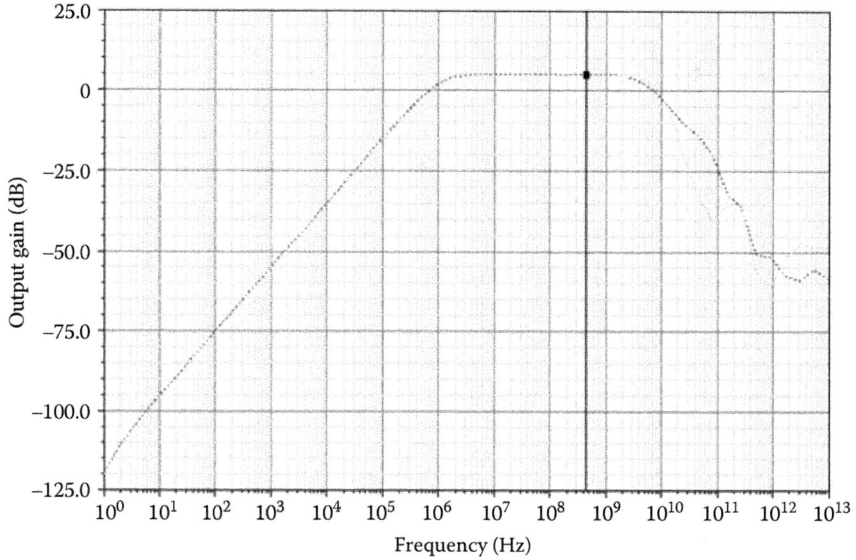

FIGURE 9.4
Gain versus *BW* (AC) simulation. *Source:* T. Noulis et al., "CMOS 65 nm 'on chip' broadband real time substrate noise measurement," *Electronics Letters*, vol, 51, no. 21, 1710–1711, October 2015.

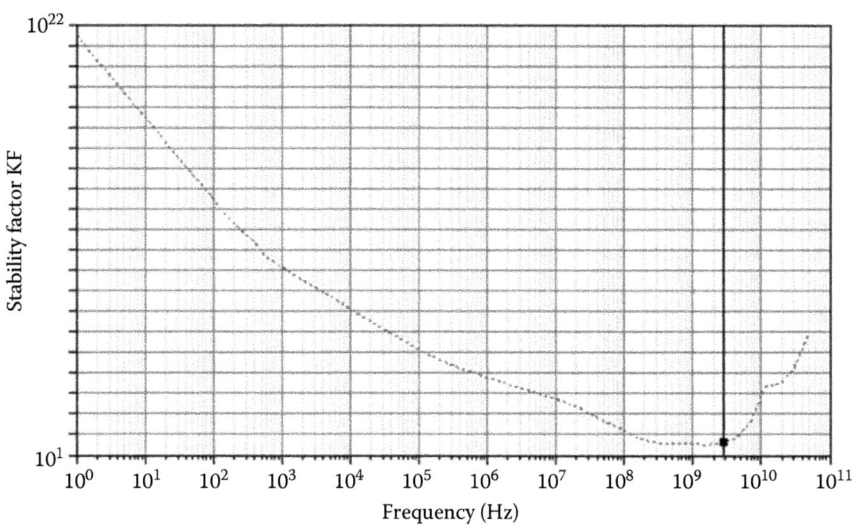

FIGURE 9.5
Stability Factor versus Frequency (s-parameter) simulation. *Source:* T. Noulis et al., Wide-band substrate crosstalk sensor for wireless SoC application, *Sensors and Actuators A: Physical*, vol. 239, 144–152, March 1, 2016.

9.4 Noise Aggressor Architecture

The digital noise transmitter circuit (shown in Figure 9.6a) consists of 768 Loop Shift Registers (LSRs) arranged in 32 rows and 24 columns and controlled and operate through a Serial Peripheral Interface (SPI) bus. A digital circuit with high and externally adjustable switching activity is required. The LSR circuit has been chosen for the test chip because it is a simple circuit used often in a lot of applications and its switching activity is directly controllable through the data written to the shift register. For a specific clock frequency, we can program its data in order to have a degree of selectivity on the produced switching activity. Each LSR circuit occupies an area of 28 um × 28 um in TSMC's 65 nm Low Power (LP) technology. The array structure of LSR circuits occupy a total area of 672 um × 896 um. Enabling and/or disabling the operation of each LSR provide us with the choice of controlling spatial and temporal correlations of the produced switching activity in relation to the analog noise victim circuit. The enabling/disabling of each LSR's operation (shifting their contents in loop mode with the externally adjustable CLK clock speed [see Section 9.1]) can be made either independently, row-wise or column-wise. Row and column selectivity operating control for the array of LSRs can be performed for one or multiple rows/columns of the array. The contents of the shift registers can be programmed randomly and independently. Similarly, they can be read back for testing purposes. If a write operation reaches the end of a file, the new data written will overwrite the old ones in a cyclic manner. While, if a read operation reaches the end of a file, the data will start read all over again from the beginning of the file. We chose to implement this kind of control on the LSRs by using SPI bus because of its simplicity and the restricted number of I/O pads we have available. An LSR of 32-bit length, capable of rotating its contents at 1 GHz frequency and with the functionality described above, occupies about 28 m × 28 um in the 65 nm CMOS technology. The noise transmitter consists of 32 rows and 24 columns of LSR circuits occupying an area of 700 um × 896 um.

The SPI bus is a synchronous serial data link standard named by Motorola (Figure 9.6b) that operates in full duplex mode. Devices communicate in master/slave mode where the master device initiates the data frame. Multiple slave devices are allowed with individual slave select (chip select) lines. The test chip is a slave and thus an external master controller is necessary for the communication.

The SPI bus specifies four logic signals that will be I/O pads in our test chip: Serial clock (SCLK), where the output comes from the master; Master Output, Slave Input (MOSI), where the output comes from the slave; Master Input, Slave Output (MISO), where the output comes from the slave; and Slave Select (SS), which is active low and has output come from the master.

The SPI bus can operate with a single master device and with one or more slave devices. In our case, it is expected that our chip will be the single slave. Selecting a different programming protocol with pre-defined number of bytes for every command, for example, the SS signal is not necessary for a single slave device. In this case, the SS pin can be fixed to logic low internally in the chip, and omitted from the I/O pads. In the case of many slaves, devices without tri-state outputs can't share SPI bus segments with other devices. Only one such slave can talk to the master and only its chip select could be activated. Not all transmissions require all four of these operations

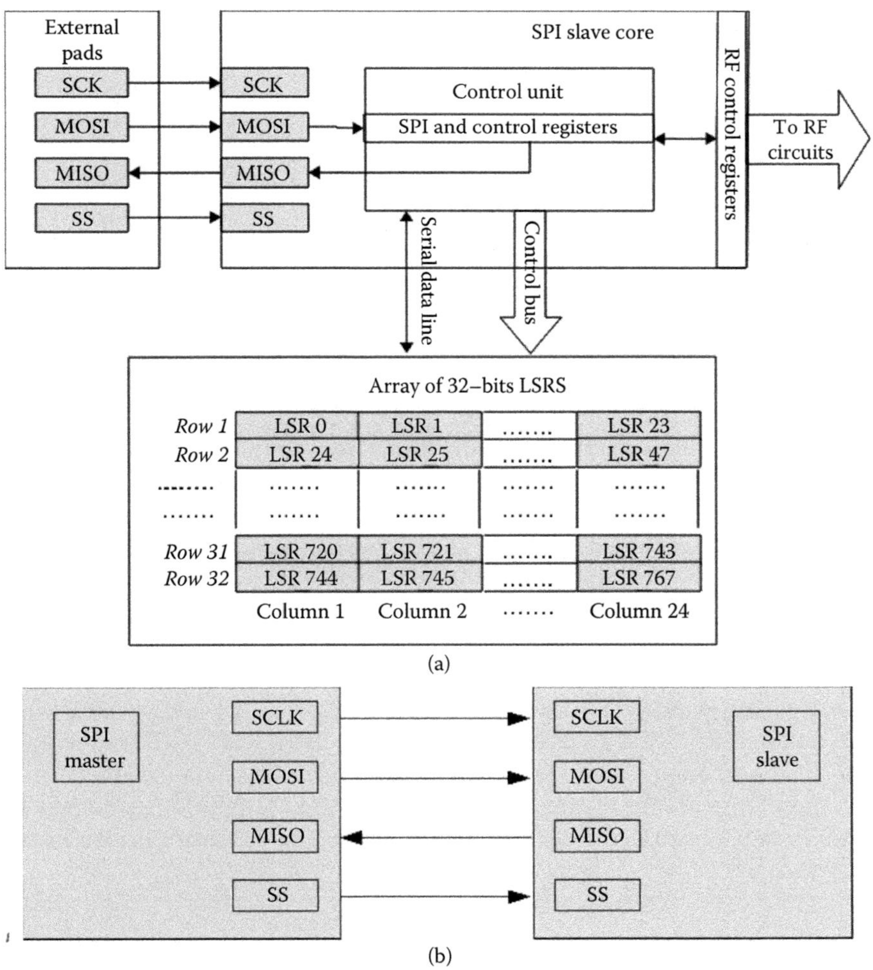

FIGURE 9.6
(a) Digital Noise Transmitter. *Source:* T. Noulis et al., Wide-band substrate crosstalk sensor for wireless SoC application, *Sensors and Actuators A: Physical*, vol. 239, 144–152, March 1, 2016. (b) SPI Bus with single master and single slave.

to be meaningful, but they do happen. Transmissions normally involve two shift registers of a given word size, such as eight bits, one in the master and one in the slave; they are connected in a ring. Data is usually shifted out with the most significant bit first, while shifting a new least significant bit into the same register. After that register has been shifted out, the master and slave have exchanged register values. Then each device takes that value and does something with it, such as writing it to memory. If there is more data to exchange, the shift registers are loaded with new data and the process repeats. Transmissions may involve any number of clock cycles. When there is no more data to be transmitted, the master stops toggling its clock. Normally, it then deselects the slave. Transmissions often consist of 8-bit words and a master can initiate multiple such transmissions if it wishes/needs.

9.5 Substrate Noise Sensing on a Wireless Communication SoC

The silicon prototype photograph for the CMOS wireless communication (ZigBee application) SoC, fabricated in a 65 nm Low Power (LP) CMOS process commercially available by TSMC, is shown in Figure 9.7a. Two noise sensors were implemented for substrate noise sensing. Noise sensor a in Figure 9.7a has the substrate input tap 2 μm and noise sensor b in Figure 9.7a has it 10 μm away from the digital logic. Noise sensor c was also implemented as a standalone circuit just for performance characterization. This particular version had his input available externally and a high frequency signal generator was used for the performance characterization. Since a Zigbee transceiver was implemented, several other RF circuitries are available on the silicon, such as a Low Noise Amplifier (LNA), a Voltage Control Oscillator (VCO), a Power Amplifier (PA), and several other peripheral blocks (Figure 9.7a). The LNA was the most sensitive to substrate crosstalk noise because of its close proximity to the digital control logic.

Concerning the substrate noise sensor standalone performance, sensor c was used for standalone characterization purposes. The sensor provides a gain equal to 6.5 dB in an operation where *BW* has a low –3 dB frequency of 1.05 MHz and high –3 dB frequency of 4.47 GHz. The –1 dB gain compression point appears for input signal amplitude of 124 mV and the output *rms* noise voltage is equal to 129 μV. In Table 9.2, the noise sensor performance is summarized.

The proposed design appears advantageous and advanced compared to the available state-of-the art designs. A table containing all of the available substrate crosstalk sensing techniques, with respect to the design requirements presented in Section 9.2, is provided in Table 9.3. The proposed sensor is the only one that can operate in the high GHz region above 4 GHz and, at the same time, satisfy all of the rest of the design requirements, such as not to affect noise propagation, being able to capture switching activity

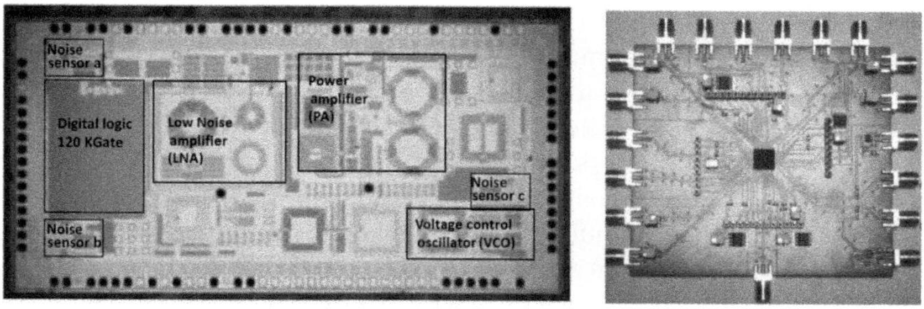

(a) (b)

FIGURE 9.7
(a) CMOS 65 nm wireless communication SoC Silicon footprint with noise sensors, digital logic noise transmitter, and other RF components highlighted, and (b) SoC PCB test board. *Source:* T. Noulis et al., Wide-band substrate crosstalk sensor for wireless SoC application, *Sensors and Actuators A: Physical*, vol. 239, 144–152, March 1, 2016.

TABLE 9.2
Noise Sensor Performance Characteristics

Noise Sensor Characterization Metrics	
Gain	6.5 dB
f-3 db low	1.05 MHz
f-3 b high	4.47 GHz
Vin_ampl@ –1dB compression point	124 mV

Source: T. Noulis et al., Wide-band substrate crosstalk sensor for wireless SoC application, *Sensors and Actuators A: Physical*, vol. 239, 144–152, March 1, 2016.

TABLE 9.3
Comparison of the Proposed Noise Sensor to the Available Substrate Noise Sensing Techniques

Design Specification	**Sensor Type**					
	[18,19]	**[20,21]**	**[22,23]**	**[24,25]**	**[26,27]**	**Proposed Sensor**
Do not affect noise propagation	√	×	√	√	×	√
GHz region upper frequency	×	×	×	×	×	√
MHz region low frequency	√	√	√	√	√	√
Compact and small size	√	×	√	√	√	√
Multiplexing capability	×	√	√	√	√	√

Source: T. Noulis et al., Wide-band substrate crosstalk sensor for wireless SoC application, *Sensors and Actuators A: Physical*, vol. 239, 144–152, March 1, 2016.

from the MHz region until the GHz region, being compact and optimized area-wise, and providing multiplexing capability for PIN number minimization. In addition, going to a lower technology node, and in a ball grid array package (BGA—Solder balls package) with much lower inductive parasitics compared to the PGA approach, the upper –3 db frequency can really be optimized to a much higher value.

The sensor's substrate noise capturing capabilities are depicted in Figure 9.8. For both noise sensors a and b, and in a frequency span from 100 MHz to 4 GHz, the noise spurs, as measured by the spectrum analyzer, are provided. They have the 120 k Gate digital logic with only the IO logic enabled and have both the IO and the Core logic fully enabled. All the crosstalk spurs captured by both the noise sensors at 2 μm and at 10 μm from the NT are given in Table 9.4. Only spurs with amplitude higher than –92 dBm are provided in Table 9.4. When the NT was fully disabled (IO and Core logic "OFF") both noise sensors did not capture any crosstalk spur with amplitude higher than –92 dBm.

The respective spectrum analyzer silicon measurements with disabled logic, IO logic "ON" and Core logic "OFF," and both IO and Core "ON" are provided in Figure 9.8 for both sensors.

TABLE 9.4

Substrate Noise Captured Signals with Different Enable and Disable Modes for the NT

	Substrate Noise Spurs—Amplitude (dBm)			
Frequency (GHz)	**Noise Sensor at 2 μm from NT**		**Noise Sensor at 10 μm from NT**	
	IO "ON" & Core "OFF"	**IO & Core "ON"**	**IO "ON"/Core "OFF"**	**IO & Core "ON"**
0.194	–44.83	–44.17	–	–59.07
0.201	–	–44.33	–64.97	–
0.396	–63.95	–	–73.86	–73.17
0.591	–46.57	–46.05	–73.12	–63.96
0.786	–68.76	–	–89.14	–88.99
0.981	–56.94	–54.68	–69.99	–69.95
1.176	–69.17	–	–82.09	–78.72
1.371	–65.57	–65.47	–	–75.84
1.379	–	–65.74	–75.52	–
1.574	–73.21	–69.49	–88.86	–82.51
1.964	–	–		–86.73
1.769	–73.33	–	–84.34	–
2.362	–	–	–88.95	–86.37
3.392	–	–74.31	–	–
3.594	–70.92	–70.72	–	–86.41

Source: T. Noulis et al., Wide-band substrate crosstalk sensor for wireless SoC application, *Sensors and Actuators A: Physical*, vol. 239, 144–152, March 1, 2016.

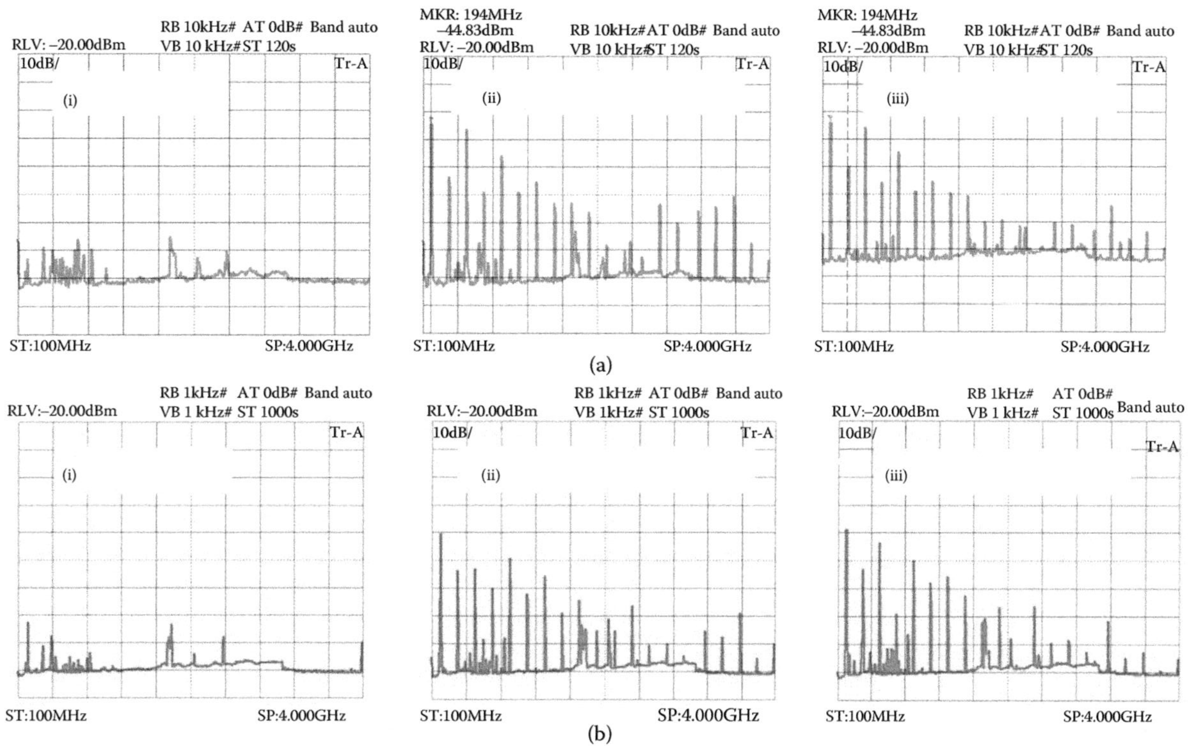

FIGURE 9.8

Substrate noise signal captured using the noise transmitter in different activity modes (a) Noise sensor spurs captured at 2 μm away from the NT with (i) the full *logic* OFF, (ii) IO logic ON and Core logic OFF, (iii) IO and Core logic ON. (b) Noise sensor spurs captured at 10 μm away from the NT with (i) the full logic OFF, (ii) IO logic ON and Core logic OFF, (iii) IO and Core logic ON. *Source:* **T. Noulis et al., "CMOS 65 nm 'on chip' broadband real time substrate noise measurement,"** *Electronics Letters*, **vol, 51, no. 21, 1710–1711, October 2015.**

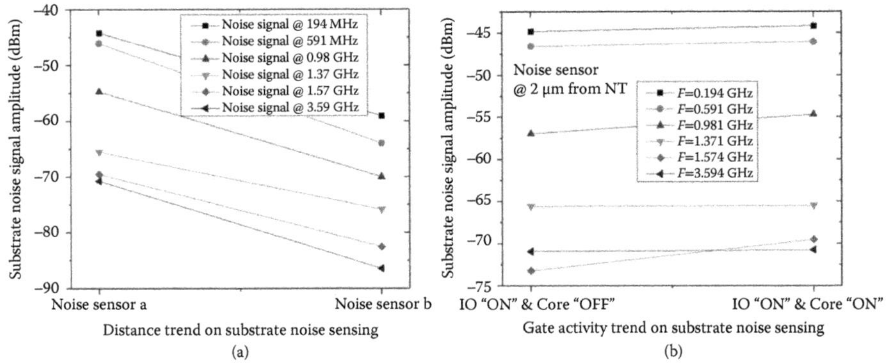

FIGURE 9.9
Substrate noise sensing trends (noise amplitude) on (a) digital noise transmitter distance and (b) on switching activity level. *Source:* T. Noulis et al., Wide-band substrate crosstalk sensor for wireless SoC application, *Sensors and Actuators A: Physical*, vol. 239, 144–152, March 1, 2016.

From Table 9.4 and Figure 9.8, it is extracted that noise sensor a, when in exactly the same frequency values as noise sensor b, captures noise signals of higher amplitude values, which makes sense since its substrate input tap is only 2 μm away from the digital logic instead of 10 μm for noise sensor b. In addition, when full logic is "ON" for both cases (2 μm and 10 μm away from the NT), the captured crosstalk spurs have higher amplitudes (on the same frequencies) compared to the lower switching activity case (only IO logic "ON,") which is also makes sense. This is presented for the sensor at 2 μm away from the NT in Figure 9.8b.

The related noise transmitter distance and gate switching activity level trends are depicted in Figure 9.9, where the agreement of the substrate noise signal trends sensed to the general substrate coupling phenomenon performance behavior [15,19], are confirmed.

Furthermore, the noise sensors can capture substrate noise signals with low amplitudes almost equal to –86.5 dBm, which corresponds to a 14.96 μV voltage peak amplitude, and can also capture really high amplitude signals in the range of 120 mV, achieving a quite satisfactory substrate signal sensing dynamic range, considering the high frequency and broadband operating *BW*.

9.6 Summary

Substrate crosstalk sensing was performed in a 65 nm CMOS ZigBee wireless communication SoC. All the related crosstalk measurement requirements were addressed by implementing a novel CMOS architecture for sensing substrate noise signals. The particular architecture can directly sense substrate crosstalk noise signals in a wide frequency region from the

MHz region until 4.5 GH. It is ideal for both wireless communication product level substrate coupling sensitive path detection and substrate crosstalk flow-noise integrity analysis validation. The silicon measurements obtained confirm the substrate coupling phenomenon trends in terms of the noise aggressor distance and the switching activity level of the noise aggressor (noise transmitter). A detailed comparison with the state of the art crosstalk sensing techniques was performed, showing the superiority of the proposed design, which appears to be optimum and suitable for use in wireless communications SoCs for respective sensitive crosstalk paths identification and, in general, for real time substrate crosstalk measurements.

While the proposed noise sensing architecture is currently the most advanced, however, considering the mobile communications SoC applications and the mmWave and 5G perspectives, the above approach has the disadvantage of the relatively large silicon area and it reaches only until the 4.5 GHz frequency region. Innovative circuit design is required to fabricate compact SoC level noise sensors that can operate in decades of GHz frequency bandwidth while occupying very little active silicon area. This can be achieved by designing a similar architecture in technology nodes below 28 nm with even greater Ft (unity current gain frequency) and Fmax (unity power gain frequency) performance values. In addition, the usage of a Ball Grid Array package, instead of a Pin Grid Array, with way less parasitics could also provide an a lot of room for improvement in terms of the operation bandwidth.

Acknowledgments

This work was co-financed by Hellenic Funds and by the European Regional Development Fund (ERDF) under the Hellenic National Strategic Reference Framework (ESPA) 2007–2013, according to Contracts no. MICRO2-15/E-II and MICRO2-15/E-III.

In addition, the contribution of Dr. Sotiris Bantas is greatly appreciated, since he was the one that initiated, inspired this project, and technically guided the full work. His contribution is greatly aknowledged.

References

1. T. Noulis, P. Merakos, E. Lourandakis, S. Stefanou, and Y. Moisiadis, Wide-band substrate crosstalk sensor for wireless SoC application, *Sensors and Actuators A: Physical*, vol. 239, 144–152, March 1, 2016.
2. T. Noulis, E. Lourandakis, S. Stefanou, and P. Merakos, "CMOS 65 nm 'on chip' broadband real time substrate noise measurement," *Electronics Letters*, vol, 51, no. 21, 1710–1711, October 2015.

3. A. Muhtaroglou, G. Taylor, and T. Rahal-Arabi, On-die droop detector for analog sensing of power supply noise, *IEEE Journal of Solid-State Circuits*, vol. 39, no. 4, 651–660, April 2004.
4. M. Felder and J. Ganger, Analysis of ground-bounce induced substrate noise coupling in a low resistive bulk epitaxial process: Design strategies to minimize noise effects on a mixed signal chip, *IEEE Transaction on Circuits and Systems II*, vol. 46, no. 11, 1427–1436, November 1999.
5. E. Charbon, P. Miliozzi, L. P. Carloni, A. Ferrari, and A. Sangiovanni-Vincentelli, Modeling digital substrate noise injection in mixed signal IC's, *IEEE Transactions on Computer Aided Design*, vol. 18, no. 3, 301–310, March 1999.
6. R. Gharpurey and G. Meyer, Modeling and analysis of substrate coupling in integrated circuits, *IEEE Journal of Solid state circuits*, vol. 31, no. 3, 125–128, March 1996.
7. B. R. Stanisic, N. K. Verghese, R. A. Rutenbar, L. R. Carley, and D. J. Allstot, Addressing substrate coupling in mixed-mode IC's: Simulation and power distribution synthesis, *IEEE Journal of Solid State Circuits*, vol. 29, no. 3, 226–238, March 1994.
8. R. Senthinathan and J. L. Prince, Simultaneous switching ground noise calculation for packaged CMOS devices, *IEEE Journal of Solid State Circuits*, vol. 26, no. 11, 1724–1728, November 1991.
9. K. Joardar, A simple approach to modeling cross-talk in integrated circuits, *IEEE Journal of Solid State Circuits*, vol. 29, no. 10, 1212–1219, October 1994.
10. A. L. L. Pun, T. Yeung, J. Lau, J. R. Clement, and D. K. Su, Substrate noise coupling through planar spiral inductor, *IEEE Journal of Solid State Circuits*, vol. 33, no. 6, 877–884, June 1998.
11. J. M. Casalta, X. Aragones, and A. Rubio, Substrate coupling evaluation in BiCMOS technology, *IEEE Journal of Solid State Circuits*, vol. 32 no. 4, 598–603, April 1997.
12. A. Helmy and M. Ismail, *Substrate Noise Coupling in RFICs*, Berlin: Springer Science, 2008.
13. M. Badaroglu, S. Donnay, H. De Man, Y. Zinzius, G. Gielen, T. Fonden, and S. Signell, Modeling and expiremental verification of substrate noise generation in a 220k-gates WLAN system-on-chip with multiple supplies, *IEEE Journal of solid State Circuits*, vol. 38, no. 9, 1250–1260, July 2003.
14. M. Nagata, "On-Chip Measurements Complementary to Design Flow for Integrity in SoCs," in *Proceedings of the 44th annual Design Automation Conference*, pp. 400–403, 2007.
15. N. Azumam, S. Shimazaki, N. Miura, M. Nagata, T. Kitamura, S. Takahashi, M. Murakami, K. Hori, A. Nakamura, K. Tsukamoto, M. Iwanami, E. Hankui, S. Muroga, Y. Endo, S. Tanaka, and M. Yamaguchi, "Measurements and Simulation of Substrate Noise Coupling in RF ICs with CMOS Digital Noise Emulator," in *9th International Workshop on Electromagnetic Compatibility of Integrated Circuits (EMC Compo)*, Nara, Japan, pp. 42–46, December 15–18, 2013.
16. S. Bronckers, G. Van der Plas, G. Vandersteen, and Y. Rolain, Substrate noise coupling mechanisms in lightly doped CMOS transistors, *IEEE Transactions on Instrumentation and Measurement*, vol. 59, no. 6, 1727–1733, May 2010.
17. M. Shen, J. Mikkelsen, O. K. Jensen, and T. Larsen, "A Compact P+ Contact Resistance Model for Characterization of Substrate Coupling in Modern Lightly Doped CMOS Processes," in *7th European Microwave Integrated Circuits Conference (EumIC)*, Amsterdam, Nederland, pp. 492–495, October 29–30, 2012.

18. D. Ozis, T. Fiez, and K. Mayaram. "A Comprehensive Geometry-Dependent Macromodel for Substrate Noise Coupling in Heavily Doped CMOS," *IEEE Custom Integrated Circuits Conference*, Orlando, FL, May 15, 2002.
19. P. Baumgartner, Substrate crosstalk analysis flow for submicron CMOS system on chip, *IET Electronics Letters*, vol. 51, no. 12, 953–954, November 2015.
20. D. K. Su, M. J. Loinaz, S. Masui, and B. A. Wooley, Experimental results and modeling techniques for substrate noise in mixed-signal integrated circuits, *IEEE Journal of Solid-State Circuits*, vol. 28, 420–430, April 1993.
21. T. Blalack, J. Lau, F. J. R. Clément, and B. A. Wooley, "Experimental Results and Modeling of Noise Coupling in a Lightly Doped Substrate," in *IEDM '96 Technical Digest*, pp. 623–626, December 1996.
22. M. Nagata, Y. Kashima, D. Tamura, T. Morie, and A. Iwata, "Measurements and Analyzes of Substrate Noise Waveform in Mixed-Signal IC Environment," in *Proceeding of 1999 IEEE Custom Integrated Circuits Conference*, pp. 575–578, 1999.
23. M. Nagata and A. Iwata, Substrate noise simulation techniques for analog-digital mixed 1si design, *IEICE Transactions on Fundamentals*, vol. E82–A, no. 2, 271–277, February 1999.
24. M. Nagata, T. Okumoto, and K. Taki, A built-in technique for probing power supply and ground noise distribution within large-scale digital integrated circuits, *IEEE Journal of Solid-State Circuits*, vol. 40, no. 4, 813–819, April 2005.
25. M. Nagata, M. Nagata, J. Nagai, K. Hijikata, T. Morie, and A. Iwata, Physical design guides for substrate noise reduction in CMOS digital circuits, *IEEE Journal of Solid-State Circuits*, vol. 36, no. 3, 539–549, March 2001.
26. M. van Heijningen, J. Compiet, P. Wambacq, S. Donnay, and I. Bolsens, "A Design Experiment for Measurement of the Spectral Content of Substrate Noise in Mixed-Signal Integrated Circuits," in Proceedings of the 1999 Southwest Symposium on Mixed-Signal Design, Tucson, AZ, pp. 27–32, April 11–13, 1999.
27. F. D. Ferraiolo, A. Sehgal, P. Song, and M. A. Sperling, "On-Chip Detection of Power Supply Vulnerabilities," Assignee: International Business Machines Corporation, Patent No. US7,355,435 B2, April 8, 2008.
28. H. W. Wagner, "Supply Voltage Characteristic Measurement," Assignee: Intel Corporation, Patent No. US 7,378,833 B2, May 27, 2008.
29. Y. Tsividis, *Operation and Modeling of the Transistor MOS*, 2nd edition, Oxford University Press, 2003.
30. VeloceRFTM v1.7 User Manual, Helic, 2011.
31. SPI Block GuideV03.06, Motorola, Inc., Document No. S12SPIV3/D, Original Release Date: January 21, 2000–Revised: February 4, 2003.

10

System-on-Chip Substrate Crosstalk Measurement Techniques

Konstantinos Moustakas, Thomas Noulis, and Stylianos Siskos

CONTENTS

10.1 Introduction

The consumer electronics market is driven by cost reduction and applications such as telecommunications, the Internet, and mobile and portable devices. Exponentially increasing functionality must be implemented in smaller areas and with higher performance and power efficiency. To meet these demands, the semiconductor industry has significantly advanced in the past decades, characterized by continuous technology scaling to deep sub-micron and nano-scale dimensions. Approaching the limits of silicon semiconductor technology, Moore's law is coming to an end. Until new electronic devices are introduced in the beyond-CMOS era, predicted after 2020, the industry will continue to improve integration and reduce cost by encompassing the functionality of complete systems in a single chip, a so-called system-on-chip (SoC). Such chips include, for example, WiFi, mobile 3G and 4G capabilities, GPS, cameras, and various sensors on the same die. Technology scaling is mostly beneficial for digital circuitry, improving performance and density and lowering power consumption. Analog/RF circuits benefit less from shrinking beyond nano-scale and it makes their design more challenging due to the low supply voltage, increased mismatches, and more complex device behavior and modeling. Although most of the functionality is implemented by the digital and DSP part, as software radio is emerging, analog/RF circuits are needed at the interface between the digital processing and the analog world. Higher levels of integration have several advantages, but also disadvantages and design complications [1]. The advantages are

1. Reduced package count and die area, significantly reducing costs.
2. Fewer pads and interconnect lines are driven, reducing the associated parasitic inductance and capacitance and the number of output buffers, thus improving power efficiency.
3. Speed and high frequency response is also improved by the reduced interconnect parasitics.
4. Increased reliability.

These benefits come with tremendous challenges due to the high circuit density, such as coupling interaction between blocks, increased switching noise, thermal limitations and design, and verification difficulties. The biggest disadvantage, coupling interaction, can occur due to the significant mutual inductance (magnetic coupling) and capacitance that exists between any two bond wires and pins in a package and through the common substrate shared by the circuits. An ever increasing number of transistors in the digital circuitry is switching with high speed and fast clocks, producing switching noise, which propagates through the substrate and degrades the performance of the sensitive analog/RF circuits, as depicted in Figure 10.1. The effect is

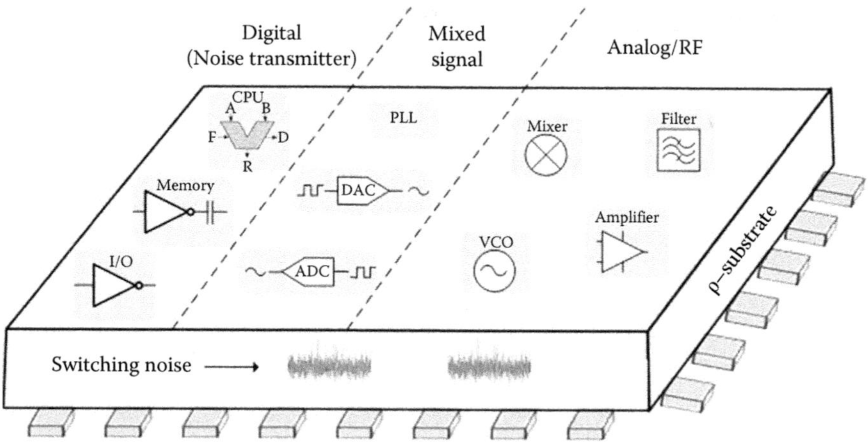

FIGURE 10.1
Substrate noise coupling on a SoC.

aggravated as technology substrate doping is increased and threshold voltage is decreased. Therefore, a high conductivity path is created through the substrate, increasing noise coupling to sensitive devices with reduced headroom and voltage swings.

As smaller technology nodes are predicted and vast integration into SoC's continues, the effect of substrate noise coupling is becoming more important. One way to deal with the substrate noise coupling, is to over-engineer analog/RF blocks. But, this strategy leads to increased time to market, higher cost, and less efficient designs. Finding ways to predict, measure, and suppress substrate noise is the key to developing successful integrated systems. Measuring the substrate noise is essential in identifying the critical coupling paths and taking the appropriate measures to isolate sensitive blocks. It can also be used in model verification to develop more accurate and reliable EDA tools to predict the noise propagation and distribution during the design phase.

10.2 Substrate Noise Coupling Mechanism

The substrate noise coupling mechanism can be divided into three phases: (a) Noise generation and injection through the digital switching circuits, (b) propagation through the medium (substrate), and (c) reception and impact on the sensitive analog blocks [2]. In Figure 10.2, an example of noise produced by a CMOS inverter (digital) affecting an NMOS transistor through substrate coupling is illustrated [2]. The mechanisms of injection, propagation, and reception are depicted in detail. The nature of propagation depends on the substrate resistivity and isolation structures implemented as well as frequency operation.

10.2.1 Noise Generation and Injection

Digital switching noise is generated during the simultaneous transition of multiple gates, producing large transient current spikes, which flow through the chip, package, and PCB power distribution network [2]. Noise current in the substrate is induced through several mechanisms. The most dominant are diffusion and interconnect wire capacitive coupling, impact ionization, and coupling through substrate contacts due to power grid fluctuations.

10.2.1.1 Capacitive Coupling of Interconnects and Passive Devices

The capacitive coupling between two interconnect wires and the substrate is illustrated in Figure 10.3 and modeled with three capacitors (C_1, C_2, C_3) [3].

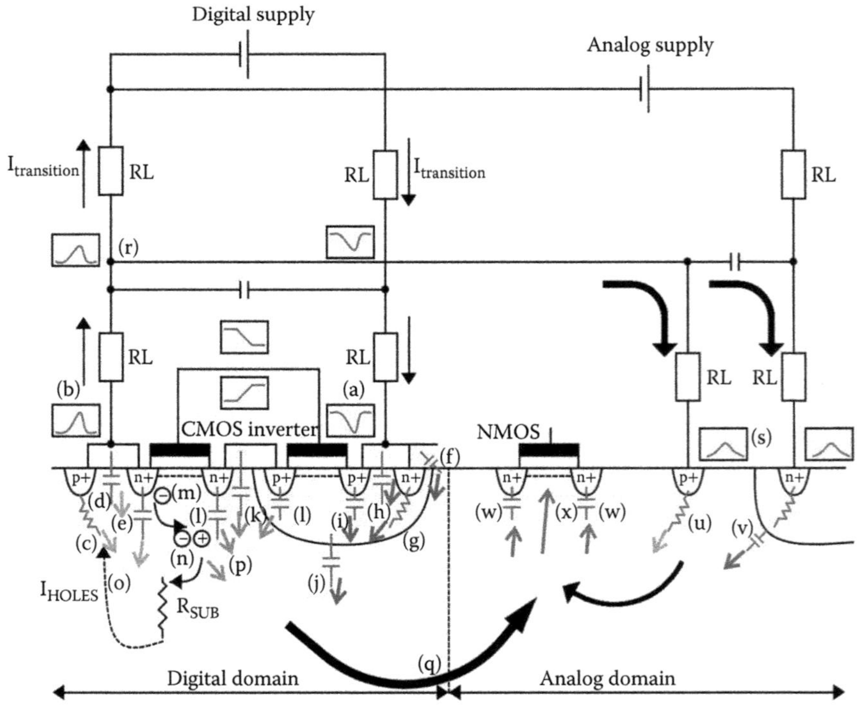

FIGURE 10.2
Substrate noise coupling mechanism.

The coupling coefficients depend on the metal layer, the dimensions of the interconnect wire, and the distance to other areas, such as other wires, junctions, etc. Interconnects in the lower metal layers are more coupled to the substrate than interconnects in the upper metal layers. Passive devices like resistors, capacitors and inductors can also capacitively induce current into the substrate. Poly resistors have a smaller capacitance than diffusion resistors. Thus diffusion resistors inject more noise current into the substrate. For a resistor grounded at one end to an AC ground, the noise current injected due to an input voltage V_{in} applied to the other end is [4]

$$I = \sqrt{\frac{j\omega C}{R}}\tanh\left(\frac{\sqrt{j\omega RCL}}{2}\right)V_{\text{in}} \tag{10.1}$$

where C is the per unit length capacitance of the resistor, R is the per unit length resistance and L is the length of the resistor. This equation is obtained by modeling the resistor as a dissipative transmission line. Capacitors types can be poly-to-poly, metal-to-poly, metal to metal and poly-to-substrate. Metal-to-metal capacitors have the largest parasitic capacitance to the substrate to nominal capacitance ratio.

10.2.1.2 Coupling through Diffusion Capacitance

Noise currents are injected to the substrate through p–n junction depletion capacitance C_{js} of drain and source regions [1].

$$C_{js} = \sqrt{\frac{q\varepsilon}{2(\psi + V_b)}}\left(\sqrt{\frac{N_1N_2}{N_1 + N_2}}\right) \tag{10.2}$$

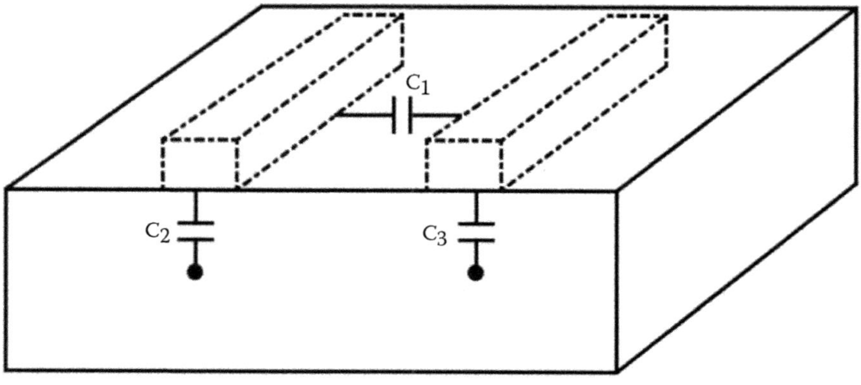

FIGURE 10.3
Capacitive coupling of interconnect wires.

where q is the electron charge, C_{js} the junction capacitance per unit area, ε the silicon dielectric constant, ψ the junction built in potential, V_b the reverse bias voltage, and N1 and N2 the p and n region doping concentrations. The coupling is dependent on the device size and switching frequency. Increased doping in modern technologies leads to higher depletion capacitance and, therefore, stronger coupling effects.

10.2.1.3 Coupling though Substrate Contacts

The body of a CMOS transistor is biased to a well-defined voltage to avoid forward biased parasitic p-n junctions. Typically, the bulk contact of a PMOS transistor is tied to the power supply voltage V_{dd} and the bulk contact of an NMOS transistor is connected to ground. The biasing contacts of the NMOS transistor in a typical CMOS process is directly connected to the substrate. In each digital gate on-chip, there is at least one substrate contact. Therefore, there are a large number of substrate contacts providing a low resistivity path for ground bounce noise to be injected [5]. In the case of PMOS transistors, power supply ringing can be injected into the n-well resistively and then to the substrate through the depletion capacitance between the substrate and the n-well. In lower frequencies, the power injected by the NMOS substrate contacts is a significant source of noise current.

10.2.1.4 Noise Injection through Impact Ionization

As the transistor channel shrinks beyond the deep-submicron region, the electric field in the channel increases and impact ionization currents become significant. When MOS devices are biased in the saturation region, a high electric field is formed near the drain. Electrons accelerated by this field constitute "hot" carriers that create, via collisions, electron-hole pairs. In the case of NMOS transistors in a p-type substrate, the generated holes are taking a path into the substrate, generating a noise voltage through the substrate resistance. For the case of PMOS transistors in an n-well, the impact ionization current is lower due to the lower hole mobility and indirect capacitive coupling through the n-well to p-substrate capacitance. The impact ionization substrate current is given by Equation 10.3 [1].

$$I_{\text{sub}} = K_1\left(V_{\text{DS}} - V_{\text{Dsat}}\right) \cdot I_d \cdot e^{\left(\frac{-K_2}{V_{\text{DS}} - V_{\text{Dsat}}}\right)} \tag{10.3}$$

where I_d is the drain current, V_{DS} the drain to source voltage, and V_{Dsat} the drain to source saturation voltage. K_1 and K_2 are empirical constants. Hot electron induced current is the dominant cause of substrate noise in NMOS devices up to one MHz. Impact ionization current occurs in both digital transitions with the same polarity. As a result, it contains large even harmonics of the fundamental switching frequency.

10.2.1.5 Noise Injection through Power Grid Fluctuations

The power distribution path includes on-chip interconnects, bond and package wires and PCB interconnects to the analog power supply, which is assumed to be clean. Large transient current spikes caused by the simultaneous switching of multiple digital gates produce transient voltage ringing, which flows through the resistive and inductive impedances of this path. The power supply voltage drop is given by Equation 10.4 [2].

$$V_{\text{ddbounce}} = R_{V_{\text{dd}}} \cdot I_{\text{DD}} + L_{V_{\text{dd}}} \cdot \left(\frac{dI_{\text{DD}}}{dt} \right) \tag{10.4}$$

The ground bounce voltage drop is given by Equation 10.5 [2]:

$$V_{\text{SSbounce}} = R_{V_{\text{SS}}} \cdot I_{\text{SS}} + L_{V_{\text{SS}}} \cdot \left(\frac{dI_{SS}}{dt} \right) \tag{10.5}$$

where R_{Vdd} and R_{Vss} are the power supply and ground wire resistance and L_{Vdd} and L_{Vss} are the power supply and ground network inductance. The sum of parasitic capacitors and, usually, decoupling capacitors located between the on-chip positive power supply and ground can be modeled as a lumped capacitor C_{eff}. C_{eff} and the sum of the parasitic inductance L_{eff} form a resonant circuit with resonance frequency [3]:

$$f_{osc} = \frac{1}{2\pi\sqrt{L_{\text{eff}} C_{\text{eff}}}} \tag{10.6}$$

When a current spike occurs, damped voltage fluctuations follow the initial voltage drop. On-chip decoupling capacitors are used to provide a low impedance path for reducing power supply fluctuations, but these capacitors also lowers the resonance frequency [3].

As shown in Figure 10.2, a positive large current spike of I_{DD} generates a negative voltage spike on the V_{dd} node (a) and a positive large current of I_{SS} generates a positive voltage spike on the V_{SS} node (b). These spikes are larger for large devices and fast digital clocks. As a result of scaling the number of gates is increased and faster current transients dI/dt occur. Thus, the generation of switching noise is expected to increase further [2]. V_{dd} and V_{SS} fluctuations couple into the substrate via the resistance of substrate contacts and through interconnect and junction capacitances.

10.2.1.6 Noise Injection Example of a CMOS Inverter

In Figure 10.2, the coupling of ground ringing through substrate contacts is depicted by (c) and is significant in low frequencies due to RC filtering by parasitic capacitances. Coupling through interconnect wires is illustrated in

(d) and through junction depletion capacitances by (e). V_{dd} ringing couples directly to the substrate through interconnect capacitance (f). V_{dd} noise is also injected to the N-well through contacts (g), interconnect capacitance (h), and depletion capacitance (i). The sum of (g), (h), and (i) is coupled to the substrate through the N-well capacitance (j).

Coupling of gate output transitions through interconnect capacitance is illustrated by (k) and through depletion capacitance is illustrated by (l). Impact ionization is depicted by hot carriers (m), producing electron–hole pairs (n). The hole current I_{HOLES} (o) produces voltage fluctuations on the substrate due to the resistance R_{SUB} (p).

10.2.2 Noise Propagation through the Substrate

The substrate acts as the medium for the sum of the noise injected to propagate, shown in Figure 10.2 as (q), and affects sensitive analog circuits. The propagation properties and their dependence on technology parameters are given by a function of frequency of operation. The current in the substrate can be given be a derivative of Maxwell's equations [1]:

$$J = (\sigma + j\omega\varepsilon_{si})E \tag{10.7}$$

where J is the current density in the substrate (A/cm^2), E the electric field (V/cm), σ the conductivity (S/cm), ε_{si} the dielectric permittivity of silicon, and Ω the frequency in rad/s. For $\sigma \gg \omega\varepsilon_{si}$, the dominant noise propagation nature is resistive. At low frequencies, the substrate can be modeled as distributed resistive network. This assumption is valid below f_c.

$$f_c = 1/2\pi\rho_{sub}\varepsilon_{si} \tag{10.8}$$

where ρ_{sub} is the substrate resistivity ($1/\sigma$).

For a typical value of $\rho_{sub} = 10\ \Omega$cm, the substrate can be considered resistive for a frequency below 15 GHz. Thus, for most RF and mixed signal applications, the substrate can be considered as a resistive mesh, as illustrated in Figure 10.4 [1]. Above $f_{c,}$ the substrate model becomes a mesh of R_s and C_s.

Although separate pins are used for the analog and digital power supply, switching noise is coupled into the analog power supply via the package parasitic capacitance. Ground fluctuation noise is also injected into analog V_{SS} through a shared ground path on the PCB. Voltage fluctuations produced on the shared impedance ([*r*] in Figure 10.2), couple into the analog ground V_{SSA} ([s] in Figure 10.2) and analog power supply V_{DDA} ([t] in Figure 10.2) [2]. V_{SSA} noise couples into the substrate through the contact resistance ([*u*] in Figure 10.2) and noise on V_{DDA} couples through the N-well depletion capacitance ([*v*] in Figure 10.2) [2].

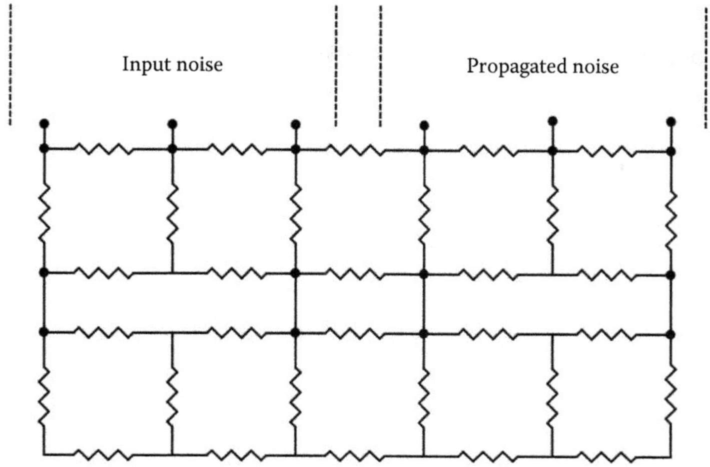

FIGURE 10.4
Substrate model at low frequencies.

10.2.3 Noise Reception and Impact

For most devices, the most common noise reception mechanism is capacitive sensing. These devices include BJTs, resistors, capacitors and interconnect lines. Well junction capacitance is the main coupling path for devices such as diffusion resistors and well-to-poly capacitors. Metal interconnects and poly resistors are coupled via the oxide capacitance to the substrate [1].

For MOS devices, noise is coupled through the source and drain junction capacitances and more severely due to the body effect. In MOS transistors, the threshold voltage V_t is a function of the substrate potential, given by [1].

$$V_T = V_{T0} + \frac{\sqrt{2q\varepsilon N_A}}{C_{ox}}\left(\sqrt{2\varphi_f + V_{SB}} - \sqrt{2\varphi_f}\right) = V_{T0} + \gamma\left(\sqrt{2\varphi_f + V_{SB}} - \sqrt{2\varphi_f}\right) \tag{10.9}$$

where ε is the substrate dielectric permittivity, N_A the substrate doping, C_{ox} the unit oxide capacitance, φ_f the surface inversion potential, and V_{SB} the source-to-body potential. The effect on the drain current is represented in the small-signal model by a linearized parameter g_{mb} [1].

$$g_{mb} = \frac{\partial I_d}{\partial V_{BS}} = g_m \frac{\gamma}{2\sqrt{2\varphi_f + V_{SB}}} \tag{10.10}$$

where g_m is the gate-to-source transconductance and g_{mb} the body-to-source transconductance. Typically g_m/g_{mb} = 0.1 to 0.3. While the capacitive noise reception is important only in high frequencies, the body effect is a significant issue at low frequencies also, making MOS devices more susceptible to substrate noise. In Figure 10.2, capacitive sensing by an NMOS is illustrated by (w) and noise reception due to body effect is illustrated in (x).

10.3 Substrate Noise Measurement

10.3.1 Overview of Substrate Crosstalk Measurement Techniques

Substrate noise coupling emerges as the most crucial problem in modern SoC applications, especially when technology multi-sourcing is used, due to complete radio spectrum usage from 3 KHz to 30 GHz and wireless applications demanding SNR below background noise. Things will become more severe as wireless SoCs move to millimeter wave and from the 4G to the 5G era [6]. Significant effort has been invested in recent years to reduce the effect of digital switching noise coupling on analog circuits. It focused on modeling and simulation, measurement, and techniques of suppression and isolation.

Various modeling techniques and approaches have been proposed to predict the propagation and impact of substrate noise during the design phase [1,4,7,8]. Substrate noise generation by large digital circuits has been also simulated using behavioral models due to the complexity of equivalent circuit models [9]. Recent studies combine models produced from EM simulators with circuit simulation to more accurately predict the effect and analyze the dependence of physical parameters [10–13]. Regarding EDA tool integration in the design flow, there exists very limited options [13,14] with questionable accuracy [6]. It is extremely important to validate the accuracy and effectiveness of the proposed modeling and simulation flows by comparing the results with directly measured substrate signals. Measurement is not only important for model and tool verification, but also for understanding noise coupling mechanisms and identifying critical paths across the chip. Therefore, we can have a deeper insight into the underlying physical parameters that affect noise propagation and injection and improve and evaluate noise suppression methodologies. It is also very significant to be able to separate magnetic and substrate noise coupling during the design phase by having the ability to measure and correlate the substrate noise across the die with the output signal spurs. By identifying the greatest noise transmission contributor, appropriate measures can be taken and the design phase can be sped-up, avoiding the testing of a lot of experimental prototypes.

Several substrate noise measurement techniques have been proposed, such as implementing sensor circuits that are connected to off chip laboratory or on-chip instrumentation (on-chip digitizers) [2]. The main challenge in mobile and wireless SoCs substrate noise measurement is developing a sensor that has wideband gain in order to be able to amplify all the intermodulation's products noise signals in the GHz region [6]. A really good quality broadband sensor for modern mobile applications must be able to measure at frequencies beyond 10 GHz.

The simplest sensing technique involves the use of p+ substrate taps through ohmic contact diffusions and sense lines, but has many disadvantages. The 50 Ω load of the spectrum analyzer practically shunts the substrate to ground, affecting the propagation of the signal through it. In Figure 10.5, the equivalent substrate noise voltage source drives the 50 Ω transmission live through a source resistance R_s that represents the resistance of the probing substrate contact. The measured signal has an amplitude of

$$V_{\text{measured}} = V_{\text{noise}} \frac{Z_{\text{in}}}{Z_{\text{in}} + R_{\text{source}}} \tag{10.11}$$

Thus, the signal is attenuated and may fall below the background noise threshold. The measured signal represents the noise in reference to the oscilloscope ground, which is equal to the PCB ground [15]. Therefore, the measurement is corrupted and includes noise in the power supply network. A more accurate technique that solves this problem involves measuring the difference between the substrate and ground (V_{SS}) signals. Also, in complex VLSI chips, the interconnect crosstalk that couples into single ended wires does not preserve signal integrity. Since each sense line requires dedicated pins on the chip, this technique becomes impractical and is limited only to small experimental chips with a few probing points [15].

Another alternative is using NMOS transistors that sense the substrate noise through the body effect and capacitive coupling to its gate, drain, and

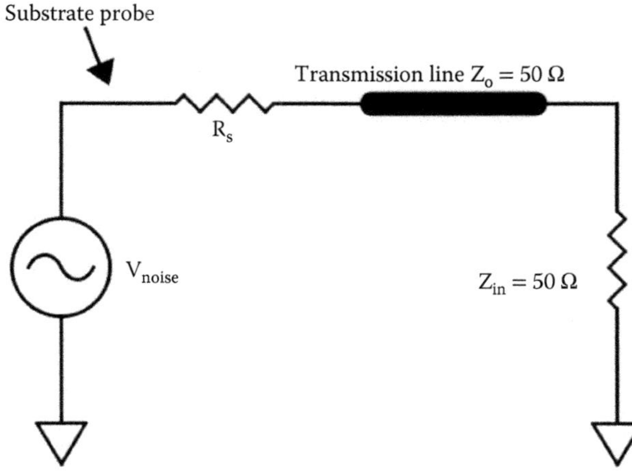

FIGURE 10.5
Noise measurement via substrate tap.

source nodes [14,16,17]. Their output current is routed to off-chip instruments (spectrum analyzers). This is an indirect measurement technique that does not measure the substrate noise voltage, but the impact it has on the NMOS operation and current. These type of sensors are effective in simple experimental designs, but, similar to the p+ substrate tap sensing, they are affected by the ground bounce and interconnect crosstalk and require dedicated pins. Also, the gain is limited by the small value of g_{mb} transconductance. As a consequence, they do not provide good results for complex chips.

A simple and direct measurement topology is proposed in [8,18,19]. It consists of a source follower as a level shifter to sense the noise voltage and a MOSFET transconductance stage (g_m) to convert the measured voltage to a current signal that can be amplified via a current mirror and multiplexed. Its disadvantages are measurement corruption from ground bounce and interconnect crosstalk since the measurement is single ended and high consumption respective to its gain. It is also limited by a moderate bandwidth.

To reduce the effects of signal contamination, on-chip samplers have been developed [2,15,18,20–26]. The sampler in [18,20,24] uses a differential latched comparator that is controlled by three-phase clocks and the sampler in [22] uses a single ended chopper-type comparator that is controlled by a four-phase clock [26]. They both measure the input noise by comparing it to a single-ended-type reference voltage. A solution to reduce signal contamination due to the on-chip sampler noise coupling involves a sampler controlled by one-phase differential clock that measures the differential input, comparing it to a differential input voltage to improve signal integrity [2,15,26]. In [25], a Threshold Inverter Quantization (TIQ) comparator is used to sense noise via the threshold voltage dependence of an inverter to the substrate voltage. On chip samplers provide a digital output signal that can be easily multiplexed and processed without expensive laboratory equipment, but take a lot of area on the chip to limit the probe points. To achieve a high resolution, a large gain must be implemented by the comparator, leading to low bandwidth. This type of measurement is also indirect, sensing the influence of substrate voltage on the output state of comparators.

A sensing technique that uses a PMOS-based differential sensor DC-coupled to the substrate is proposed in [2,15,26]. The DC coupling allows the frequency response to extend down to DC and the differential architecture reduces contamination from ground bounce and interconnect crosstalk. The sensors are also compact and can be multiplexed. The disadvantage when connecting the output to a spectrum analyzer is the need for a buffer to match the output impedance of the sensor to 50 Ω and very low gain. It can achieve good bandwidth performance but this is true due to the low gain.

Another direct measurement technique involves using differential amplifiers to remove common mode noise coupling through the ground node [6,27,28]. Differential signaling also improves immunity to interconnect crosstalk. The inputs are ac coupled using capacitors and the bandwidth cannot extend down to DC. A differential sensor with improved bandwidth using sub-threshold biasing is proposed in [6].

10.3.2 Substrate Noise Sensor Requirements

Accurately measuring the switching noise propagating through the substrate is a very challenging task because of the nature of the application. Low magnitude voltages in the order of μV have to be sensed and amplified over a very wide bandwidth. Signal contamination from ground bounce and interconnect crosstalk, both in-chip and off-chip, in the connection to the measurement equipment must be greatly reduced. In complex VLSI chips requiring a lot of probe points, sensors occupying a lot of area are impractical. Therefore, the exotic requirements of a compact sensor circuit, such as being capable of measuring very low amplitude signals from DC up to 10 Ghz and beyond without contaminating the measurement, have to be met. In more detail, the requirements that a state-of-the art substrate sensor must incorporate are as follows:

1. The sensor must not affect the noise propagation through the substrate or inject noise itself. The purpose is the measurement of substrate noise at probe points across the die. Because the sensor itself is not part of the main function of the chip, noise propagation must be the same as if the sensor was not built-in. On-chip digitizers, which are digital circuits, must be physically designed with great caution and measures have to be taken to not inject additional noise. Care must be also taken in substrate contacts to prevent latch-up of the sensor's transistors. They shunt the substrate to ground and alter noise propagation [2,15].
2. The signal must not be contaminated from interconnect crosstalk on the on-chip and off-chip path. This is more important in complex SoC chips where noise from surrounding circuits can couple to the interconnect wires. Therefore, the sensor must use differential signaling to cancel out the common-mode noise between the substrate signal and ground. Immunity to V_{dd} fluctuations is also important and can be achieved using decoupling capacitors or symmetrical circuit design techniques. The power supply and ground pins of the sensor must be separate and as isolated as possible [2,15].
3. The measured bandwidth must extend down to DC and high enough to be able to measure all the intermodulation products [6]. In modern wireless and mobile applications, bandwidth above 10 GHz is very important for an accurate measurement. The low cutoff frequency is usually set by the input coupling capacitors and the high frequency by the technology, the load capacitance, and the type of the amplifier used. Wideband techniques in the amplifier design are, therefore,

essential to achieve the high frequency response without using area-consuming inductors [2,15].

4. The sensors need to be compact and occupy as small area an as possible for an easy implementation in complex SoC applications where lots of probe points are needed [2,15].
5. The sensor output is essential for multiplexing in order to reduce the number of pins needed, especially in large VLSI designs [2,15].
6. The sensor gain, in the case of using off-chip instrumentation, does not need to be high. A value of about 6 dB is adequate because a high performance spectrum analyzer can measure signals down to approximately –170 dbm. Apart from voltage amplification, the sensor acts as a buffer between the substrate tap and the 50 Ω load of the spectrum analyzer and prevents shunting the substrate to ground, which would affect noise propagation.

10.3.3 Analysis and Simulation of Substrate Measuring Techniques

The most important substrate noise sensing techniques reported in literature will be analyzed and simulated in 65 nm technology using Cadence Virtuoso™. Designing sensor circuits in a nanometer process node is challenging due to the low supply voltage V_{dd} = 1.1V and the strong influence of small channel effects, such as Channel Length Modulation (CLM), Drain Induced Barrier Lowering (DIBL), and threshold voltage variation on transistor operation.

10.3.3.1 Bulk-Driven NMOS Transistors

The dominant mechanism of noise injection in MOS devices is the body effect, as mentioned in Section 10.2.3. Due to the small drain and source depletion capacitances and the gate oxide capacitance, the noise injection though them is important only in very high frequencies. NMOS transistors fabricated on the p-substrate can be used to sense substrate noise via the drain current modulation due to the source-body transconductance g_{mb} (Equation 1.10). Figure 10.6 illustrates the technique [15].

The NMOS transistor is connected to the spectrum analyzer 50 Ω input trough a bias tee. A bias tee is used to set the DC bias point independently of the AC signal. The high impedance of the inductor does not allow the AC signal to flow through it, but there is a short-circuit at DC that sets the bias point of the drain at 0V. The capacitor is an open circuit at DC and a short circuit at AC, providing a low impedance path from the AC signal to the spectrum analyzer input. Since the input is connected to ground, typically the NMOS has to be biased from negative power supplies [15]. This is easy in simple test chips, but is impractical is production environments. The drain potential is set to 0V to achieve the maximum V_{DS}, so the gain is not

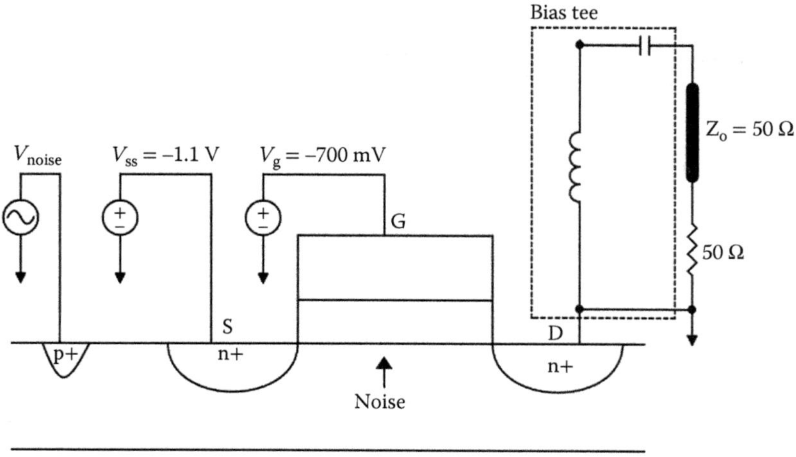

FIGURE 10.6
Substrate noise measurement using bulk-driven NMOS.

decreased by CLM. The gain depends on the bias current and the transistor size, as seen by Equation 10.12.

$$g_{mb} \cong 0.1\text{–}0.3 g_m, \quad g_m = \sqrt{2\mu_n C_{ox} \frac{W}{L} I_D} \tag{10.12}$$

where g_m is the gate-to-source transconductance, g_{mb} the body-to-source transconductance, μ_n the electron mobility, C_{ox} the gate oxide capacitance, I_D the drain current, W the width, and L the length of the transistor. The values of the gate-source and body-source transconductance calculated via simulation are: g_m = 100 mS, g_{mb} = 16.1 mS, therefore g_{mb}/g_m = 0.161. Because the measurements are high-frequency, S-parameter analysis is used to characterize the sensor performance. The NMOS dimensions are given by Table 10.1 and the simulation results by Table 10.2. In Figure 10.7, the S_{21} parameter characterizing the frequency response and the gain of the sensor is illustrated for process and temperature variations. In Figure 10.8, the intrinsic noise voltage in the output due to the sensor is also depicted.

The negative gain of the sensor is due to the small transconductance g_{mb} compared to g_m. An optimal combination of channel length (which affects output resistance), width, and bias gate overdrive voltage (which affect bias

TABLE 10.1

Device Size and Values of the Bulk-Driven NMOS Sensor

Transistor	Dimensions (W/L)
M1	190 μm/90 nm

TABLE 10.2
Simulation Results of the Bulk-Driven NMOS Sensor

Specification	Value
Gain	–3.75 dB
Gain PT variation	–3 dB to –3.7 dB
Bandwidth	0 Hz to 7.18 GHz
Bandwidth PT variation	0 Hz–7.18 GHz to 0 Hz–7.6 GHz
Power consumption	8.3 mW
1dB compression point	980 mV
Area	180 μm²

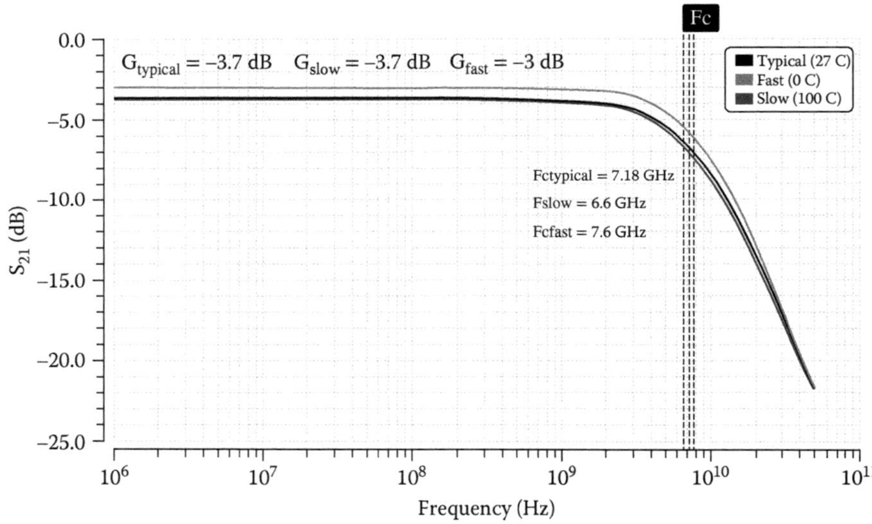

FIGURE 10.7
S_{21} parameter for PT variations of the bulk-driven NMOS sensor.

current) was selected. The gain can be higher with a larger bias current, but the power consumption will be too high. To be able to carry a large current, interconnects must be made wide, increasing their capacitance and decreasing the bandwidth due to parasitics in the fabricated chip. The bandwidth is extended down to DC and is relatively high, but this response is due to the low gain that limits the accuracy and the applications of the NMOS bulk-driven sensor. Because it is a transistor-only sensor, it is very compact and has a very high 1 dB compression point. Also the intrinsic noise in the output is very low, below the spectrum analyzer noise floor. As mentioned in Section 10.3.2, because the output is single ended, the output signal integrity is compromised by interconnect and ground bounce crosstalk. It needs dedicated pins for each transistor and the outputs cannot be multiplexed. Therefore its applications are limited to experimental designs [15].

10.3.3.2 Source Follower–Transconductor (SF+Gm) Sensor

The sensor schematic is illustrated in Figure 10.9. The circuit is composed of a source follower (M1–M2), a transconductance stage (M3), and a current mirror with current amplification (M4,M5) [8,18,19]. A bias tee is used in the

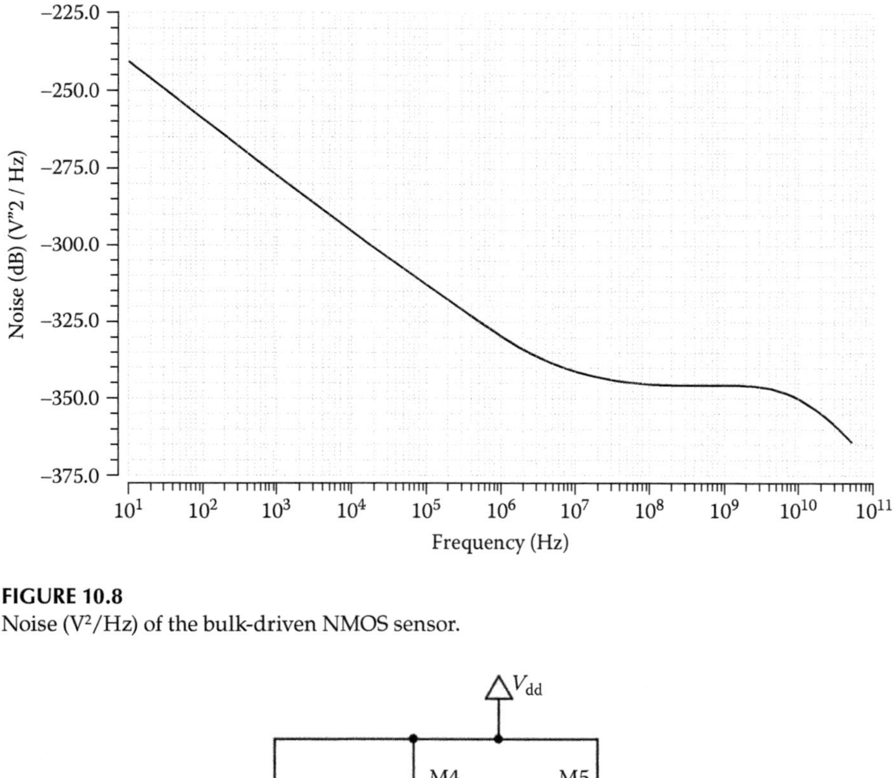

FIGURE 10.8
Noise (V^2/Hz) of the bulk-driven NMOS sensor.

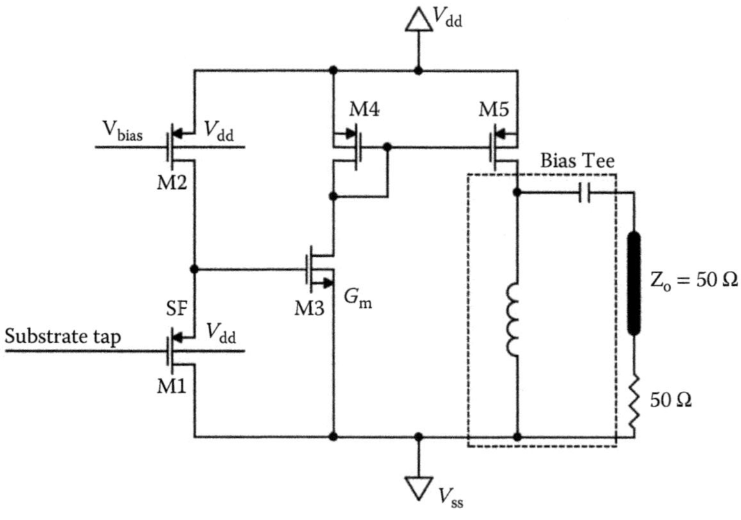

FIGURE 10.9
SF+Gm substrate noise sensor.

output as in the previous case. The noise voltage signal at the input of the source follower is level-shifted by V_{GSM2} and driven to the gate of M3. This voltage is translated to drain current I_{DM3} and mirrored by M4–M5 to the output.

The gain is given by the following equation:

$$A_{vSF} = \frac{g_{m1}}{g_{m1} + G_{s1}} = \frac{g_{m1}}{g_{m1} + g_{mb1} + g_{ds1} + g_{ds2}} = g_{m1}\left(\frac{1}{g_{m1}} \,||\, \frac{1}{g_{mb1}} \,||\, r_{ds1} \,||\, r_{ds2}\right) \tag{10.13}$$

where g_m is the gate-to-source transconductance, G_{s1} the total conductance at the source node of M1, $g_{ds} = 1/r_{ds}$ the drain-source conductance, and r_{ds} the drain-to-source resistance.

$$r_{ds} = \frac{1}{\lambda I_D} \tag{10.14}$$

The gain of a source follower is always less than unity. Conductances g_{ds1} and g_{ds2} are usually much smaller than g_{mb}, so the major source of gain reduction is the body effect. $A_{vSF} = 0.815$ as calculated by simulation and g_{m3} depends on M3 size and by the biasing voltage V_{GSM3} and is calculated equal to 7.3 mS. Because $V_{GSM4} = V_{GSM5}$ and, neglecting CLM for simplicity, from the drain current equation:

$$I_D = \frac{\mu_n C_{ox}}{2}\frac{W}{L}(V_{GS} - V_T)^2 \tag{10.15}$$

where μ_n is the carrier mobility, C_{ox} the gate oxide capacitance per unit area, W and L the dimensions of the MOS device, V_{GS} the gate-to-source voltage, and V_T the MOSFET threshold voltage, the current gain of the current mirror can be calculated as

$$A_{iCM} = \frac{\left(W/L\right)_{M5}}{\left(W/L\right)_{M4}} = 6 \tag{10.16}$$

Due to the CML effect, the simulated current gain is $A_{iCM} = 3.66$. The sensor gain therefore is

$$\frac{V_o}{V_{in}} = A_{vSF} \cdot g_{mM3} \cdot A_{iCM} \cdot Zo = 0.815V \cdot 0.0073\frac{A}{V} \cdot 3.66 \cdot 50\Omega = 1.1 \cong 1.1\,\mathrm{dB} \tag{10.17}$$

which is equal to the simulated gain. Transistor dimensions are given by Table 10.3 and the simulation results by Table 10.4. In Figure 10.10, the S_{21} parameter is illustrated for process and temperature variations and, in Figure 10.11, the intrinsic noise voltage of the sensor is also depicted.

This sensor has a low but adequate gain. To achieve it the power consumption is high due to the high bias currents of the current mirror and g_m stages. It suffers from signal contamination, as previously mentioned, because it is single-ended. It is compact and can be used on complex chips because the output current can easily be multiplexed using pass transistors to activate or deactivate the current mirror. The bandwidth is relatively high, but cannot be further extended without impractical power consumption. There is also a significant decrease in performance in the slow process corner. The output noise is very low and does not pose a problem in the performance of the sensor.

10.3.3.3 DC-Coupled PMOS Sensor

The sensor schematic is illustrated in Figure 10.12. This measurement technique uses PMOS-based differential sensors DC coupled to the substrate taps

TABLE 10.3

Device Size and Values of the SF+Gm Sensor

Transistor	Dimensions (W/L)
M1	48 μm/120 nm
M2	48 μm/120 nm
M3	16 μm/120 nm
M4	12 μm/120 nm
M5	50 μm/120 nm

TABLE 10.4

Simulation Results of the SF+Gm Sensor

Specification	Value
Gain	1.1 dB
Gain PT variation	–2.1 dB to 1.6 dB
Gain mismatch variation	0.94 dB to 1.26 dB
Bandwidth	0 Hz–7 GHz
Bandwidth PT variation	0 Hz–4.7 GHz to 0 Hz–8.2 GHz
Bandwidth mismatch variation	0 Hz–6.85 GHz to 0 Hz–6.9 GHz
Power consumption	11 mW
1dB compression point	130 mV
Area	570 μm2

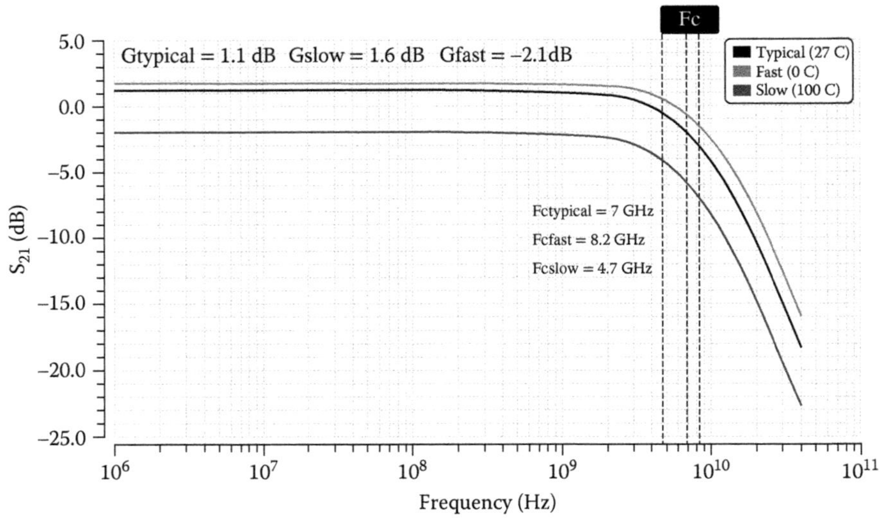

FIGURE 10.10
S_{21} parameter for PT variations of the SG+Gm sensor.

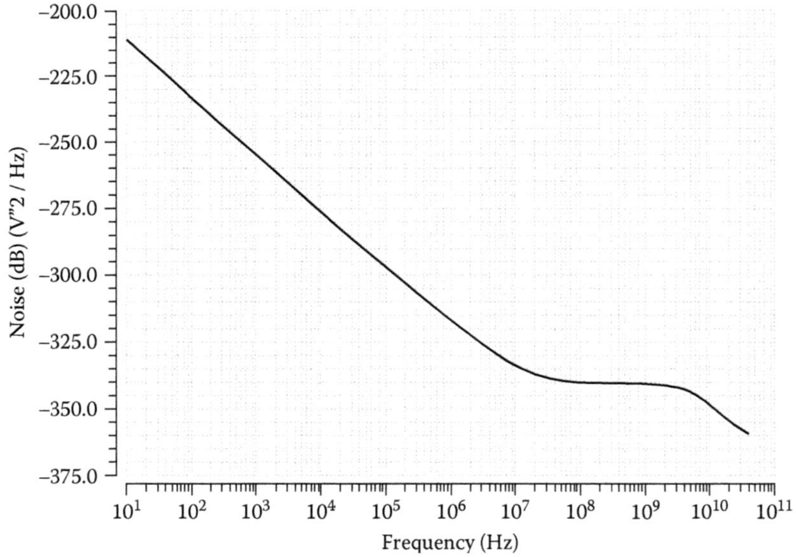

FIGURE 10.11
Noise (V^2/Hz) of the SF+Gm sensor.

[2,15,26]. Since no input capacitor is used, the measurement bandwidth can expand down to DC. The main advantage of this method over the previous is that the sensor measures the signal in differential mode. The differential-to-single action of the balun used at the output removes the common mode noise of the ground and the signal coming from the substrate. Therefore,

the signal integrity from the measurement point to the input of the measurement instrumentation is preserved. In physical layouts of circuits containing NMOS and PMOS, transistor substrate shunting contacts are often used to prevent latch-up. A careful layout reduces this problem, but a circuit with only PMOS transistors, such as the sensor mentioned here, guarantees no interaction with the substrate noise propagation. The output can be connected to a spectrum analyzer (off-chip) or an on-chip digitizer. In the first case, buffers must be used in the output to match the sensor's output resistance with the 50 Ω impedance of the spectrum analyzer.

The sensor circuit consists of two branches made of transistors M_1, M_3, M_2, and M_4 [2,15,26]. The bias generator consists of transistors M_5 and M_6. M_7 and M_8 are NMOS source followers used as buffers to match the output impedance of the sensor to the transmission line. The gates of M_1 and M_2 are connected to the same DC potential, so, for equal bias currents, if $M_1 = M_2$ and $M_3 = M_4$, $V_{GS1} = V_{GS2}$. Therefore, the output voltages of the source followers made from M_1,M_3 and M_2,M_4 are equal for DC. Using small signal analysis of the circuit of Figure 10.12, the sensor gain can be calculated [2,15,26]. The differential output voltage V_{out} can be expressed by the equation [2,15,26]:

$$V_{OUT} = V_{IN} \frac{g_{m1}(r_{o1} \,||\, r_{o3})}{1 + (g_{m1} + g_{mb1})(r_{o1} \,||\, r_{o3})} \tag{10.18}$$

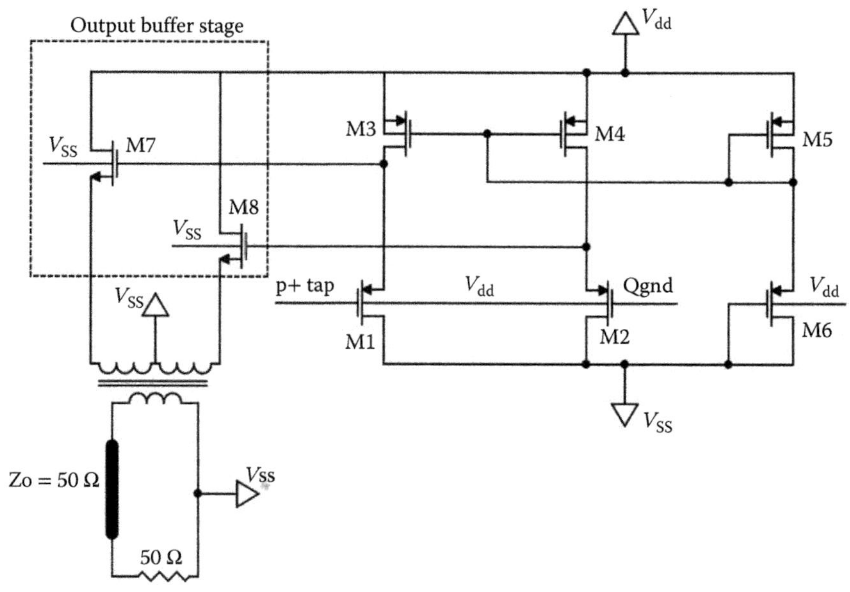

FIGURE 10.12
DC-Coupled PMOS substrate noise sensor.

where V_{in} is the substrate noise voltage, g_m the gate-to-source transconductance, g_{mb} the body-to-source transconductance, and r_o the MOSFET drain-to-source resistance. Because the following condition is typically true,

$$(g_{m1} + g_{mb1})(r_{o1} \,||\, r_{o3}) >> 1 \tag{10.19}$$

The sensor differential gain is calculated as

$$\frac{V_{OUT}}{V_{IN}} = \frac{g_{m1}}{g_{m1} + g_{mb1}} \tag{10.20}$$

Considering g_{mb} to be approximately 0.15 g_m, $V_{out}/V_{in} \approx 0.87$. Buffer transistors M_7 and M_8 were scaled for output resistance of about 50 Ω to match the transmission line impedance and avoid oscillations and reflections. The buffer output resistance, $R_{obuffer}$, forms a resistive divider with the transmission line impedance Z_o.

$$V_o = V_{OUT}\frac{Z_o}{R_{obuffer} + Z_o} \cong \frac{V_{OUT}}{2} = V_{OUT}(\text{dB}) - 6\,\text{dB} \tag{10.21}$$

Therefore, the gain of this amplifier is less than unity if only buffer transistors are used. An additional gain stage can be used, but will significantly affect the bandwidth. The output of this sensor can be easily multiplexed, because the output common mode voltage is in the input range of analog multiplexers. The DC common mode output voltage is given by [2,15,26]:

$$V_{CM} = \sqrt{\frac{2I_B}{\mu_p C_{ox}\frac{W}{L}}} + |V_{TP}| \tag{10.22}$$

where μ_p is the hole mobility, C_{ox} the gate oxide capacitance, V_{TP} the PMOS threshold voltage, I_B the bias current, and W and L are transistor dimensions. The sensor design is symmetrical. Variations of the bias current through M_5 and M_6 caused by V_{dd} fluctuations are mirrored on the M_1–M_3 and M_1–M_3 branches. Because they are identical, the current variation is cancelled as a common mode ac signal. Transistor dimensions are given by Table 10.5 and the simulation results by Table 10.6. In Figure 10.13, the S_{21} parameter is illustrated for process and temperature variations and, in Figure 10.14, the intrinsic noise voltage of the sensor is also depicted.

The measurement assumes an ideal balun with no insertion loss. A real balun will reduce the gain further by about –3 dB depending on its type. The CMRR of this sensor is, essentially, the CMRR of the balun because it performs the subtraction passively. A typical value for a good quality balun is a CMRR of 25–30 dB.

TABLE 10.5

Device Size and Values of the DC-Coupled PMOS Sensor

Transistor	Dimensions (W/L)
M1	48 µm/120 nm
M2	20 µm/120 nm
M3	48 µm/120 nm
M4	20 µm/120 nm
M5	2 µm/60 nm
M6	6 µm/60 nm

TABLE 10.6

Simulation Results of the DC-Coupled PMOS Sensor

Specification	Value
Gain	–6.2 dB
Gain PT variation	–7 dB to –5.6 dB
Gain mismatch variation	–6.25 dB to –6.2 dB
Bandwidth	0 Hz to 9 GHz
Bandwidth PT variation	0 Hz–7.8 GHz to 0 Hz–12 GHz
Bandwidth mismatch variation	0 Hz–9.3 GHz to 0 Hz–9.7 GHz
Power consumption	6.8 mW
1dB compression point	280 mV
Area	928 µm2

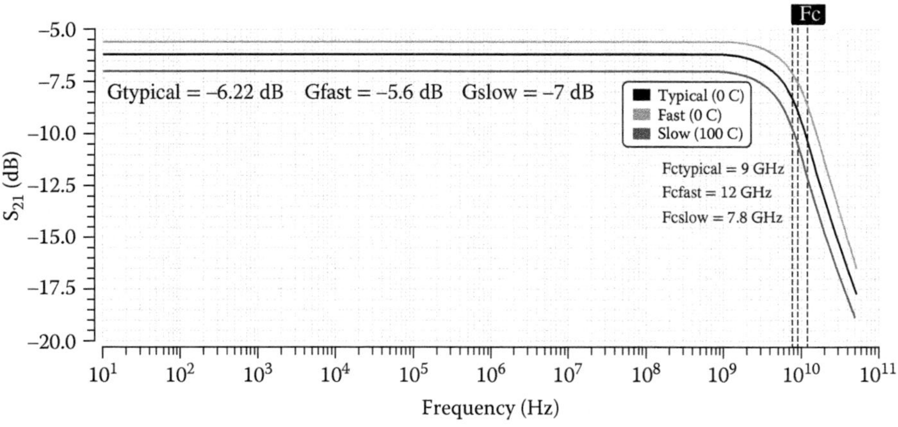

FIGURE 10.13
S_{21} parameter for PT variations of the DC-Coupled PMOS sensor.

As mentioned, the gain of this sensor is below unity because it is actually composed of two source followers on each signal (substrate and ground). One source follower is used to sense the differential signal and one as buffer to match the 50 Ω impedance of the transmission line. Therefore, no stage with

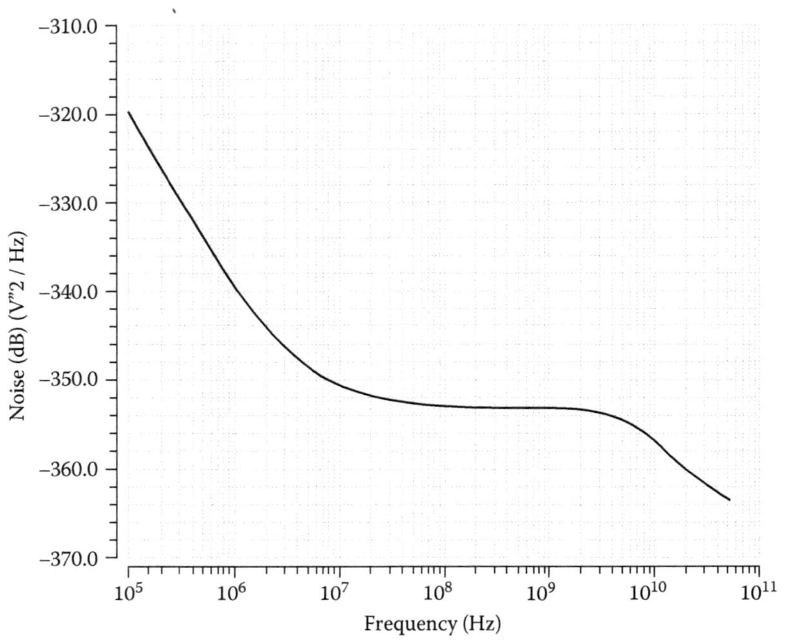

FIGURE 10.14
Noise (V^2/Hz) of the DC-Coupled PMOS sensor.

gain>1 is used. This is the most significant disadvantage of this measurement method. The bandwidth is high partially due to the low gain. Because it is a symmetrical design, statistical mismatch variations are subtle. The output intrinsic noise is very low and does not pose a significant measurement limitation. Without the buffers, it consumes little area. The buffers need to be wide to achieve an output resistance of 50 Ω and, therefore, take more area. The sensor circuit can be deployed across the die at various measure points and the outputs can be multiplexed to the buffers or an on-chip digitizer.

10.3.3.4 Differential Amplifier Based Substrate Noise Sensor

Differential amplifiers can be used to measure the differential voltage between two probes. Connecting one input to a substrate contact and the other to ground, the amplified differential output signal, which is less sensitive to interconnect and ground bounce crosstalk, can be measured by a spectrum analyzer [6,27,28]. The input voltage DC level cannot be equal to the V_{SS} potential because the differential pair transistor would be cut off. Therefore, capacitive AC coupling must be implemented at the input. In Figure 10.15, a typical differential amplifier based sensor is illustrated [27,28].

M_1, M_2 constitutes the differential pair transistors, which are resistively loaded by R_2 and R_3. Resistive loading is used instead of an active device

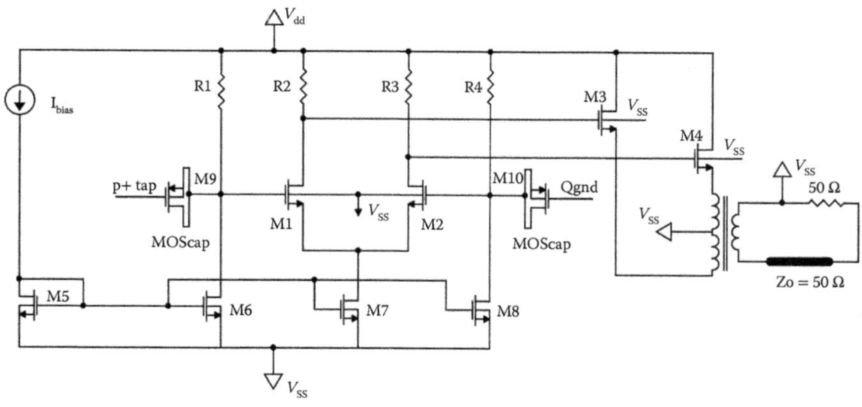

FIGURE 10.15
Differential amplifier substrate noise sensor.

to limit the parasitic capacitance at the differential pair output nodes that would degrade frequency performance. M_7 is the current source transistor of the differential pair and M_5, which is driven by a bias current generator, is used to create a bias voltage for M_5, M_6 and M_7. The drain currents of M_6, M_8 cause a voltage drop to R_1 and R_4, creating the bias potential for M_1 and M_2. The differential outputs are driven by two NMOS buffer transistors for impedance matching and current amplification and are converted to single ended mode by a balun.

From an AC perspective, a high pass filter is present in the input composed of the input capacitance implemented by MOScap and the parallel combination of R_1, r_{ds6} and R_4, r_{ds8} respectively. The MOScap is a capacitance formed by the gate oxide of a MOSFET. A PMOS transistor is used because it is shielded by an N-Well and its body can be connected to its drain and source, forming a parasitic capacitor. The gate capacitance is given by the following equation:

$$C_{gg} = \frac{\varepsilon_{ox}}{t_{ox}} WL \tag{10.23}$$

where ε_{ox} is the silicon dioxide dielectric permittivity, t_{ox} the oxide thickness, and W, L the MOSFET dimensions. The cutoff frequency is calculated assuming a first order high pass filter as:

$$f_c = \frac{1}{2\pi R_i C_i} = \frac{1}{2\pi\left(r_{ds1} \,||\, R_1\right) C_{gg9}} = \frac{1}{2\pi\left(r_{ds8} \,||\, R_4\right) C_{gg10}} \tag{10.24}$$

where R_i, C_i are the filter total resistance and capacitance, r_{ds} the drain-to-source resistance, and C_{gg} the gate capacitance. The gain and frequency response can be estimated from the AC mode half circuit depicted in Figure 10.16. R_s is the parallel combination of R_1 and r_{ds6} and $R_d = R_2$. The half circuit

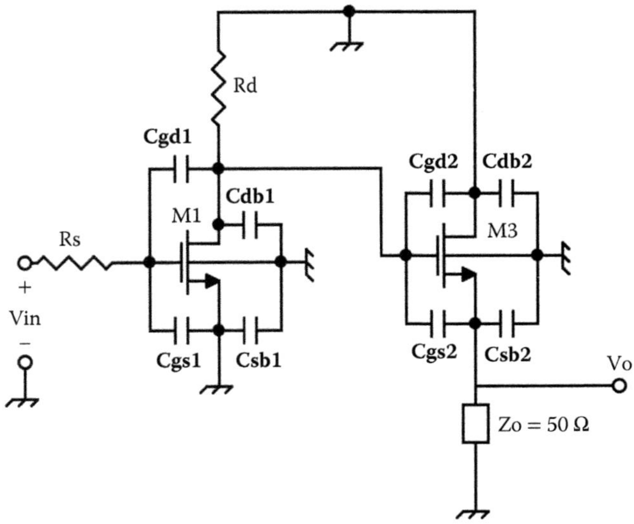

FIGURE 10.16
AC mode half circuit of the differential amplifier substrate noise sensor.

DC gain is easily calculated by multiplying the current gain of M_1 with the resistance seen on the drain node of M1 and with the source follower voltage gain. The source follower was sized to have output resistance of approximately 50 Ω, therefore, its gain, A_{vSF}, is about 1/2.

$$A_{vhc} = \frac{g_{m1}}{2} \cdot R_d \cdot A_{vSF} \tag{10.25}$$

where g_m is the gate-to-source transconductance. The differential gain is double the half circuit gain:

$$A_{vhc} = g_{m1} \cdot R_d \cdot A_{vSF} \tag{10.26}$$

To make a simplified calculation of the frequency response at first order the zero-value time constant analysis is used [29]. The resistance seen from each capacitor in the circuit is calculated, neglecting all the other capacitors and setting all independent sources to zero. The R-C time constants that arise are then summed. The $\Omega_{-3\ \mathrm{dB}}$ frequency is then

$$\omega_{-3\mathrm{dB}} = \frac{1}{\sum \tau_k} = \frac{1}{\sum R_k C_k} \tag{10.27}$$

where τ_k is the time constant, C_k the corresponding capacitance, and R_k the calculated resistance. The time constant for C_{gs1} is

$$t_k = C_{gs1} R_S \tag{10.28}$$

C_{gd2} and C_{db1} are parallel and the time constant of their sum is

$$t_k = \left(C_{gd2} + C_{db1}\right) R_d \tag{10.29}$$

C_{db2} and C_{sb1} are short circuited to ground and are neglected. The time constant of C_{sb2} is

$$t_k = C_{sb2} Z_o \tag{10.30}$$

where Z_o is the transmission line characteristic impedance. The time constant of C_{sb2} is very small, so it can also be neglected. C_{gd1} is a feedback capacitor and its equivalent miller capacitor value is $(1 + A)C_{gd1}$ where $A = g_{m1}R_d$. Its time constant is

$$t_k = \left[R_s\left(1 + g_{m1}R_d\right) + R_d\right] C_{gd1} \tag{10.31}$$

C_{gs2} is also a feedback capacitor, for an amplifier with a positive gain less than unity. Therefore, its time constant is:

$$t_k = \left[R_d\left(1 - A_{vSF}\right) + Z_o\right] C_{gs2} \tag{10.32}$$

The cutoff frequency therefore is

$$\omega_{-3dB} = \frac{1}{C_{gs1}R_s + \left(C_{gd2} + C_{db1}\right)R_d + \left[R_s\left(1 + g_{m1}R_d\right) + R_d\right]C_{gd1} + \left[R_d\left(1 + A_{vSF}\right) + Z_o\right]C_{gs2}} \tag{10.33}$$

where C_{gs} is the gate-to-source capacitance, C_{gd} the gate-to-drain capacitance, C_{db} the bulk-to-drain capacitance, A_{vSF} the source follower gain, Z_o the transmission line characteristic impedance, and R_d the drain load resistance. This equation can be simplified by keeping only the dominant terms.

$$\omega_{-3\mathrm{dB}} = \frac{1}{R_d\left[C_{gd2} + C_{db1} + \left(g_{m1}R_s + 1\right)C_{gd1}\right]} \tag{10.34}$$

Therefore, the value of R_d and the size of the transistors determine the first order dominant pole. To increase the gain, R_d or g_m must be increased. But, increasing R_d or increasing the transistor width to increase g_m also increases the time constants, shifting the dominant pole to a lower frequency. The constant gain-bandwidth (GBW) is indicative of the frequency response and speed of voltage amplifiers and depends on the technology parameters, topology, biasing, and parasitic capacitances. R_d can be used to control the gain-frequency tradeoff. GBW product is generally given by

$$\begin{pmatrix} A_v \approx g_m R_d \\ \omega_{-3dB} \approx \dfrac{1}{R_d C_{total}} \end{pmatrix} \Rightarrow GBW \approx \frac{g_m}{C_{total}} \tag{10.35}$$

where A_v is the gain, g_m the gain stage transconductance, R_d the gain stage total output resistance, and C_{tot} the total capacitance at the output node. In Table 10.7, transistor dimensions, resistors, and bias current values are shown. R_1, R_4 are poly resistors without silicide to reduce the area they consume and take advantage of the high resistance per square. A lot of current is drawn through R_1 and R_2 and their value is low, so poly resistors with silicide, which have low resistance per square and can support a high current without being damaged, were used. Simulation results are given in Table 10.8. In Figure 10.17, the S_{21} parameter is illustrated for process and temperature variations and, in Figure 10.18, the intrinsic noise voltage of the sensor is also depicted.

The advantage of this sensor is the high gain and immunity from crosstalk and ground bounce coupling due to differential signaling. Because the DC potential at the inputs is not compatible with the substrate potential, capacitive ac coupling must be used to limit the bandwidth at low frequencies. The sensor takes up more area compared to previous solutions and has a relatively low 1 dB compression point, which is, nevertheless, adequate for the application given the very low magnitude of the signals measured. The sensor output DC voltage is within the range of analog multiplexers and the

TABLE 10.7

Device Size and Values of the Differential Amplifier Based Substrate Noise Sensor

Transistor	**Dimensions (W/L)**
M1	50 μm/60 nm
M2	50 μm/60 nm
M3	53.1 μm/60 nm
M4	53.1 μm/60 nm
M5	2.1 μm/60 nm
M6	1 μm/120 nm
M7	29.4 μm/60 nm
M8	1 μm/120 nm
M9	160 μm/1 μm
M10	160 μm/1 μm
Resistor	**Value**
R1	9 KΩ
R2	650 Ω
R3	650 Ω
R4	9 KΩ
Current Source	**Value**
Ibias	200μA

TABLE 10.8

Simulation Results of the Differential Amplifier Based Substrate Sensor

Specification	Value
Gain	6.2 dB
Gain PT variation	5.8 dB to 7 dB
Gain mismatch variation	4.65 dB to 6.6 dB
Bandwidth	9 MHz–7 GHz
Bandwidth PT variation	10.9 MHz–4.2 GHz to 8.3 MHz–11 GHz
Bandwidth mismatch variation	9.2 MHz–6.9 GHz to 9 MHz–7.8 GHz
Power consumption	7 mW
1dB compression point	91 mV
Area	2002 μm2

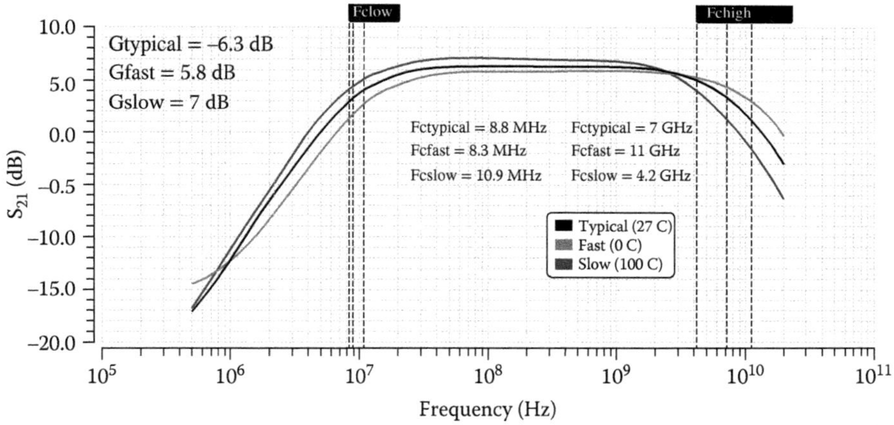

FIGURE 10.17
S_{21} parameter for PT variations of the differential pair based substrate noise sensor.

output signals can be easily multiplexed. The output of this amplifier is differential. The two output signals are

$$V_{d+} = \frac{A_v}{2}\left(V_{\text{substrate}} - V_{\text{qgnd}}\right)$$
$$V_{d-} = -\frac{A_v}{2}\left(V_{\text{substrate}} - V_{\text{qgnd}}\right) \quad (10.36)$$

where A_v is the differential gain, $V_{\text{substrate}}$ the substrate noise voltage, and V_{qgnd} the analog quiet ground voltage. These two signals are subtracted by the balun to create the single ended output signal:

$$V_o = A_v\left(V_{\text{substrate}} - V_{\text{qgnd}}\right) \quad (10.37)$$

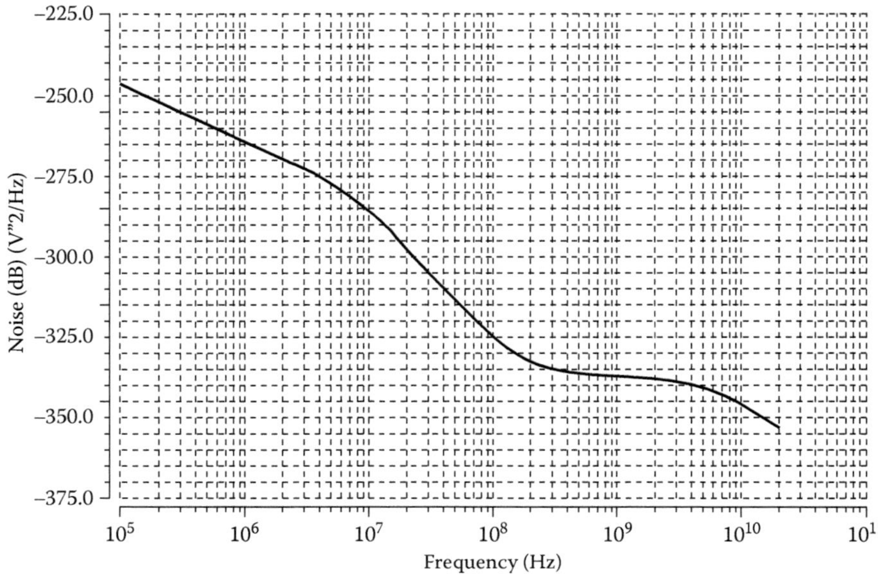

FIGURE 10.18
Noise (V^2/Hz) of the differential pair based substrate noise sensor.

Therefore, the total CMRR is the CMRR of the differential amplifier plus the CMRR of the balun and is higher than other implementations that use only the balun for signal subtraction. In Figure 10.19, the CMRR of the differential amplifier is illustrated.

10.3.3.5 Subthreshold Biased Differential Amplifier Substrate Noise Sensor

A variation of the circuit in Section 2.3.4 was proposed in [6] that employs subthreshold biasing of the differential pair and open drain output buffers to enhance performance. It can achieve higher bandwidth or the same bandwidth as the previous sensor, but with reduced power consumption. To compare the two circuits, the subthreshold differential amplifier was designed for the same bandwidth and the sensor efficiency was evaluated. The circuit is shown in Figure 10.20.

Typically, subthreshold circuits are used in low power applications and the currents in the circuit are small. These circuits are slow in frequency response. But, for the same high current as in full inversion biasing, similar or better frequency performance can be obtained with careful design. This is true because of the exponential relationship of transconductance in subthreshold operation. The drain current of an n-channel MOSFET in the subthreshold region is [30]

$$I_D = \frac{W}{L} I_0 e^{\frac{V_{GS}-V_{TH}}{mV_T}\left(1-e^{-\frac{V_{DS}}{V_T}}\right)} = \frac{W}{L} I_0 e^{\frac{V_{GS}-V_{TH}}{mV_T}} \tag{10.38}$$

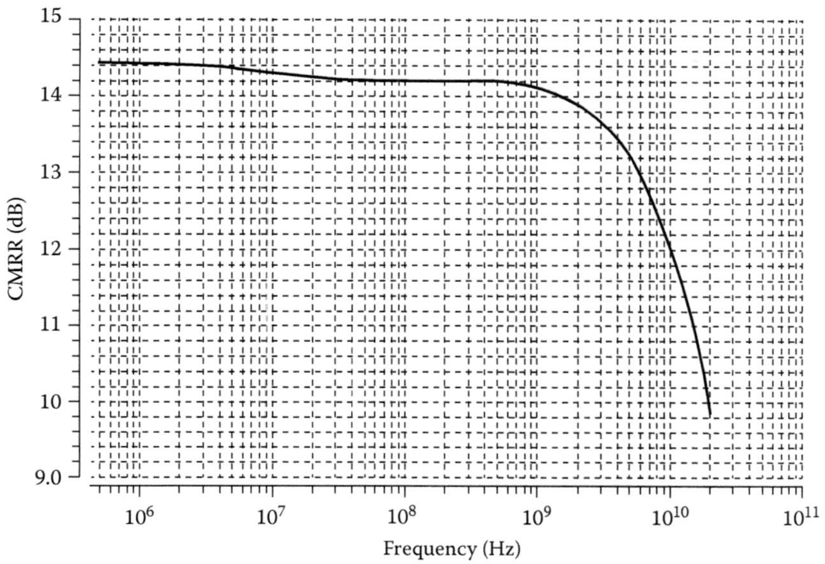

FIGURE 10.19
CMRR of the differential pair based substrate noise sensor.

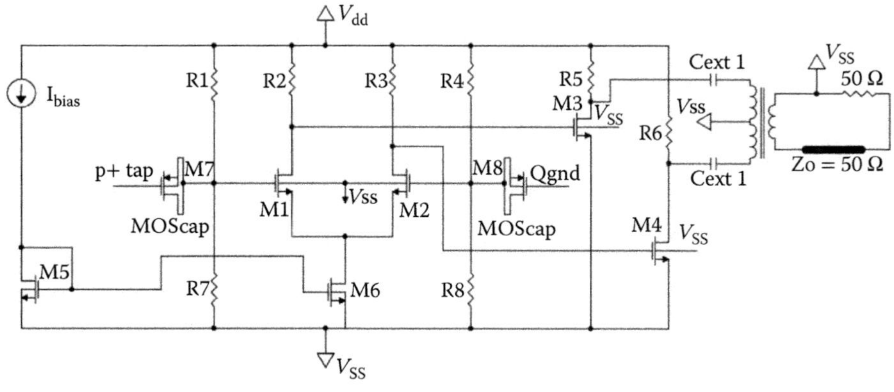

FIGURE 10.20
Subthreshold biased differential pair substrate noise sensor.

where *W/L* is the aspect ratio of the transistor, V_{GS} and V_{DS} the gate-to-source and drain-to-source voltages, V_{TH} the threshold voltage, V_T the thermal voltage (26 mv), I_0 the current for $V_{GS} = V_{TH}$, and m the slope factor ranging from 1,2 to 2 depending on technology. The subthreshold transconductance is given by:

$$g_{msub} = \frac{I_D}{mV_T} \tag{10.39}$$

For a given bias, current is higher than the strong inversion transconductance:

$$g_m = \frac{2I_D}{V_{GS} - V_{TH}} \tag{10.40}$$

To achieve high gain in the differential amplifier, without increasing the value of the load resistance too much, the transconductance of the differential pair transistors must be as high as possible. The tranconductance in strong inversion in relation to the drain current and $V_{eff} = V_{GS}\text{-}V_{TH}$ is

$$g_m = \frac{2I_D}{V_{eff}} \tag{10.41}$$

To increase g_m, V_{eff} must be decreased and the transistor is forced to operate at moderate inversion just above the threshold. With a small increase in the transistor width, in order to draw the bias current, and adjusting the biasing point to subthreshold, we can take advantage of the higher transconductance in this region, but with the disadvantage of a small increase in the capacitance. The transconductance-to-capacitance ratio is increased and, therefore, the GBW value from equation 10.34 is increased too.

Apart from subthreshold biasing, the circuit operation is almost the same as previously described in Section 2.3.4, but with two more modifications. First, open drain outputs are used instead of source follower buffers. The output resistance is the parallel combination of the drain-source resistance r_{ds} and the drain resistance. The major advantage over using a source follower topology is that the body effect is eliminated and does not reduce the gain. The second modification is using purely resistive dividers for adjusting the differential pair bias point instead of a mosfet current source and a resistor. The advantage is less parasitic capacitance from the input to ground, which lowers the signal magnitude by leakage. Finally, external capacitors are used to AC couple the balun because at DC it is a short circuit to ground that would shunt the drain DC potential. Transistor dimensions and resistor and bias current values are given by Table 10.9 and the simulation results by Table 10.10. In Figure 10.21, the S_{21} parameter is illustrated for process and temperature variations and in Figure 10.22 the intrinsic noise voltage of the sensor is also depicted. Finally, in Figure 10.23 CMRR of the subthreshold differential pair sensor is illustrated.

It can be seen from Table 10.10 that this sensor performs better than the differential amplifier based substrate noise sensor in Figure 10.15 in terms of efficiency. For the same bandwidth, it consumes less power and, if designed for the same power, has larger bandwidth. The large variation of the gain in PT variation simulation is due to the variation of the resistor R_5 and R_6 values.

TABLE 10.9

Device Size and Values of the Subthreshold Differential Amplifier Substrate Noise Sensor

Transistor	Dimensions (W/L)
M1	52 μm/60 nm
M2	52 μm/60 nm
M3	15 μm/60 nm
M4	15 μm/60 nm
M5	2.1 μm/60 nm
M6	33.6 μm/60 nm
M7	160 μm/1 μm
M8	160 μm/1 μm
Resistor	**Value**
R1	10 KΩ
R2	720 Ω
R3	720 Ω
R4	10 KΩ
R5	500 Ω
R6	500 Ω
R7	8 KΩ
R8	8 KΩ
Current Source	**Value**
Ibias	200μA

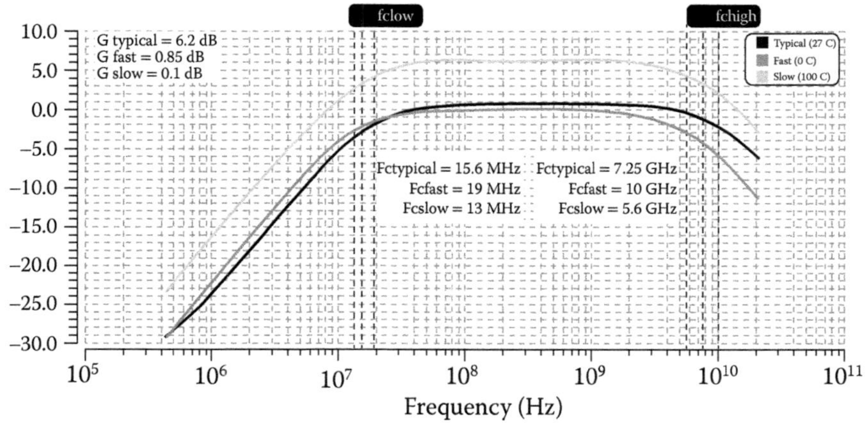

FIGURE 10.21

S_{21} parameter for PT variations of the subthreshold differential pair substrate noise sensor.

10.3.3.6 On-Chip Digitizer

The on-chip samplers are used as on-chip instrumentation to convert the analog signal sensed to digital data by successive comparisons with an external

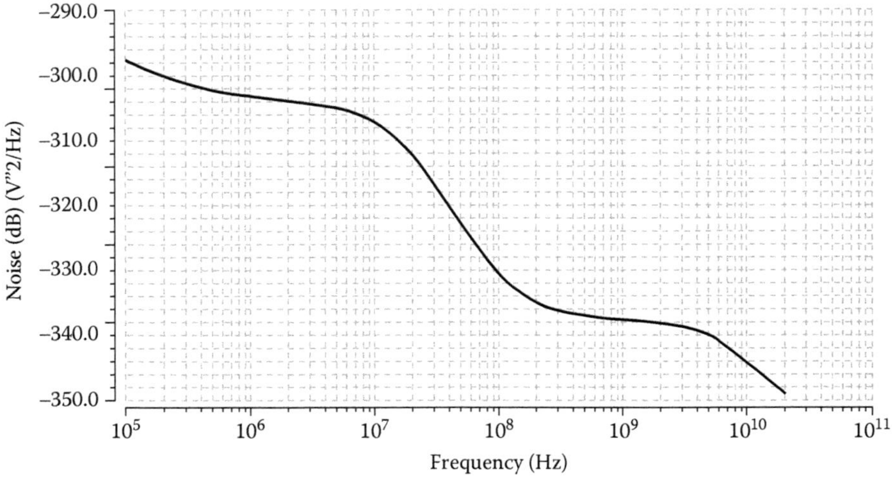

FIGURE 10.22
Noise (V^2 /Hz) of the subthreshold differential pair substrate noise sensor.

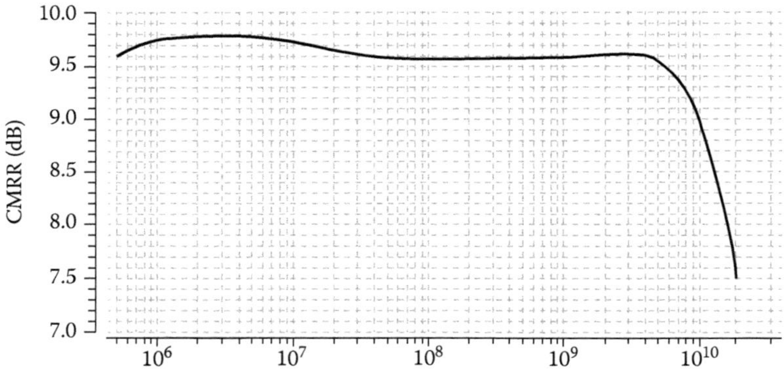

FIGURE 10.23
CMRR of the subthreshold differential pair substrate noise sensor.

programmable reference voltage [2,15,18,20–26]. A single bit sequence of samples is generated for variable delay between the sampling time and the digital switching clock that injects noise. For each delay time, a reference voltage is swept at the reference input of a comparator. Change in the comparator output indicates the magnitude of the sampled signal within some resolution limits. This method requires the sampled signal to be periodic. Because the measured substrate noise of a mixed-signal IC is generated mainly by the digital core clock, it can assumed to be periodic and synchronized with it [2,15,26]. From a frequency domain perspective, this method is using undersampling. Reconstruction of the signal is possible because, due to periodicity,

TABLE 10.10

Simulation Results of the Differential Amplifier Based Substrate Sensor

Specification	Value
Gain	6.1 dB
Gain PT variation	0.1–6.2 dB
Gain mismatch variation	6 dB to 6.3 dB
Bandwidth	15.6 MHz to 7.25 GHz
Bandwidth PT variation	13 MHz–5.6 GHz to 19 MHz–10 GHz
Bandwidth mismatch variation	15.6 MHz–7.2 GHz to 15.8 MHz–7.3 GHz
Power consumption	4.4 mW
1dB compression point	91 mV
Area	2204 μm2

the extended Shannon-Nyquist sampling theorem is applicable. The digital data produced by the sampler are much less sensitive to interconnect crosstalk and ground bounce and can be routed off chip without compromising signal integrity [2,15,26]. Furthermore, the various signals from probes across the die can be multiplexed to the input of the on-chip sampler. Only one pin is needed for the data to be routed off-chip.

The type of comparator usually used is a single flip-flop comparator composed of two latches [15]. The positive feedback latch illustrated in Figure 10.24 is a simple type of latch with good performance and high gain.

The operation of the positive feedback latch is divided into two modes. When CLKP is LOW (CLKN is HIGH), the outputs QP and QN are proportional to the input signals DP and DN and the latch operates in transparent mode. In transparent mode the latch is operating as a differential pair amplifier with a differential gain of

$$A_{\text{vcomp}} = g_{m3,4} \cdot R_{\text{L}} \tag{10.42}$$

where g_m is the gate-to-source transconductance and R_L is the load resistance.

When CKLP switches high and CLKN low, the latch enters latched mode [15]. The positive feedback action amplifies QP-QN differential signal exponentially to the full swing levels: V_{dd} and V_{DD}-$R_L I_{Bias}$. These levels also called CML levels must be converted to CMOS levels again at the output. In latched mode the differential QP-QN output becomes a digital signal with its value depending on the polarity of the differential input. If a second latch is connected to the outputs of the first with inversed clocks, it holds the output value during the transparent mode of the comparator. The two latches together form a flip-flop [15]. The complete circuit of the on-chip digitizing sensor is show in Figure 10.25.

To reduce the switching noise injected by clock distribution circuits, a single phase clock is used and both latches use differential clocks [2,15,26].

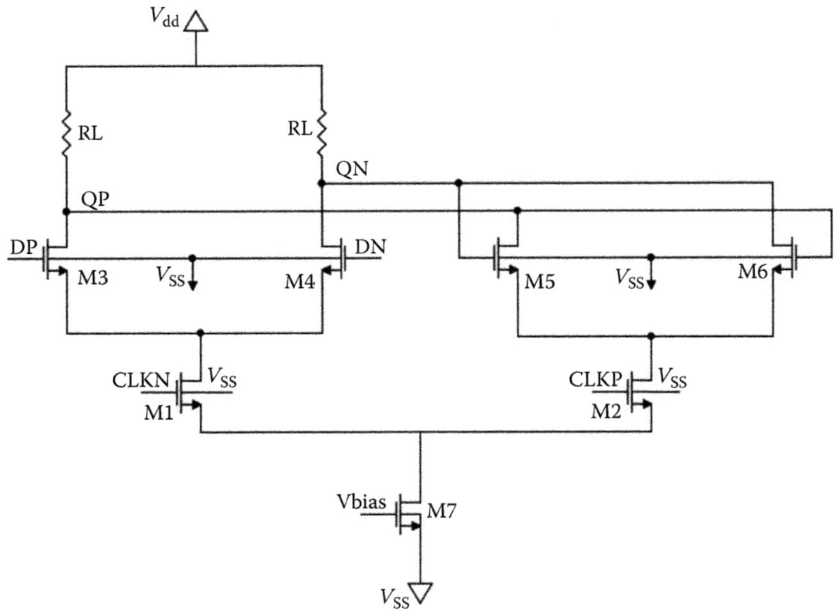

FIGURE 10.24
Positive feedback latch.

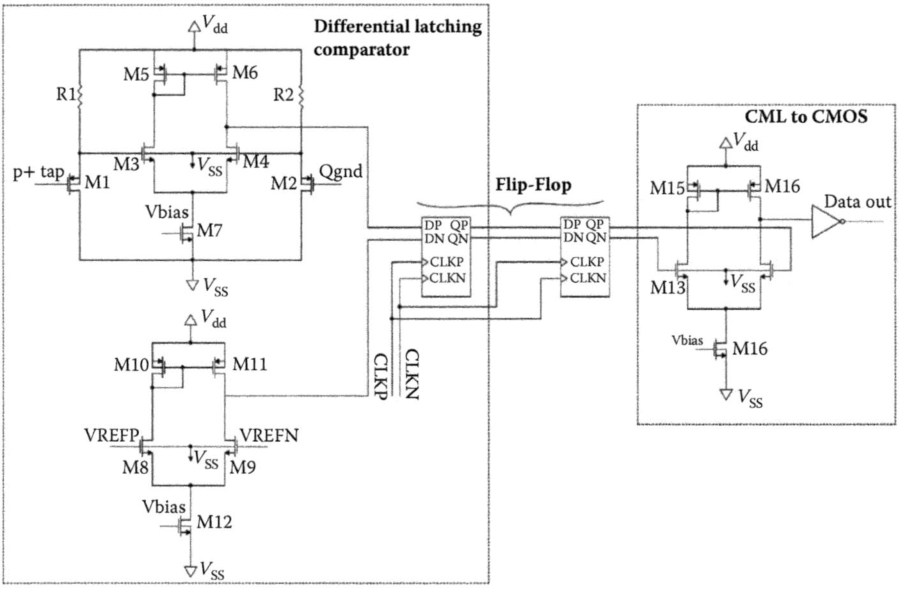

FIGURE 10.25
On-chip digitizing sensor.

Both the input signal and the reference voltage are differential to reduce contamination from crosstalk coupling. Differential amplifiers with active load and single outputs are used to amplify the input and reference signals (for symmetry) and convert them from differential to single ended. The substrate noise and quiet ground signals are level-shifted to a compatible with the amplifier input DC potential using PMOS level shifters with resistive source degeneration. The sampling flip-flop converts the analog signal to digital and a CML to CMOS converter is used in the output to make the output levels compatible with most digital devices. In transparent mode the output of the comparator is equal to

$$V_{\text{ocomp}} = A_{\text{vcomp}}\left[(DP - DN)A_{\text{vin}} - (VREFP - VREFN)A_{\text{ref}}\right] \tag{10.43}$$

where

$$A_{\text{vin}} = A_{vSF} \cdot g_{m1} \cdot r_{ds4} \tag{10.44}$$

$$A_{\text{ref}} = g_{m6} \cdot r_{ds9} \tag{10.45}$$

where A_{vin} is the gain of the substrate noise differential input stage (M1–M6, R1–R2), A_{vSF} the input stage source follower gain (M1–R1, M2–R2), A_{ref} the gain of the reference voltage input stage (M8–M12), g_m the gate-to-source transconductance, and r_{ds} the drain-to-source resistance.

The resolution depends on the minimum voltage able to activate the positive feedback in the first latch. For better resolution, higher gain is needed in the input, which limits the bandwidth. The timing diagram of the sampling sensor is illustrated in Figure 10.26. The analog input is compared to V_{REF} and the result is sampled on the rising edge of SCLK, which is a divided version of the digital clock delayed by T_d [2,15]. The output is LOW if V_{in} is lower than V_{REF} and high if the opposite is true. V_{IN} is sampled at time T_S while performing successive comparisons sweeping V_{REF}, and monitoring the digital output. The procedure is repeated of multiple T_S values and the signal is reconstructed at the post-processing phase [2,15]. Transistor dimensions and resistor values of the on-chip digitizing sensor and the feedback latches are given by Tables 10.11 and 10.12 accordingly. The simulation results are given by Table 10.13. In Figure 10.27 the S_{21} parameter of the sensor analog front end (from the input to the comparator transparent mode) is illustrated for process and temperature variations and in Figure 10.28 the simulation to determine the sensor resolution is depicted.

To achieve a resolution of 1.5 mV, a gain of 17 dB had to be used, which limits the bandwidth significantly. The on-chip sampler requires only one pin. Inputs and outputs can be multiplexed and, due to differential signaling, the common mode noise is removed. The output is digital and laboratory

TABLE 10.11

Device Size and Values of the on-Chip Digitizing Sensor

Transistor	Dimensions (W/L)
M1, M2	8 μm/60 nm
M3, M4, M8, M9	37.05 μm/60 nm
M5, M6, M10, M11	30 μm/60 nm
M7, M12	12.35 μm/60 nm
M13, M14	14.82 μm/60 nm
M15, M16	24 μm/60 nm
M17	9.88 μm/60 nm
Resistor	**Value**
R1, R2	435 Ω

TABLE 10.12

Device Size and Values of Positive Feedback Latches

Transistor	Dimensions (W/L)
M1, M2	24.7 μm/60 nm
M3, M4	14.82 μm/60 nm
M5, M6	22.23 μm/60 nm
M7	9.88 μm/60 nm
Resistor	**Value**
RL	900 Ω

TABLE 10.13

Simulation Results of the on-Chip Digitizing Sensor

Specification	Value
Gain	17 dB
Bandwidth	0 Hz–2.5 GHz
Resolution	1.5 mV
Power consumption	7.7 mW
1dB compression point	90 mV
Area	3000 μm2

equipment, such as a spectrum analyzer, is not needed. The disadvantages are the large area, the complexity, and low BW. Very high caution has to be taken during the physical layout to isolate the sensor so it does not inject additional switching noise [2].

10.3.4 Comparison of the Simulated Designs

In Table 10.14, the sensors that analyzed and simulated in Section 10.3.3 are compared in light of the requirements for high performance substrate noise sensors as discussed in Section 10.3.2. In Table 10.15, the simulation results for all the sensors are presented for easier comparison.

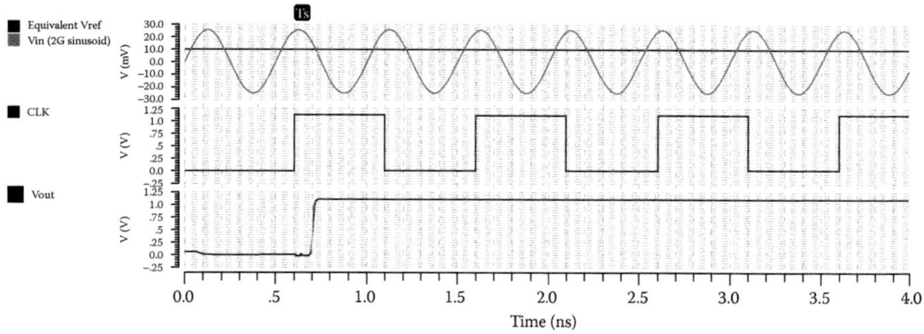

FIGURE 10.26
Timing diagram of the on-chip digitizing sensor.

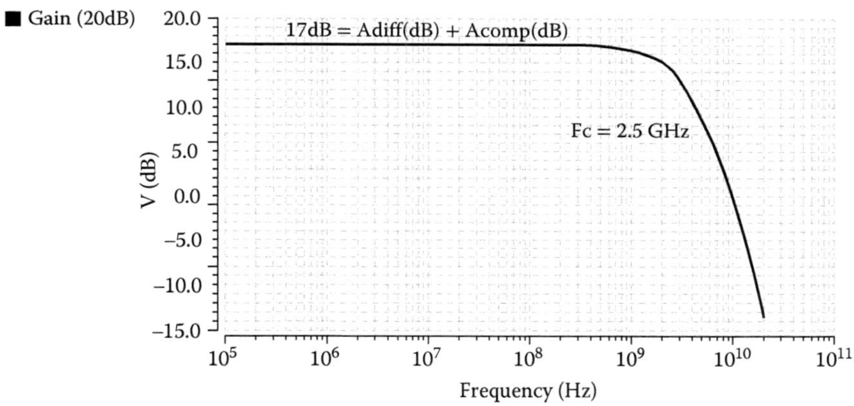

FIGURE 10.27
S_{21} parameter of the analog front end of the on-chip digitizing sensor.

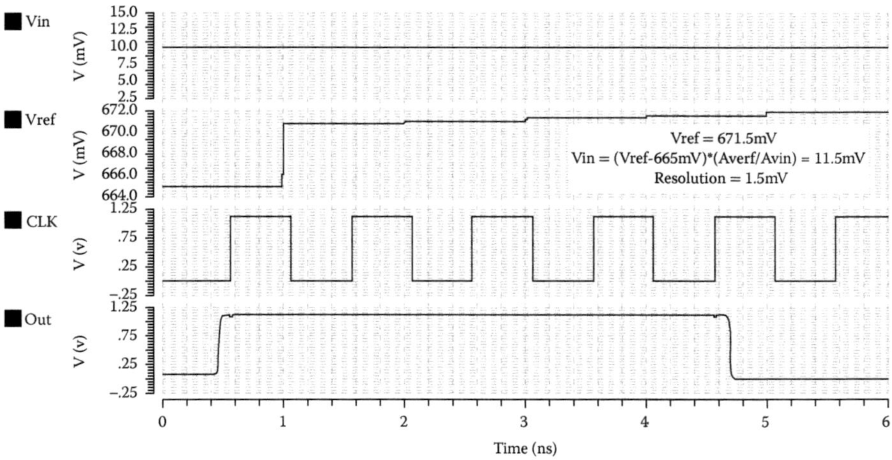

FIGURE 10.28
Resolution of the on-chip digitizing sensor.

TABLE 10.14

Compliance of the Simulated Sensors with the Requirements of a High-Performance Substrate Noise Sensor

	Bulk-Driven NMOS	SF+GM	DC Coupled PMOS	Differential Amplifiers	Subthreshold Diff. Amps	On-Chip Digitizer
Does not affect noise	✔	✔	✔	✔	✔	
Immune from interconnect crosstalk			✔	✔	✔	✔
High BW	✔	✔	✔	✔	✔	
Expands down to DC	✔	✔	✔			✔
Compact	✔	✔	✔			
Multiplexed		✔	✔	✔	✔	✔

Depending on the application, some sensors are more suitable than others. DC coupled sensors may satisfy the requirements, but they have gain less than unity. For modern wireless and mobile SoC applications, the bandwidth must be as high as possible to be able to measure all of the intermodulation products injecting noise into the substrate. As mm-wave and 5G are emerging as the future in wireless communications, even higher bandwidth will be required from substrate noise sensors and accurate substrate noise sensing will be more crucial than ever to avoid problematic designs and reduce design time. None of the simulated sensors can measure up to 10 GHz and beyond. Therefore, there is a a lot of interest in designing a high performance substrate noise sensor fitting the requirements of Section 10.3.2 that can also achieve very high bandwidth.

Acknowledgment

This work was financed by the Aristotle University Research Committee and the Special Account for Research Grants in the framework of the "Strengthening Research Activity in Aristotle University of Thessaloniki—2015" action, and under the project "SoCs NOISE & SIGNAL INTEGRITY"—Project Code Number 93296.

TABLE 10.15

Comparison of the Simulated Substrate Crosstalk Noise Sensors

	Bulk-Driven Sensor	SF-Gm Sensor	DC-Coupled Sensor	Diff. Amplifier Sensor	Subthreshold Diff. Amp. Sensor	On-Chip Digitizer
Gain	–3.75 dB	1.1 dB	–6.2 dB	6.2 dB	6.1 dB	17 dB
Gain PT variation	–3 dB to –3.7 dB	–2.1 dB to 1.6 dB	–7 dB to –5.6 dB	5.8 dB to 7 dB	0.1 dB to 6.2 dB	–
Gain mismatch variation	–	0.94 dB to 1.26 dB	–6.25 dB to –6.2 dB	4.65 dB to 6.6 dB	6 dB to 6.3 dB	–
Bandwidth	0 Hz–7.18 GHz	0 Hz–7 GHz	0 Hz–9 GHz	9 MHz–7 GHz	15.6 MHz –7.25 GHz	0 Hz–2.5 GHz
Bandwidth PT variation	0 Hz–7.18 GHz to 0 Hz–7.6 GHz	0 Hz–4.7 GHz to 0 Hz–8.2 GHz	0 Hz–7.8 GHz to 0 Hz–12 GHz	10.9 MHz–4.2 GHz to 8.3 MHz –11 GHz	13 MHz–5.6 GHz to 19 MHz –10 GHz	–
Bandwidth mismatch variation	–	0 Hz–6.85 GHz to 0 Hz–6.9 GHz	0 Hz–9.3 GHz to 0 Hz–9.7 GHz	9.2 MHz–6.9 GHz to 9 MHz –7.8 GHz	15.6 MHz–7.2 GHz to 15.8 MHz –7.3 GHz	–
Power consumption	8.3 mW	11 mW	6.8 mW	7 mW	4.4 mW	7.7 mW
1dB compression point	980 mV	130 mV	280 mV	91 mV	91 mV	90 mV
Area	180 μm^2	570 μm^2	928 μm^2	2002 μm^2	2204 μm^2	3000 μm^2
Resolution	–	–	–	–		1.5 mV

References

1. A. Helmy and M. Ismail, *Substrate Noise Coupling in RFIC's*, New York, NY: Springer, 2008.
2. C. Iorga, "Measurement, Suppression, and Prediction of Digital Switching Noise Coupling in Mixed-Signal System-on-Chip Applications," PhD Diss., Stanford University, 2007.
3. E. Backenius, "On Reduction of Substrate Noise in Mixed-Signal Circuits," PhD Diss., Linköpings University, 2005.
4. E. Charbon, R. Gharpurey, P. Miliozzi, R. G. Meyer, and A. Sangiovanni-Vincentelli, *Substrate Noise: Analysis and Optimization for IC Design*, Dordrecht: Kluwer Academic Publishers, 2001.
5. M. Van Heijningen, J. Compiet, P. Wambacq, S. Donnay, M. G. E. Engels, and I. Bolsens, Analysis and experimental verification of digital substrate noise generation for epi-type substrates, *IEEE Journal of Solid-State Circuits*, vol. 35, 1002–1008, 2000.
6. T. Noulis, P. Merakos, E. Lourandakis, E. Stefanou, and Y. Moisiadis, Wide-band substrate crosstalk sensor for wireless SoC applications, *Sensors and Actuators, A: Physical*, vol. 293, 144–152, 2016.
7. M. Badaroglu, S. Donnay, H. De Man, Y. Zinzius, G. Gielen, T. Fonden, and S. Signell, Modeling and experimental verification of substrate noise generation in a 220k-gates WLAN system-on-chip with multiple supplies, *IEEE Journal of Solid-State Circuits*, vol. 38, 1250–1260, 2003.
8. M. Nagata, "On-chip Measurements Complementary to Design Flow for Integrity in SoC," in *Paper presented at the 44th annual Design Automation Conference*, San Diego, CA, June 04–08, 2007.
9. M. Van Heijningen, M. Badaroglu, S. Donnay, M. Engels, and I. Bolsens, "High-Level Simulation of Substrate Noise Generation Including Power Supply Noise Coupling," in *Paper presented at the 37th annual Design Automation Conference*, Los Angeles, CA, June 5–9, 2000.
10. N. Azuma, S. Shimazaki, N. Miura, M. Nagata, T. Kitamura, S. Takahashi, M. Murakami, K. Hori, A. Nakamura, K. Tsukamoto, M. Iwanami, E. Hankui, S. Muroga, Y. Endo, S. Tanaka, and M. Yamaguchi, "Measurements and Simulation of Substrate Noise Coupling in RF ICs with CMOS Digital Noise Emulator," in *Paper presented at the 9th International Workshop on Electromagnetic Compatibility of Integrated Circuits*, Nara, Japan, December 15–18, 2013.
11. S. Bronckers, G. Van der Plas, G. Vandersteen, and Y. Rolain, Substrate noise coupling mechanisms in lightly doped CMOS transistors, *IEEE Transactions on Instrumentation and Measurement*, vol. 59, 1727–1733, 2010.
12. M. Shen, J. Mikkelsen, O. K. Jensen, and T. Larsen, "A Compact P+ Contact Resistance Model for Characterization of Substrate Coupling in Modern Lightly Doped CMOS Processes," in *Paper presented at the 7th European Microwave Integrated Circuits Conference*, Amsterdam, Netherlands, October 29–30, 2012.
13. T. Noulis and P. Baumgartner, Substrate crosstalk analysis flow for submicron CMOS system on chip, *IET Electronics Letters*, vol. 51, 953–954, 2015.
14. D. K. Su, M. J. Loinaz, S. Masui, and B. A. Wooley, Experimental results and modeling techniques for substrate noise in mixed-signal integrated circuits, *IEEE Journal of Solid-State Circuits*, vol. 28, 420–430, 1993.

15. C. Iorga, *Noise Coupling in Integrated Circuits: A Practical Approach to Analysis, Modeling, and Suppression*, Newbury Park, CA: NoiseCoupling.com, 2008.
16. T. Blalack, J. Lau, F. J. R. Clement, and B. A. Wooley, "Experimental Results and Modeling of Noise Coupling in a Lightly Doped Substrate," in *Paper presented at the 1996 International Electron Devices Meeting*, San Francisco, CA, December 8–11, 1996.
17. X. Aragones and A. Rubio, Experimental comparison of substrate noise coupling using different wafer types, *IEEE Journal of Solid-State Circuits*, vol. 34, 1405–1409, 1999.
18. M. Nagata, T. Okumoto, and K. Taki, A built-in technique for probing power supply and ground noise distribution within large-scale digital integrated circuits, *IEEE Journal of Solid-State Circuits*, vol. 40, 813–819, 2005.
19. M. Nagata and A. Iwata, Substrate noise simulation techniques for analog–digital mixed LSI design, *Analog Integrated Circuits and Signal Processing*, vol. 25, 209–217, 2000.
20. M. Nagata, D. Kashima, T. Morie, and A. Iwata, Measurements and analyses of substrate noise waveform in mixed-signal IC environment, *IEEE Transactions on Computer-Aided Design of Integrated Circuits and Systems*, vol. 19, 671–678, 2000.
21. M. Nagata, J. Nagai, H. Hijukata, T. Morie, and A. Iwata, Physical design guides for substrate noise reduction in CMOS digital circuits, *IEEE Journal of Solid-State Circuits*, vol. 36, 539–549, 2001.
22. K. Makie-Fukuda, T. Anbo, T. Tsukada, T. Matsuura, and M. Hotta, Voltage comparator-based measurement of equivalently sampled substrate noise waveforms in mixed-signal integrated circuits, *IEEE Journal of Solid-State Circuits*, vol. 31, 726–731, 1996.
23. X. Weize and E.G. Friedman, "A Substrate Noise Circuit for Accurately Testing Mixed-Signal Ics," in *Paper presented at the 2002 IEEE International Symposium on Circuits and Systems*, Scottsdate, AZ, May 26–29, 2002.
24. M. Nagata and A. Iwata, "Substrate Crosstalk Analysis in Mixed Signal CMOS Integrated Circuits," in *Paper presented at the 22nd Asia and South Pacific Design Automation Conference*, Yokohama, Japan, January 25–28, 2000.
25. B. T. Kang, N. Vijaykrishnan, M. J. Irwin, and I. D. Duarte, "The Substrate Noise Detector for Noise Tolerant Mixed-Signal IC," in *Paper presented at the 2003 IEEE System-on-Chip Conference*, Portland, OR, September 17–20, 2003.
26. C. Iorga, Y. C. Lu, and R. W. Dutton, A built-in technique for measuring substrate and power-supply digital switching noise using PMOS-based differential sensors and a waveform sampler in System-on-Chip applications, *IEEE Transactions on Instrumentation and Measurement*, vol. 56, 2230–2237, 2007.
27. M. Van Heijningen, J. Compiet, P. Wambacq, S. Donnay, and I. Bolsens, "A Design Experiment for Measurement of the Spectral Content of Substrate Noise Inmixed-Signal Integrated Circuits," in *Paper presented at the 1999 Southwest Symposium on Mixed-Signal Design*, Tucson, AZ, April 11–13, 1999.
28. B. E. Owens, S. Adluri, P. Birrer, R. Shreeve, S. K. Arunachalam, K. Mayaram, T. S. Fiez, Simulation and measurement of supply and substrate noise in mixed-signal ICs, *IEEE Journal of Solid-State Circuits*, vol. 40, 382–391, 2005.
29. T. C. Carusone, D. Johns, and K. Marti, *Analog Integrated Circuit Design*, 2nd edition, Hoboken, NJ: John Wiley & Sons, 2012.
30. Y. Tsividis, *Operation and Modeling of the Transistor MOS*, New York, NY: McGraw-Hill, 1987.

11

3D IC Floorplanning Based on Thermal Interactions

Boris Vaisband and Eby G. Friedman

CONTENTS

11.1 Introduction

Heterogeneous applications are an important focus of the integrated circuit (IC) community. Collaborative efforts to increase the functionality of heterogeneous systems are driven by the increasing demand for application diversity. Different types of circuits (e.g., optical and temperature sensors, analog and digital CMOS, and L1, L2, and L3 memory) are combined within these systems, generating greater integration complexity that produces a variety of significant design and manufacturing obstacles [1]. A three-dimensional (3D) structure has been demonstrated to be an efficient platform for heterogeneous integration [2,3]. Different layers are individually manufactured, and stacked to form a heterogeneous system. Each layer can be optimized according to the specific requirements of the circuits on that layer (e.g., different substrates, individually optimized power networks, local synchronization). The different layers are connected by through-silicon vias (TSVs). TSVs within a 3D structure are short vertical interconnects (~ 20 μm [4]) that replace the long, high impedance global interconnect, greatly alleviating global signaling issues.

With scaling and increasing on-chip power densities, on-chip temperatures are rising. This issue has drawn a great deal of attention within the IC community, and remains a major concern in high performance circuits [3,5]. Higher on-chip temperatures lead to degradation in the electrical characteristics of the on-chip devices; for example, lower mobility carriers [6,7] and threshold voltage drift in transistors [8,9], as shown in Figure 11.1. In 3D structures, thermal congestion is greatly exacerbated [3,10,11] since heat is trapped between the different layers, and the thermal path(s) to the heat

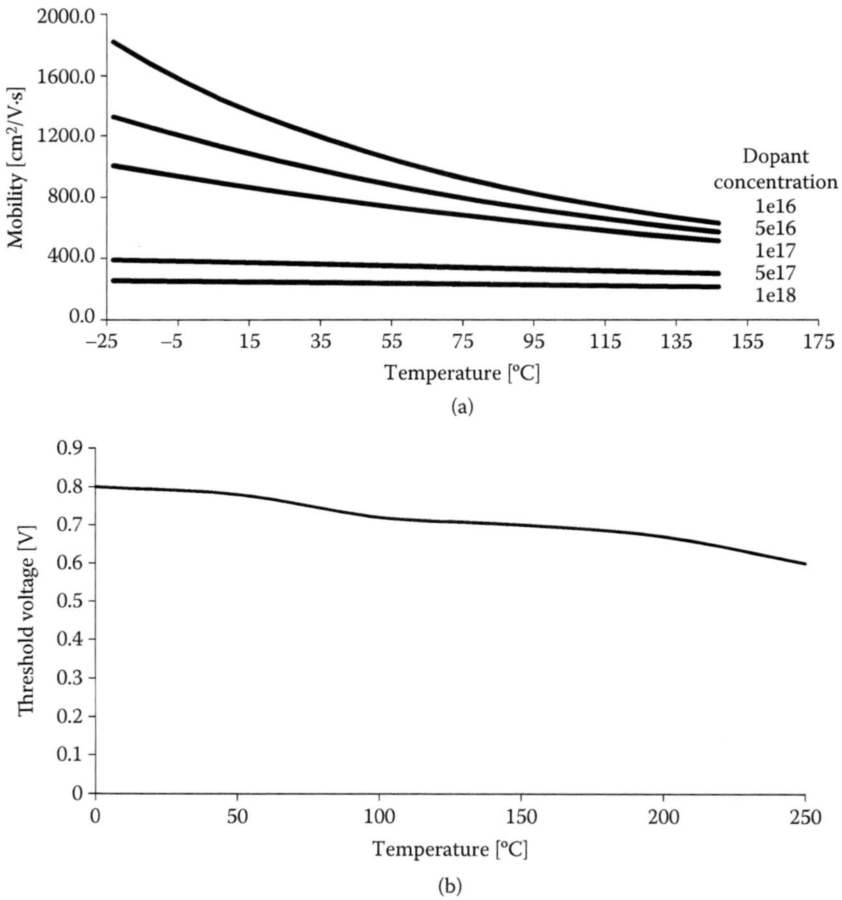

FIGURE 11.1
Effect of increasing temperature on (a) mobility, and (b) threshold voltage.

sink(s) is (are) significantly more complicated [3]. The issue of moving the heat towards the heat sink has been shown to be a practical obstacle in next generation high performance 2D and 3D ICs [3,12].

To address issues of thermal congestion, a comprehensive solution at all design levels is required. Many solutions at the manufacturing level have been proposed (e.g., thermal TSVs [13,14], microfluidic channels [15], thermoelectric cooling [16]). Thermal aware floorplan algorithms have also been developed to address the issue of thermal effects at the physical placement level [17–20]. A thermal aware methodology is required to produce a high performance thermal aware 3D system. Though previously manufactured 3D structures are mostly comprised of similar layers (e.g., memory cube structures [21–23]), future circuits will include heterogeneous integrated circuits composed of a diverse variety of material substrates, each manufactured in different processes and integrated within a single three-dimensional system (e.g., photonic IC, magnetic RAM, analog and digital CMOS, MEMS). In this chapter, a thermal aware floorplan methodology for heterogeneous three-dimensional systems, and a thermal interaction driven floorplan algorithm are proposed.

The rest of the chapter is composed of the following sections. Thermal interaction is proposed as the cost function within the proposed algorithm, as presented in Section 11.2. A system level floorplan methodology is described in Section 11.3. The proposed algorithm is reviewed in Section 11.4. Evaluation of the algorithm on benchmark circuits is described in Section 11.5. Some conclusions are offered in Section 11.6.

11.2 Thermal Interaction

Lowering the peak temperature of a system has classically been the focus of thermal aware algorithms. Thermal interactions among the different modules within a system are not considered when optimizing for this metric because temperature is the measure of heat at a specific location and does not describe the effect of this heat on other locations. Lowering the maximum temperature of the circuit, therefore, does not guarantee the functionality and performance of all of the modules within a circuit. As exemplified in Figure 11.2, the maximum temperature, denoted as T_{max}, is generated in module A, while a temperature denoted as T_b ($T_b < T_{max}$) is generated in module B, placed in close proximity to module C. Module C contains a temperature sensitive analog sense amplifier which is greatly affected by the heat generated in module B. The thermal interaction between modules B and C is not considered by existing thermal aware algorithms that aim to lower the peak temperature. The result is functional failure or degradation in performance of the circuit in module C.

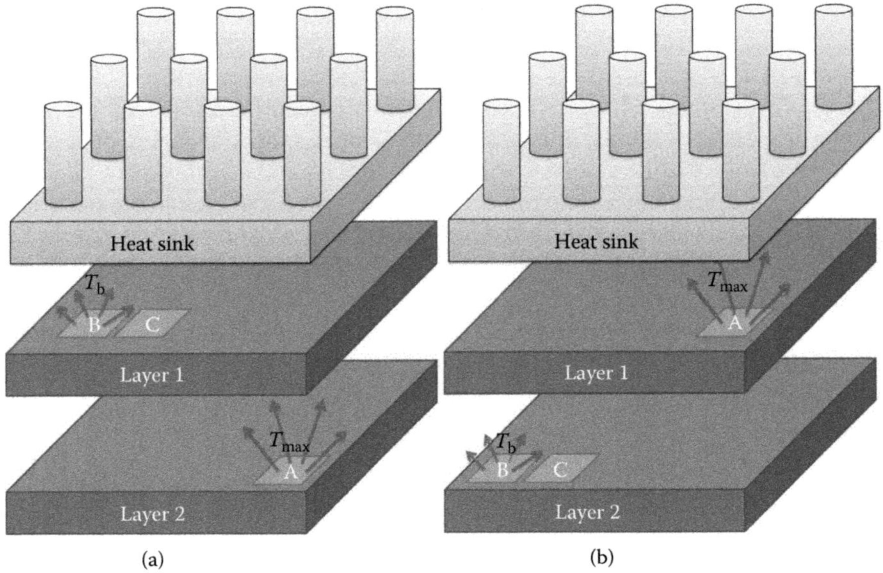

FIGURE 11.2
Illustration of placement of three modules A, B, and C, (a) initial state, and (b) after application of a temperature driven floorplan algorithm. T_{max} and T_b are the temperatures generated, respectively, in modules A and B.

Note that the thermal interaction between the different modules is the key issue, and an algorithm is proposed that minimizes these interactions.

A unique floorplan consisting of differently sized rectangular modules is typically placed on each layer of a heterogeneous 3D structure. To develop circuits in a thermally efficient manner, each module is assigned two parameters: thermal aggressiveness t_a and thermal sensitivity t_s. The thermal aggressiveness is based on the heat flux generated within each module as well as the material properties of the module and surrounding area. This parameter characterizes the thermal effect of a module on neighboring modules. The thermal sensitivity is determined according to the electrical nature of a specific module (e.g., sensitivity of a critical data path to degradation of mobility of carriers due to an increase in temperature). Issues such as delay uncertainty, noise margin, threshold voltage variations, and dynamic range can also determine the thermal sensitivity of a module. This parameter therefore characterizes the sensitivity of a module to thermal coupling from neighboring modules. The parameters t_a and t_s range from 0 to 1 and are integrated within the floorplanning algorithm to determine the most thermally efficient location of each module. The thermal influence between modules i and j is defined in Equation 11.1,

$$T_{\text{inf}}^{i,j} \equiv t^i{}_a t^j{}_s + t^j{}_a t^i{}_s \qquad (11.1)$$

The thermal influence therefore ranges from 0 to 2, where the lower bound considers those modules that do not thermally influence another module (and may be placed in close proximity to each other) and the upper bound considers those modules that exhibit a significant influence on another module (and must be placed far from each other). Note that thermal influence is not a vector, therefore $T_{\text{inf}}^{i,j} = T_{\text{inf}}^{j,i}$. Each module can be both thermally aggressive and thermally sensitive depending upon the nature of the circuit and other thermal characteristics. The thermal influence is therefore the mutual influence of the modules on each other and not the influence of a certain module on another module. (This is expressed as the superposition of the individual effects in Equation 11.1).

Another important component of the thermal interaction is the characteristics of the thermal path between modules i and j described by the thermal resistance $R^{i,j}{}_{th}$ [°C/W] [3,24]

$$R^{i,j}{}_{th} = \Delta x / (k \cdot A) \tag{11.2}$$

The thermal conductivity k [W/(m°C)] is a property of the material, Δx [μm] is the length of the thermal path between two modules, and A [m^2] is the surface area of the thermal path through which the heat is transferred. Similar to thermal influence, the thermal resistance is not a vector, therefore $R^{i,j}{}_{th} = R^{j,i}{}_{th}$. Finally, the thermal interaction between modules i and j is

$$T^{i,j}{}_{int} \equiv T^{i,j}{}_{inf} / R^{i,j}{}_{th} \tag{11.3}$$

Thus, $T^{i,j}{}_{int} = T^{j,i}{}_{int}$ as the result of the division of two scalars. Thermal interaction [W/°C] is similar to thermal conduction adjusted according to the thermal influence between modules. Substituting Equation 11.1 into 11.3,

$$T^{i,j}{}_{int} = (t^{i}{}_{a} t^{j}{}_{s} + t^{j}{}_{a} t^{i}{}_{s}) / R^{i,j}{}_{th} \tag{11.4}$$

The cost function of the proposed algorithm minimizes $max(T_{int})$ and therefore minimizes the conduction of heat between thermally aggressive and thermally sensitive modules. This algorithm does not necessarily lower the maximum temperature, but rather generates a 3D IC floorplan less susceptible to functional failure and/or performance degradation.

11.3 3D Floorplan Methodology

A floorplan methodology for 3D systems is proposed in this section. The proposed methodology is applicable to heterogeneous 3D systems comprised of multiple layers with different properties. Each layer may be manufactured from different material substrates and individual processes and composed of a variety of different circuits (e.g., processor, MEMS structures, and photovoltaic devices). Due to the heterogeneous nature of the different layers (e.g., different substrate materials), modules may not be shifted among different layers within a 3D system. Based on the proposed thermal interactions, the different layers of the 3D structure are ordered in a thermal-aware manner followed by a thermal interaction-driven placement of the modules within each layer. The floorplan methodology is composed of two steps, as described in Sections 11.3.1 and 11.3.2.

11.3.1 Global Floorplan

The different layers of the 3D stack are vertically ordered according to previously determined thermal parameters and application-specific constraints (*e.g.*, the image sensors need to be placed on the top layer). Some of the layers may be thermally passive (neither thermally aggressive nor sensitive) and act as thermal insulators to separate the thermally aggressive planes from the thermally sensitive layers. An example of a global floorplan is illustrated in Figure 11.3a. In this figure, the photonic IC layer, which is thermally passive, is used as a thermal insulator and placed in the middle of the 3D stack to separate the thermally aggressive processor from the thermally sensitive RF layer.

11.3.2 Local Floorplan

The second step consists of moving different modules within each layer to separate the thermally aggressive modules from the thermally sensitive modules. Each module is placed to minimize thermal interactions with the remaining modules, including modules on other layers. Based on the heterogeneity concept, modules may be shifted within their original layer only, but the thermal interactions for each module are determined with respect to all other modules, including those modules on other layers of the 3D system. A local floorplan is illustrated in Figure 11.3b. In this figure, the thermally aggressive clock generating module is placed at the bottom right corner of the layer, while the thermally sensitive analog circuit is placed at the top left corner of the layer. The sensor modules in this example are assumed to be constrained by location and cannot be shifted. A similar process is performed on the other layers to minimize thermal interactions within a 3D system.

11.4 Proposed Algorithm

The proposed thermal interaction-driven 3D floorplan algorithm generates a thermally efficient placement for all component modules making up a 3D

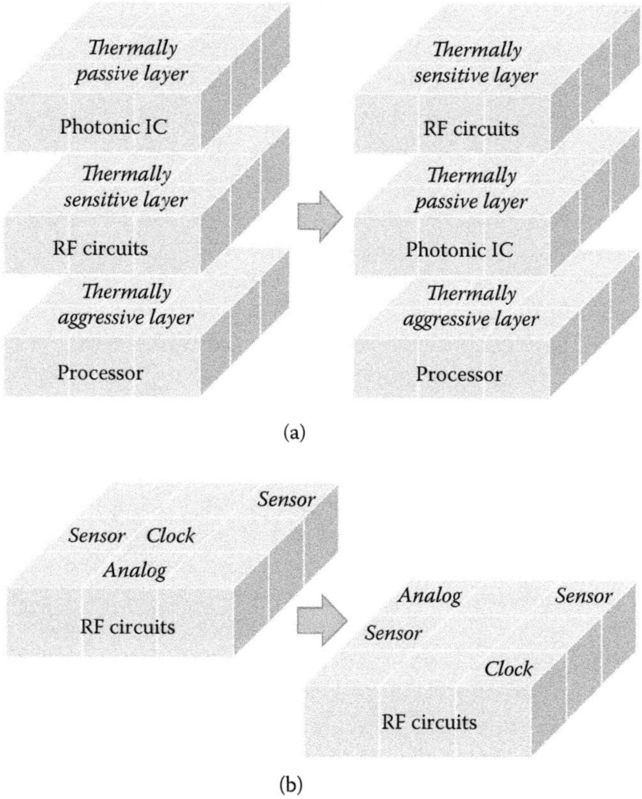

FIGURE 11.3
Thermal-aware methodology based on (a) a global floorplan, and (b) a local floorplan. The clock generating module is thermally aggressive, while the analog circuit is thermally sensitive.

system. The algorithm is based on the TCG-S floorplanning algorithm [25] proposed by Lin and Chang. The algorithm is a combination of the TCG (transitive closure graph) and the packing sequence of the SP (sequence pair) representations. Both horizontal and vertical transitive closure graphs (C_h, C_v) and a packing sequence ($\Gamma\_$) are used in TCG-S to represent a floorplan. Simulated annealing [26] is the iterative engine of the proposed algorithm.

The cost function of the new algorithm is driven by both area and thermal interactions using weighted coefficients for each of the two components W_{Area} and W_{Tint} for, respectively, area and thermal interactions. To compensate for the different magnitudes of the cost function components, each weight function is normalized according to the average cost, as described in [27]. The overall cost function is therefore

$$\mathrm{Cost} = W_{\mathrm{Area}}\left(\mathrm{Area}/\mathrm{Area}^{\mathrm{avg}}\right) + W_{T\,\mathrm{int}}\left(T_{\mathrm{int}}/T_{\mathrm{int}}^{\mathrm{avg}}\right) \tag{11.5}$$

where $W_{Area} + W_{Tint} = 1$. The area and thermal interaction normalization parameters are, respectively, $Area^{avg}$ and T_{int}^{avg}. These parameters are extracted over multiple algorithm evaluations for each circuit. The optimization is performed on a per layer basis for the area, and for the complete 3D system for the thermal interaction, i.e. the thermal interaction is optimized between modules on all layers, while the area is optimized individually on each layer. The thermal interaction for intra-layer modules is determined according to the thermal influence from Equation 11.1 and the thermal resistance from Equation 11.2, where Δx is determined according to the distance between modules within a layer (after each perturbation of the floorplan). The thermal interaction for the inter-layer modules is determined similarly where the thermal resistance, in this case, is determined according to the different thermal conductivity characteristics of the different layers, and the geometric and thermal properties of the connecting TSVs. To accommodate a multiple layer structure, during each cycle of the simulated annealing process, a single perturbation of a floorplan is performed on each layer and the resulting area is extracted prior to determining the thermal interaction of the overall system. The area is therefore minimized for each individual layer while thermal interactions are considered for all of the layers. After each cycle, the execution conditions are checked for each layer, and, if a layer has converged to a satisfactory solution, the algorithm is no longer executed on that layer, but continues to run only on those layers that have not yet converged. The algorithm either converges to a solution or reaches a cooling point, the execution is terminated, and the coordinates of all modules are produced. Pseudo-code for the algorithm is provided in Appendix 11.1.

11.5 Evaluation Results

The algorithm has been evaluated on a suite of existing MCNC benchmark circuits. The algorithm is evaluated on an Intel Core i5-2410M CPU @ 2.30 GHz machine with Windows 7 64-bit OS and implemented in C++. The thermal resistance is determined according to the thermal conductivity of silicon at 50°C (k = 138 [W/m°C]), size of the modules, and distance between the different modules placed within a layer. The thermal aggressiveness and thermal sensitivity in the benchmark circuits are randomly generated. To assess the quality of the resulting floorplan, a quality figure of merit η is used,

$$\eta \equiv T_{int}^{wc} / T_{int}^{max} \tag{11.6}$$

where T_{int}^{max} is the maximum thermal interaction, defined as the greatest thermal influence over the smallest thermal resistance, and T_{int}^{wc} is the worst case thermal interaction for a generated floorplan. A smaller value of η indicates a better result.

11.5.1 Suite of MCNC Benchmark Circuits

MCNC benchmark circuits are evaluated for different number of layers. Three to five layers are assumed for each benchmark circuit and the area of the floorplan is determined by the largest layer. Each test circuit is evaluated for different weight function ratios of W_{Area} and $W_{T\,int}$. The worst case thermal interaction, the quality parameter (η) from Equation 11.6, and computation run time are listed in Table 11.1 for each test case. All of the evaluations are compared to a baseline case ($W_{Area} = 1$ and $W_{T\,int} = 0$) describing a traditional area only driven floorplan algorithm. As the weight of the thermal interactions in the cost function increases, the worst case thermal interaction of the circuit decreases at the expense of larger area. This tradeoff between area and thermal interaction is exemplified in Figure 11.4 for the ami33 circuit (from the suite of MCNC benchmark circuits). For small weights of thermal interaction, the improvement in η is rapid while the increase in total area is slow. It is, therefore, not necessary to sacrifice significant area to greatly lower the thermal interactions among the modules of the circuit. Similar trends are exhibited by all of the MCNC benchmark circuits. The weight of the thermal interactions used by the floorplanning algorithm is an important design parameter that may be controlled to satisfy the area and thermal budgets of a system.

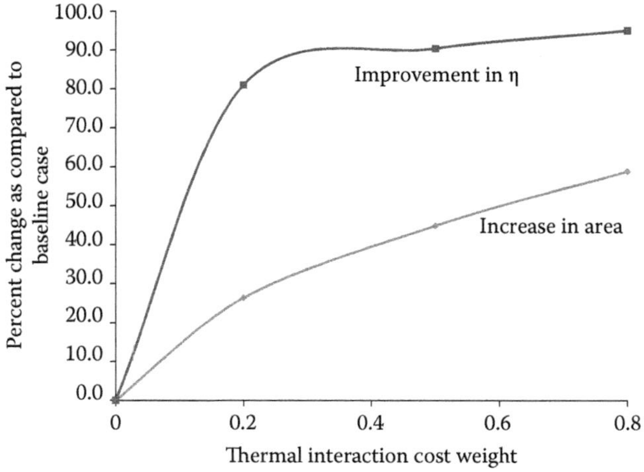

FIGURE 11.4
Percent increase in area and improvement in η as compared to the baseline case evaluated for the ami33 benchmark circuit.

TABLE 11.1

Application of Proposed Algorithm on MCNC Benchmark Circuits

Benchmark	Layers	Area Part	Thermal Interaction Part	Floorplan Area [mm^2]	T_{int}^{wc} [10^{-3}· W/°C]	η	Run Time [s]	% Improvement in η as Compared to Baseline Case
ami33	3	1	0	1.29	1.84	0.021	50.4	Baseline case
		0.8	0.2	1.63	0.36	0.004	25.7	81
		0.5	0.5	1.87	0.21	0.002	48.7	90.5
		0.2	0.8	2.05	0.12	0.001	37.6	95.2
ami49	3	1	0	38.4	2.43	0.028	125.1	Baseline case
		0.8	0.2	45.1	0.98	0.011	91.9	60.7
		0.5	0.5	45.3	0.59	0.007	94.1	75
		0.2	0.8	66.4	0.23	0.003	67.3	89.3
apte	4	1	0	52.3	6.1	0.085	3	Baseline case
		0.8	0.2	52.6	4.2	0.059	2.9	30.6
		0.5	0.5	54.5	0.52	0.007	2.8	91.8
		0.2	0.8	54.6	0.32	0.004	2.6	95.3
hp	4	1	0	14.2	3.11	0.037	4.5	Baseline case
		0.8	0.2	15.3	1.8	0.022	3.1	40.5
		0.5	0.5	16.1	0.88	0.011	4	70.3
		0.2	0.8	17.2	0.68	0.008	3.8	78.4
xerox	5	1	0	22.1	4.86	0.06	6.1	Baseline case
		0.8	0.2	23.1	1.12	0.014	4.1	76.7
		0.5	0.5	23.5	0.75	0.009	4.9	85
		0.2	0.8	24.2	0.4	0.005	4.4	91.7

Similar circuits are used on each layer to evaluate the application of the algorithm to homogeneous systems. To emphasize the applicability of the algorithm to heterogeneous systems, a 3D system consisting of different benchmark circuits on each layer is evaluated in Section 11.5.3.

11.5.2 Temperature Evaluation Using COMSOL

Temperature evaluations of thermally sensitive modules are provided in this section. A 3D system consisting of three layers with the Xerox circuit (from the suite of MCNC benchmark circuits) on each layer is thermally evaluated using COMSOL Multiphysics [28]. Three test cases with different cost function weights are evaluated. The power density ranges from 10^5 to 10^7 [W/m^2] and is generated according to the thermal aggressiveness and area of each module. The temperature is determined at the center of the nine most thermally sensitive modules. Simulation results are shown in Figure 11.5. A significant reduction of up to 41.5°C is exhibited between the baseline case ($W_{Area} = 1$ and $W_{Tint} = 0$) and test case 1 ($W_{Area} = 0.5$ and $W_{Tint} = 0.5$), and up to 56.3°C between the baseline case and test case 2 ($W_{Area} = 0.5$ and $W_{Tint} = 0.8$). This investigation emphasizes the importance of optimizing the physical floorplan for thermal interactions among modules rather than minimizing the peak temperature of a system.

The area penalty versus improvement in thermal interaction (η) data are listed in Table 11.2. The randomness of the assumed thermal characteristics adds uncertainty to the evaluated temperature in test case 2 (a thermal sensitivity of 0.67). The remaining temperatures decrease as the thermal interaction is emphasized (a larger weight in the cost function). Good correlation is demonstrated between minimization of thermal interactions and lower temperatures.

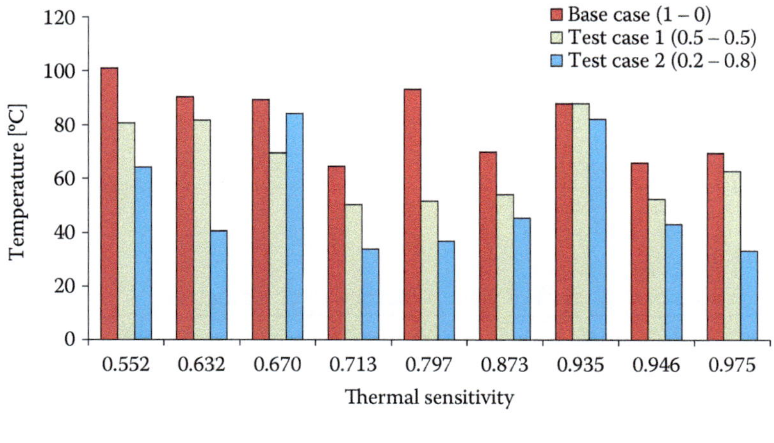

FIGURE 11.5
Temperature evaluation using COMSOL for three different test cases. Cost function weights are denoted in the label for each test case in the following format: area weight–thermal interaction weight.

11.5.3 Heterogeneous 3D Systems

A heterogeneous system is evaluated in this section. This system consists of three layers with a different MCNC benchmark on each layer (apte, hp, and xerox). TSVs are integrated on each layer, occupying 10% of the area. Different substrate materials are assigned to each layer to emulate different circuit types; for example, Si for digital CMOS, GaAs for RF circuits, and Ge for photovoltaic cells [29]. Various TSV materials with different thermal conductivities are also evaluated. The materials and thermal conductivities of the substrate and TSVs on each layer are listed in Table 11.3.

Results of the evaluation are listed in Table 11.4. The area of the system is recorded on a per layer basis, although the maximum area determines the actual die size (assuming that all layers must be the same size). The test circuit is evaluated for different weight function ratios of W_{Area} and $W_{T\,int}$ in terms of the worst case thermal interaction, quality parameter (η), and computational run time. All of the evaluations are compared to a baseline case ($W_{Area} = 1$ and $W_{T\,int} = 0$) similar to Section 11.5.1. Improvement in η of up to 73.9% is exhibited at the expense of an increase of 39.8% in total area. The tradeoff between die area and worst case thermal interaction is shown in Figure 11.6. Note that the apte benchmark on the first layer of the 3D heterogeneous benchmark circuit requires the largest area among all layers; the total die area is therefore based on layer 1. Similar to the homogeneous case, the improvement in η is rapid as compared to the increase in total die area for small thermal interaction weights.

TABLE 11.2

Area and η for Three Test Cases Evaluated for Temperature Using COMSOL

	Area (mm²)	η
Baseline case (W_{Area}=1 and $W_{T\,int} = 0$)	21.95	0.031
Test case 1 ($W_{Area} = 0.5$ and $W_{T\,int} = 0.5$)	40.8	0.01
Baseline case ($W_{Area} = 0.2$ and $W_{T\,int} = 0.8$)	67.63	0.005

TABLE 11.3

Heterogeneous 3D System Test Case Structure and Parameters

Layer	Benchmark	Number of Modules	Substrate Material/ Thermal Conductivity [W/m°C]	TSV Material/ Thermal Conductivity [W/m°C]
1	apte	9	Si/138	Tungsten/174
2	hp	11	GaAs/40	Carbon nanotube/3000
3	xerox	10	Ge/45	Copper/390

TABLE 11.4

Heterogeneous 3D System Test Case

Area Part	Thermal Interaction Part	Layer	Layer Area (mm²)	T_{int}^{WC} [10^{-3}· W/°C]	η	Run Time [s]	% Improvement in η as Compared to Baseline Case
1	0	1	47.5	1.6	0.023	1.8	Baseline case
		2	10.3				
		3	22				
0.8	0.2	1	47.5	1.16	0.017	1.6	36.1
		2	11.6				
		3	22.3				
0.5	0.5	1	51.1	0.84	0.012	1.6	47.8
		2	13.8				
		3	23				
0.2	0.8	1	66.4	0.41	0.006	1.5	73.9
		2	14.2				
		3	23.8				

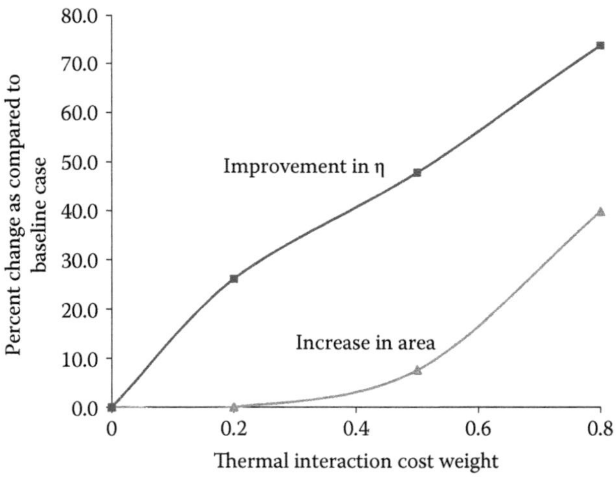

FIGURE 11.6
Percent increase in area and improvement in worst case thermal interaction as compared to baseline case, evaluated for the heterogeneous 3D structure.

11.6 Conclusions

Rising temperatures in 3D ICs may lead to circuit failure and degradation in performance. Alleviating thermal effects between aggressive and sensitive modules within heterogeneous 3D systems is the focus of this chapter.

A floorplan methodology and algorithm are proposed to mitigate heat congestion. The primary concept behind the proposed methodology is to lower thermal interactions between different thermally aggressive and thermally sensitive parts of a 3D IC.

A thermal interaction floorplan algorithm for heterogeneous 3D systems is developed and implemented in C++. The cost function of the algorithm considers both area and the thermal interactions among the modules within a 3D IC. Evaluation of this algorithm on several benchmark circuits is provided. The results are compared to the baseline case ($W_{Area} = 1$ and $W_{T\,int} = 0$), and a significant improvement in η is achieved, ranging from 30% to 95%. The floorplans generated by the proposed algorithm have also been evaluated for temperature using COMSOL. A maximum improvement of 56.3°C is achieved as compared to the baseline case. Evaluation of the algorithm on a heterogeneous 3D system is also presented and, as compared to the baseline case, reveals improvement in η from 26% to 74%.

Acknowledgment

This research is supported by the Ministry of Education, academic research fund (AcRF), TIER 2, Singapore under Grant No. MOE2014-T2-2-105.

Appendix 11.1

Characterization and Pseudo-Code of Algorithm

Algorithm Inputs

- 3D floorplan netlist
- Thermal aggressiveness of all modules
- Thermal sensitivity of all modules
- Thermal characteristics of layers (substrate and TSVs)
- Cost function weights (area and thermal interaction)

Algorithm Outputs

- 3D floorplan
- Maximum thermal interaction T_{int}^{wc}

Algorithm Pseudo-Code

```
Thermal interaction driven heterogeneous 3D floorplan
Generate random initial solution
while simulated annealing break conditions not met do
    for each non-converged layer do
        Perturb modules in layer
        Determine current area cost
    end for
    Determine worst case thermal interaction across all layers
    Determine total cost for each layer (area and thermal interaction)
    Evaluate simulated annealing conditions for each layer
    for each non-converged layer do
        if break condition met for layer i then
            Record area
            Set layer i as converged
            if layer i is last then
                Record thermal interaction
                Break
            end if
        end if
    end for
end while
```

References

1. E. Salman and E. G. Friedman, *High Performance Integrated Circuit Design*, New York, NY: McGraw-Hill Publishers, 2012.
2. V. F. Pavlidis and E. G. Friedman, *Three-Dimensional Integrated Circuit Design*, Burlington, MA: Morgan Kaufmann, 2009.
3. B. Vaisband, I. Savidis, and E. G. Friedman, "Thermal Conduction Path Analysis in 3-D ICs," in *Proceedings of the IEEE Symposium on Circuits and Systems*, pp. 594–597, June 2014.
4. I. Savidis and E. G. Friedman, Closed-form expressions of 3-D via resistance, inductance, and capacitance, *IEEE Transactions on Electron Devices*, vol. 56, no. 9, 1873–1881, September 2009.
5. N. Nelson, G. Briggs, M. Haurylau, G. Chen, H. Chen, D. H. Albonesi, E. G. Friedman, and P. M. Fauchet, "Alleviating Thermal Constraints While Maintaining Performance Via Silicon-Based On-Chip Optical Interconnects," in *Proceedings of the Workshop on Unique Chips and Systems*, pp. 45–52, March 2005.
6. S. S. Li and W. R. Thurder, The dopant density and temperature dependence of electron mobility and resistivity in N-type silicon, *Solid-State Electronics*, vol. 20, no. 7, 609–616, July 1977.

7. N. D. Arora, J. R. Hauser, and D. J. Roulston, Electron and hole mobilities in silicon as a function of concentration and temperature, *IEEE Transactions on Electron Devices*, vol. 29, no. 2, 292–295, February 1982.
8. J. J. Tzou, C. C. Yao, R. Cheung, and H. Chan, The temperature dependence of threshold voltages in submicrometer CMOS, *IEEE Electron Device Letters*, vol. 6, no. 5, 250–252, May 1985.
9. H. Wong, C. Liang, and N. W. Cheung, On the temperature variation of threshold voltage of GaAs MESFETs, *IEEE Transactions on Electron Devices*, vol. 39, no. 7, 1571–1577, July 1992.
10. B. Vaisband and E. G. Friedman, "3-D Floorplanning Algorithm to Minimize Thermal Interactions," in *Proceedings of the IEEE International Symposium on Circuits and Systems*, pp. 2133–2136, May 2015.
11. M. Jie, K. Kawakami, and A. K. Coskun, "Optimizing Energy Efficiency of 3-D Multicore Systems with Stacked DRAM under Power and Thermal Constraints" *Proceedings of the ACM/IEEE Design Automation Conference*, pp. 648–655, March 2012.
12. W. Huang, M. R. Stan, K. Skadron, K. Sankaranarayanan, S. Ghosh, and S. Velusamy, "Compact Thermal Modeling for Temperature-Aware Design," in *Proceedings of the IEEE/ACM Design Automation Conference*, pp. 878–883, June 2004.
13. M. G. Farooq, T. L. Graves-Abe, W. F. Landers, C. Kothandaraman, B. A. Himmel, P. S. Andry, C. K. Tsang, E. Sprogis, R.P. Volant, K.S. Petrarca, K. R. Winstel, J.M. Safran, T. D. Sullivan, F. Chen, M. J. Shapiro, R. Hannon, R. Liptak, D. Berger, and S. S. Iyer, "3-D Copper TSV Integration, Testing and Reliability," in *Proceedings of the IEEE International Electron Devices Meeting*, pp. 7.1.1–7.1.4, December 2011.
14. M. Nihei, A Kawabata, T. Murakami, M. Sato, and N. Yokoyama, "Improved Thermal Conductivity by Vertical Graphene Contact Formation for Thermal TSV," in *Proceedings of the IEEE International Electron Devices Meeting*, pp. 33.5.1–33.5.4, December 2012.
15. Z. Li and M. S. Bakir, "Electrical and Fluidic Microbumps and Interconnects for 3D-IC and Silicon Interposer," in *Proceedings of the IEEE SOC Conference*, pp. 159–164, September 2012.
16. G. J. Snyder, M. Soto, R. Alley, D. Koester, and B. Conner, "Hot Spot Cooling Using Embedded Thermoelectric Coolers," in *Proceedings of the IEEE Semiconductor Thermal Measurement and Management Symposium*, pp. 135–143, March 2006.
17. Z. Pingqiang, M. Yuchun, L. Zhouyuan, R. P. Dick, S. Li, Z. Hai, H. Xianlong, and Z. Qiang, "3D-STAF: Scalable Temperature and Leakage Aware Floorplanning for Three-Dimensional Integrated Circuits," in *Proceedings of the IEEE/ACM International Conference on Computer-Aided Design*, pp. 590–597, November 2007.
18. J. Cong, J. Wei, and Y. Zhang, "A Thermal-Driven Floorplanning Algorithm for 3D ICs," in *Proceedings of the IEEE/ACM International Conference on Computer Aided Design*, pp. 306–313, November 2004.
19. W. L. Hung, G. M. Link, Y. Xie, N. Vijaykrishnan, and M. J. Irwin, "Interconnect and Thermal-Aware Floorplanning for 3D Microprocessors," in *Proceedings of the IEEE International Symposium on Quality Electronic Design*, pp. 98–104, March 2006.

20. A. K. Coskun, A. B. Kahng, and T. S. Rosing, "Temperature- and Cost-Aware Design of 3D Multiprocessor Architectures," in *Proceedings of the IEEE Euromicro Conference on Digital System Design, Architectures, Methods and Tools*, pp. 183–190, August 2009.
21. C. L. Bertin, D. J. Perlman, and S. N. Shanken, Evaluation of a three-dimensional memory cube system, *IEEE Transactions on Components, Hybrids, and Manufacturing Technology*, vol. 16, no. 8, 1006–1011, December 1993.
22. J. Jeddeloh and B. Keeth, "Hybrid Memory Cube New DRAM Architecture Increases Density and Performance," in *Proceedings of the IEEE Symposium on VLSI Technology*, pp. 87–88, June 2012.
23. "Micron Hybrid Memory Cube," 2014. Available at: http://www.micron.com/products/hybrid-memory-cube.
24. W. Huang, "HotSpot—A Chip and Package Compact Thermal Modeling Methodology for VLSI Design," PhD Thesis, University of Virginia, January 2007.
25. J. M. Lin and Y. W. Chang, "TCG-S: Orthogonal Coupling of P*-Admissible Representations for General Floorplans," in *Proceedings of the IEEE Design Automation Conference*, pp. 842–847, June 2002.
26. S. Kirkpatrick, C. D. Gelatt, and M. P. Vecchi, Optimization by simulated annealing, *Science*, vol. 220, no. 4598, 671–680, 1983.
27. T. C. Chen and Y. W. Chang, "Floorplanning," in *Electronic Design Automation: Synthesis, Verification, and Testing*, Chap.10, L. T. Wang, Y. W. Chang, and K. T. Cheng (Eds.), pp. 575–634, Burlington, MA: Morgan Kaufmann, 2008.
28. "COMSOL Multiphysics," 2013. Available at: http://www.comsol.com/.
29. E. U. Onyegam, J. Mantey, R. A. Rao, L. Mathew, M. Hilali, S. Saha, D. Jawarani, S. Smith, D. A. Ferrer, S. V. Sreenivasan, and S. K. Banerjee, "Exfoliated Sub-10 μm vThin Germanium for Cost-Effective Germanium Based Photovoltaic Applications," in *Proceedings of the IEEE Photovoltaic Specialists Conference*, pp. 271–273, June 2011.

12

A Unified Method for Calculating Parasitic Capacitive and Resistive Coupling in VLSI Circuits

Alkis A. Hatzopoulos and Michael G. Dimopoulos

CONTENTS

12.1 Introduction

Substrate coupling prediction and modeling has become a really challenging problem in the recent years, mainly due to the continuous scaling of CMOS technology, which makes feasible the integration of more devices at closer distances. Thus, the demand for techniques accurately predicting the behavior of the chip that consider substrate coupling is constantly rising. Various approaches have been reported and can be classified into two major categories, the comparison of which shows a tradeoff between speed of simulation and accuracy of results. More specifically, methods of the first category try to directly solve the differential equations, which describe the problem using numerical techniques. Some of the most commonly used are the Finite-Difference Method (FDM), the Finite Element Method (FEM), or the Boundary Element Method (BEM)—the term Method of Moments (MoM) is sometimes used instead [1–5]. The above techniques exhibit high accuracy, but at the expense of simulation time and memory requirements. In fact, simulation time grows exponentially with the increase of the geometric complexity of the problem, while in some cases the simulation totally fails to converge.

Methods of the second category [6–13], solve the problem using fitting techniques, which offer increased speed, but with less accuracy and a limited range of validity. The major drawbacks of both are twofold. First, they can only be applied after tape-out, therefore becoming technology dependent and failing to provide a priori insight to the designer. Secondly, most of the methods of this category, for simplicity reasons, treat the coupling mechanism as a 2D problem, resulting, in some cases, in a faulty outcome. These methods cannot be applied, in general, in pure 3D problems, such as capacitive coupling. Considering all the above, it seems no coincidence that little may be found in literature [13–16] regarding analytical formulas of capacitive coupling. Most of the suggested formulas are semi-empirical in the best case and with a limited range of validity. Taking, additionally, into account that, for a few GHz and above, capacitive coupling becomes the dominant coupling mechanism and the importance of accurate capacitive models becomes apparent.

Accurate doping profile determination is only possible by means of specialized techniques and instrumentation [17–19]. For this reason, simpler techniques are necessary. In [20], the authors try to develop a rather complex methodology that relies on a series of many different simulators to avoid detailed information about doping. In [21], the resistivity is extracted by measurement data from test structures. Although, sufficiently accurate, this method requires a significant amount of time and a fair knowledge of EM simulations. [22] is also based on the measurement data of test structures, but relies on a simulator to calibrate first an assumed doping profile. The substrate noise is then computed using a fitting model the parameters of which are extracted by the calibrated profile. Therefore, it is highly dependent on the accuracy of the simulator while, due to the fitting techniques employed, its range of validity is expected to be limited.

In this chapter, a unified method for calculating Resistive and Capacitive Coupling exploiting Geometry constraints (RCCG) is described, which was initially proposed in [23,24]. The resulting model is directly obtained from the physical understanding of the problem. Neither fitting techniques nor measurements are required for the evaluation of its components. Hence, it is fast, accurate, scalable, and technology independent. The only information required is the geometry of the problem. For this reason, the developed model is characterized by generality and, although initially developed for determining the substrate coupling—current flow, it is by no means limited to that. It is a general method that may be also applied for predicting the flow of other quantities, for example, fluids.

A new approach for extracting the substrate resistance, initially proposed in [25], is also proposed in this chapter. It is fast, general, easy-to-use, requires a small number of simple DC measurements, and does not rely on fitting techniques. The method for resistance extraction relies on the geometric interpretation and the prediction of the current streamlines between two coupled structures and it is an extension of the RCCG method. The proposed scheme has been tested by utilizing data from both simulations and measurements based on a fabricated chip and also results from literature.

This chapter is organized as follows. First the general model description is presented in Section 12.2. The RCCG method formulation and the algorithm for computing the resistance or capacitance of a structure are presented in Section 12.3 together with typical case studies. The formulation for the geometric substrate resistance computation method and the proposed algorithm for extracting the substrate resistance are described in Section 12.4. Experimental results of the application of the RCCG method to several cases and simulation and measurement results for the substrate resistance computation method are presented in Sections 12.5 and 12.6, respectively. Finally, a summary concludes the chapter.

12.2 General Model Description

Various analytical modeling methods for the calculation of resistive or capacitive coupling have been presented in the literature. However, to solve the studied problem, ideal cases are assumed and several restrictions posed by an actual case are ignored, the most common of which is the structures in vicinity. In all these cases, a system free of other structures is assumed. Without doubt, this is not the case if other metal structures do exist in the vicinity or if the examined structure is located near some kind of PEC (Perfect Electric Conductor) or insulator wall, as it is, for example, on the outer sides of the chip. In every case, the electric field lines do not spread to infinite distance as assumed, but are confined to a limited range according to the restrictions of each problem. This approximation may result in a predicted resistance value up to 50% lower than the actual one and a predicted capacitance up to 50% higher.

Thus, a general formula that can accurately predict the resistance or capacitance, including the special restrictions of each problem, is needed. The analysis followed in this work is based on simple geometric calculations regarding the evaluation of the resistance or capacitance of various shapes.

For this reason, the RCCG method can, due to its generality, successfully deal with both resistive and capacitive coupling effects while also taking into account the phenomena other methods ignore.

The main idea of RCCG method underlies one of the most prevalent principle of physics, the principle of the least energy, according to which every system tries to come to the state which requires the least energy. In the examined case, this principle takes the form of the well-known Law of Least Resistance (LLR). It is an experimental fact that the current always choose to flow to the path of the minimum possible resistance. Hence, the idea is to determine, in every case, the current flow shape that exhibits the minimum resistance. The process of finding the resistance (the same analysis may be applied for calculating the capacitance) of any structure is then reduced to just a geometric calculation. In addition, by exploiting symmetries unveiled by a closer observation of the current flow lines, the complexity of the problem may be reduced significantly, as will be made clear next.

The RCCG method presented in the next section [24] is an enhanced version of the method previously reported [23], where the simulation time and the memory consumption are improved and, in addition, the method is extended to predict capacitive coupling. The RCCG method for the case of the resistance can be outlined as follows: Let current flow in a medium with resistivity ρ forming an arbitrary shape. The latter is approximated by a combination of simple geometric shapes described altogether by a set

$$P = \{p_1, p_2, ... p_n\} \tag{12.1}$$

of N geometrical parameters. This shape will be called the *assumed shape*. The resistance R of the assumed shape is a function $f(P)$ of those parameters multiplied by the resistivity of the medium:

$$R = \rho f(P) \tag{12.2}$$

The process of computing the coupling resistance in every case is reduced for determining the correct assumed shape. According to LLR, this shape needs to exhibit the minimum resistance value. Hence, a set of possible assumed shapes is produced and the resistance values of them are calculated. Then, the minimum of those values is selected as the resulted coupling resistance.

The set

$$\hat{P} = \{\hat{p}_1, \hat{p}_2, ... \hat{p}_n\} \tag{12.3}$$

of the parameter values that correspond to the assumed shape of the minimum resistance are called the *minimizing parameters*. In other words, $\hat{P}$ satisfies the equation,

$$\rho f(P = \hat{P}) = R_{\min} \tag{12.4}$$

The number of parameters in Equation 12.4 can be reduced if the physical depiction of the flow is taken into consideration. Here, the well-known Rayleigh-Maxwell method [26] is used for calculating the resistance. Other methods can be used equivalently.

The calculation of the capacitance by means of the RCCG method is straightforward, if we consider the definition of resistance and capacitance between two given structures:

$$R = \rho \frac{d}{A} \tag{12.5}$$

$$C = \varepsilon \frac{A}{d} \tag{12.6}$$

Regarding the geometric properties of any structure, that is, surface area A and distance d, the capacitance is the reciprocal of the resistance. The only difference lies in the multiplication factors ρ and ε, which depend, respectively, on the conductive and dielectric properties of the material of the medium. Therefore, the capacitance of a structure can be computed by the inverse of the function giving the minimum resistance

$$C_{\max} = \varepsilon \left[f(P = \hat{P}) \right]^{-1} \tag{12.7}$$

The resulting capacitance will be the maximum one since it has dimensions of conductance.

One of the strong points of the RCCG method is the fact that resistivity takes the form of a simple multiplication factor in the resulting model (see Equation 12.2). Since the shape itself does not depend on the resistivity value, this model is independent of the technology parameters and universal.

12.3 RCCG Method Formulation

In this section the RCCG method is described along with the elementary structures that are used to partition the structure under study. These elementary structures exhibit symmetries, thus leading to simpler equations and increased speed of computation. The resistance of the given structure is computed by the superposition of the resistances of the elementary structures. Likewise, the capacitance of the given structure is computed by the superposition of the resistances capacitances of the elementary structures.

The RCCG method is based on two general principles. First, theory states that electric field lines do not cross each other. Secondly, the LLE *imposes* field lines to choose the path that will result in the minimum total energy of the system. By exploiting these facts, any 3D structure, in general, may be treated as the superposition of simpler cases and the minimum resistance (or maximum capacitance) of each may be easily calculated. Two elementary structures are used in each case, the circular Disk with radius r_p at height h above an Infinite Plane (DIP) and the Square of side length w_p at height h above an Infinite Plane (SIP). The assumed shape of the first consists of a truncated cone of height h_p followed by a cylinder of height $h - h_p$ while the second consists of a truncated pyramid of height h_p followed by a orthogonal parallelepiped of height $h - h_p$. In addition, the Rectangle above an Infinite Plane (RIP) is also used. The latter, although not an elementary one, may be considered primary regarding the computation of line tracks. These structures are described in more detail in Section 12.3.1. Due to their symmetry, the equations that describe them are simple, thus resulting in fast computation times.

The bottleneck of the previously reported method [23] is the computation of the integral included in the SIP case. Its computation using typical techniques, like Newton-Raphson, requires more than the half of the overall simulation time. At this work, the simplicity of the equations of the circular disk are exploited to eliminate integrals and, consequently, to greatly reduce the simulation time. Particularly, the proposed algorithm of the method is a three-phase process and can be summarized as follows.

In the first phase, a virtual infinite plane is assumed between the coupled structures and then each one of them is decomposed (divided) into the elementary SIP and DIP structures. Obviously, if the resistance (or capacitance) of a single structure to the backplane is required, the SIP and DIP structures directly correspond to the required one. Then every SIP structure is approximated with its corresponding DIP, that is, squares are replaced by circular disk(s). The DIPs that replace the SIP structures will be called rDIPs in order to be distinguished from the initial DIPs. Next, in phase two, a search domain range is defined and the minimum resistance for every initial DIP and the minimizing parameters for every rDIP are computed respectively. By the end of this phase, the resistance of the assumed shape is evaluated with a relative error of about 1%–2% in respect to the value obtained with SIP equations. If greater accuracy is not needed, the algorithm may stop here. Otherwise, it continues to the next phase (third) where the initial SIPs replaced by rDIPs, are restored, and their resistance is calculated. The minimizing parameters found in previous step are used as a starting point for the construction of the search domain range of SIP structure parameters. The final resistance is obtained by the parallel combination of all the elementary ones. If the capacitance is required instead, it is obtained by the inverse of the final resistance.

Before describing the RCCG algorithm some notations are given as follows:

w_c	is the width of a given square.
r_c	is the radius of a given circle or circular disk.
h	is the height of either the square or the circular disk above the infinite plane.
s	is the separation distance between two coupled structures.
w_p	is the parameter describing the width of the truncated pyramid of the assumed shape for the SIP case.
r_p	is the parameter describing the radius of the truncated cone of assumed shape for the DIP case.
h_p	is the parameter describing the height of either the truncated pyramid or the truncated cone of the assumed shape for the SIP and DIP cases respectively.
x_p	describes any parameter.
$\hat{r}_p$	is the minimizing parameter describing the radius of the truncated cone of assumed shape for the DIP case.
$\hat{h}_p$	is the minimizing parameter describing the height of the truncated cone of assumed shape for the DIP case.
$\hat{x}_p$	describes any minimizing parameter.
D	is the one-dimensional array of N elements describing the search domain for the r_p parameter. D_i is the i-th value of r_p.
H	is the one-dimensional array of M elements describing the search domain for the h_p parameter. H_i is the i-th value of h_p.
R	is the resistance matrix and R_{ij} is the element of the i-th row and j-th column.
$R_{\min}$	is the minimum value of the resistance matrix.
R_k	is the resistance of the k-th region, when the substrate is divided in K uniform regions.
R_f	is the final resistance of the structure or the coupled structures.
$f(p_1, p_2, \ldots p_n)$	is the function giving the resistance (or capacitance) of an assumed shape described by parameters $(p_1, p_2, \ldots p_n)$.
e	is the relative error between the resistance value calculated by rDIP and DIP.

In the following, the algorithm phases along with the steps they consist of are described for the computation of the resistance. As was mentioned earlier, the capacitance is obtained by the reciprocal of the final "resistance" value. The equations used for both the DIP and SIP cases are given in Section 12.3.1.
Phase 1 – Initial Phase:

1. **If** the coupling resistance between two structures is calculated **then** assume an infinite plane between them and set $s = s/2$
2. Divide the structure into elementary ones, namely SIP and DIP.
3. **For each** SIP (***replace every SIP with the corresponding rDIP***) set its width w_c to $2r_c = 2w_c/\sqrt{\pi}$

Phase 2 – DIP Calculation:

1. **If** the substrate profile is non uniform **then**
   Divide substrate into K regions with a uniform doping profile each.
2. **For each** structure (*both DIP and rDIP*)
   construct the search arrays **D** and **H** as $\mathbf{D} = \{r_c, 100r_c\}$ $\mathbf{H} = \{0, h\}$ for its parameters r_p and h_p respectively
3. **For each** structure (*both DIP and rDIP*)
   For each D_i, $i = 1$ to N
   For each H_j, $j = 1$ to M
   For each substrate region, $k = 1$ to K ($k = 1$ *if substrate is uniform*)
   Calculate the resistance of each region $R_k = \rho_k f(D_i, H_j)$
   End
   Calculate the total resistance (*of all regions*)
   $$R_{ij} = \sum_{k=1}^{K} R_k$$
   End
   End
   End
4. **For each** structure (*both DIP and rDIP*) search matrix **R** and find its minimum value $R_{\min}$ that corresponds to indices values $i = i_0$ and $j = j_0$ respectively.
5. **For each** structure (*both DIP and rDIP*)
   If structure is **Not** an rDIP **then**
   store its $R_{\min}$
   Else (*structure is a rDIP*)
   get and store the minimizing parameters $\hat{r}_p = D_{i_0}$ and $\hat{h}_p = H_{j_0}$
6. **If the required relative error is greater than *e* then goto phase 3 – Step 3 (*final step*)**
   Else continue to next phase

Phase 3 – SIP Calculation:

(*The substrate segmentation of K regions performed in the previous phase is retained here.*)

1. **For each** SIP construct, its search arrays **D** and **H** are $\mathbf{D} = \{0.75 \times 2\hat{r}_p, 1.25 \times 2\hat{r}_p\}$ and $\mathbf{H} = \{0.75 \times \hat{h}_p, 1.25 \times \hat{h}_p\}$ respectively.

2. Repeat steps 3 and 4 of phase 2.
3. **Final step**: Calculate the total resistance R_t by the parallel combination of all R_{min}, computed in the previous steps, namely

$$R_t = \left[\sum 1 / R_{min}\right]^{-1}$$

If the coupling resistance between two structures is required, **then** final resistance $R_f = 2R_t$ **else** $R_f = R_t$

The algorithm of the RCCG method will be illustrated with the simple case of two coupled identical cubes. The electric field lines between them are depicted in Figure 12.1. As can be seen, they start on one side of the left cube and terminate at the same side of the right cube. Therefore, each cube may be treated as the superposition of six square sides every single of which is coupled with the same side of the other cube. Since the coupling resistance is required, an infinite plane is assumed between the two cubes, according to step 1 of phase 1. Then (step 2) each cube is divided into six SIPs, and each SIP is replaced by an rDIP (step 3). Next, in phase 2, the resistance of each DIP is calculated. If the required relative error is greater than e, the process stops here. The whole process up until this point is very fast with an average completion time of a few msec. If better accuracy is required, it continues to the third phase where the six initial SIPs are restored and their resistance is

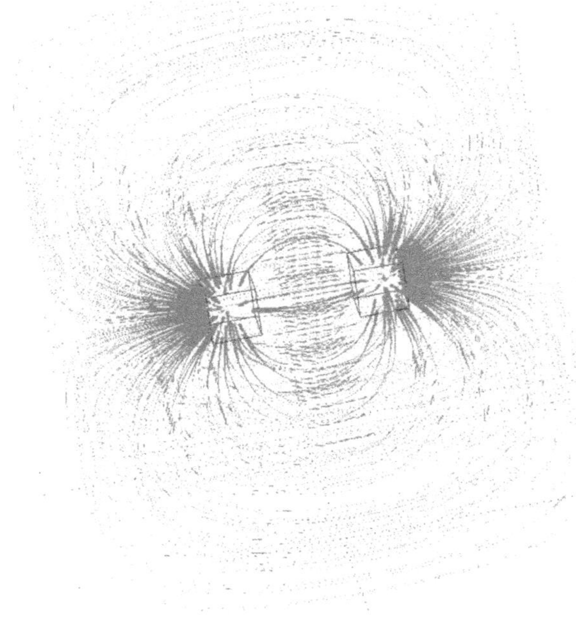

FIGURE 12.1
Electric field lines between a pair of identical cubes.

calculated. The final resistance is the double (one half was computed) of the parallel combination of the six SIPs.

It must be mentioned that the most time-consuming calculation in the SIP equations is the integral, the calculation of which requires over the half of the overall run time. Particularly, for its computation to have a sufficient accuracy, roughly 10^3 mathematic operations need to be performed [23]. As a result, the overall process time is increased with an additional time $T_q = (10^3 NM)T_o$, where T_o is the time required for each operation and N and M the number of rows and columns, respectively, of the largest resistance matrix that appears during calculation of the SIP cases. This is a noticeable period of time even for a modern computer. It is instructive to illustrate that the reduction in simulation time succeeded with a small example. Let $N = 500$ and $M = 600$ and assume that $T_o = 10\text{n sec}$. Then, the total additional time required for the calculation of all integrals within search range is equal to $T_q = 3\,\text{sec}$.

Here, by eliminating the integral and exploiting the simple formula of DIP instead, the complexity of calculations can be greatly reduced. What's more important, the use of DIP equations has small impact on the accuracy of results. The relative deviation between the values obtained from SIP and DIP equations are practically zero, apart from the cases—not commonly encountered in practice—where the ratio $w_p/h_p > 10$, the relative deviation is practically zero.

Run time can be further improved by applying typical optimization techniques to phases 2 and 3. For example, any common optimized search algorithm technique, such as the binary search, can be used for step 4 of both phase 2 and phase 3. Binary search for example has a complexity of $O(\log_2 n)$, which is low compared to the $O(n)$ complexity of the exhaustive search.

One more optimization step that may be implemented when considering large circuits is based on the uniformity of the distances. In a typical CMOS circuit, the distances by which the structures of a circuit block are separated from each other are more or less of the same order. We can exploit this fact to reduce the total calculations. That is, if a resistance (or capacitance) of a structure has been already calculated, the search domain of another similar structure need not to be constructed from scratch. Instead, it is built around the minimizing parameters already calculated in the same away as shown in phase 3, step 1.

As discussed earlier, one actual common effect other methods ignore is the structures in the vicinity. Our method can successfully take into account the restrictions imposed by this effect very easily by properly setting the search domain arrays at step 2 of phase 2 and step 1 of phase 3. In more detail, the upper limit of array **D**, describing the width of the assumed shape, is set to the maximum distance allowed by the restrictions of every problem.

12.3.1 Elementary Structures of the RCCG Method

The heart of the RCCG method is comprised of the calculation for the minimum resistance (or maximum capacitance, correspondingly) concerning the

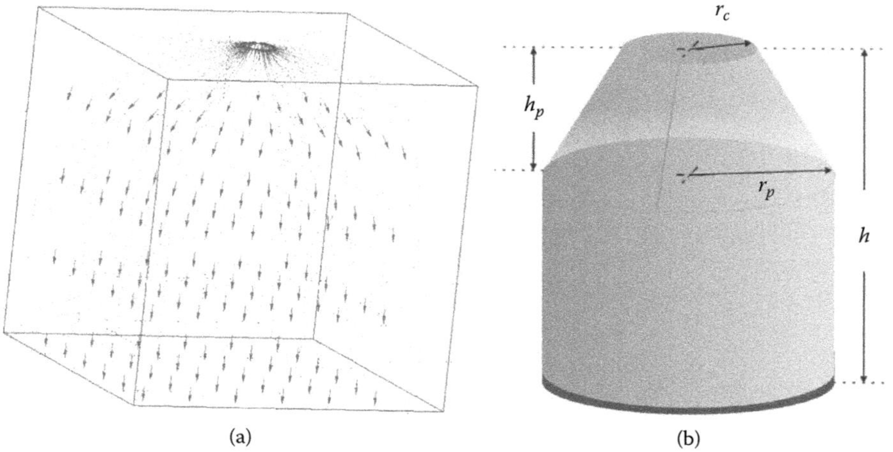

FIGURE 12.2
Circular disk over an infinite plane. (a) Electric field lines. (b) Assumed shape.

elementary structures of a circular disk and of a square both above an infinite (ground) plane, as described earlier, called DIP and SIP for short respectively. These structures are described in detail below, starting with the DIP, which is simpler.

Let a circular disk of radius r_c reside at height h over an infinite ground plane, as shown in Figure 12.2a. Observing the shape of the current flow lines it can be seen that it resembles that of a cone followed by a cylinder. Hence, a sufficient approximation of the current flow shape will be a truncated cone of height h_p and radius of the upper and lower bases r_c and r_p, respectively, followed by a cylinder of radius r_p, as shown in Figure 12.2b. Consequently, the set of parameters required to be determined for this simple case is consisted only of two, namely, the radius r_p and the height h_p.

Using the Rayleigh-Maxwell method the resistance of this shape will be

$$R_{\min} = \frac{R^+ + R^-}{2} \tag{12.8}$$

where

$$R^- = \rho \frac{1}{\pi} \frac{h_p}{r_c r_p} \tag{12.9}$$

$$R^+ = \rho \frac{1}{\pi r_c} \frac{(r_p - r_c)^2}{h_p r_p} \frac{1}{\ln\left[1 + (r_p - r_c)^2 / h_p^2\right]} \tag{12.10}$$

The case of a square of width w_c at height h_p over an infinite (ground) plane, SIP for short, is examined next. Clearly, this case corresponds in practice to square metal contacts over substrate. Note, that the resulting equations can be used to directly calculate the substrate resistance or capacitance to backplane of such contacts. The shape of the current flow lines, illustrated in Figure 12.3a, can be approximated by a truncated pyramid of height h_p and side width of lower base w_p, followed by an orthogonal parallelepiped of the same base width, shown in Figure 12.3b. Again, only two geometric parameters are necessary to fully define the structure of the shape, that is, w_p and h_p.

The resulting resistance value for this case is given by the following set of equations:

$$R_{\min} = \frac{R^+ + R^-}{2} \tag{12.11}$$

$$R^- = \rho \frac{h_p}{w_c w_p} + R_{op} \tag{12.12}$$

$$R^+ = \rho \left[\int_0^{w_t/2} \frac{1 - \lambda h_p}{h_p \lambda \sqrt{\lambda^2 x^2 + 1}} \tan^{-1}\left(\frac{\lambda w_p}{\sqrt{\lambda^2 x^2 + 1}} \right) dx \right]^{-1} + R_{op} \tag{12.13}$$

where

$$R_{op} = \rho \frac{h - h_p}{w_p^2} \tag{12.14}$$

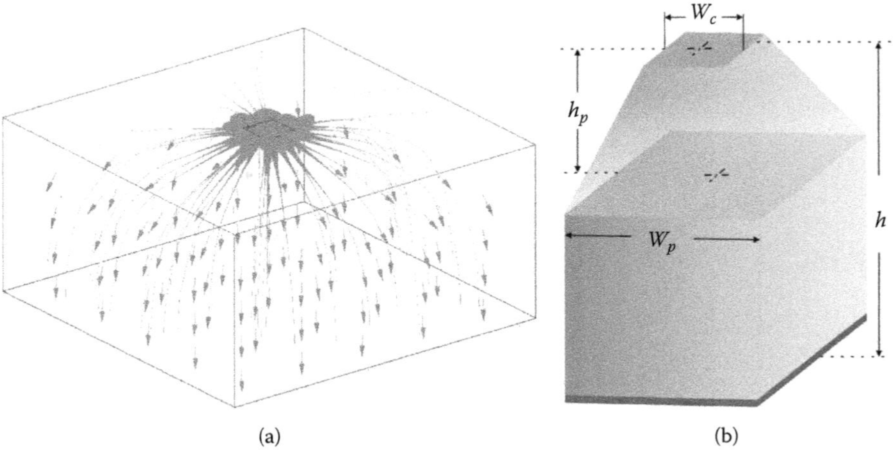

FIGURE 12.3
Square over an infinite plane. (a) Electric field lines. (b) Assumed shape.

$$\lambda = (1 - w_c / w_p) / h_p \tag{12.15}$$

The case of an orthogonal parallelogram, that is rectangle, of length L and width w_c over an infinite (ground) plane, RIP for short, is studied next. Rectangular structures like this one correspond to line tracks or microstrips and striplines. Although, this case may not be considered as an elementary one since it includes a SIP case, it is primary regarding the track lines or microstrips. For this reason, the steps followed for the computation of the RIP case may be considered as an additional phase following the basic ones of the RCCG algorithm.

In Figure 12.4, the assumed shape for the RIP case is depicted. The latter is divided into three regions: S_1, S_2, and S_3. Apparently, S_1 and S_2 constitute the two halves of the SIP case. For this reason the RCCG method is directly applied to the union S_1 and S_2. At region S_3, current is assumed to flow in planes perpendicular to the rectangle, as shown in Figure 12.4. At each plane, current flow lines form the same shape, which is composed of a trapezoid followed by an orthogonal parallelepiped. The width of the latter is w_p, while the upper and lower bases of the trapezoid are w_c and w_p, respectively, and its height h_p.

Applying the Rayleigh-Maxwell technique to this plane, the overall resistance of region S_3 is

$$R_{S_3} = \frac{R_{\bar{S}_3} + R_{\bar{S}_3}}{2} \tag{12.16}$$

where

$$R_{\bar{S}_3} = \rho \frac{h_p}{(w_p - w_c)(L_c - w_c)} \ln\left(\frac{w_p}{w_c}\right) + R_{op} \tag{12.17}$$

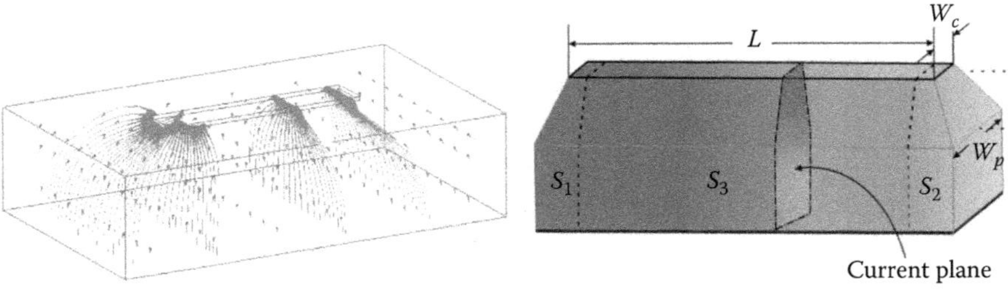

(a) (b)

FIGURE 12.4
Rectangular line over an infinite plane. (a) Electric field lines. (b) Assumed shape.

$$R_{S_3}^{+} = \rho \frac{\ln\left(\dfrac{w_t}{w_c}\right)}{2(L_c - w_c)\tan^{-1}\left(\dfrac{w_p - w_c}{2h_p}\right)} + R_{op} \tag{12.18}$$

with R_{op} to be given from

$$R_{op} = \rho \frac{h - h_p}{w_p(L_c - w_c)} \tag{12.19}$$

Consequently, the steps followed for the calculation of the resistance of the RIP case can be summarized as described below:

1. Divide the assumed shape into three regions S_1, S_2, S_3, so that the union of S_1 and S_2 constitutes the SIP case.
2. Apply all three phases of RCCG method to this union to calculate its resistance $R_{S_{12}}$ and the minimizing parameters $\hat{h}_p$ and $\hat{r}_p$.
3. Call phase 3 using Equation 12.16 through 12.19 as the function of the assumed shape to calculate the resistance R_{S_3} of region S_3.
4. Take the total resistance R_L by the parallel combination of the resistances $R_{S_{12}}$ and R_{S_3}, namely

$$R_L = \frac{R_{S_{12}} R_{S_3}}{R_{S_{12}} + R_{S_3}} \tag{12.20}$$

12.3.2 Implementation on Typical Cases

In this section, the RCCG method is demonstrated on commonly encountered in practice structures. The steps need to be followed at every case are thoroughly described. The analysis in each case is performed either for the calculation of the capacitance or for the resistance. As discussed earlier, both the analysis and the results are the same with the only difference being that the final capacitance is given by the reciprocal of the minimum resistance and vice versa. The first case that is studied regards the computation of the capacitance between a pair of two identical cubes. The subcase where one dimension of the structure is very small in respect with the other two is also examined. Next the coupling (substrate) resistance between two contacts is investigated, which is significant in the CMOS design.

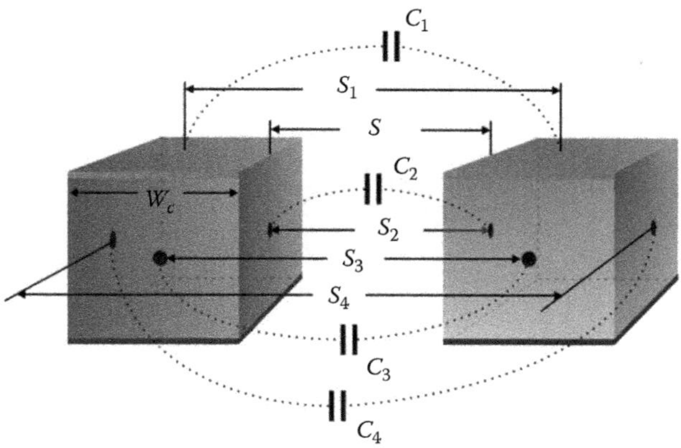

FIGURE 12.5
Formulated capacitances between two identical cubes.

12.3.2.1 Coupling Capacitance

Let us consider the general case of two cubes of side width w_c separated at distance s measured between the two closest sides, as shown in Figure 12.5. The electric field lines between the cubes are illustrated in Figure 12.1. Without a loss of generality, it is assumed that both cubes reside at the same plane. Six capacitances, one per coupled pair of sides, are defined (phase 1, step 2).

The total capacitance C_t is the parallel combination of the six capacitances, namely

$$C_t = \sum_{t=1}^{6} C_i \tag{12.21}$$

The average length of the electric field lines between each coupled pair is considered in calculations given by the following equations.

$$s_1 = s + w_c \tag{12.22}$$

$$s_2 = s \tag{12.23}$$

$$s_3 = s + w_c \tag{12.24}$$

$$s_4 = s + \frac{5}{2} w_c \tag{12.25}$$

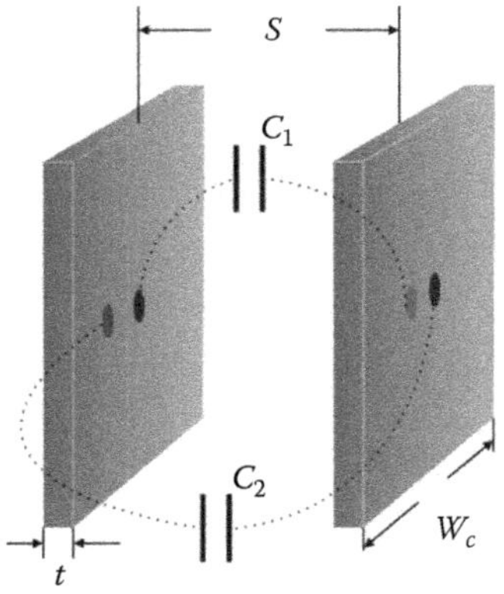

FIGURE 12.6
Formulated capacitances for the small thickness approximation case.

The distance of the below and the back pair is obviously the same as the top and the front respectively, that is, s_1 and s_3.

Now, consider the case where the dimension lateral to the closest side pair is very small in respect with the others, as shown in Figure 12.6. This case is very common in practice if one takes into account that in modern technologies the thickness of any but the top metal is roughly half a micrometer. It follows that every structure with a minimum length (or width) over 5 μm or more can be considered as belonging in this case.

Although each cube may be decomposed again to six sides, it is sufficient here to take the two big sides only. This approximation will be called the small thickness approximation (STA). This way, the total number of the capacitances needs to be computed is reduced to only two, resulting in a faster computation. However, it should be noted, that the contribution of the other four sides to the overall capacitance is by no means ignored. In fact, it is incorporated with the other two by considering for each one an effective area slightly bigger. Specifically, the effective width w_e of each assumed square is given by the sum of its initial width and the thickness of each side, as implied by Equation 12.26 below. The average length of the electric field lines s_1 and s_2 regarding capacitances C_1 and C_2 are given in Equations 12.27 and 12.28, respectively.

$$w_e = w + t \tag{12.26}$$

$$s_1 = s \tag{12.27}$$

$$s_2 = s + w / 2 \tag{12.28}$$

The total capacitance C_t is given by the parallel combination of the C_1 and C_2, that is, $C_t = C_1 + C_2$.

12.3.2.2 Coupling Resistance between Two Contacts

The case of two coupled square contacts over a conductive medium is investigated below. This case corresponds to the coupling resistance between metal contacts over the conductive substrate. It can be also used for modeling the resistive (or capacitive) coupling between lines, N-/P-wells and N-/P-contacts, or other passive regions. Although, it may be considered as a special case for the two coupled cubes presented previously, it is worth to paying special attention to this case due to its great importance in the VLSI design.

The current flow lines between the two square contacts are depicted in Figure 12.7a and their assumed shape is shown in Figure 12.7b. The average length of all different current flow paths between the two coupled contacts is given by

$$\overline{s} = w_c + s \tag{12.29}$$

As it can be seen from Figure 12.7, the assumed shape corresponds to the superposition of two SIP structures with a virtual infinite plane between them, as referred in phase 1, step 1 of the RCCG algorithm. For the calculation of each SIP, half of the average length $\overline{s}/2$ is used (phase 1, step 1) and the final resistance is the double of the single SIP structure (phase 3, step 3).

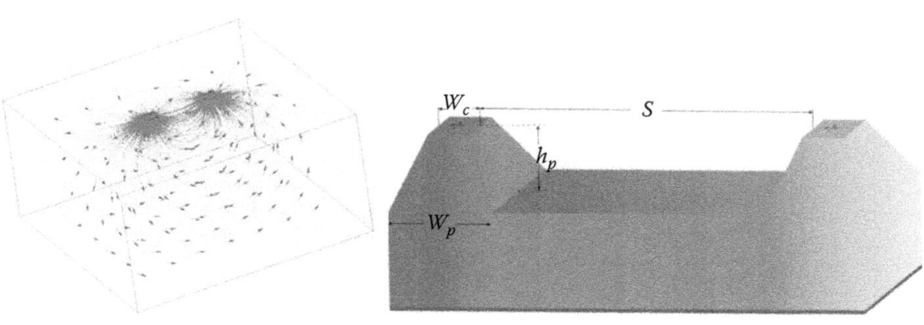

(a) (b)

FIGURE 12.7
A pair of identical square contacts above conductive media. (a) Electric field lines. (b) The assumed shape.

12.4 Substrate Resistance Extraction Method

For an accurate resistance extraction method based on measurements, an accurate and scalable formula is needed to relate the resistance of a given structure to the resistivity of the medium. Fitting techniques are not considered as an option since their range of validity is limited to certain values of their parameters. For this purpose, the Substrate Resistance computation based on Geometry constraints (SRG) will be described. The SRG method [25] is an extended version of the RCCG method [24] in order to improve the resistance extraction accuracy in epitaxial lightly doped substrates where the error percentage of the RCCG method is in the neighborhood of 15%, a value not adequate for an accurate resistance determination. The SRG method relies on the geometric interpretation and the prediction of the current streamlines between two coupled structures. It exploits the principle of the least energy to provide the correct geometric shape that the current flow follows. A brief description of the substrate resistance computation based on the RCCG method will be given next followed by the detailed modifications made in the SRG method in order to improve the accuracy of doping profile determination in epitaxial lightly doped substrates.

In epitaxial lightly doped substrates, the truncated cone approximation breaks down. In this type of substrate, a thin layer of approximately 2 μm thickness exists above the bulk substrate, which has a thickness of about 300 μm. The resistivity of this thin layer varies from 1 to 1000 times less than the bulk resistivity value. These extreme ratios of the thickness and resistivity values between the two layers leads to a very large spreading of the current streamlines in the epi-layer [27]. The assumed shapes of the RCCG method are properly modified in the SRG method to capture the behavior above with the accuracy required for doping profile extraction.

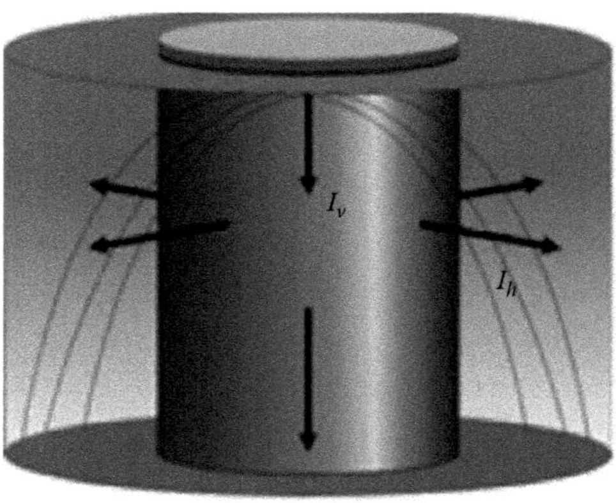

FIGURE 12.8
The assumed current flow used in the SRG method for the upper boundary resistance. A derived circular upper contact is shown at the top of the cylinder comprising the current lines.

The new approach followed in the SRG method for the computation of the substrate resistance by taking into account the upper and lower boundary resistances R_m^+ and R_m^- will be detailed next. In the computations involved, a square contact of side width w is approximated with a circular disk of the same surface area; its radius is $r_c = w/\sqrt{\pi}$. This approximation, which is also followed in [24], although not necessary for the SRG method, simplifies the involved equations.

The upper boundary resistance R_m^+ will be determined first. The current is resolved into a horizontal (I_h) and a vertical (I_v) component, as shown in Figure 12.8. The vertical current flow is approximated with a cylinder with the same cross section as the circular disk (an orthogonal parallelepiped for the square contact). The horizontal component is assumed to flow between two coaxial cylindrical shells, the first of which is the vertical one with the same cross section as the disk (two coaxial orthogonal shells for the square contact).

The resistance corresponding to the vertical resultant is given by

$$R_v = \frac{1}{\sigma}\frac{h_p}{\pi r_c^2} \tag{12.30}$$

whereas the resistance of the horizontal resultant can be obtained by applying Gauss law, which after some calculations gives

$$R_h = \frac{1}{\sigma}\frac{1}{2\pi h_p}\ln\left(\frac{r_c}{r_p}\right) \tag{12.31}$$

where the same notation used in Figure 12.8 is followed.

The two resistances are assumed to be in series. Thus the final resistance value of the upper boundary is defined as

$$R_m^+ = R_h + R_v \tag{12.32}$$

The lower boundary resistance R_m^- will be determined next. The equipotential surfaces are approximated as spheroid sectors parallel to the surface of the contact, as shown in Figure 12.9b. The spheroids become increasingly oblate as they reach the backplane where they become flat. The resistance between the spherical sector located at the contact and the spherical sector located at the backplane can be expressed by the definition formula of the resistance as

$$R_s = \frac{V_1 - V_2}{I} \tag{12.33}$$

where V_1, V_2 is the potential at the contact and at the backplane, respectively, and I the current flowing from the first to the second. Approximating

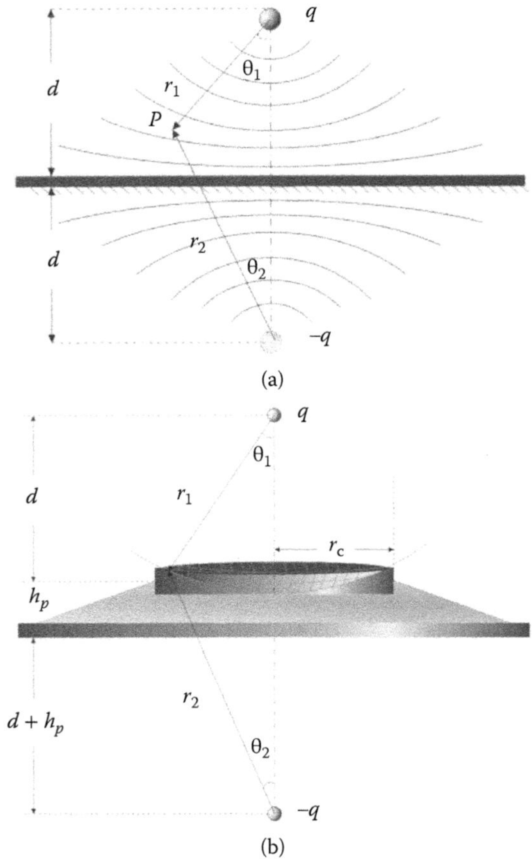

FIGURE 12.9
Method of images (a) for the charge above a perfect conductor and (b) the application for computing the lower boundary of the resistance.

these surfaces systematically is a cumbersome task. Instead, the method of images [26] can be utilized for this purpose.

According to the method of images, the problem of a current source (CS) above a ground (or a conductive) plane can be easily solved by placing an equal and opposite CS to the symmetric position of the first CS about the plane, as shown in Figure 12.9a. The potential at any given point above the plane can be then calculated as

$$V = \frac{I}{4\pi\sigma}\left(\frac{1}{r_1} - \frac{1}{r_2}\right) \tag{12.34}$$

The equipotential surfaces of this problem match that of the initial problem of the contact above a ground plane (except when very close to the contact).

Therefore, by substituting Equation 12.34 into 12.33 we get for the resistance between contact and backplane:

$$R_s = \frac{1}{4\pi\sigma}\left[\left(\frac{1}{r_1} - \frac{1}{r_2}\right) - \left(\frac{1}{r_1 + h_p} - \frac{1}{r_2 - h_p}\right)\right] \tag{12.35}$$

Eventually, the lower boundary resistance R_m^- is computed by

$$R_m^- = R_s / (a\beta) \tag{12.36}$$

where the coefficients α, β are used to normalize the area of the spherical sector to that of the contact by the following way. The values of α and β are determined as follows. A CS is placed above the contact at a relatively large distance in comparison to the dimensions of the contact. This ensures that the curvature of the corresponding sphere becomes negligible in respect to the dimensions of the contact and the base circle of the spherical sector will better approximate the flat contact. The particular value used in this work is $d = 4r_c$. The spherical sector corresponds to a sphere of radius d centered at the CS having a total surface area equal to $A_{SP} = 4\pi d^2$. The surface area of the sector is $A_{SCT} = 2\pi d^2(1 - \cos\theta)$. Thus, the ratio a of the latter to the area of the sphere is

$$a = \frac{A_{SCT}}{A_{SP}} = \frac{2\pi d^2(1-\cos\theta)}{4\pi d^2} = \frac{1-\cos\theta}{2} \tag{12.37}$$

The surface area of the contact (circular disk) is $A_{CON} = \pi r_c^2$. So, the ratio β of the spherical sector to this area is

$$\beta = \frac{A_{SCT}}{A_{CON}} = \frac{2\pi d^2(1-\cos\theta)}{\pi r_c^2} = \frac{2d^2(1-\cos\theta)}{r_c^2} \tag{12.38}$$

The final resistance of the contact to the backplane is:

$$R_{tr} = \frac{R_m^+ + R_m^-}{2} \tag{12.39}$$

One of the strong points of the SRG method is that it is independent of the resistivity. The latter takes the form of a simple multiplication factor to the unit resistance of the assumed shape (see Equation 12.39). Therefore, the calculation of the resistance of the assumed shape for each set of its parameters needs to be performed only once. Then, the resistance for every resistivity value is obtained by a simple multiplication with the unit resistance. Also, the SRG method may be directly used without any modification for the doping profile extraction of an arbitrary multi-layer substrate. As it will be shown

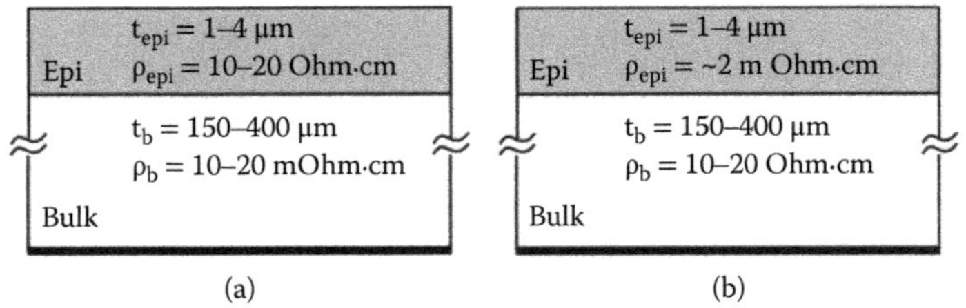

FIGURE 12.10
Types of substrate profiles: (a) heavily doped and (b) lightly doped.

later, the average difference of the SRG with respect to simulation and measurement results is within 5%.

12.4.1 Substrate Resistance Extraction Algorithm

The two most commonly used substrate doping profiles are the heavily doped and the lightly doped, both shown in Figure 12.10. The substrate doping profile is not uniform in general [22]. However the common approach, also followed in this work, is to stratify it in a number of uniform layers of specific resistivity and thickness. Every layer is then characterized by a pair of parameters at most, its thickness and its equivalent resistivity value.

The proposed method for the extraction of the substrate resistance is depicted in Figure 12.11. Measurement data are first obtained by test structures on the chip under study. The user introduces next into the extractor mainly the number of layers and, optionally, some additional parameters that will be described below. Once the resistivity has been determined, the substrate coupling resistance between any two structures can be defined by employing SRG with the determined resistivity. The extractor, with the SRG approach implemented, generates a data set based on its input values, which is then compared to the measurement data set. If the stop criteria, comprising mainly the relative error between the two data sets, are not met the extractor parameters are automatically refined and the procedure of generating new data set and comparing their values with the corresponding measurement data is repeated.

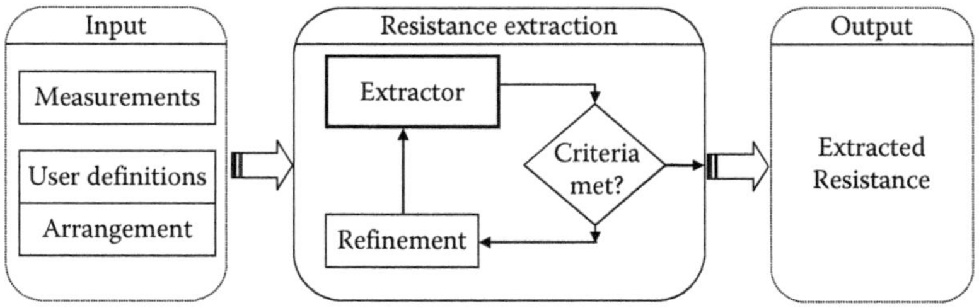

FIGURE 12.11
Block diagram of the method for resistance extraction.

The measurement data used in this work correspond to the measurement of the contact to backplane resistance of each structure. There are two main reasons for this choice. First, the separation of the test structures is not included as a parameter for the process, which otherwise would increase the complexity of the latter (several different separation distances would be needed). The second reason is related to the area overhead due to the test structures, which is a crucial parameter for any measurement method. Here, the measurement is performed in DC, implying that the structure under study is not capacitively affected by the rest of the structures in its vicinity. Therefore, the separation among them may be reduced to the smallest possible. Additionally, since the formula is independent of the dimensions of the structure, accurate characterization may be accomplished by the minimum area allowed by the process for any individual structure. Thus the area overhead, attributed to the overall area occupied by the block of test structures, can be kept very small (e.g., 0.004 mm^2 for double layer stratification).

Let us consider now the extractor tool that comprises the heart/kernel of the proposed method in more detail. The extractor takes as input the geometry of the test structures along with a set of definitions by the user and provides in each iteration a new set of resistance values. Specifically, user definitions consist primarily of the number of layers that will be used for the substrate stratification. Optionally, for each layer parameter (thickness, resistivity) specific search ranges may be defined and also an initial guess value may be given. Instead of ranges, single values may also be provided. This may be used for example to set fixed heights for each layer. Fixed values would significantly reduce the search space and, consequently, the overall execution time. If the initial guess value is not provided by the user, then the mean average of the upper and lower bounds of each range is used.

In the proposed resistance extraction method, two stopping criteria are employed. The first is the difference between the extracted resistance value and the measured one, which should not exceed the TOL value (a user defined parameter). The second one is the maximum number of iterations of MAX_ITER (another user defined parameter) are needed for the method to converge. After extensive experimentation, a value of MAX_ITERS=15 has been found to suffice.

12.5 RCCG Method Validation and Applications

The RCCG method has been validated after examining various test cases. In the following, the RCCG method is compared against measurement data and data obtained from simulations. For the measurement data, a test chip was fabricated in the UMC 0.18 lightly doped CMOS process and, in addition, measurement data from the literature [7] has been used.

12.5.1 Measurement Results

In this section the RCCG results are compared with measurement data from a chip fabricated in a 0.18 lightly doped CMOS process. The substrate is about 500 μm thick with a bulk resistivity of 20 Ωcm. An epitaxial layer exists over the bulk substrate with a resistivity ranging from a few mΩ.cm to the bulk resistivity value. Measurement data from [7] were also used to validate the proposed method.

A die photo of the test chip is illustrated in Figure 12.12, showing the several contact arrangements it includes. The coupling resistance between square contacts of three different side widths $W = [20, 30, 40]\,\mu m$ has been studied. Figure 12.13 shows the comparative results for the coupling resistance as a function of their separation. Next, the RCCG method was applied to line tracks, that is, rectangular contacts, of side length $L = 80\,\mu m$ and two different side widths $W = [20, 40]\,\mu m$. In Figure 12.14, the coupling resistance as a function of their separation is depicted. As can be seen, the RCCG results are, in both cases, in close agreement with measurement data with the typical error falling below 5% with a single exception of about 7%.

As far as the comparison with the measurement data obtained from [7] is concerned, an identical substrate profile is used in order for the comparison to be fair. The substrate is about 250 μm thick with a bulk resistivity of about 13 Ωcm. The coupling resistance of two different cases has been examined. First, between square contacts of various side widths $W = [18, 24, 30, 36, 42]\,\mu m$ and secondly, between a square contact of side $W = 6\,\mu m$ and a rectangular contact of the same side width and various lengths $L = [18, 24, 42]\,\mu m$. The

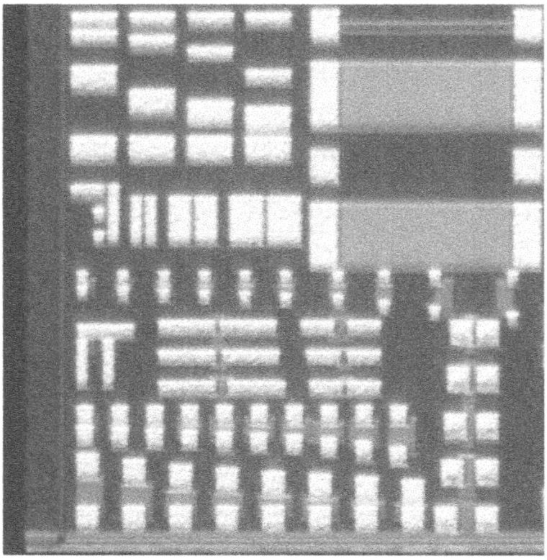

FIGURE 12.12
Die photo of the test chip in the UMC 0.18 lightly doped process containing various contact arrangements.

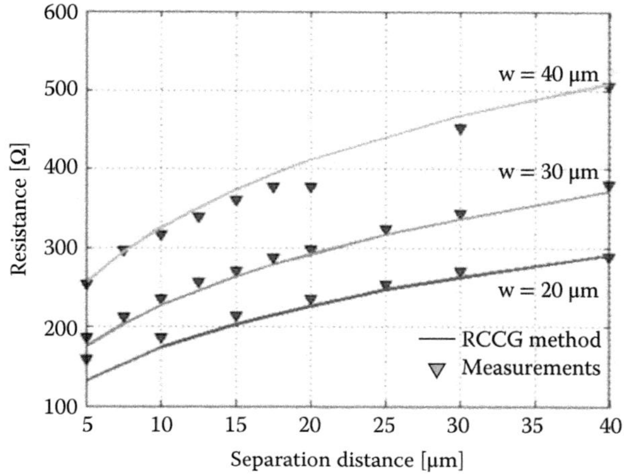

FIGURE 12.13
Comparison between the RCCG method and measurement results for the case of the coupling resistance between two square contacts of width (20,30,40) µm as a function of its separation distance.

results, which were presented in [23], are again in close agreement, suggesting the validity of the method.

Note that the slope of the resistance curve is greater in measurement data from the first chip. This is attributed to the existence of the small resistivity

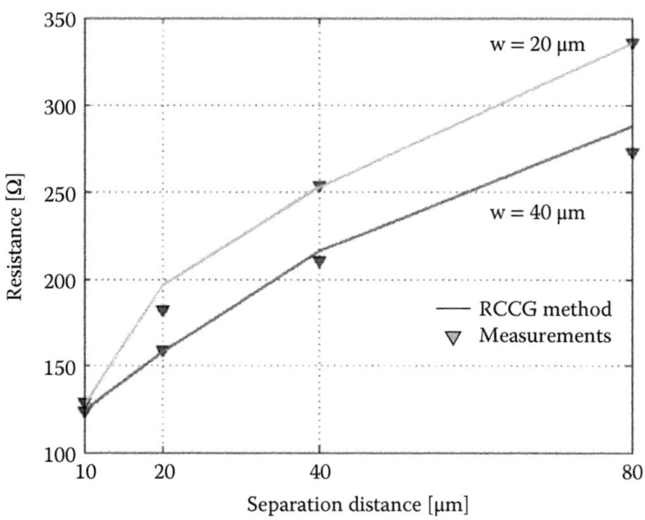

FIGURE 12.14
Comparison between the RCCG method and measurement results for the case of the coupling resistance between rectangular contacts of length L = 20 µm and two different widths (20,40) µm as a function of its separation distance.

thin epitaxial layer in the specific process of the first chip. The RCCG method captures this behavior very well.

12.5.2 Simulation Results

The RCCG results are compared here with the results of two commercial simulators, COMSOL [28], which uses the FEM technique and ADS [29], which uses the MoM technique. Several experiments commonly encountered in practice arrangements have been setup to test the validity of the method. The resistive coupling effect for both lightly and heavily doped substrates has been thoroughly examined in [23] using the ADS simulator. For all the cases in [23], the results have a typical error which falls within 1%–5%. Concerning particularly the case of the resistance to backplane of a line track above a substrate, the RCCG method was compared also with simulation results obtained by COMSOL and ADS, as shown in Figure 12.15. It must be noted that, when the ratio of the line length to the substrate height is close or greater than unity, the data of the RCCG method almost matches data obtained with both COMSOL and ADS. When the above ratio becomes less than unity, there is a discrepancy between the results obtained by ADS and the results obtained by the RCCG method and COMSOL. This becomes more apparent as the width of the line becomes smaller and may be attributed to some limitations of the MoM technique in the specific range of values. The

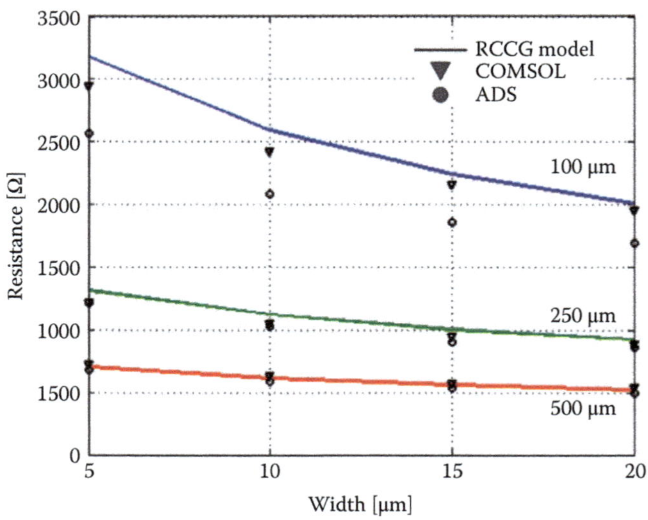

FIGURE 12.15
Comparison between the the RCCG method and simulation results for the case of the resistance to backplane of a line track as a function of its width for different lengths (100, 250, 500) µm. The substrate height is 500 µm.

TABLE 12.1

Performance Comparison between the Proposed Model and Simulators Using Analytical Meshing Techniques

	Resistance between Contact-Contact			Resistance between Contact-Line		
	Simulation Setup Time	Simulation Run Time (sec)	Relative Error	Simulation Setup Time (sec)	Simulation Run Time (sec)	Relative Error
ADS Simulator	Moderate	20	~5%	Moderate	20	~5%
COMSOL Simulator	Long	25	~5%	Long	30	~5%
RCCG Method	Fast	<1 sec	2%–10%	Fast	<1 sec	2%–10%

Dimensions of Contact and Line are 30 × 30 μm and 30 × 150 μm respectively.

results of the RCCG method are, in every case, in close agreement with the results obtained by COMSOL.

Finally, a performance comparison table is presented in Table 12.1, since one of the main targets of the RCCG method is the fast computation speed of the substrate coupling effect. In Table 12.1, the performance of the RCCG method is compared with both COMSOL and ADS simulators. The mesh density of both simulators is set to the minimum possible so that the relative error falls below 5% in comparison with measurement data. The simulation setup time is the typical time required for an experienced designer to setup from scratch the experiment under simulation, that is, the input of the simulator. It is clear that, in terms of total simulation time (run time and setup time) the RCCG method outperforms both simulators.

12.6 Experimental Results for the Resistance Extraction

Before applying the proposed algorithm for substrate resistivity extraction the accuracy of the SRG method is first investigated by means of a commercial simulator [28]. Finally, the proposed algorithm is validated with measurement data from a test chip.

12.6.1 Simulation Results

The computed (by the SRG method) substrate resistance value of a contact to the backplane is compared against simulation results using the FEM method as obtained from the COMSOL commercial simulator [28]. First, several contact dimensions are studied for three different substrates of different resistivity and height. The results, as a function of the contact dimensions, are

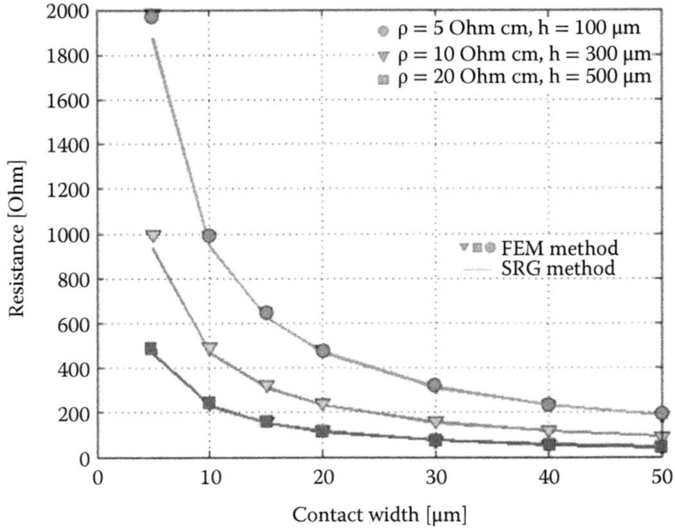

FIGURE 12.16
Comparative results between the proposed formula and COMSOL for the substrate resistance of several contact widths and for different substrates.

depicted in Figure 12.16. As shown, simulation data are in close agreement with the SRG results.

Next, the SRG results are compared with COMSOL data in terms of the substrate thickness and the dimensions of the test structures for the two common substrate profiles. The dimensions of the test structures vary from 5 μm to 50 μm and the substrate thickness ranges between 100 μm and 500 μm, which covers the typical range.

Various geometries (circular, rectangular) and values are examined. The comparison reveals, in general, that the relative difference is small, on average about 5%. For the heavily doped substrate, the biggest difference is observed for the small dimensions and falls practically below 5% for a side width larger than 20 μm. For the lightly doped substrate, the average relative difference is 5%.

12.6.2 Measurement Results

In this section, the proposed resistance extraction method is validated against measurement data obtained by a test chip (Figure 12.12) fabricated in a UMC 0.18 lightly doped CMOS process.

The test structures consist of three arrays of square contacts of side widths 20, 30, and 40 μm, as was shown in Figure 12.12.

The measurements are performed in DC, thus de-embedding is not needed for removing parasitics (inductive or capacitive coupling). The calibration of the multimeter is performed on the chip under test and, specifically, by

TABLE 12.2

Extracted Resistance of Different Test Structures Obtained by Measurement Data and SRG

Test Structure w × w (μm²)	Test structure separation w (μm)	Measurement Data (Ohm)	SRG Data (Ohm)	SRG Error (%)
20 × 20	5.0	268.63	278.43	3.65
	7.5	309.29	301.71	2.45
	10.0	342.41	362.03	5.73
	12.5	369.70	397.91	7.63
	15.0	393.85	384.44	2.39
	17.5	414.22	434.06	4.79
	20.0	433.62	404.65	6.68
	30.0	492.57	516.61	4.88
	40.0	535.75	555.95	3.77
30 × 30	5.0	187.75	197.06	4.96
	7.5	213.76	207.18	3.08
	10.0	238.90	224.09	6.20
	12.5	258.76	275.92	6.63
	15.0	277.11	260.71	5.92
	17.5	292.82	308.25	5.27
	20.0	307.00	288.92	5.89
	25.0	333.22	345.05	3.55
	30.0	353.99	374.91	5.91
	40	391.05	411.89	5.33
40 × 40	5	161.19	148.96	7.59
	10	189.02	179.61	4.98
	15	216.65	228.74	5.58
	20	237.97	221.48	6.93
	25	260.38	277.33	6.51
	30	276.19	292.82	6.02
	40	306.38	326.11	6.44

touching a rectangular contact designed on the top metal of dimensions 25 × 75 μm. A simple calculation reveals that, for these dimensions, the total resistance sensed by the probe tips would be less than 50 mΩ, far less than the expected values to be measured, which typically vary above 100 Ω.

Table 12.2 presents the measurement results corresponding to the fabricated test chip in comparison with the results obtained by SRG. As can be seen from the results, the prediction of the resistive coupling is in close agreement with the measurement results with 5.3% as the average error.

In Figure 12.17, which is based on the measurement data from the fabricated test chip, the case examined involves the coupling resistance of a pair of contacts with widths $W = \{20, 30, 40\}$ μm. For both cases, uniform and non-uniform, the proposed method correctly predicts the resistance

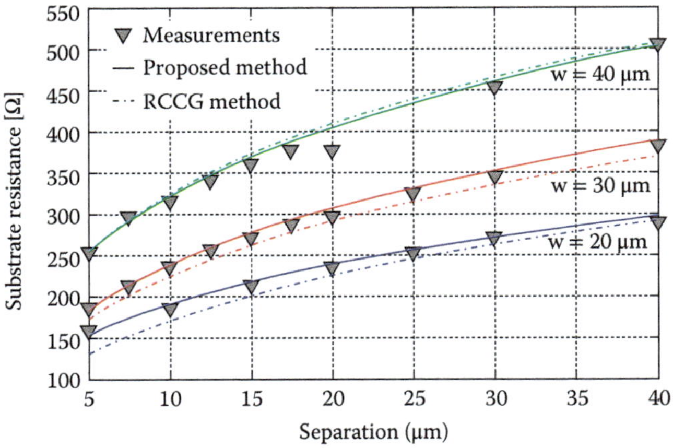

FIGURE 12.17
Comparison of the proposed method with measurement data for the coupling resistance between two square contacts as a function of their separation. Three different side widths are studied.

coupling with the average error being approximately 5%. Also, in this figure, the proposed method is compared against the RCCG method [24].

12.7 Summary

In this chapter, a unified method for calculating resistive and capacitive coupling exploiting geometry constraints is presented. This method is based on the geometry of the electric field lines produced by the coupled structures and on one of the primary principles of physics, the Law of Least Energy. The shape of the current flow lines is investigated in each case. The presented method exploits the underlying symmetry in these shapes for simplifying both the analysis process and the resulting equations used for describing them. The superposition of three simple structures, that is, square, circular disk, and rectangle over an infinite (ground) plane, is used for modeling the coupling mechanism in every case. The shape of the electric field lines in each of these simple structures is approximated by simple geometric shapes (cylinders, orthogonal parallelepipeds, etc.), whose resistance (or capacitance) may be computed explicitly. Since the proposed method is based on geometric calculations, it is generally scalable and independent from technology parameters. Comparative results have been presented where the proposed method exhibits high accuracy and fast computation speed in comparison to other methods from literature.

An automated method for extracting the substrate resistance has also been presented. It uses a geometric formulation to derive an analytical model for the substrate resistance for an arbitrary substrate doping profile. The presented formulation does not rely on fitting techniques and is general. Extensive simulation experiments on various substrates heavily doped, lightly doped, and uniform have been performed. Also, measurements based on a fabricated chip have been used to validate the proposed method. Both simulation and measurement results show that the average error in resistance computation is within 5%, supporting the validity of the proposed method.

References

1. M. Badaroglu, M. van Heijningen, V. Gravot, J. Compiet, S. Donnay, G. G. E. Gielen, and H. J. De Man, Methodology and experimental verification for substrate noise reduction in CMOS mixed-signal ICs with syn- chronous digital circuits, *IEEE Journal of Solid-State Circuits*, vol. 37, no. 11, 1383–1395, November 2002.
2. E. Schrik and N. P. van der Meijs, "Combined BEM/FEM Substrate Resistance Modeling," in *Proceeding of IEEE Design Automation Conference (DAC'02)*, New Orleans, LA, pp. 771–776, June 2002.
3. M. van Heijningen, J. Compiet, P. Wambacq, S. Donnay, M.G.E. Engels, and I. Bolsens, Analysis and experimental verification of digital substrate noise generation for epi-type substrates, *IEEE Journal of Solid-State Circuits*, vol. 35, no. 7, 1002–1008, July 2000.
4. E. Charbon, R. Gharpurey, R. G. Meyer, and A. Sangiovanni-Vincentelli, Substrate optimization based on semi-analytical techniques, *IEEE Transactions on Computer-Aided Design*, vol. 18, no. 2, 172–190, February 1999.
5. R. Gharpurey and R. G. Meyer, Modeling and analysis of substrate coupling in integrated circuits, *IEEE Journal of Solid-State Circuits*, vol. 31, no. 1, 344–353, March 1996.
6. A. Samavedam, A. Sadate, K. Mayaram, and T. S. Fiez, A scalable substrate noise coupling model for design of mixed-signal IC's, *IEEE Journal of Solid-State Circuits*, vol. 35, no. 6, 895–904, June 2000.
7. H Lan, T. W. Chen, C. O. Chui, P. Nikaeen, J. W. Kim, and R. W. Dutton, Synthesized compact models and experimental verifications for substrate noise coupling in mixed-signal ICs, *IEEE Journal of Solid-State Circuits*, vol. 41, no. 8, 1817–1829, August 2006.
8. N. K. Verghese, D. J. Allstot, and M. A. Wolfe, "Fast Parasitic Extraction for Substrate Coupling in Mixed-Signal ICs," in *IEEE Custom Integrated Circuits Conference*, Santa-Clara, California, May 1995, pp. 121–124, 1995.
9. D. Ozis, T. Fiez, and K. Mayaram, "A Comprehensive Geometry-dependent Macromodel for Substrate Noise Coupling in Heavily Doped CMOS Processes," in *Proceedings of the IEEE Custom Integrated Circuits Conference*, Orlando, FL, pp. 497–500, September 2002.

10. S. Ahyoune, J. Sieiro, J. M. López-Villegas, and M. N. Vidal, "Substrate Coupling Modeling in Integrated Circuits Using Analytical Green's Function," in *Conference on Design of Circuits and Integrated Systems*, Madrid, Spain, 2014.
11. A. Anvar, S. Karmalkar, R. Gokul, and C. Akhil, "A Quasi-Static Model for the Coupling Impedance between Coplanar Rectangular Contacts on a Bulk Substrate," in *29th International Conference on VLSI Design and 15th International Conference on Embedded Systems (VLSID)*, Kolkata, India, pp. 338–342, 2016.
12. M. Babic and M. Krstic, "A Coarse Model for Estimation of Switching Noise Coupling in Lightly Doped Substrates," in *IEEE 18th International Symposium on Design and Diagnostics of Electronic Circuits & Systems*, Belgrade, Serbia, pp. 217–222, 2015.
13. G. Shomalnasab and L. Zhang, New analytic model of coupling and substrate capacitance in nanometer technologies, *IEEE Transactions on Very Large Scale Integration (VLSI) Systems*, vol. 23, no. 7, 1268–1280, July 2015.
14. F. Stellari and A. L. Lacaita, New formulas of interconnect capacitances based on results of conformal mapping method, *IEEE Transactions on Electron Devices*, vol. 47, no. 1, pp. 222–231, January 2000.
15. A. Bansal, B. Paul, and K. Roy, An analytical fringe capacitance model for interconnects using conformal mapping, *IEEE Transactions on Computer-Aided Design of Integrated Circuits and Systems*, vol. 25, pp. 2765–2774, 2006.
16. M. I. Elmasry, Capacitance calculation in MOSFET VLSI, *IEEE Electron Device Letters*, vol. EDL-3, 6–7, 1981.
17. P. Eyben, D. Alvarez, M. Jurczak, R. Rooyackers, A. De Keersgieter, E. Augendre, and W. Vandervorst, Analysis of the two-dimensional-dopant profile in a 90 nm complementary metal–oxide–semiconductor technology using scanning spreading resistance microscopy, *Journal of Vacuum Science & Technology*, vol. 22, no. 1, 364–368, 2004.
18. P. Eyben, M. Xu, N. Duhayon, T. Clarysse, S. Callewaert, and W. Vandervorst, Scanning spreading resistance microscopy and spectroscopy for routine and quantitative two-dimensional carrier profiling, *Journal of Vacuum Science & Technology*, vol. 20, no 1, 471–478, 2002.
19. S. Morita, *Roadmap of Scanning Probe Microscopy (NanoScience and Technology)*, Berlin: Springer, p. 18, November 2006.
20. S. Bronckers, K. Scheir, G. V. Plas, G. Vandersteen, and Y. Rolain, A methodology to predict the impact of substrate noise in analog/RF systems, *IEEE Transactions on Computer-Aided Design of Integrated Circuits and Systems*, vol. 28 no. 11, pp. 1613–1626, November 2009.
21. A. Sharma, P. Birrer, S. K. Arunachalam, C. Xu, T. S. Fiez, and K. Mayaram, Accurate prediction of substrate parasitics in heavily doped CMOS processes using a calibrated boundary element solver, *IEEE Transactions on Very Large Scale Integration (VLSI) Systems*, vol. 13 no. 7, 843–851, July 2005.
22. B. Peterson, K. Mayaram, and T. S. Fiez, "Automated Extraction of Model Parameters for Noise Coupling Analysis in Silicon Substrates," in *IEEE Custom Integrated Circuits Conference, CICC '07*, San Jose, CA, pp. 853–856, September 2007.
23. Y. Bontzios and A. Hatzopoulos, "A New Method for Calculating Resistive Losses on Lightly and Heavily Doped Substrates", *in Proceeding of IFIP International Conference Very Large Scale Integration*, Rhodes, Greece, pp. 301–305, October 2008.

24. Y. Bontzios and A. Hatzopoulos, A unified method for calculating capacitive and resistive coupling exploiting geometry constraints on lightly and heavily doped CMOS processes, *IEEE Transactions on Electron Devices*, vol. 57, no. 8, 1751–1760, August 2010.
25. Y. Bontzios, M. Dimopoulos, and A. Hatzopoulos, Automated substrate resistance extraction in nanoscale VLSI by exploiting a geometry-based analytical model, *International Journal of Microelectronics and Computer Science*, vol. 2, no. 3, 81–87, 2011.
26. A. F. Peterson, *Computational Methods for Electromagnetics*, Hoboken, NJ: Wiley-Blackwell, pp. 24–34, 1997.
27. E. Charbon, R. Gharpurey, P. Miliozzi, R. G. Meyer, and A. Sangiovanni-Vincentelli, *Substrate Noise: Analysis and Optimization for IC Design*, Berlin: Springer, April 2001.
28. Comsol, "Femlab 3.1," Available at: http://www.comsol.com.
29. Keysight Technologies, Advanced Design System (ADS), http://www.keysight.com

13

Coupling through Substrate for Millimeter Wave Frequencies

Vasileios A. Gerakis and Alkis A. Hatzopoulos

CONTENTS

13.1 Introduction

As the market's demand increases for advanced applications such as supercomputers, IoT devices, autonomous vehicles and aircrafts, machine learning, and big data, the CMOS technology pushes its scaling and the Integrated Circuits' (IC) operation frequency reaches up to 100 GHz. This way, faster computing cycles and data transmission is accomplished. In this band of Millimeter Wave (MW) frequencies, the coupling between devices through the substrate maximizes, often leading to observations opposite to what a designer would expect. The substrate isn't left untouched by the increased penetration of MW frequency electromagnetic signals. The effect of the substrate of the on-chip devices, the crosstalk between on-chip devices through the substrate, and means to attenuate the crosstalk are topics that have been attracting focus ever since the ICs' operating frequency reached the GHz range [1]. Furthermore, as the frequency increases through the MW range and up to 100 GHz, phenomena such as skin and proximity effect [2], significantly alter the shape of the paths through which the currents flow inside the conductors and the semi-conductors, thus affecting the electrical behavior of all devices, components, and interconnections on chip. Design rules and best practices in lower frequencies cannot be taken for granted in the MW frequency range. Every design case should be studied thoroughly, simulated, and tested in its operation frequency early, before the design process and again prior to fabrication. This would be prerequisite to reassure the substrate's effect on chip will be taken into consideration and even attenuated by employing a means of protection to critical parts to maximize the design's efficiency and minimize unpredictable and undesired results of the crosstalk through the substrate.

The effect of the substrate profile on the coupling between devices in the MW band has yet to receive attention. In this chapter, a typical 65 nm technology is leveraged to study this effect. Interconnections and inductors are designed and simulated to describe the effects of the substrate signals that pass from one component to another. Means to attenuate it are also studied, such as guard rings and shields placed around and below both aggressor and victim components, respectively. The resistivity of the substrate considered is 10 Ohm-cm and its thickness approximately 780 μm. The oxide and metal layer stack up is designed as in a typical 65 nm technology. The higher—considerably thicker—metal layers were employed to design the inductors, typically offering lower resistivity and higher quality values. A collection of design guidelines to reduce the on-chip substrate noise coupling between interconnections and inductors in high frequency applications are presented. Moreover, the influence of the substrate and the crosstalk reduction methods to the quality (Q) and the self-resonance frequency (SRF) of the inductors are explored.

13.2 Interconnections Coupling through Substrate

Typically, metal paths' models in the literature include coupling through substrate [3]. In this section, two parallel metal paths are designed in ANSYS HFSS, leveraging typical 65 nm design rules to study the effect of this coupling at MW frequencies up to 100 GHz. The paths' width is 4 μm and length 100 μm. They were designed on the second metal layer and at various separation distances in between them from 24 μm up to 74 μm. The distance is measured between their facing edges. To better model the damascene copper process that would be employed at fabrication, the paths' cross-section is designed not as rectangular, but as a trapezoid with the top base larger than the bottom one by 10%, as shown in Figure 13.1a. This trapezoidal hypothesis is especially important in higher than 1 GHz frequencies through which the skin depth becomes relatively small. Four wave ports were utilized to excite the interconnections and s-parameters were extracted for frequencies up to 100 GHz.

The magnitude of the S13 parameter represents the signal attenuation from the first wave port of the second interconnection path to the first wave port

FIGURE 13.1
Layout of (a) the parallel paths and a zoom in the cross section of the metal and (b) parallel paths and grounded path of the same geometry in the middle of the distance between them.

of the first interconnection path. Due to the symmetry of the layout, the S13 parameter is identical to the S31 one. Using the S13 parameter from the simulation results, the S13 difference (%) between the case with silicon substrate and without one is calculated, for all distance and frequency values, as in

$$\Delta S13_{Si-Interconnects}(\%) = 100 \cdot \frac{S13_{With\text{-}Si} - S13_{Without\text{-}Si}}{S13_{Without\text{-}Si}} \tag{13.1}$$

Also, the S13 difference (%) between the case with a grounded metal interconnection path and without one is calculated as in

$$\Delta S13_{Ground-Interconnects}(\%) = 100 \cdot \frac{S13_{With\text{-}Ground} - S13_{Without\text{-}Ground}}{S13_{Without\text{-}Ground}} \tag{13.2}$$

In Equation 13.1 and Equation 13.2, only S13 in magnitude values are used. The reason for not utilizing S13 in dB values is that they are not able to properly represent the magnitudes' differences when subtracted. Difference of 1 dB corresponds to larger magnitude difference values when it is derived from greater dB values (e.g., 40 dB–39 dB) than lower ones (e.g., 10 dB–9 dB). So, for the same 1 dB difference we would have different magnitude differences and this could lead to false conclusions. Furthermore, it should be noted that a 100% ΔS13 value means doubling the first S13 value compared to the second from the nominator. On the contrary, −50% ΔS13 value means reducing by half the first S13 value compared to the second of the nominator and −100% means reducing the first S13 value to zero.

In Figure 13.2a, a 3D plot of Equation 13.1 is pictured for all frequency and distance values. The difference (%) of the magnitudes of S13 between the structure simulated with and the ones without silicon substrate appears to be positive throughout the whole frequency range and separation distance variation. So, the structure with silicon substrate has greater S13 values and, thus, the attenuation between the two interconnections is less in this case. The obvious deduction is that the substrate offers parasitic conductive and capacitive paths for the alternating currents to pass from the one interconnection to the other, reducing the isolation in between them and, thus increasing, crosstalk. The separation distance of the signal interconnections is a critical variable affecting the mutual coupling of the paths [4]. Especially at higher separation distance values between them, the substrate seems to improve the crosstalk up to 50%.

In Figure 13.2b, the S13 magnitude difference $\Delta S13_{Ground\text{-}Interconnects}$ (%) of the structures with and without grounded metal interconnection in the middle, as pictured in Figure 13.1b and defined in Equation 13.2, is plotted. The middle metal grounded path is of the same geometry and dimensions and lies exactly at the center of the in-between separation distance of the two paths. As presented in 13.2b, the values of $\Delta S13_{Ground\text{-}Interconnects}$ are negative, which means that adding a metal grounded path in the middle of the two interconnections offers an electrically ground wall as it significantly reduces the S13 magnitude values by approximately 50%.

To summarize this section's deductions:

- Figure 13.2a shows that the existence of substrate increases the S13 parameter and, thus, crosstalk.
- The maximum influence of the silicon substrate on S13 and on crosstalk is at lower frequencies for all distances except the shortest ones.
- Regarding the interconnections' separation distance, the influence of the substrate maximizes its influence at the highest distance values, reaching 50% increase, as pictured in Figure 13.2a. This happens because the interconnections need to be separated at a minimum distance for the parasitic capacitances and resistances to be enabled and to increase ΔS13 (%) and crosstalk.

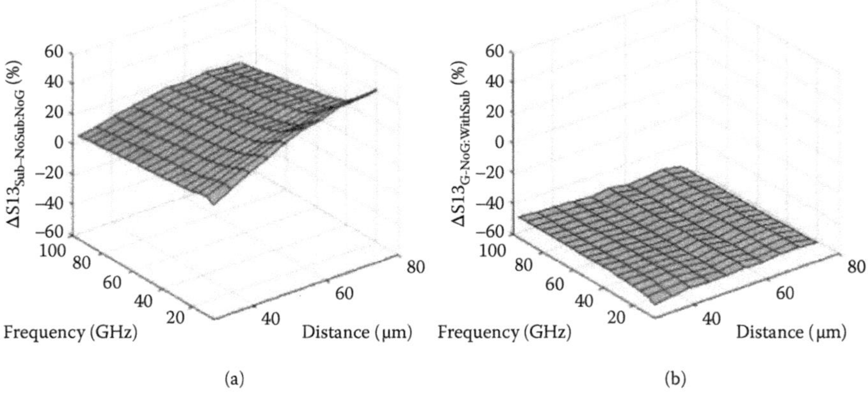

FIGURE 13.2
(a) $\Delta S13_{Si\text{-}Interconnects}$ (%) difference between the structures with and without silicon substrate, as defined in Equation 13.1 and (b) $\Delta S13_{Ground\text{-}Interconnects}$ (%) difference between the structures with and without ground interconnect in the middle, as defined in Equation 13.2. Separation distance is 24–74 μm and frequency from 1 to 100 GHz.

- For a given interconnections' separation distance, the influence of the substrate increases when lowering the signal's frequency.
- Grounded metal lines placed in the middle between interconnections act as a wall blocking parasitic signals and reducing coupling at all distances and frequencies up to 50%, as shown in Figure 13.2b.

13.3 Inductors Coupling through Substrate

One widely investigated passive device is the integrated inductor utilized for RF applications. The inductor is the device that creates the most intense electromagnetic flux due to the nature of its operation. It induces AC signals in nearby devices, altering their electrical performance. This aggression is increased by the silicon substrate that offers an alternative conductive and capacitive path for the coupling ac noise currents to reach nearby victims. The interconnection path is formed through the epi and bulk layers of the silicon substrate due to its conductivity and permittivity. Noise signals are increased to affect devices through the substrate that lie with even greater distances between them through the MW frequency range. Previous works show that high resistivity substrates are reducing substrate losses and keep the self-resonance frequency of the inductor as high as possible [5]. In Figure 13.3a, the commonly used model of an integrated inductor is presented, taking into consideration the substrate effect [6]. The resistance and inductance of the inductor are represented by R and L, and the fringing capacitance between the ports of the inductor is represented by C. The capacitance through the oxide layers are represented by C_{ox} and the capacitance and resistance through the silicon substrate to the ground by C_{si} and R_{si}.

However, this model is proven to be less accurate in the MW frequency range and other more complicated models are needed. Especially for the modeling of the eddy current under an on-chip inductor, induced inside the substrate, more lumped elements need to be incorporated at the model to offer more accuracy in predicting the inductor's behavior. Iv. Lai and M. Fujishima have proposed a model with 5 more elements, depicted in Figure 13.3b, which includes an inductance produced by the substrate's eddy currents L_{sub} in series with a respective resistance R_{sub} and a mutual inductance, without coupling coefficient k, that is formed between the inductance of the substrate L_{sub} and the inductor L [7]. Optimization analysis leveraging HFSS Optimetrics tools and Quasi Newton Optimizer was conducted to fit the 9 and 14-element models of Figures 13.3a and b to the s-parameters matrix extracted by simulating the octagonal inductor used in this chapter. The fitting frequency is from 1 to 100 GHz. The optimizer was not able to fit the 9-element model to the simulated s-parameter matrix of the inductor. On the other hand, it was successful for the 14-element model of Figure 13.3b. In Figures 13.3c through e the inductance, quality and resistance are presented, extracted from the y-parameters from the simulated in HFSS structure, expressed in Equation 13.13, Equation 13.4, and Equation 13.5, respectively.

$$L = \frac{im\left(\frac{1}{Y_{11}}\right)}{2\pi \cdot f} \tag{13.3}$$

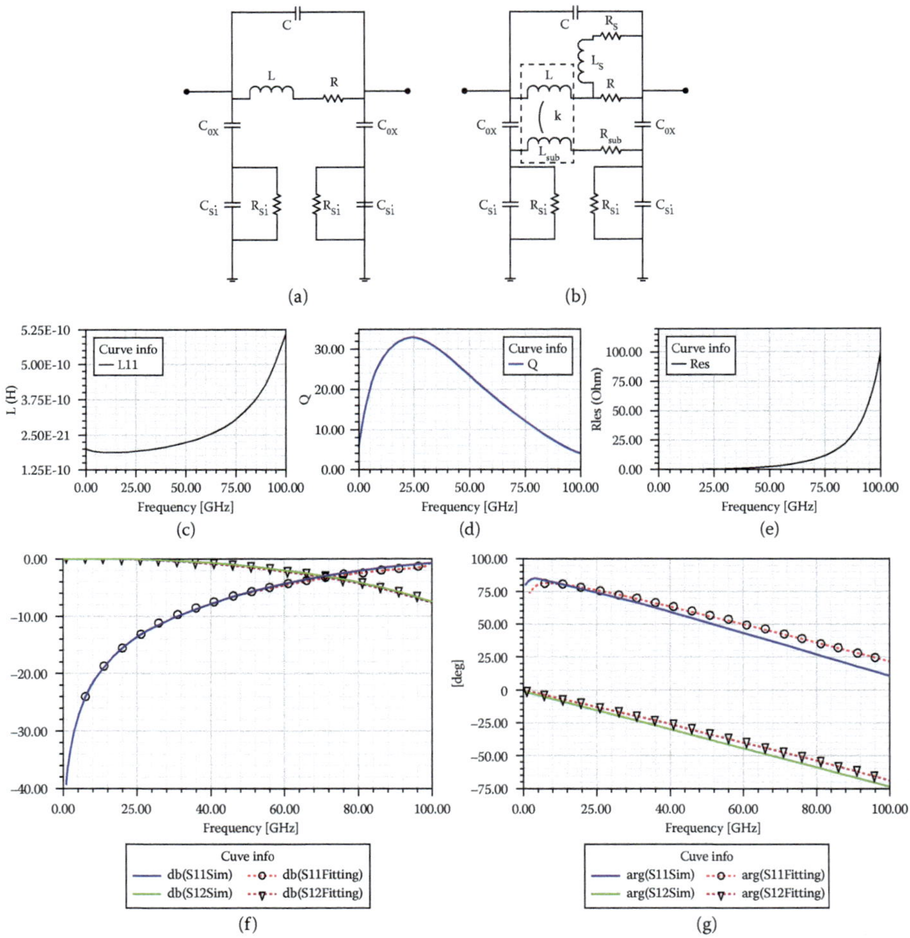

FIGURE 13.3
(a) 9 element and (b) 14 element circuit model of the inductor. (c) Inductance, (d) quality, and (e) resistance of the inductor up to 100 GHz. This shows the fitting of the 14-parameter lumped element model of (b) to the s-parameters of the inductor simulated in this chapter (diameter 130 μm, path width 20 μm, single turn, uppermost metal layer).

$$Q = \frac{im\left(\frac{1}{Y_{11}}\right)}{re\left(\frac{1}{Y_{11}}\right)} \tag{13.4}$$

$$R = re\left(\frac{1}{Y_{11}}\right) \tag{13.5}$$

In Figures 13.3f and g, the fitting of the 14-element model (red lines) reprsenting (b) to the simulated s-parameters of the HFSS (blue and green lines) is presented. In (f) the magnitude and in (g) the arguments of the s-parameters are pictured. The optimizer seems to fit satisfactorily the model of the s-parameters, and, thus, it is concluded that the 14-element model is successful at describing the inductor's behavior up to 100 GHz.

Table 13.1 presents the values of the lumped elements' model of Figure 13.3b extracted by the optimizer during the fitting procedure. These values are a mathematical solution and their only physical meaning is to prove that a model of this complexity may be enough to describe the s-parameters of an inductor up to 100 GHz. It should be also noted that a symmetrically designed octagonal inductor, as in this chapter, is expected to have identical C_{ox}, C_{si}, and R_{si} values below both ports, reducing the parameters of input to the optimizer to 11 and reducing the complexity of the fitting procedure. However, inductors with more than one turn would not be symmetrical and would be expected to need 14 parameters to be optimized correctly.

In this chapter, integrated inductors are mainly used to study the substrate coupling phenomena in the MW frequency range. Several structures representing various inductor topologies were simulated utilizing ANSYS HFSS simulation software. To observe the coupling phenomena through the substrate, two co-planar inductors were designed following typical 65 nm design technology rules, as in the previous section. To sweep both the frequency range up to 100 GHz efficiently and fast and to run several parametric analyses for distances between the inductors up to 1 mm, Aristotle University of Thessaloniki and Hellasgrid supercomputing infrastructures were leveraged, a node of 64 CPU cores and 200 GB RAM. For the inductors to reach a high operation and resonance frequency, the inductors were chosen to have only one turn, as depicted in Figure 13.3a. They were also placed at the uppermost metal layer, as the greater metal thickness in this layer provides less resistivity and higher quality to the inductors. Each inductor is designed so as its terminals to have coplanar lumped ports that are configured for s-parameter extraction. The attenuation of the signal that passes from the first port of the second inductor to the respective port of the opposing first inductor is represented by the S13 parameter. In this study, all structures were symmetrical, so defining which inductor was the aggressor and which the victim was meaningless as

TABLE 13.1

Elements Extracted by Fitting the Model in Figure 13.3b to the Extracted S-Parameters of the Inductor Studied in This Chapter

Element	Value	Units
R	0.507	Ohm
R_s	1.25	mOhm
R_{sub}	4.361	Ohm
R_{si}	0.613	Ohm
L	0.416	nH
L_s	63.855	nH
L_{sub}	91.4	nH
k	0.772	
C	7.7	fF
C_{ox}	0.92	fF
C_{si}	7.1	fF

they both serve both roles. This was also confirmed by the identical S13 and S31 values at all cases.

To picture the effect of the silicon substrate on the coupling, the structure was simulated at a substrate-less geometry too. In Figure 13.4a, the layout of the inductor is presented. The inductor is octagonal, designed with a single turn for its operation frequency to easily reach the MW range. The path width is 20 μm and the inductors' outer diameter is, in all cases, 130 μm. The process considered in this chapter is the Cu damascene process, in which all metal lines' cross-sections are generally trapezoidal rather than rectangular. Since, practically, the cross-sections of the paths are less ideal than the rectangular shapes usually considered and to improve the accuracy of the models designed, a more realistic tapered form of the paths was designed, as in the previous section, in which the bottom smaller base of the trapezoid was approximately 10% less than the top one.

A first indication of the effect of the silicon substrate on the inductors' operation is their reduced quality value Q, as pictured in Figure 13.4b. The inductor's quality is reduced by almost 25% due to the silicon substrate. The inductor quality Q was extracted from the y-parameter simulation results, using Equation 13.4.

To discover the coupling between two inductors and how silicon and other parameters affect it, the S13 parameters were extracted by simulating two opposing inductors, designed as in Figure 13.5a. The magnitude of the S13 parameter indicates the attenuation of the noise signal that passes from the first port of the second inductor to the first port of the opposing first inductor. Lower S13 values imply greater attenuation of the signal that passes from the aggressor to the victim inductor and thus improved isolation between them. To compare two structures' S13 parameter, a difference (%) is calculated for all frequencies and distances, leveraging the Mathworks Matlab environment.

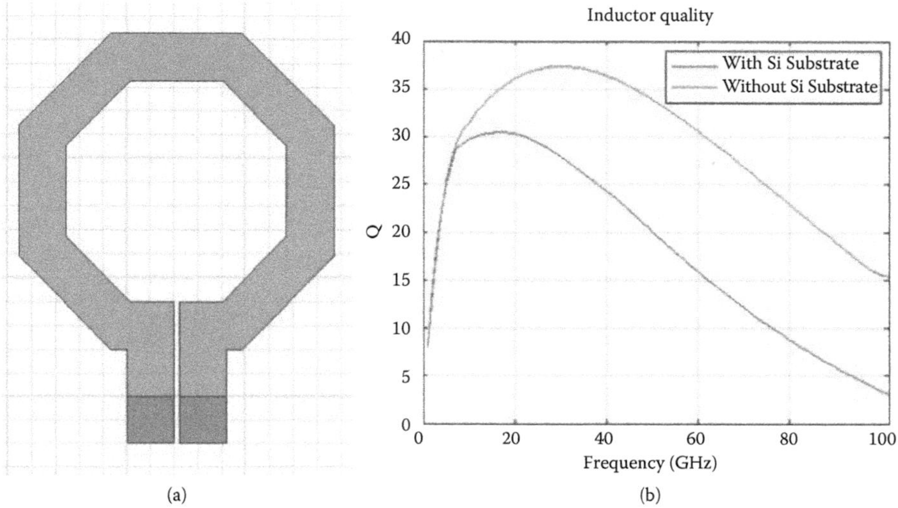

FIGURE 13.4
(a) Layout of the octagonal inductor. (b) Quality of the inductor at a structure with and without Si Substrate.

When this difference is positive, the subtracter is lower and, thus, offers better attenuation and isolation as a structure. When it is negative, the subtrahend is lower, offering better isolation. To offer a better understanding of the impact at each case, the result of this difference is divided by the subtracter to present the difference in a relative percentage. So, the (%) difference of the S13 parameters' magnitude in this case, is derived in Equation 13.6.

$$\Delta S13_{\text{Si-Inductors}}(\%) = 100 \cdot \frac{S13_{\text{With-Si}} - S13_{\text{Without-Si}}}{S13_{\text{Without-Si}}} \tag{13.6}$$

In Figure 13.5b, a 3D plot of Equation 13.6 is presented. All simulations are performed for frequencies up to 100 GHz and separation distances between the outer edges of the two inductors from 20 up to 620 μm. As pictured, the difference $\Delta S13_{\text{Si-Inductors}}$ of the S13 parameters, with and without Si substrate is always positive, so the S13 parameter of the structure with substrate is greater and offers increased coupling and worse isolation. The silicon substrate significantly increases the dissemination of noise coupling, reaching almost 400% greater values at the frequency of 100 GHz and at the distance of 600 μm, as depicted in the yellow areas that occur in the 3D plot in Figure 13.5b. This implies that, in the MW frequency range, inductors' coupling is increased especially at greater separation distance. For the blue plateau areas of lower distance values below 500 μm and lower frequency values below 70 GHz, the ΔS13 (%) remains below 20%, showing that, for those frequency-distance combination values, the substrate effect is kept low.

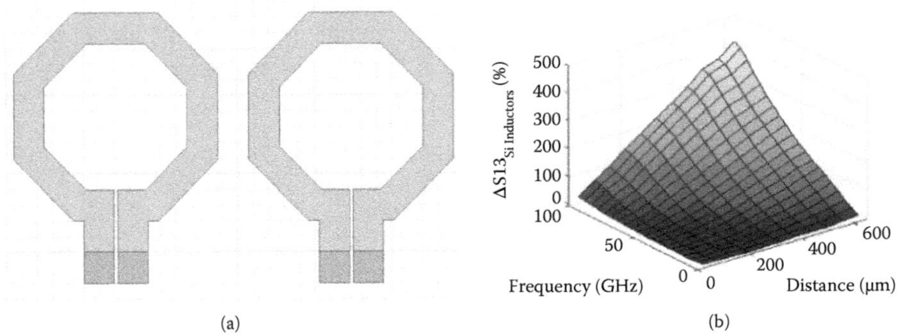

FIGURE 13.5
(a) The layout of the two opposing inductors simulated. (b) $\Delta S13_{\text{Si-Inductors}}$ (%) difference of the structures with and without silicon substrate, as defined in Equation 13.6 for frequencies of 1–100 GHz and with a separation distance between the inductors from 20 to 620 μm.

To summarize this section's deductions:

- The model in Figure 13.3b can describe the inductor's behavior and the substrate's effect up to 100 GHz.
- Silicon substrate deteriorates the inductor's quality, as pictured in Figure 13.4b.
- Silicon substrate increases crosstalk between inductors, especially in higher frequencies and at greater separation distances, up to 400%.
- For lower combinations of frequency-distance values, the substrate affects the crosstalk by less than 20%.

13.4 Guard Rings

A widely-used method to isolate devices and especially inductors from nearby aggressors or victims is to surround them with guard rings. Guard rings serve to block the electromagnetic interference of the inductors through the substrate to nearby noise sensitive devices. They may be designed in various layouts in the substrate as p+ diffusion areas in high resistivity substrates [8–10] or n+ [11], or both p+ and n+ [12], or p-minus areas alone or combined with p+ in lightly doped CMOS substrates [13]. Guard rings are more often designed above the substrate at the polysilicon or at a metal layer [14], or in all of them, either connecting the various rings with metal vias or not. They may be either grounded or floating [15] or even induced with currents [16] to cancel the coupling effects at nearby components. They may be formed by direct paths, cut in one or several parts, of broken by paths or even several smaller metal tiles to negate eddy currents formation [17]. However, isolating an inductor is not the only purpose of leveraging guard rings. Many studies investigate the positive impact of a guard ring to the quality of the inductor.

Guard rings utilized in the MW frequency range, if not pre-studied early in the design process, simulated, and tested carefully may lead to undesirable effects. It is often stated that for frequencies below 50 GHz the distance of the ring from the inductor is a critical factor [14]. Placing the ring too close to the inductor may reduce the coupling isolation and the quality factor of the inductor. To study the effect of the distance of the ring from the inductor, 3 cases were simulated, as pictured in Figure 13.6a through c, with the metal square rings placed at the uppermost metal layer and encircling both opposing inductors and at a close, a wide, and a far separation distance between the ring and each inductor. The rings' metal width is 5 μm. In order for the comparison to be fair, the inductors in all 3 setups were placed at the same separation distance range between their outer edges, from 350 μm to 1350 μm. The area that each inductor and its surrounding ring occupies is 0.03 mm^2, 0.07 $mm^{2,}$ and 0.2 mm^2 for the close, wide, and far inductor ring separation distances, respectively.

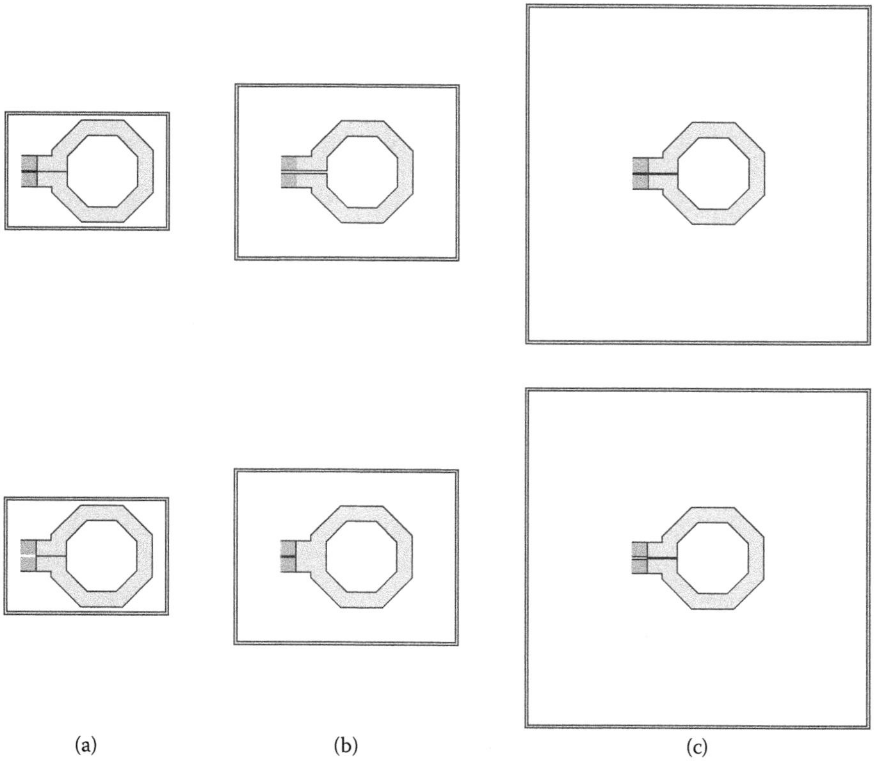

FIGURE 13.6
Layout of the inductor structures with (a) close square rings, (b) wide square rings, and (c) far square rings.

The simulations' S13 parameters without and without rings differences (%) were calculated for these three cases:

$$\Delta S13_{RingsClose}(\%) = 100 \cdot \frac{S13_{Rings\ close\ to\ inductors} - S13_{Without\ rings}}{S13_{Without\ rings}} \tag{13.7}$$

$$\Delta S13_{RingsWide}(\%) = 100 \cdot \frac{S13_{Rings\ wide\ to\ inductors} - S13_{Without\ rings}}{S13_{Without\ rings}} \tag{13.8}$$

$$\Delta S13_{RingsFar}(\%) = 100 \cdot \frac{S13_{Rings\ far\ from\ inductors} - S13_{Without\ rings}}{S13_{Without\ rings}} \tag{13.9}$$

The 3D plots in the first column of Figure 13.7 represent the S13 differences (%) as expressed in Equations 13.7 through 13.9. As shown in Figure 13.7c, when the rings are placed far from the inductor at a distance of 150 μm, as in Figure 13.6c, the ΔS13(%) difference is minimized compared to Figures 13.7a and b, which correspond to the close and the wide setup. Thus, the layout with the rings at the far distance from the inductor has in all frequencies and at all distances, lower S13 values and, thus, improved isolation. For the square type of guard rings, sacrificing area by placing the ring at a greater distance from the inductor leads to slightly improved isolation.

Furthermore, in Figure 13.7d, the quality of the inductors in the four cases is presented. It is apparent that a ring close to the inductor square slightly reduces the quality of the inductors. On the other hand, leveraging square metal rings at greater distances improves the quality of the inductors in a similar way. One other deduction is that utilizing a metal guard ring shifts the quality curve to higher frequency values, leading to better performance at higher frequencies. This is a significant advantage of guard rings compared to the shields that will be studied in the following section. However, the positive effects of improved isolation and quality for the inductors are offered only when, in exchange, the designer sacrifices area, due the fact that the distance of the ring from the inductor is comparable to the outer diameter of the inductor. All the following simulations and analysis in this section were conducted for two different values of distances of the rings from the inductor. The inductors surrounded by "wide" rings were placed at a distance twice the path's width from the inductor, similarly to Figure 13.6b. The inductors surrounded by "far" rings were placed far from the inductor, approximately at a distance equal to the diameter of the inductor, similarly to Figure 13.6c. These two layouts were selected for further analysis as the "close" ring case reduces slightly the inductors' quality.

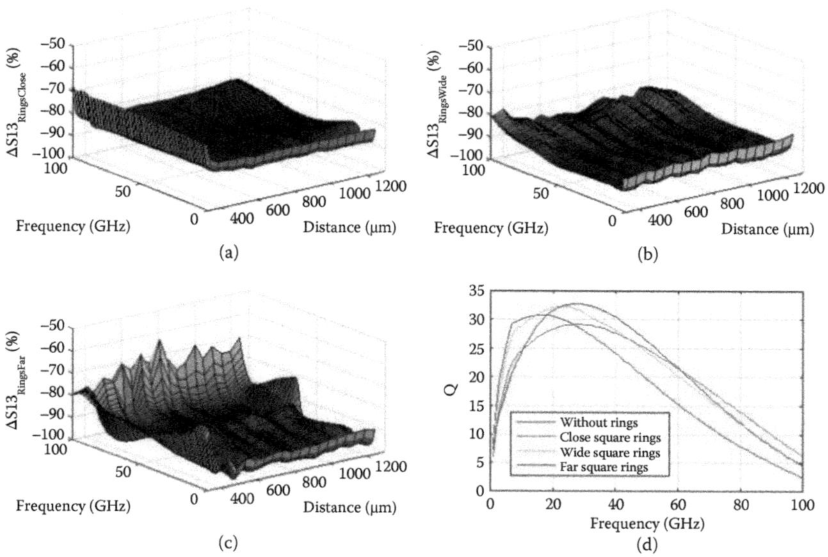

FIGURE 13.7
ΔS13 (%) difference of the structures with (a) close rings as in Equation 13.7, (b) wide rings as in Equation 13.8, and (c) far rings as in Equation 13.9. The inductors are at a separation distance between their outer edges of 350 μm to 1350 μm. (d) Comparison of the quality of the case without a ring and the close, wide, and far rings cases.

As presented in Figure 13.8, various geometries of rings were designed around the two inductors and simulated for frequencies up to 100 GHz and separation distances between the outer edges of the inductors up to 1350 μm. Firstly, square and octagonal metal rings that occupy the greatest amount of area were simulated. As shown in Figure 13.9, they offer improvement in terms of the quality factor Q of the inductors, in both wide and far rings setup. The critical advantage of the octagonal metal rings is the considerably reduced area they occupy. In the far rings setup, the metal rings shift the quality curve of the inductor to greater values, for about 5 GHz. Next a p+ ring was designed and simulated on the substrate, as pictured in Figure 13.8c, which improves slightly the quality of the inductor in both wide and far rings setup cases. The advantage of leveraging this kind of rings is that the area occupied is considerably less than the metal rings, as they are formed underneath the metal layers. Lastly, a structure in which both metal rings around and p+ rings below the inductors were designed and simulated, as in Figure 13.8d. This setup's intersection is pictured in Figure 13.8e. As shown in Figure 13.9, this kind of setup improves by almost 10% the quality of the inductors in the cases of both the close and far rings setup.

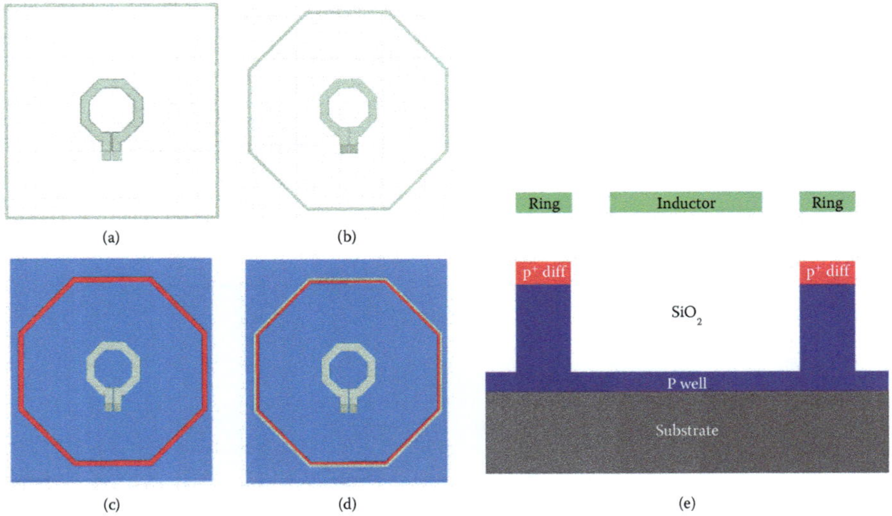

FIGURE 13.8
Various geometries of guard rings. The green color is the highest metal layer, where the inductors and the rings lie, blue is the is p-well, and red is the p+ diffusion paths that create the p+ rings. The figures depict (a) a square metal ring, (b) an octagonal metal ring, (c) an octagonal p+ ring, (d) both metal and p+ octagonal rings far from the inductor, and (e) the intersection of the layout in (d).

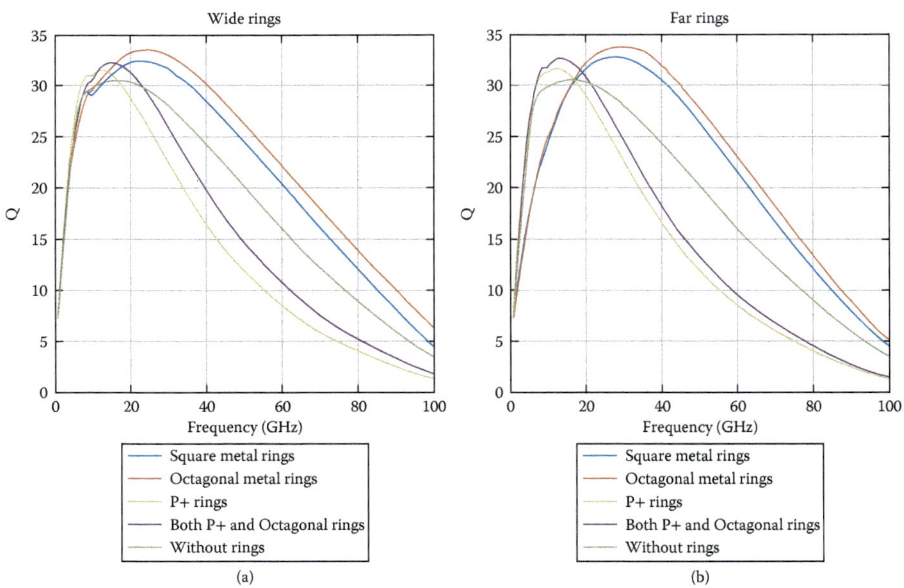

FIGURE 13.9
Comparison of the quality Q of the inductors with and without various kinds of guard, such as (a) wide rings and (b) rings placed far from the inductor.

To compare the effectiveness of the different kind of rings, the S13 magnitude difference (%) is calculated compared to the un-ringed case. So, these differences are expressed in Equations 13.10 through 13.13:

$$\Delta S13_{\text{Squarerings}}(\%) = 100 \cdot \frac{S13_{\text{Square rings}} - S13_{\text{Without rings}}}{S13_{\text{Without rings}}} \tag{13.10}$$

$$\Delta S13_{\text{Octagonalrings}}(\%) = 100 \cdot \frac{S13_{\text{Octagonal rings}} - S13_{\text{Without rings}}}{S13_{\text{Without rings}}} \tag{13.11}$$

$$\Delta S13_{P+\text{rings}}(\%) = 100 \cdot \frac{S13_{\text{P+ rings}} - S13_{\text{Without rings}}}{S13_{\text{Without rings}}} \tag{13.12}$$

$$\Delta S13_{P+\&\text{Octagonalrings}}(\%) = 100 \cdot \frac{S13_{\text{P+\&Octagonalrings}} - S13_{\text{Without rings}}}{S13_{\text{Without rings}}} \tag{13.13}$$

Equations 13.10 through 13.13 have been depicted twice, first in Figure 13.10a through d for the wide ring setup cases, and, second, at Figure 13.11a through d for the cases where the rings are placed far from the inductor. In both the

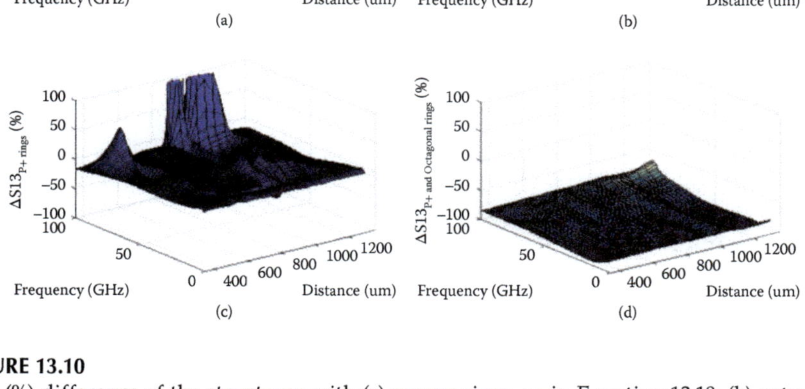

FIGURE 13.10
ΔS13 (%) difference of the structures with (a) square rings, as in Equation 13.10; (b) octagonal rings, as in Equation 13.11; (c) p+ rings, as in Equation 13.12; and (d) both p+ and octagonal rings, as in Equation 13.13. All rings are at a "wide" setup, as in Figure 13.6b. Separation distance between the outer edges of the inductors is from 350 μm to 1350 μm.

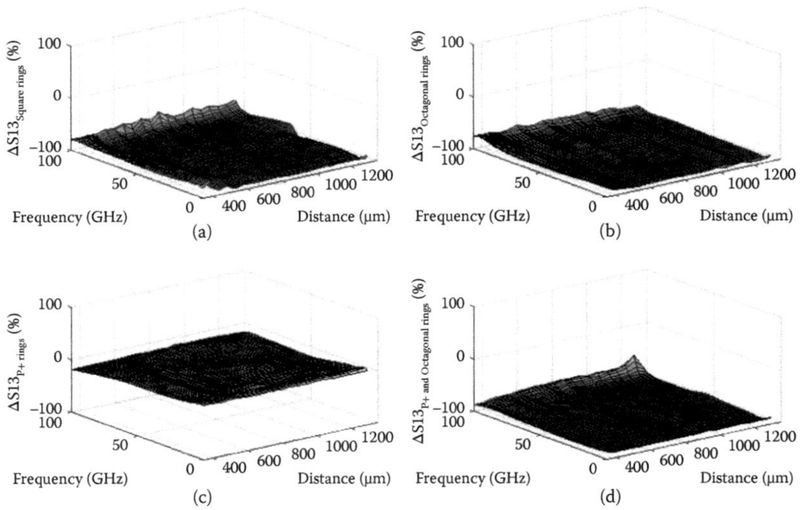

FIGURE 13.11
ΔS13 (%) difference of the structures with (a) square rings, as in Equation 13.10; (b) octagonal rings, as in Equation 13.11; (c) p+ rings, as in Equation 13.12; and (d) both p+ and octagonal rings, as in Equation 13.13. All rings are placed far from the inductors, as in Figure 13.6c. Separation distance between the outer edges of the inductors is from 350 μm to 1350 μm.

wide and far ring setup cases, the best improvement, in terms of isolation, is in cases where the 3D surface of the 3D plots is lower. This happens in the case where octagonal metal rings are implanted and where octagonal metal rings were implemented with p+ rings cases, as shown in Figure 13.10b and d and, in Figure 13.11b and d, reaching almost −90%. This means that the magnitude of the S13 parameter is reduced by almost 90% when surrounded by an octagonal metal grounded ring for all frequencies and separation distance values.

It must be noted that for the cases of inductors with p+ rings underneath them, at specific combinations of higher frequencies and distances, peaks that reach even above 100% occur, as depicted in Figures 13.10c, in the case of the wide rings. These peaks are the result of resonance frequency points for specific separation distances for the inductors and created by the incorporation of p+ rings on top of a p-well area inside the substrate. So, utilizing p+ rings may slightly reduce the crosstalk and not widen the occupied metal layer area, but it reduces the quality of the inductors and reveals an extra risk by present resonance peaks at higher frequencies.

To summarize this section's deductions:

- Metal rings improve both isolation and inductor quality, as pictured in Figure 13.9, but occupy a great amount of area around the inductor.
- Octagonal metal rings offer the maximum quality improvement to the inductors and at the same time achieve the best attenuation of crosstalk, as shown in both Figures 13.10b and 13.11b for the "wide" and the "far" setup, respectively.

- P+ rings may not occupy extra metal area, but the improvement of isolation is slight compared to the metal rings. Also, it slightly reduces the quality of the inductors and the risk of resonance points caused by P+ rings may result in a significant increase in crosstalk even above 100%, as presented in Figures 13.10c.

13.5 Shields

Planar shields are often placed underneath the inductor to block the inductor's parasitic currents from reaching the substrate and through it nearby devices, reducing the level of penetration of electric field into the substrate. Although their effectiveness has been controversial, this section tries to demonstrate what their behavior would be at MW frequencies utilizing a typical 65 nm technology. Shields may be designed at the poly-silicon [18] or the first three metal layers [19] and could be either grounded [20] or floating [21,22]. Shields come in various layouts: planar, patterned grounded or floating [20,23], finger floating [24], and others less widely adopted [21]. Due to the eddy currents induced inside a shield by the inductor, the quality of the inductor is sometimes reduced and the resonant frequency moves to higher values [24]. To cancel the effect of reducing the quality factor, the shields are often cut or patterned to leave gaps in between the conducting paths. So, leaving gaps in the shields increases the quality of the inductor for lower frequencies, but reduces the isolation compared to the ungapped 1-piece planar shield. These gaps are supposed to prevent eddy currents from forming and must be as narrow as possible for them to block the signals that try to pass from the inductor to the substrate. In Figure 13.12, the layouts of the shields chosen to be designed and simulated in this study are presented. They are placed underneath the inductors in the second metal layer. In Figure 13.12a, the 1-piece planar shield without gaps is pictured, and the next two, b and c, are the gapped ones. Figure 13.12d and e shows the patterned ground shields, grounded and floating, respectively. Lastly, Figure 13.12f, represents the finger floating shield, which is the only one that lies in two metal layers, the second and the third ones. The percentage of the area of the metal to the whole area of the shield for the b, c, d, and e cases is approximately 90%.

The effectiveness of the planar shields is examined in terms of the isolation of the inductors. First, two opposing inductors, as in Figure 13.5a, are designed with shields placed underneath them in six different cases corresponding to the 6 different layouts of Figure 13.12. These 6 structures are simulated for frequency of 1 GHz to 100 GHz and separation distance values between the outer edges of the inductors 20 μm to 620 μm. Then, as in previous sections, the S13 parameters are extracted. All structures are compared

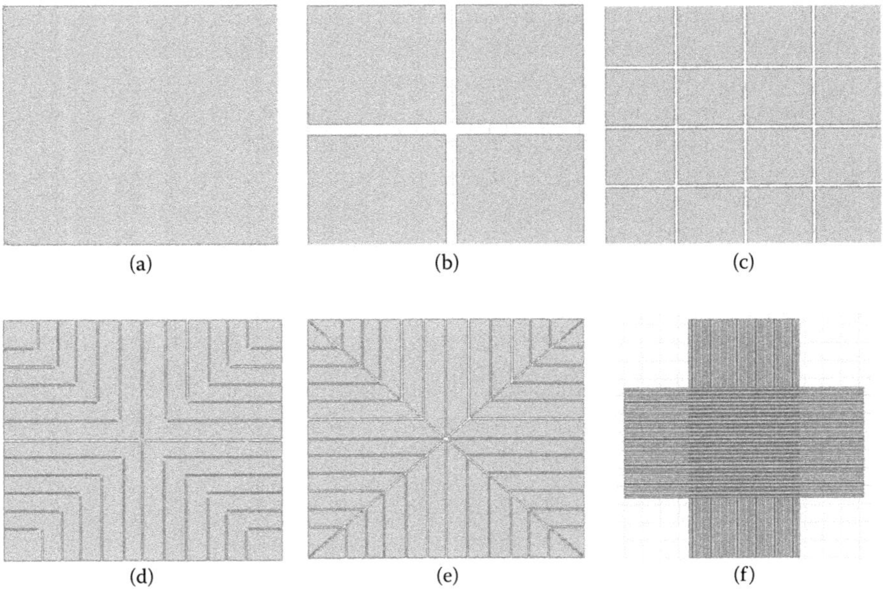

FIGURE 13.12
Shield layouts designed in ANSYS HFSS: (a) 1-piece shield, (b) 4-piece shield, (c) 16-piece shield, (d) patterned ground shields (PGS), (e) patterned floating shields (PFS), and (f) finger floating shields (FS).

to the un-shielded case, calculating the ΔS13(%) difference of each case, as in Equations 13.14 through 13.19.

$$\Delta S13_{\text{1-piece shield}} = 100 \cdot \frac{S13_{1-\text{pieceshield}} - S13_{\text{WithoutShields}}}{S13_{\text{WithoutShields}}} \tag{13.14}$$

$$\Delta S13_{\text{4-piece shield}} = 100 \cdot \frac{S13_{4-\text{pieceshield}} - S13_{\text{WithoutShields}}}{S13_{\text{WithoutShields}}} \tag{13.15}$$

$$\Delta S13_{\text{16-piece shield}} = 100 \cdot \frac{S13_{16-\text{pieceshield}} - S13_{\text{WithoutShields}}}{S13_{\text{WithoutShields}}} \tag{13.16}$$

$$\Delta S13_{\text{PGS}} = 100 \cdot \frac{S13_{\text{PGS}} - S13_{\text{WithoutShields}}}{S13_{\text{WithoutShields}}} \tag{13.17}$$

$$\Delta S13_{\text{PFS}} = 100 \cdot \frac{S13_{\text{PFS}} - S13_{\text{WithoutShields}}}{S13_{\text{WithoutShields}}} \tag{13.18}$$

$$\Delta S13_{FS} = 100 \cdot \frac{S13_{FS} - S13_{WithoutShields}}{S13_{WithoutShields}} \tag{13.19}$$

These figures are pictured in 3D plots in Figure 13.13. The isolation is mostly enhanced when the planar shield is designed in one piece without gaps, beneath each inductor. This is shown in Figure 13.13a, as the surface of this 3d plot is the lowest of all 6 plots, representing the maximum attenuation of the signal that passes from the first port of the second inductor to the first port of the opposite first one. The gaps in the shields reduce the isolation significantly in higher frequencies as the ΔS13 surface shifts up with the cutting of the shield increasing to 4, 16, and more pieces through the 3D plots in Figures 13.13a through f. Furthermore, finger floating shields offer, for most distance and frequency values, worse isolation as compared to patterned ground and floating shields, as shown in Figure 13.13d through f.

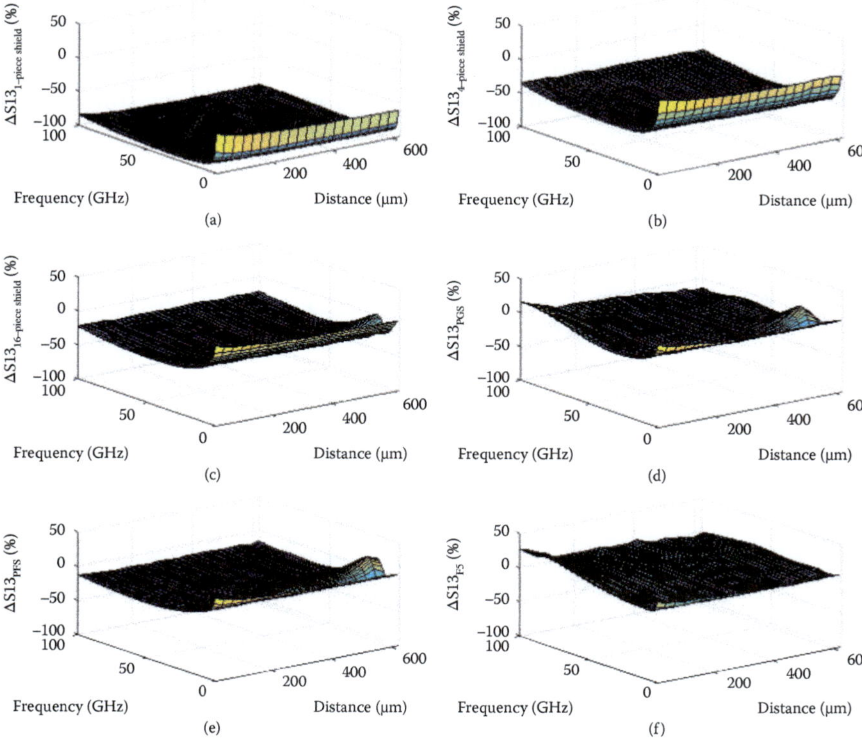

FIGURE 13.13
ΔS13 (%) difference of the structures with (a) 1-piece shields, (b) 4-piece shields, (c) 16-piece shields, (d) patterned ground shields, (e) patterned floating shields, and (f) finger floating shields. Separation distance between the outer edges of the inductors is from 20 µm to 620 µm.

So, the 1-piece grounded, un-gapped shield offers the best isolation compared to all other shield types, as it completely blocks the passing of the parasitic signals to the substrate as a metal grounded wall. However, for frequencies lower than 25 GHz, the utilization of 1-piece shields significantly reduces the quality factor Q of the inductor, as shown in Figure 13.14a. For frequencies higher than 45 GHz though, 1-piece shields, apart from offering the best isolation, offers the best quality factor and, for frequencies greater than 65 GHz, it offers better quality even when compared to the unshielded case. Finger floating shields may offer the worst isolation compared to every other kind of shield, but have the least negative impact on the quality of the inductors below 50 GHz, as shown in Figure 13.14a.

In Figure 13.14b, the inductor is designed with a smaller diameter of 100 μm. This 30 μm reduction in the diameter causes the curves of the quality of the inductor and its' resonance frequency at the various cases to be shifted to higher frequencies and greater quality values. This is a useful method to optimize the performance of the inductor at a higher frequency without increasing the crosstalk between nearby inductors.

The reduction in the quality of the inductors by utilizing shields has been noted previously [24]. Placing a shield underneath the inductor may dramatically decrease the inductance and increase the resistance of the inductor. This is due to the close distance between the layers of the shield

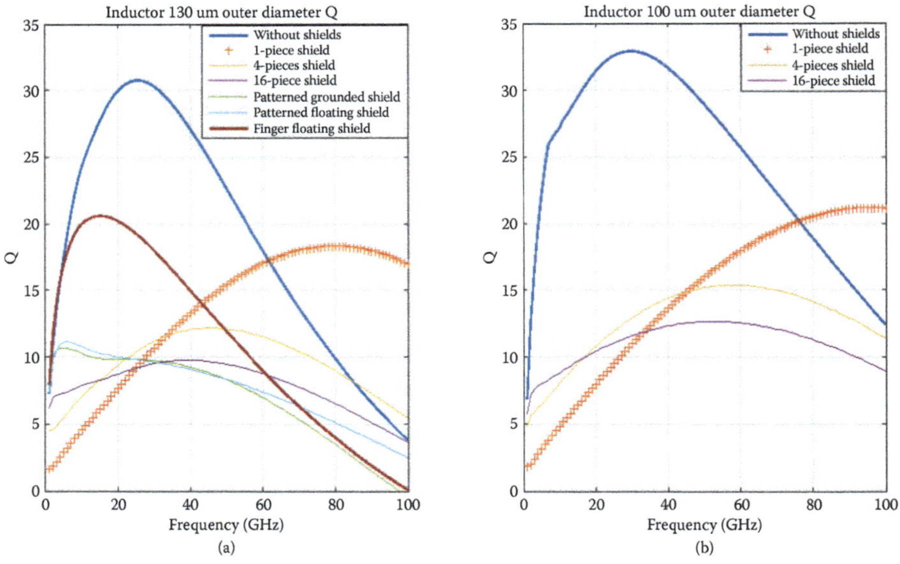

FIGURE 13.14
Inductors' quality Q for various types of shields (a) for the typical inductor diameter in this chapter of 130um and (b) for inductor diameter of 100um.

and the inductor, which is approximately 4.2 μm for the cases studied in this chapter. At this close distance, the proximity effect occurs, which alters the paths in which current flows inside the inductor and causes greater resistance and reduced inductance values, leading to the reduction of the quality factor of the inductor. The proximity effect is increased by decreasing the distance between the metals. To test this, the 1-piece shield was placed at the fifth and the seventh metal layer, instead of the second one, expecting to increase the proximity effect, due to the smaller distance between the inductor and the shield, and lead to higher resistance and lower inductance values for the inductor. After simulating these two cases, the anticipated result was observed and led to the reduction of the maximum quality factor value for the two cases, from 18.2 for the inductor with the shield at the second metal layer to 15.4 at the second metal layer, and 12.6 at the seventh layer.

To summarize this section's deductions:

- Metal shields may be used to block the crosstalk through the substrate without sacrificing area, but in the case of a typical 65 nm technology they significantly reduce the quality of the inductor.
- The 1-piece planar metal grounded shield offers the best isolation as it completely blocks the conductive paths through the substrate. Leaving gaps in the shield, may improve the quality of the inductor at lower frequencies, but deteriorates the isolating ability of the shield.
- Decreasing the diameter of the inductor and the shield shifts the quality curve of the inductor to higher frequency values and slightly improves it due to the reduction of resistance that occurs.
- Placing a shield even closer to the inductor reduces its quality.
- Patterned and finger shields do not offer the best isolation between the examined shield cases and shift the quality curve to lower frequency values. At frequencies below 25 GHz, they offer the best quality to the inductor when compared to the 1, 4, and 16-piece shields.

13.6 Conclusions

Table 13.2 summarizes all the deductions of this chapter. It refers to the simulations performed and a typical 65 nm CMOS technology. The symbol "+" means improvement, "++" means significant improvement and "–" and "–" the opposite.

TABLE 13.2

Summary of the Performance of the Structures in This Chapter

	Area	Isolation	Quality of Inductor	Unwanted Resonance Points
Rings				
Metal square	−	++	+	+
Metal octagonal	−	++	+	+
P+	+	+	+	−
P+ and metal octagonal	−	++	+	−
Shields				
1-piece	+	++	−	+
4-piece	+	++	−	+
16-piece	+	+	−	+
Patterned ground	+	+	−	−
Patterned floating	+	+	−	−
Finger floating	+	−/+	−	−

References

1. A. Helmy and M. Ismail, A Design Guide for Reducing Substrate Noise Coupling in RF Applications, *IEEE Circuits and Devices Magazine*, September/October 2006.
2. H. Johnson and M. Graham, *High-Speed Signal Propagation-Advanced Black Magic*, Upper Saddle River, NJ: Prentice Hall PTR, 2003.
3. M. Datta, S. Sahoo, and R. Kar, Bandwidth Modelling for Distributed On-Chip RLCG Interconnect Considering Coupling Effects, in *International Conference on Devices and Communications (ICDECom)*, 2011.
4. D-W. Kim, C. Li, P. Lo Guo Qiang, Signal integrity and crosstalk analysis of the transmission lines on SOI substrate for high-speed up to 50 GHz, *IEEE 18th Electronics Packaging Technology Conference (EPTC)*, 2016.
5. A. S. Royet, J. P. Michel, B. Reig, J. L Pornin, M. Ranaivoniarivo, B. Robain, P. de Person, G. Uren, Design of optimized high Q inductors on SOI substrates for RF ICs, *IEEE International Conference on Electronics, Circuits and Systems (ICECS)*, 2016.
6. N. A. Talwalkar, C. P. Yue, and S. S. Wong, Analysis and synthesis of on-chip spiral inductors, *IEEE Transactions on Electron Devices*, vol. 52, no. 2, February 2005.
7. I. C. H. Lai and M. Fujishima, Simple modeling expressions for substrate network of on-chip inductors, *IEEE Internationa Conference on Microelectronic Test Structures*, vol. 18, April 2005.
8. M. Ben Salah, D. Pasquet, F. Voiron, P. Descamps, J-L. Lefebvre, D. Lesenechal, Reducing Substrate noise coupling in a 3D-PICS integrated passive device by localized P+ guard rings, *IEEE Topical Conference on Power Amplifiers for Wireless and Radio Applications (PAWR)*, 2013.

9. R. R. Reddy, S. Tanna, S. G. Singh and O. K. Singh, TSV Noise Coupling in 3D IC Using Guard Ring, *IEEE International 3D Systems Integration Conference*, 2015.
10. M. Shen, J. H. Mikkelsen, K. Zhang, O. K. Jensen, T. Tian, and T. Larsen, Modeling and design guidelines for P+ guard rings in lightly doped CMOS substrates, *IEEE Transactions on Electron Devices*, vol. 60, no. 9, September 2013.
11. L. Zhang, X-P. Yu, and E-P. Li, The investigation of the substrates noise suppression using guard rings in CMOS technology, *IEEE International Conference on Microwave and Millimeter Wave Technology (ICMMT)*, 2016.
12. C-T. Dai and M-D. Ker, Optimization of guard ring structures to improve latchup immunity in an 18 V DDDMOS process, *IEEE Transactions on Electron Devices*, vol. 63, no. 6, June 2016.
13. E-P. Li, L. Zhang, and X-P. Yu, P-minus substrate guard ring modeling for the purpose of noise isolation in CMOS substrates, *IEEE Electrical Design of Advanced Packaging and Systems Symposium (EDAPS)*, 2015.
14. M. H. M. Ali, C-L. Ler, S. C. Rustagi, Y. M. Yusof, N. D. Arora, B. Y. Majlis, The Impact of Electromagnetic Coupling of Guard Ring Metal Lines on the Performance of On-chip Spiral Inductor in Silicon CMOS, in *2nd Asia Symposium on Quality Electronic Design (ASQED)*, 2010.
15. P-L. You and T-H. Huang, A switched inductor topology using a switchable artificial grounded metal guard ring for wide-FTR MMW VCO applications, *IEEE Transactions on Electron Devices*, vol. 60, no. 2, February 2013.
16. S-S. Yoo, K-Y. Lee, and H-J. Yoo, Coupling-shielded inductor for high isolation between PA and LC-based DCO, *IEEE Electron Device Letters*, vol. 38, no. 1, January 2017.
17. Taiwan Semiconductor Manufacturing Company (TSMC), www.tsmc.com.
18. J. Y. Hasani, M. Kamarei, and F. Ndagijimana, Transmission line Inductor Modeling and Design for Millimeter Wave Circuits in Digital CMOS Process, *IEEE International Workshop on Radio-Frequency Integration Technology*, 2007.
19. S. Rasidah, M. H. Siti Maisurah, E. F. Nazif, K. Norhapizin, and A. I. A. Rahim, The Design of Ground Shield Spiral Inductor Using 0.13 μm CMOS Technology for Millimeter-Wave Radio over Fiber Applications', *IEEE Regional Symposium on Micro and Nanoelectronics (RSM)*, August 2015.
20. A. Sahu, B. Grayczyk, M. Almalkawi, and V. Devabhaktuni, High-Q Spiral Inductors with Multilayered Split-Ring Resonator (SRR) Patterned Ground Shields, *IEEE Antennas and Propagation Society International Symposium (APSURSI)*, 2014.
21. C-C. Lim, H. Ramiah, J. Yin, P-I. Mak, and R. P. Martins, A High-Q Spiral Inductor with Dual-Layer Patterned Floating Shield in a Class-B VCO Achieving a 190.5-dBc/Hz FoM,
22. R. L. Haner and S. Krishnan, Spiral inductors with floating low-loss RF shields for LC resonator applications, *IEEE Transactions on Components, Packaging and Manufacturing Technology*, vol. 5, no. 9, September 2015.
23. J. C. Rautio, J. D. Merrill, and M. J. Kobasa, Efficient Electromagnetic Analysis of Spiral Inductor Patterned Ground Shields, *IEEE International Conference on Microwaves, Communications, Antennas and Electronics Systems (COMCAS)*, 2013.
24. C-H. Jeong and J-P. Hong, Analysis of substrate shielding methods for sub-THz on-chip inductors, *Electronics Letters*, vol. 50, no. 22, 2014.

14

Paradigm Shift of On-Chip Interconnects from Electrical to Optical

Swati Joshi, Amit Kumar, and Brajesh Kumar Kaushik

CONTENTS

14.1 Introduction

In modern day very-large-scale integration (VLSI) chips, the complexity due to dense interconnects and a high data transmission rate presents several communication problems within the integrated circuit. The interconnect technologies are not able to keep up with the performance expected from high-performance computing systems and microprocessor chips at further scaled technology nodes. This is a critical factor in the design of high-performance electronic systems. The international technology roadmap for semiconductors (ITRS) has highlighted the existence of interconnect problems with scaling [1]. The fundamental paradigm shift takes place when shifting to quantum concepts from conventional laws of governing and explaining the behavior of interconnects. For VLSI and ultra-large-scale integration (ULSI) circuits, ITRS reports suggest that there will be a material limit imposed by conventional metal interconnects beyond a technology node of 8 nm. Thus, once we reach this limit, different materials, techniques, and technologies would have to be explored that can offer the projected performance requirements compatibility with existing CMOS technology. Several emerging technologies, such as carbon nanotubes (CNTs) and graphene nanoribbons (GNRs), have been given extensive importance by researchers to meet chip performance in deep sub-micrometer and nanometer regimes. Being reliable and economical, on-chip communication is currently monopolized by electrical interconnections, however, optical interconnects dominate long distance communication due to their high bandwidth and low signal attenuation.

Optical interconnects to electronic chips are an attractive candidate that have been the subject of research for many years. With the advancements in silicon photonic technology, on-chip optical networks are possible for future high-performance systems. Novel technologies have been designed that support optical interconnects on silicon chips. Research in optoelectronic and optical devices are critical for optics to solve the central interconnect problems on silicon chips. The active research in this direction has indicated that on-chip optical interconnects can offer various advantages compared to electrical interconnects.

14.2 On-Chip Electrical Interconnects

An on-chip electrical link is shown in Figure 14.1. It consists of a driver, interconnect (Cu- based, CNT-based, or GNR-based) that allows the signal to traverse across it in electrical form and the load section, which is generally a capacitive one for such schemes. Line parameters resistance (*R* in Ω/m), inductance (*L* in H/m) and capacitance (*C* in F/m) describe the characteristics of the interconnect. The role of interconnects is to ensure the proper distribution of clock, signals, and power/ground connections to different circuit/functional blocks on a chip.

A chip has interconnects in the form of metal layers and vias. It is desirable to have a multilayer interconnection on at least three levels to obtain high packaging densities within the chip and smaller footprints. On-chip interconnects, based on their length and cross-sectional dimensions, generally have three classifications, namely: local, intermediate, and global level interconnects. Figure 14.2 shows statistical distribution of wire lengths on a chip. Local interconnects comprise very thin lines that connect transistors and gates of a functional block. These wires cover short distances, usually a few gates' span, and occupy first and sometimes second metal layers. Intermediate wires are used within adjacent blocks for inter-module communications. Intermediate interconnects have wider and longer lengths as compared to local interconnect and they provide low resistance signal paths in a functional block. Global interconnects mainly supply wires, clock, and buses, which are long wires running between functional blocks. Global interconnects range from 1 to 2 cm in length and are placed at higher level interconnect layers [2, 3].

Earlier, the speed of the integrated circuit was primarily decided by the speed of logic gates as interconnects connecting the devices were treated as

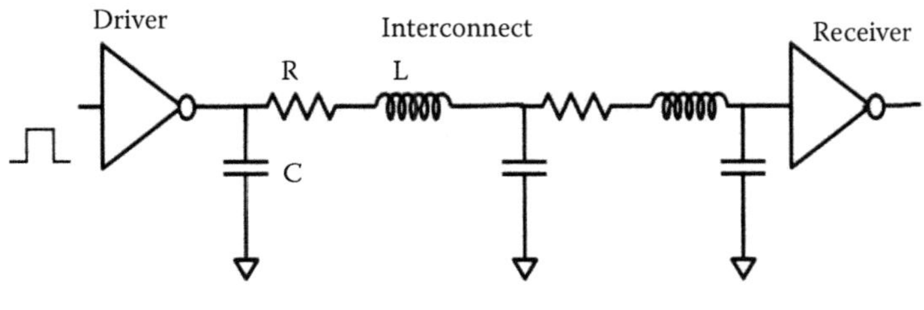

FIGURE 14.1
An on-chip electrical link.

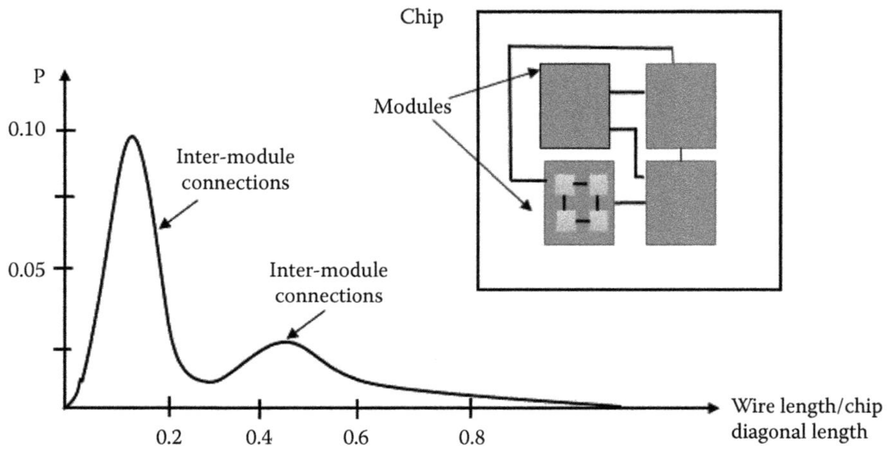

FIGURE 14.2
Statistical distribution of wire length on a chip.

short-circuited or an ideal conductor. Thus, interconnects used to have little or no effect in integrated circuits. However, as semiconductor devices in sub-micrometer and now in nanometer regime are introduced, interconnects can no longer be treated as ideal conductors, and hence the chip performance is largely decided by interconnect lines. Beginning from the contact levels with PMD's, that is, pre-metal dielectrics, the metal layer lasts up to wire bond pads, defined by steps of deposition, etch, and planarization.

14.3 Issues with Electrical Interconnects

Scaling of an on-chip electrical interconnect is shown in Figure 14.3. The effects, as a result of aggressive scaling, pose several issues in front of on-chip electrical interconnects. The most critical of them are signal reliability, bandwidth density requirements, power dissipation, latency, and crosstalk. These issues need proper attention to avoid the interconnect bottleneck on the overall device performance.

14.3.1 Scaling of Electrical Interconnects

Consider that we have a metal wire of resistivity (ρ), length (l), width (w), and thickness (t_m) running on insulator with thickness (t_{ox}) and permittivity (ε) above the ground plane, as shown in Figure 14.4.

Mathematically, resistance and capacitance of a wire are expressed as

$$R = \frac{\rho l}{w t_m} \tag{14.1}$$

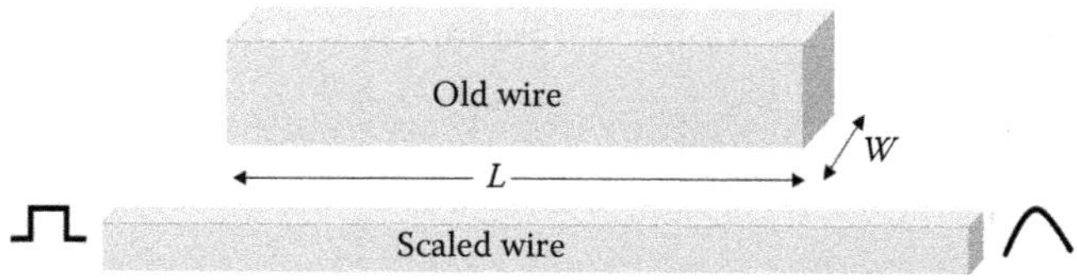

FIGURE 14.3
Scaling of interconnect.

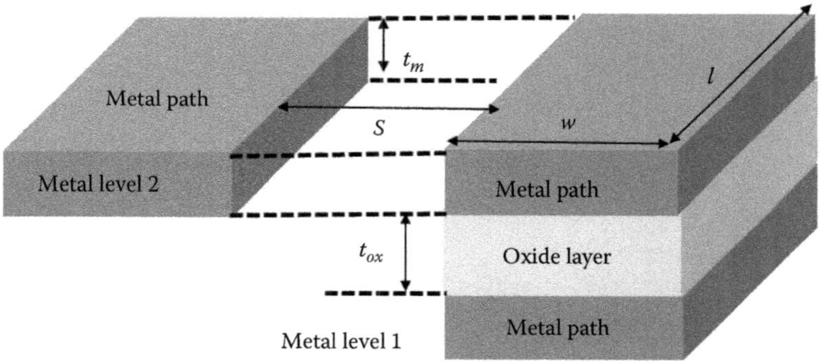

FIGURE 14.4
Interconnect dimensional parameters.

$$C = \frac{\varepsilon l w}{t_{ox}} \tag{14.2}$$

The charge time of the wire, which is proportional to the product of resistance and capacitance given by Equations (14.1) and (14.2), is therefore [3]

$$\text{Charge time} \approx RC = \frac{\rho \varepsilon l^2}{t_m t_{ox}} \tag{14.3}$$

Equation 14.3 suggests that delay in the electrical interconnect is directly proportional to the square of wire length and independent of wire width, since increasing wire width reduces the resistance and increases the capacitance in the same ratio.

$$\text{Interconnect delay} = kl^2, \quad k = \text{constant}$$

Inside a chip, wires are distributed around small length (i.e., local interconnects) and die size length (i.e., global interconnects), while technology scale down the local interconnect length scale down since the immediate devices will be nearer, however, global interconnect scales not by the transistor size but by the die size. Global interconnects include long wires running between functional blocks, delay through these lines can be comparable to or greater than the gate delay and are an area of serious concern to designers [4,5]. Signal propagation delay is negligible in local interconnects as compared to the gate delay. This delay problem is somewhat reduced by buffer insertion, however, there are difficulties with buffer insertion since it consumes power and silicon area. Global interconnects use, bidirectional lines and buffer insertion fix the direction of data flow, which requires additional directional control line.

In present scenario, the speed of an integrated circuit is limited by global interconnects. Some key design approaches to overcome this limit includes routing the global interconnects on upper metal layers with increased width and thickness of interconnects, using low κ dielectric materials, reverse scaling and use of high-resistivity substrates [6]. However, these approaches are not ultimate solution and can be effective up a to few technology nodes. Figure 14.5 shows global interconnect delay and gate delay for commonly used materials.

14.3.2 Aspect-Ratio Limit

The bit-rate capacity (B) of an electrical line is proportional to the inverse of RC, which depends on the aspect ratio that is the ratio of the length (l) to cross section area (or more strictly for a line with cross-section area A, the quantity $\sqrt{A}$). This ratio is independent of scaling and is only dimensional dependent. Bits per second send by the line remain invariant whether we use a large cross-section line at high clock speed or parallel small cross section

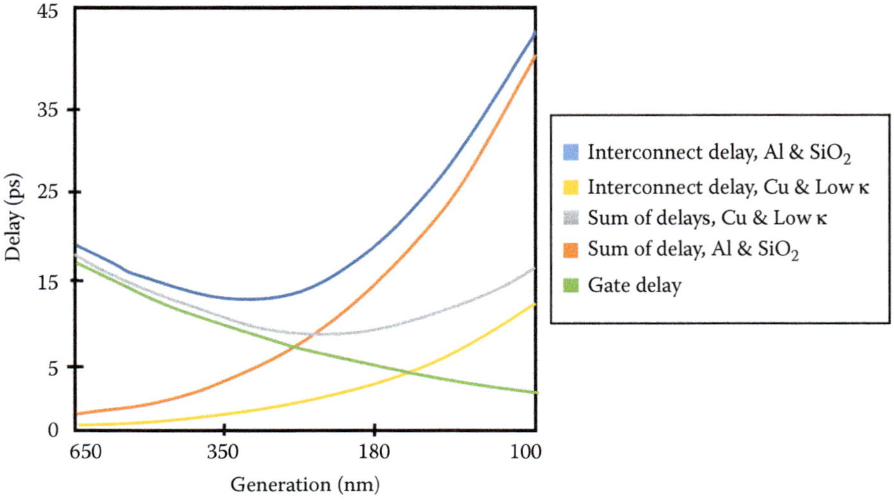

FIGURE 14.5
Gate delay and interconnect delay as a function of technology node.

line at lower clock speed and is also nearly independent of the design of the electrical lines. Lines are usually designed for optimal parameters (i.e., minimum delay, best use of cross-section area) and redesigning them can improve one characteristic while deteriorating other [5–7].

14.3.3 Scattering

At current technology nodes, increase in resistivity, especially resulting from electron scattering from the rough surface, grain boundary, and impurities, as illustrated in Figure 14.6, has become a major performance issue for on-chip interconnects [8]. Scattering phenomena due to size effects is a major concern for future lower technology nodes when the height of roughness becomes comparable to dimensions of the interconnect line or mean free path of electrons and there is a reduction in effective cross-section area of interconnect [9,10]. Also, there is an increase in density of grain boundaries. Interconnect dimension below electron mean free path (Copper, 39 nm) results in increased resistivity. Figure 14.7 shows resistivity of Cu as a function of wire width, considering scattering and barrier effect with the atomic layer deposition (ALD) barrier [10]. Due to scaling, size effects are becoming more prominent and result in increased effective resistivity of the interconnect. The resistivity of a 40-nm-wide Cu wire is reported 2.5 times higher than for bulk Cu [11,12]. Higher interconnect resistance worsens the circuit performance due to increase in power consumption and signal delay.

14.3.4 Electromigration

Electromigration (EM) is the process of metal self-diffusion as metal ions interact with moving electrons when there is high-density current flow in the interconnect. The motion of metal lattice ions is driven by momentum transfer of electrons for the metal ions under the applied electrical field. This movement results in the creation of a void where metal ions migrate and a

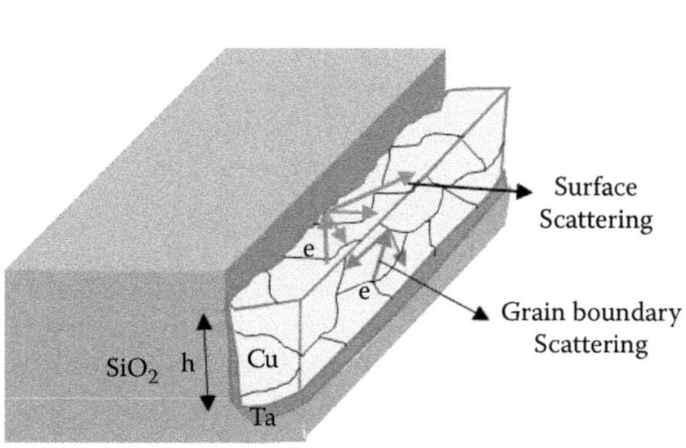

FIGURE 14.6
Schematic illustration of surface and grain boundary scattering.

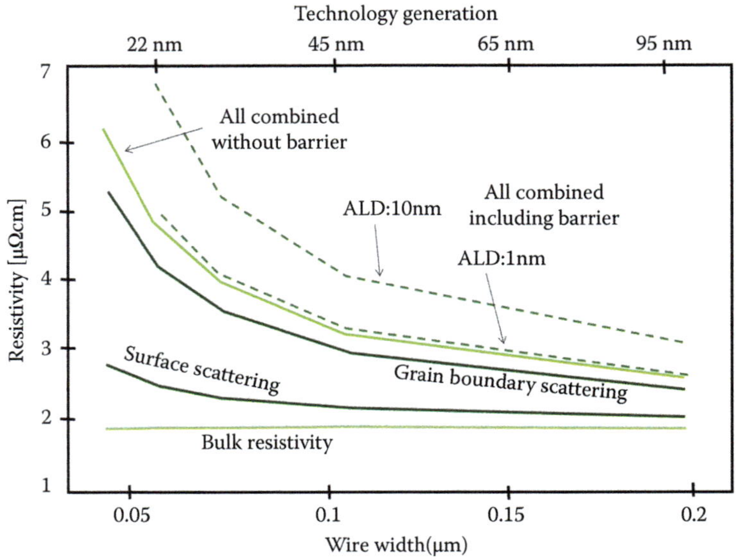

FIGURE 14.7
Cu resistivity as a function of wire width, considering the scattering and the barrier effect. (From K. H. Koo, Comparison study of future on-chip interconnects for high performance VLSI applications, Ph.D. dissertation, Dept. Electr. Eng., Stanford University, Stanford, CA, USA, 2000; K. Koo et al., *IEEE Trans. Electron Devices*, 56, 1787–1798, 2009.)

hillock where they accumulate. The void results in an open circuit or an increase in the resistance and the hillock can result in a short circuit between two adjacent interconnects [13]. Due to the increase in electron scattering events with high current density, the rate of displacement for the atoms rises. The time required to reach the minimum void size causing an electrical open, or a fixed amount of resistivity increase for the conductor line, is known as the EM lifetime [14,15].

As the technology scales down, the dimension and spacing between the interconnect keeps decreasing. Electromigration is becoming a crucial design challenge for chip reliability due to higher current densities. If high current density keeps flowing for a long time, voids and heaps can appear in interconnects and eventually result in loss of connections or interconnect failures [16,17]. Silicon substrate provides good thermal contact to the metal layer, making it withstand high current density and, hence, a pronounced electromigration effect. These effects become more prominent at the grain boundaries due to loss of symmetry as momentum transfer is strengthened in these areas. Metal ions are loosely bound and can be easily dislocated and transported in the direction of the current. However, interestingly, as dimension becomes comparable with grain size, the electromigration lifetime increases [18,19]. The failure rate of a chip is very much dependent on interconnects and electromigration effect

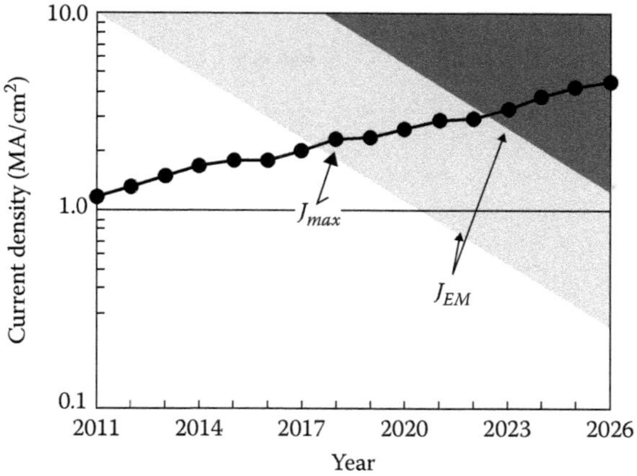

FIGURE 14.8
Projected evolution of J_{max} (from device requirement) and J_{EM} (from targeted lifetime) from ITRS road map. (From "ITRS Reports," *International Technology Roadmap for Semiconductors*, 2015. Available: http://www.itrs2.net/itrs-reports.html.)

is the dominant effect in interconnect failure. Increased temperature due to joule heating can increase this effect and decrease the lifetime of a chip. With continuous scaling and increase in clock frequency, maximum current density in a conventional copper interconnect will exceed the electromigration limit, as highlighted in the ITRS roadmap [1] shown in Figure 14.8. Major innovations are required to withstand these technological challenges.

14.3.5 Reliability Issues with Low-κ Dielectric Material

In an integrated circuit (IC), dielectric separates the conducting parts from one another. The low dielectric constant (κ) materials, with value lower than that of silicon oxide (κ = 3.9) have been used to reduce the parasitic capacitance, thus reducing the *RC* time delay, power dissipation, and crosstalk effects in VLSI circuits. A combination of Cu and low-κ dielectrics has been searched out for benefits such as low resistance, reduced electromigration effects, and low dielectric constant [20]. Low-κ dielectrics, however, suffer from reliability issues such as leakage, lower breakdown strength, time-dependent dielectric breakdown (TDDB), inferior adhesion strength, copper induced dielectric failure due to Cu outdiffusion through poor barrier confinement, and degraded performance due to moisture absorption (being hydrophilic in nature, absorbed H_2O increases dielectric polarization) [21,22]. Therefore, the introduction of low-κ dielectrics presents several challenges in meeting the required performance goals for interconnects.

14.3.6 Signal Integrity (Impedance Matching, Crosstalk, and Electromagnetic Interference)

In IC design, unintended interactions arising from parasitic effects associated with interconnects is the area of major concern. Crosstalk, in an integrated circuit, refers to undesired coupling among interconnects, which results in the transfer of unwanted perturbations among them. Scaling trends have led to wires spaced physically close together. Increased clock frequency is linked to higher parasitic capacitive (Figure 14.9) and inductive properties of the wires coming into play, leading to increased crosstalk effects. At high frequencies, line inductance is taken into account in global interconnects affecting signal propagation characteristics. Ringing due to parasitic capacitances and inductances can cause undesirable effects in the circuits, such as input voltage dropping below ground level or exceeding the supply voltage, unwanted triggering, and electromagnetic radiation. These are the most common sources of signal integrity problems in integrated circuits and should be mitigated as much as possible. Crosstalk effects can be reduced by designing layout for decrease coupling, limiting the length of lines running parallel to each other, ground shielding, and increased coupling to ground planes; however, implementing these techniques come at their own expense [23,24].

Reflections in the transmission line occur when traveling waves see discontinuities on the line. Discontinuity is defined as any change in line impedance. Changes in paths of the signal, such as vias, bends, wire crossovers, bonding wires, and so on, result in signal reflection. At high data rates, this issue needs to be taken into consideration. Electromagnetic interference (EMI) results from the external noise source or coupling between subsystems. The noise in the ground cable and power line can adversely affect circuit operation. EMI suppression uses techniques such as shielding, decoupling, and grounding [25]. All of the above effects (crosstalk, EMI, ringing, reflections in a transmission line, etc.) degrade the quality of the

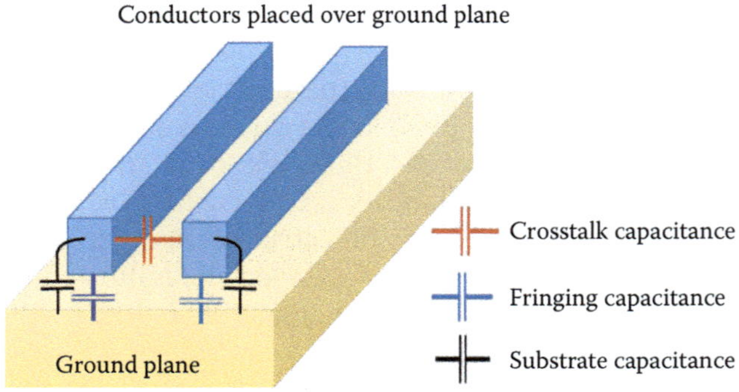

FIGURE 14.9
Wire capacitances.

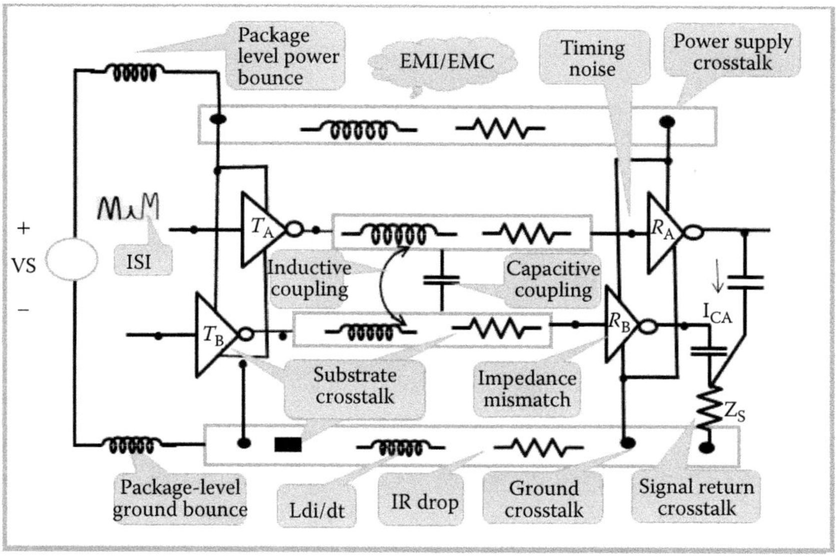

FIGURE 14.10
Illustration of different noises affecting signal integrity.

received signal, corrupt the signal, and can affect the signal traveling on the same interconnect line at a later time, resulting in inter-symbol interference (ISI), that is, energy transfer from one signal in the interconnect channel to a later unrelated signal as illustrated in Figure 14.10.

14.3.7 Synchronization Problem

Noise may affect the transition time of a signal. Due to timing noise, there could be a mismatch in the operation of active components that is triggered by signal transition because of variability in signal timing. Jitter and skew are common timing problems. Noise such as crosstalk, mismatched line lengths, and parameter variations can cause synchronization problems in the circuit.

14.3.8 Power Dissipation

Power dissipation is a critical factor in the design of a high-performance system and is a major design consideration. Global interconnect in an IC account for 40%–50% of the total dynamic power dissipation in an IC [26]. Therefore, a large portion of power dissipation is contributed by on-chip interconnects. The gate load is often dominated by interconnect capacitance. Switching of interconnection capacitances is a major cause of power dissipation. Scaling of interconnects has increased their power density, which is an area of growing concern. At high clock frequencies, skin effect in the metal and dielectric loss may cause additional losses. Uniform repeaters driving long interconnects for improved performance increase the dynamic power dissipation due to their additional input capacitance. The performance of an interconnect is measured mainly in terms of power-delay product (PDP) [27].

14.4 Conventional On-Chip Interconnects

Interconnect lines, for a long time, used aluminum as its material due to its compatibility with silicon. But, the induced electromigration problems caused by increased current density in aluminum made researchers look for alternatives since its reliability kept on decreasing due to device scaling. IBM, in 1997, declared plans to introduce copper, a material that has the lower resistivity, as a substitute for aluminum [28]. It provides current density as high as 10^7 A/cm^2, which also results in less significant electromigration effects. However, it was realized later on that even copper couldn't quite fulfil the requirements of high speed interconnects because of the following reasons:

1. The decrease in the reliability of copper was observed because of the increased current densities as the aggressive scaling in the dimensions of interconnects continued.
2. At reduced dimensions, the surface and grain boundary scattering results in increased resistivity of copper.
3. Joule heating effects rapidly increased the resistivity of this material.
4. Skin effects result in reduced conductivity at high frequencies of operation.

One of the future technology requirements of interconnects, as dimension scaling continues, is high reliability, but copper, as an interconnect material, is unable to meet this requirement. This widening gap factor inclined design engineers and researchers to focus more towards the novel interconnect materials. The promising solution to this is seen in graphene-based nano-interconnects for succeeding on-chip interconnects. The interesting and amazing properties of graphene, like its longer mean free path (on the order of several micrometers), higher current density (more than 10^9 A/cm^2), and better thermal stability than copper compelled researchers to look at this material as a possible solution to replace the conventional on-chip interconnect material in future VLSI on-chip interconnect systems. Nano-interconnects based on graphene can be divided into CNTs, that is, carbon nanotubes, and GNRs, that is, graphene nanoribbons.

14.4.1 Carbon Nanotubes (CNTs)

Carbon nanotubes are cylindrically rolled up graphene sheets with inner diameters varying in the range between 1 to 5 nm. CNTs can be classified as single-wall nanotubes (SWNTs) or multi-walled nanotubes (MWNTs), depending on the number of concentric rolled up sheets of graphene. CNTs have larger MFP (mean free path) ranging in micrometers because of ballistic transport phenomenon that results in scatter free movement of electrons along the CNT axis. On the other hand, the absence of this interesting ballistic transport in copper limits the mean free path of charge carriers only up to few nanometers [29,30]. The larger mean free path and smaller diameters of carbon nanotubes results in far better conductivity than copper. The low resistance also ensures incredibly low energy dissipation in CNTs and, hence, the dissipated power density issue can be appropriately addressed. Here, it is also important to state the interesting electrical properties due to the one-dimensional structure of carbon nanotubes, which are as follows [31–33]:

1. Due to the ballistic transport, high-quality carbon nanotubes have MFP in the range of 1 μm–5 μm.
2. The higher conductivity and better mechanical strength of carbon nanotubes are because of stronger sigma and pi bonds.
3. The electron mobility in carbon nanotubes [$\sim 2 \times 10^5$ cm^2/ (V.s)] is orders higher than electron mobility in copper [$\sim 3 \times 10^3$ cm^2/ (V.s)]. This property results in higher drift velocity.
4. Current densities in CNTs (10^9 A/cm^2) are several orders higher than current densities in copper (10^7 A/cm^2).

Chirality of carbon nanotubes is one of their critical parameters. Chirality decides the semiconducting and metallic properties of CNTs. Figure 14.11 shows the structure of a single-walled, cylindrically rolled up graphene sheet known as a single-walled carbon nanotube. Armchair, zigzag, and chiral are the different classifications for CNTs [34], which depend on the shape of the circumference of the cylinder, as shown in Figure 14.11. SWCNT with random chirality does not show any edge in terms of conduction in comparison to conventional interconnects. But, this concern can be sorted out by using MWCNTs in place of SWCNT [35]. Figure 14.12 shows a MWCNT containing concentrically rolled up graphene sheets. Statistically, a MWCNT bundle has one-third of its CNTs as conducting in nature, that is, metallic, and the remaining of the CNTs behave like semiconductors. The diameter of MWCNTs range from 2 nm to several tens of nanometers. As the energy gap in MWCNTs is inversely proportional to shell diameter, therefore, larger diameter MWCNT shells behave conductively even if they have semiconducting chirality. The energy band gap between fermi level and conduction band for a semiconducting CNT with a diameter more than 20 nm is 0.0258 eV, which can be easily overcome by environmental temperature [36].

Exceptional electrical, optical, thermal, and mechanical properties of carbon nanotubes challenge the conventional trends and researchers are still trying to describe and unravel the phenomenon of such peculiar behavior. Two interesting physical properties of carbon nanotubes are there scalability up to a single atomic layer and that an electronic device can be formed by joining two differently structured CNTs with a slight structural difference. The resultant device's properties depends on the type of carbon nanotubes that are used to form the device [37].

14.4.2 Graphene Nanoribbons (GNRs)

At present graphene nanoribbons, have been acknowledged by the researchers as a promising candidate for realizing on-chip interconnects. The electrical as well as physical properties of GNRs are similar to that of CNTs. The advantage of GNRs over carbon nanotubes is that both transistors and interconnects can be fabricated using same graphene layer. Thus, manufacturing difficulties arising out of metal-nanotube contact can be circumvented. Based on the number of stacked single layered sheets of graphene,

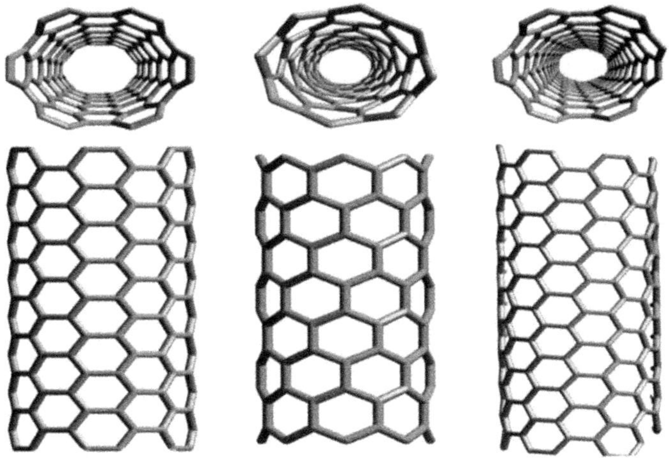

FIGURE 14.11
Illustration of three different SWCNT structures: (a) armchair nanotube, (b) zigzag nanotube and, (c) chiral nanotube. (From M. Scarselli et al., *J. Phys. Condens. Matter*, 24, 313202-1–313202-36, 2012.)

GNRs can be classified as SLGNRs, that is, single-layer graphene nanoribbons and MLGNRs, that is, multi-layer graphene nanoribbons [38]. MLGNRs are preferred over SLGNRs because of their easy fabrication process and lower resistivity. However, edge roughness has always been shown up to some extent by all MLGNRs fabricated to date. The rough edges cause the electron scattering that ultimately reduces the MFP results in lowering the conductance of the MLGNRs. This limits the performance of MLGNR interconnects [38].

At nanoscale dimensions, grain boundaries and sidewall scattering degrade the performance of copper-based interconnects. In the coming days, GNRs are likely to outpace the Cu interconnects for smaller widths. High-quality graphene nanoribbon sheet has an MFP (mean free path) ranging from 1 to 5 μm. Graphene nanoribbons have the ability to carry higher current densities, that is, more than 10^8 A/cm^2. Carrier mobility of GNRs can reach up to 3×10^3 cm^{2}V^{-1}s^{-1}. A graphene nanoribbon is so thin and so limited in width that it results in a 1D (one-dimensional) structure. High current carrying ability of MLGNR over SLGNR makes it a promising interconnect solution [39,40]. Figure 14.13

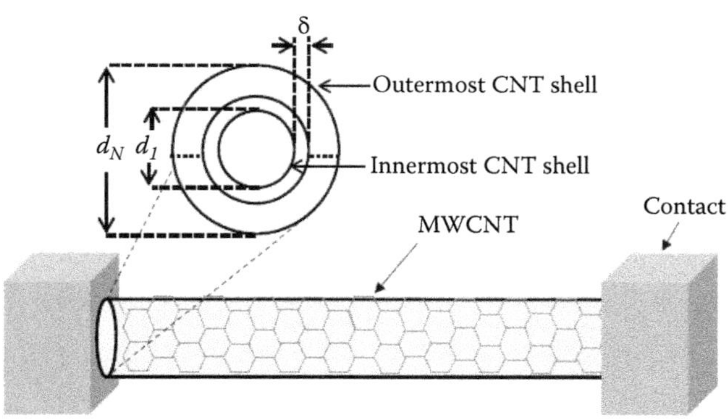

FIGURE 14.12
Schematic of MWCNT placed between the two contacts.

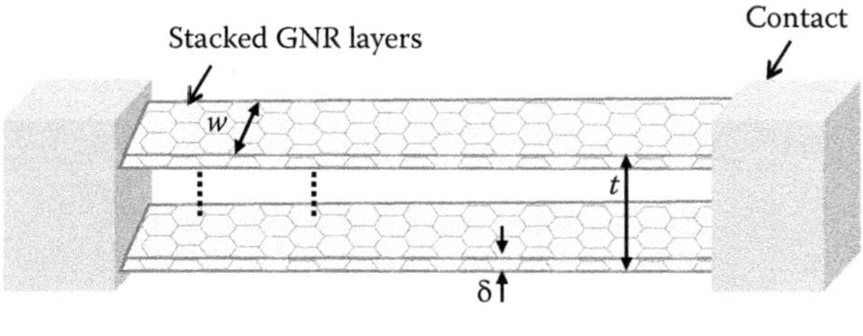

FIGURE 14.13
Structure of an MLGNR interconnect.

shows the geometric structure of a multi-layer graphene nanoribbon having N number of layers with width w, interlayer distance δ, and thickness t.

14.5 On-Chip Optical Interconnects

Optical interconnects are considered as top candidates to replace electrical interconnect for on-chip applications. Long distance communication is already dominated by optical fibers working in the wavelength band around 1550 nm with the optical loss of about 0.22 dB/km and enormous bandwidth [41]. Optical interconnects use light as a means of communication. The development in this direction requires significant efforts to develop components for on-chip optical communication including light sources, waveguides, couplers, passive devices, modulators, and detectors. An on-chip optical interconnect link is illustrated in Figure 14.14. It consists of a laser source, modulator, waveguide, and photodetector along with signal processing circuitry.

14.6 Benefits of Optical Interconnects

In the last few years, optical interconnects for electronic chips is an active area of research and several investigations have been carried out for the

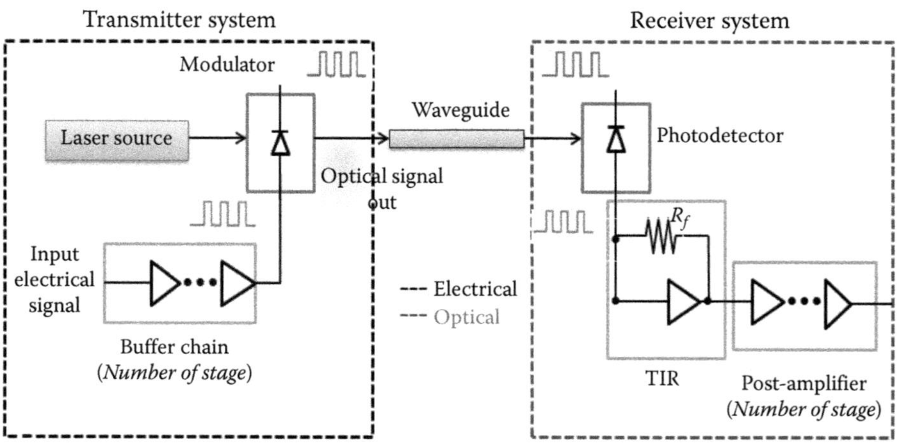

FIGURE 14.14
Optical communication link.

potential benefits of on-chip optical communication. Some possible advantages with optical interconnects are as follows:

1. Aspect ratio, which limits the bandwidth in an electrical contact, does not exist in optical interconnects [7]. The physics behind the optical interconnects is different from electrical, where it depends on the resistance, capacitance, and inductance of the wire. Signal propagation in an optical interconnect takes place through total internal reflection and the propagation characteristics are determined by waveguide geometry and index contrast between waveguide core and cladding. This communication technique can solve some of the problems limiting the performance of electrical interconnect [42]. With continuously rising clock frequency, optical interconnects with no comparable limit can be a possible solution. Optics can help in designing long distance interconnects even with continuous scaling, operating at higher clock speeds without aspect ratio problem.
2. With optical interconnects, the losses are usually associated with components and coupling at the two end and not the distance of propagation. Not having the additional requirement of drivers and resynchronization units may result in power saving as compared to electrical interconnects, provided the optoelectronic devices are well-integrated [43]. A breakthrough length is considered beyond which optical interconnects are more favorable than electrical interconnect.
3. Issues that degrade the performance of electrical interconnects, such as crosstalk effects, hot carrier scattering effects, and high-frequency attenuation [44,45] do not exist in optical interconnects. Optical signals do not produce and are immune from electromagnetic interference.
4. The electrical interconnect suffers from impedance matching and wave reflection phenomena, buses need to be redesigned with increased clock frequency for each successive generation, active impedance matching consumes power and area in the circuit. Optics can simplify design, as it uses the quantum generation and detection mechanism "Quantum impedance conversion" [46], where photons above material bandgap can generate electrons in the photodetector and a source can generate photons for absorbed electrons.
5. Electrical interconnections show filtering effect from the *RC* lines and skin effect (conduction taking place near the surface of the metal) in *LC* lines, resulting in distortion of signal at high modulation frequencies. Modulation has little effect on propagation characteristics of a high carrier frequency in optics. Dispersive effects are

seen at long distance. Also, change in modulation frequency has no effect on the loss experienced by a signal [47,48].

6. The resistance of the line is dependent on the temperature (increases at the rate of 0.004/°C, in the case of copper). Increased loss and variability in the signal rise and fall time, which affect signal propagation characteristics, can cause synchronization problems with the clock signal. Temperature variations have less effects on the propagation speed of the signal in optical systems (change in the refractive index is of the order of 10^{-5}/°C in silicon waveguides) [44].
7. Optics provide voltage isolation due to quantum nature of light generation and detection [46].

14.7 Optical Communication Link

The use of the silicon in the electronics industry is very mature, thus offering an efficient and economical solution, that is, compatible with the silicon industry, and efforts are being made to explore the use of silicon in the optical domain [49]. The motivations behind the use of silicon in optics are:

1. Silicon is an excellent optical material (Transparent in the 1.3–1.6 μm range).
2. High thermal conductivity (~10 times higher than III-V).
3. High optical damage threshold (~10 times higher than III-V).
4. High optical nonlinearities.
5. Compatibility with the CMOS fabrication technology.
6. Strong optical confinement due to the high index contrast between silicon (n = 3.45) and SiO_2 (n = 1.45).

Silicon photonics uses silicon as a medium for optical communication, that is, for generation (electrical-to-optical conversion), guidance, control, and detection (optical-to-electrical conversion) of light [50–52]. Among several platforms considered, silicon on insulator (SOI) wafers are attractive platforms for electronic, photonic, and optoelectronic integrated circuits. The refractive index contrast offered by SOI provides strong light confinement, thus allowing compact structures [53]. A designer can monolithically integrate a comprehensive set of optical and electronic functions on a single silicon substrate. Electronic-photonic integrated circuits (EPICs) on Si utilize high-speed data processing of electrons with high bandwidth data

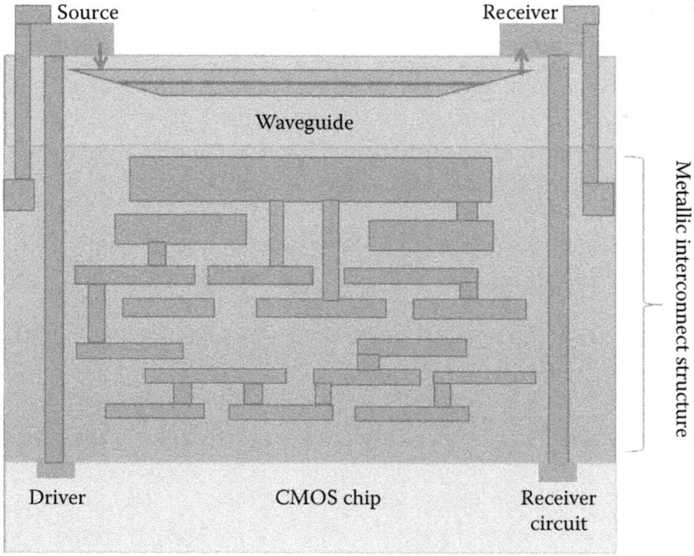

FIGURE 14.15
Hybrid interconnect structure.

transmission of photons, providing efficient synergy. A hybrid system is shown in Figure 14.15. The active research area is establishment of photonics into the silicon complementary metal oxide semiconductor (CMOS) manufacturing process. An optical link is a subject of extensive research and development with the potential of replacing the electrical link in the coming future.

14.7.1 Optical Sources

Optical interconnects are proposed to replace the global on-chip electrical links for communications in future technologies. The off-chip laser sources driving these optical interconnects can lead to higher coupling losses, packaging difficulties, and thermal mismatch between on-chip circuitry and laser source [54]. The advancement in the area of the efficient on-chip laser source is a major factor deciding the future of on-chip optical communication. The growth of an on-chip light source on Si has not kept pace with that of other photonic components, restricting the progress of optical interconnection technology due to the following reasons [55,56]:

1. Silicon has an indirect bandgap. As a result, only about one excited electron in a million can successfully release a photon. Common lasing media such as gallium arsenide feature emission efficiencies 10,000 times larger than silicon.

2. Silicon features wider, less steep-sided upper bands and, as a result, has a large tendency to support free carrier absorption.
3. A process known as Auger recombination also impedes silicon lasing. In this phenomenon, rather than emitting light, an electron in the upper band loses its energy to other electrons, which subsequently give up the excess energy as heat [57].

As a result, silicon cannot emit light efficiently when energized and, whatever radiative recombinations may occur, they are incapable of "stimulating" other photons and amplifying light into a laser beam. Recent progress in lasers on silicon includes techniques that can be used in realizing efficient light emission in silicon:

1. *III-V based Si Laser*: The widely used approach for obtaining on-chip light sources is the integration of III–V semiconductors on the silicon substrate. III–V semiconductors, due to their direct bandgap structure and high material gain, are efficient light emitters; however, due to mismatch in the lattice constant and thermal expansion coefficient dislocations, power losses occurs which hinders the growth of III–V gain materials on silicon [55]. Different integration approaches, such as direct mounting, wafer bonding, and direct hetero-epitaxial growth have been used to solve the above problem. Most of the commercially used lasers have their active layer as quantum well structure. Different laser designs include the Fabry-Perot laser, distributed Bragg reflector laser (DBR), distributed feedback laser (DFR), and vertical cavity surface emitting laser (VCSEL) [58]. A VCSEL, shown in Figure 14.16, is the most commonly used source in present-day optical communication.

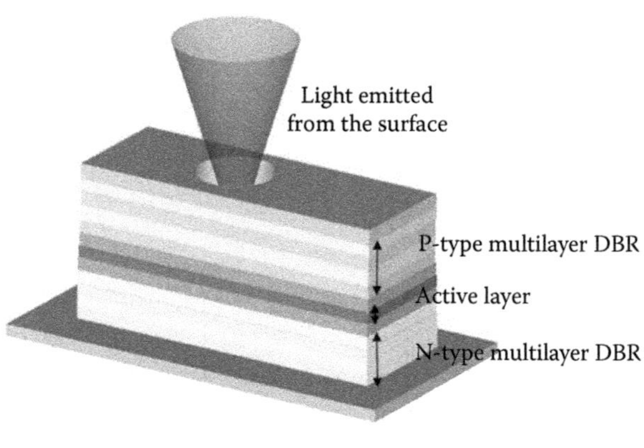

FIGURE 14.16
Simple VCSEL structure.

2. *Overcoming the indirect bandgap structure by using electron confinement*: According to the Heisenberg uncertainty principle, carrier confinement makes velocity and, hence, momentum more random, thus increasing the probability of a radiative recombination. The confinement of electrons in a silicon nanocrystal [59] increases the probability of radiative recombination once the confinement accounts for the mismatch in momentum.
3. *Introducing optically active rare-earth dopants*: One advance that exploits quantum confinement makes use of rare earth elements, such as Erbium (Er). Si-nanocrystals make use of quantum effects to localize the carriers; a nanocrystal adjacent to an erbium ion transfers its energy to the ion, which then returns to its ground state by emitting a photon, and, in this way, increasing the internal quantum efficiency (IQE) of the material [59].
4. *Germanium-On Si Laser*: The pseudo-direct gap behavior of Ge, due to small energy difference between the direct valley and indirect valleys, can be exploited for optical applications. The special band structure of Ge can be turned to the direct band gap by using any of the following methods: n-type doping, using germanium-tin (GeSn) alloy, or by introducing tensile strain [60]. This opens the possibility of achieving efficient light emission at communication wavelength 1550 nm.

14.7.1.1 Device Requirements

An ideal light source should satisfy the following requirements:

1. Emission at 1310 or 1550 nm.
2. Compatibility with CMOS manufacturing process.
3. Low threshold current.
4. Sufficient output power.
5. Narrow spot size for good coupling.
6. Narrow linewidth (less dispersion).
7. Capable of simple signal modulation (i.e., direct) over a wide bandwidth.

14.7.2 Waveguides

The growth of on-chip optical communication mostly relies on advancement in the area of silicon photonics. Silicon, being transparent in the communication wavelengths of 1.3 μm and 1.55 μm, is suitable for optical applications in this wavelength band. Waveguides are the basic component used for guiding light in optical circuits. They are used as interconnects and basic elements for other active and passive photonic components. In SOI wafers, the topmost high index Silicon (Si) form waveguide core and low refractive index silica (SiO_2) is used as waveguide cladding. The light propagates inside a waveguide through the total internal reflection confined between high index core and low index cladding. The high index contrast between silica and silicon results in strong light confinement, which results in small structures suited for dense circuits. The analysis of the waveguide is carried out by solving electromagnetic wave equations written in terms of electric field E, magnetic field H, permeability and permittivity of free space μ_0 and ε_0, and refractive index n of the medium, as given by Equations 14.4 and 14.5. The wave equations are derived from the Maxwell equations. The material and interface properties determine boundary conditions required for the solution of the wave equations [49, 53].

$$\nabla^2 E = \mu_0 \varepsilon_0 n^2 \frac{\partial^2 E}{\partial t^2} \tag{14.4}$$

$$\nabla^2 H = \mu_0 \varepsilon_0 n^2 \frac{\partial^2 H}{\partial t^2} \tag{14.5}$$

The solutions of the wave equation are called modes. Modes refer to stable field distribution in the direction transverse to the propagation direction as the confined electromagnetic field travels inside the waveguide. In terms of

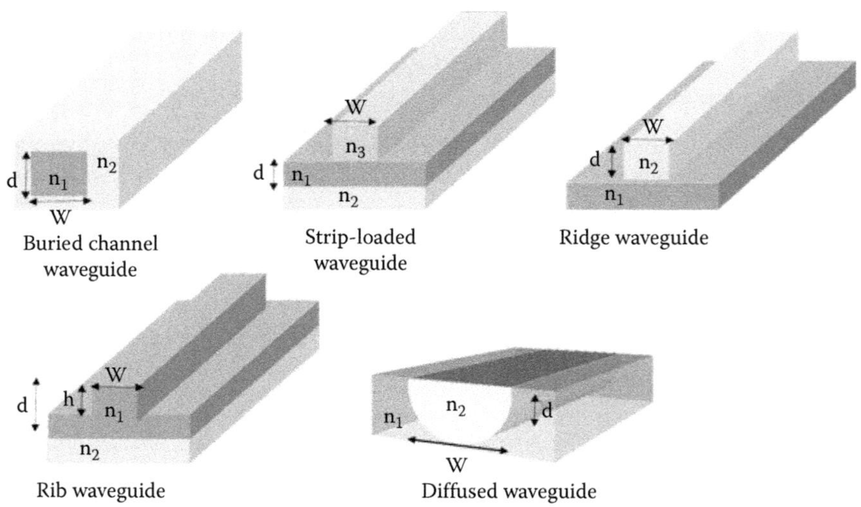

FIGURE 14.17
Different types of integrated optical waveguides.

ray theory, different modes are associated with the different angle between the wave vector of the ray and the guiding axis of the waveguide and their characteristics depend on the geometry of the waveguide, wavelength, and polarization of the light beam. Light propagating inside a waveguide is composed of different modes, each having a mode propagation constant. The dimensions of the core for the single-mode condition should be comparable or smaller than half-wavelength of a guided wave.

Light, in the form of modes, travel inside the waveguide in one of the following manners:

1. Transverse electric (*TE* modes) with field components, which have no electric field in the direction of propagation.
2. Transverse magnetic (*TM* modes) with field components, which have no magnetic field in the direction of propagation.
3. Hybrid modes with both electric and magnetic field components in the direction of propagation.

Transverse electromagnetic mode (*TEM* mode), having no electric or magnetic field in the direction of propagation, do not exist inside a dielectric waveguide. The single-mode waveguide has lots of advantages over the multi-mode design, such as high bandwidth and lower losses and it is essential in the development of optical communication. The optical circuitry is built around the fundamental guided mode. The dimensions of the waveguide are kept so as to satisfy a single-mode condition. There are different waveguides designs, as shown in Figure 14.17, each with its own set of characteristics, such as field confinement, effective index, and propagation losses. Our selection of waveguide design depends on the application that would give us the best results. A smaller bend radius (Figure 14.18) is needed for larger-scale integration, but radiation losses increase with smaller bend

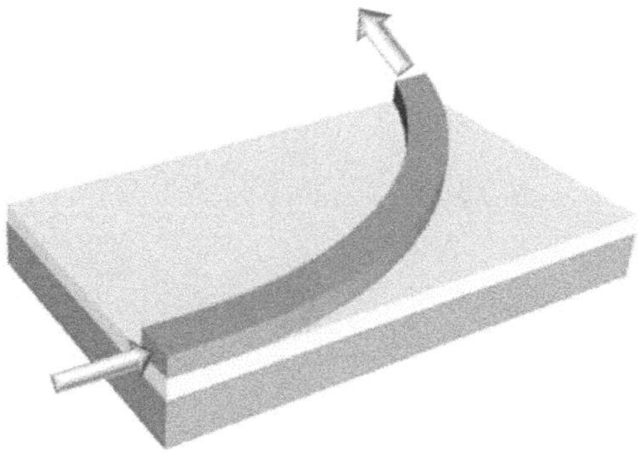

FIGURE 14.18
A SOI bent waveguide.

radii. In the case of sharp bends, mode shifts outward and there is no guiding by inner interface. A strong field on the outer interface also result in scattering losses [61]. The optimal geometries of waveguide polarization and the material to be used is a subject of intense research and is an important factor in the performance and growth of photonic integrated circuits.

14.7.3 Passive Devices

SOI waveguides have attracted attention as an emerging technology for optical interconnects in the design of densely integrated compact photonic devices and circuits [53]. The passive photonic components require no input power to function. These devices are generally used in optical signal processing to split, combine, filter or alter signals in various manners. The performance characteristics of the waveguide are key in the design of these passive devices and determine their overall performance.

14.7.3.1 Directional Coupler

Directional couplers are primitive component in the design of photonic devices and systems. They are based on coupled mode theory which deals with the interaction between two propagating modes in proximity with each other. A directional coupler consist of waveguides placed close to each other, as shown in Figure 14.19, so that the evanescent field of one waveguide "feels" the second waveguide and the gradual coupling of light takes place between waveguides. Coupling is mainly controlled by spacing and medium between waveguides. Full coupling of light is possible in the case of identical waveguides; otherwise, the coupling is usually partial [62]. Couplers find

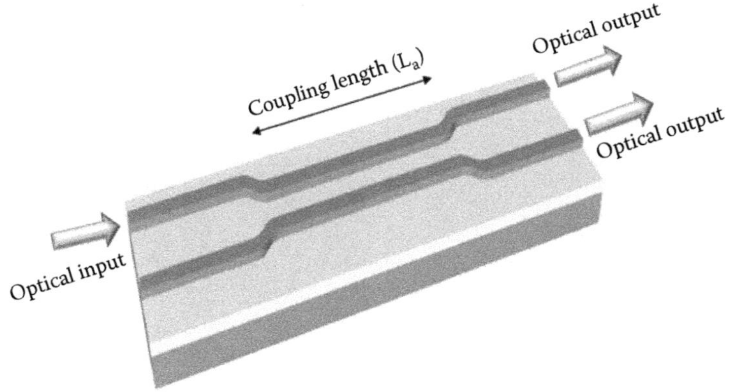

FIGURE 14.19
A SOI directional coupler.

applications in various design such as power splitter, switches, wavelength filters, and polarization selectors.

14.7.3.2 Tapers

Tapers are used to achieve the efficient coupling of light between two different cross-section's waveguides by amending the shape and size of the optical mode [63]. For efficient coupling, the modal area of the two waveguides needs to be approximately equal. A taper gradually expands or contract a mode in the transverse spatial dimensions by gently changing the size of its cross-section with hardly any mode conversion to higher-order or radiation modes, as shown in Figure 14.20. If there is gradual change in a mode shape, the mode can adiabatically adjust its parameters according to the considered position.

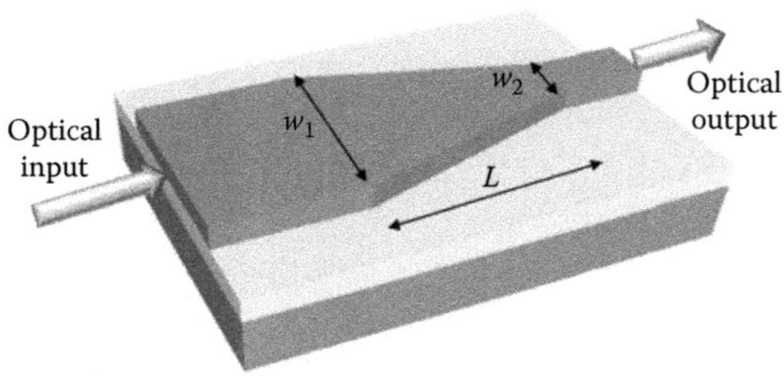

FIGURE 14.20
A SOI taper.

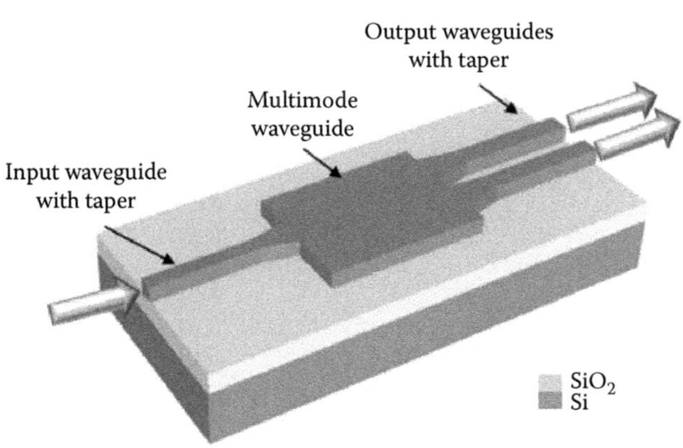

FIGURE 14.21
A SOI MMI splitter.

14.7.3.3 Multimode Interference Coupler (MMI)

MMI consists of single mode input waveguides and a large core waveguide region called the multimode region, which supports a large number of guided modes and single mode output waveguides. Through interference between the different modes in the multimode region, single or multiple images are produced at a periodic interval along the propagation direction of the waveguide according to the self-imaging principle [64]. By proper dimensioning of multimode region and output waveguides positions, characteristics of the field on each waveguide can be controlled. In a simple structure, shown in Figure 14.21, a single waveguide excites the multimode region and the output field emerges from two waveguides. It is possible to build $N \times M$ structure, consisting of N input waveguide and M output waveguides. MMI are compact and polarization independent, as compared to directional couplers, and find application in switches, couplers, modulators, and wavelength-division multiplexing circuits.

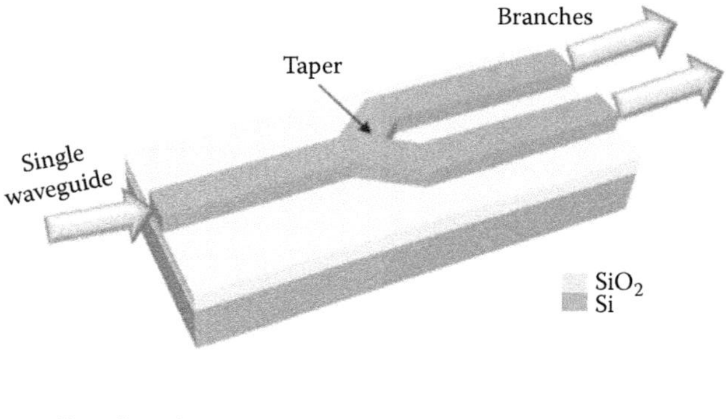

FIGURE 14.22
A SOI Y-Junction splitter/coupler.

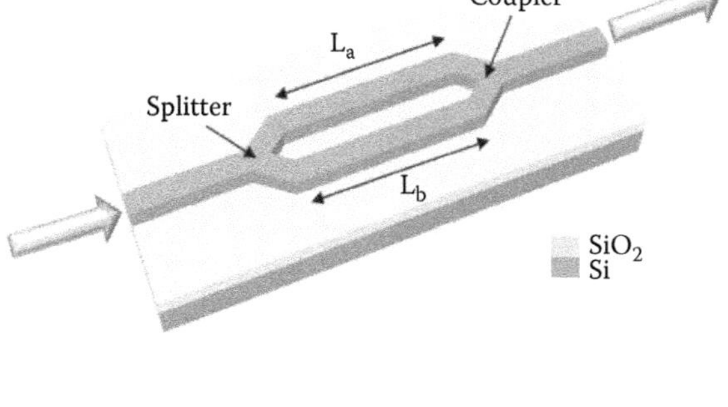

FIGURE 14.23
A SOI MZI structure.

14.7.3.4 Y-Junction

The Y-junction structure, shown in Figure 14.22, consists of a waveguide branching into two different waveguides with the same dimensions and characteristics of the input waveguide mode. The cross-over region is called the taper region. The splitting angle between the input and output waveguide is an important design parameter in terms of power loss. It mainly operates as either a power divider or a combiner according to the direction of mode propagation [65].

14.7.3.5 Mach-Zehnder Interferometer (MZI)

MZI comprises two decoupled waveguides with different lengths or physical characteristics connected at the two ends by a power splitter or combiner, normally something like a Y junction or MMI coupler, as shown in Figure 14.23. Light from input waveguides passes through the splitter resulting in two light waves traversing different optical paths L_a and L_b . It is possible to control the output characteristics by modifying the effective index of the mode in the two branches and/or by adjusting the length of the branches [66]. MZI finds applications in power divider/combiner, switching, wavelength division multiplexing, modulators, and so on.

14.7.3.6 Ring Resonator

Ring resonators are basic components in current state-of-the-art silicon photonics circuits. They consists of a closed looped waveguide such that, when the optical path length of the loop is equal to the whole number of the wavelength, constructive interference occurs inside the loop and the intensity of light builds up over multiple round trips. The light can be guided into a resonator through evanescent coupling, as shown in Figure 14.24. Compact

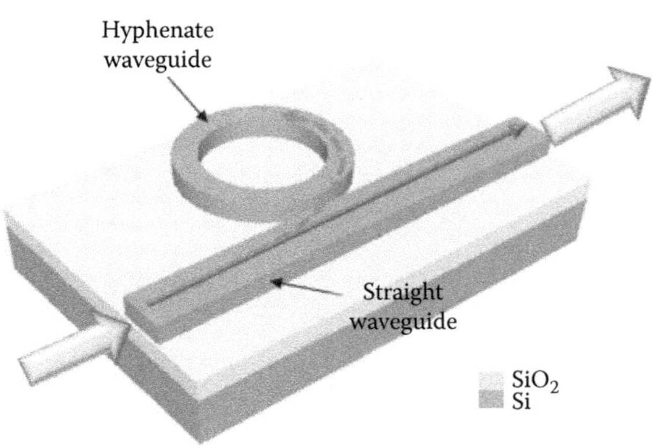

FIGURE 14.24
Schematic diagram of a SOI ring resonator.

size and high-Q resonances enable a wide range of use for these structures in devices including lasers, wavelength filters, modulators, switches, photodetectors, sensors, and delay elements [67]. These are key components for spectral filters in present WDM systems. Advancements in the design of resonators shall play an important role in the success of silicon photonics.

14.7.3.7 Arrayed Waveguide Gratings (AWG)

An AWG structure consist of a number of input and output waveguides and two free propagation regions (i.e., a wide multimode region) connected by an array of waveguide of different length, as shown in Figure 14.25. Light entering the input waveguides transverses the free propagation region, entering into the array of single mode waveguides. The length of the waveguides in the array have a constant length increment ΔL as we move from

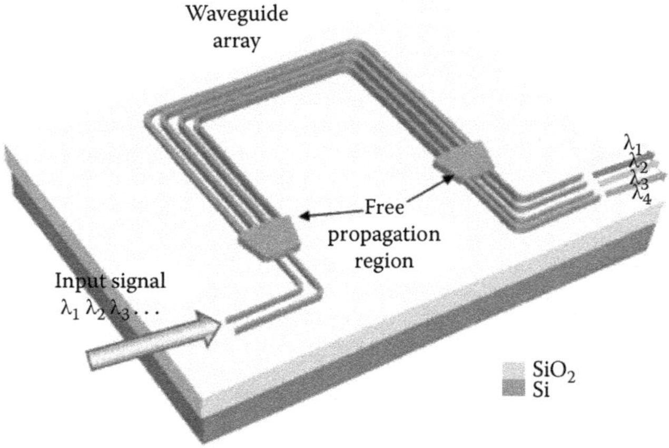

FIGURE 14.25
Schematic diagram of a SOI array waveguide grating.

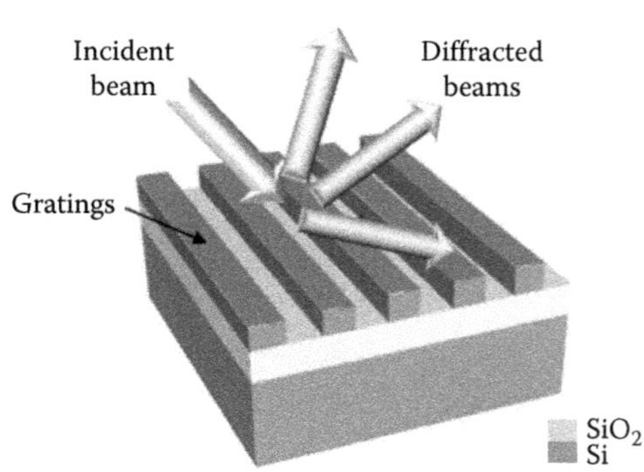

FIGURE 14.26
A SOI diffraction grating structure.

one waveguide to the other; thus, applying a different phase shift at the exit. Light then traverses another free space region and interferes at the entries of the output waveguides. By laying out the waveguides at the end of the output free space region at positions along which, at different wavelength, we have constructive interference, it is possible to separate wavelengths on to the different waveguides [68,69]. AWG finds major applications in current WDM systems as wavelength multiplexer/demultiplexer.

14.7.3.8 Diffraction Grating

A Diffraction grating consists of periodic variations in the refractive index, as shown in Figure 14.26, that diffracts and splits light into different beams by an amount varying according to the wavelength. The grating is integrated into a waveguide structure to allow in and out coupling of light outside the waveguide. When a diffracted mode matches the angle of a guided mode, strong coupling of light take place in the waveguide. These gratings can be used to couple light between independent modes and as mode converters. Gratings are also used as reflectors in state-of-the-art lasers, replacing mirrors in providing necessary feedback [70]. The advantage of using gratings instead of mirror is their wavelength selectivity, providing narrow linewidth emission. The wavelength selectivity of the grating is also used for WDM applications.

14.7.4 Optical Modulators in Silicon Photonic Circuits

Modulators are essential components in photonics circuits. They control the characteristics of light through the circuits according to an external modulating signal. There are stringent requirements for optical modulators to meet complex future demands, such as high speed, low energy per bit, compact design, low loss, large optical bandwidth, modulation depth, low-temperature sensitivity, CMOS compatible process flow, and low drive voltage. The thermo-optic effect, electro-optic effect, and microelectromechanical

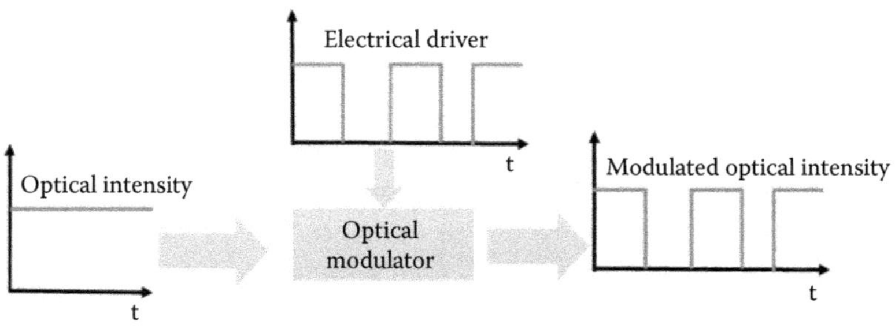

FIGURE 14.27
Electro-optical modulation scheme.

systems (MEMS) are some of the current optical modulation schemes. In an electro-optic modulator, the electro-optic effect is used to modulate a beam of light. The modulation can be in terms of change of intensity, phase, or polarization of the output beam. MEMS and thermal effects find less application in current devices due to their slow speed. An optical modulation scheme is shown in Figure 14.27.

Silicon lacks intrinsic linear electro-optic effect due to its centrosymmetric structure, and, thus, is not an ideal material to build an efficient modulator [71]. Indirect techniques are used to obtain electro-optical modulation in silicon. Application of an electric field causes change in the real part Δn (electrorefraction) and the imaginary part $\Delta\alpha$ (electro-absorption) of the refractive index. Mechanisms mostly used for optical modulation in silicon are as follows [72,73]:

1. *Free carrier concentration variations in silicon (plasma dispersion effect)*: This is the most adapted method for obtaining optical modulation in silicon. Free carrier concentration variation in silicon leads to change in both the real and imaginary parts of refractive index. Modulation (change in refractive index) is achieved through variation in the free carrier density. Figure 14.28 illustrates one such scheme. Equations 14.6 and 14.7 are used to calculate the change in the refractive index corresponding to free carrier concentration ΔN at 1.55 μm.
   At $\lambda = 1.55$ μm,

$$\Delta n = -8.8\times 10^{-22}\Delta N_e - 8.5\times 10^{-18}\Delta N_h^{0.8} \quad (14.6)$$

$$\Delta\alpha = 8.5\times 10^{-18}\Delta N_e - 6\times 10^{-18}\Delta N_h \quad (14.7)$$

   where ΔN_e, ΔN_h denotes change in free concentrations of electron and hole (cm^{-3}), respectively, and Δn and $\Delta\alpha$ denote change in the real and imaginary parts of the refractive index. A few methods used to achieve free electron and hole concentration variation in silicon are

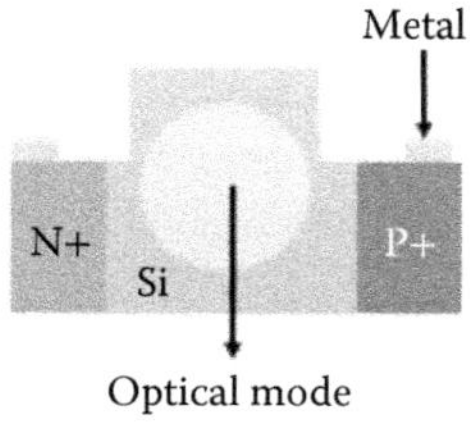

FIGURE 14.28
Optical mode in PIN diode.

- Carrier injection by forward bias in PN/PIN diode
- Carrier accumulation in MOS capacitor
- Carrier depletion by reverse bias in PN/PIN diodes

2. *Franz–Keldysh effect*: This effect alters the optical absorption of a bulk semiconductor while the electric field is applied across it. It appears, due to photon-assisted tunneling, that electrons gain energy from photons and lift up to the energy from where they can tunnel from the valance band to the conduction band of the semiconductor. Thus, the application of the electric field shifts the absorption edge to lower energy or higher wavelength due to increase in tunneling probability (large electric fields result in steeper bands, hence the reduction in barrier width) at higher wavelength.
3. *Quantum Confined Stark effect (QCSE)*: In quantum well structures, electron wave function is confined by the potential barrier. This results in discrete energy levels, barrier widening, and exciton (bound electron-hole pair) formation even at room temperature. The excitons have higher binding energy than bulk material and are not easily dissociated, even under the influence of electric field, due to strong quantum confinement. Application of an electric field results in steeper bands, increasing the probability of quantum tunneling. In the case of quantum structures, because of the step response of the absorption coefficient and the presence of excitonic peaks, the intensity change is large, which can be used in electro-absorption modulators.

The electro-optic effect in silicon can also be obtained using techniques such as

- Pockel's effect in strained silicon
- Si-Ge superlattice
- Integration of an electro-optic material on silicon

14.7.4.1 Basic Modulator Structures

MZI and ring resonator-based modulators [74] have been vigorously researched over the past few years and are commonly used in current photonic circuits. Technique like plasma dispersion effect and the integration of CMOS-compatible hybrid material are among some of the widely used methods for obtaining modulation with these structures. MZI generally needs a large drive voltage and long device length; however, it has the advantage of broadband operation and thermal stability. On the other hand, ring resonator-based structures have compact design and low voltage requirements, but suffer from high thermal sensitivity and narrow band operation. Therefore, a proper choice is required for selecting a structure according as per requirement. New approaches are required to meet the optimum goals.

14.7.5 Photodetectors

A photodetector converts light energy into electrical form and is an essential component in an optical link. It primarily works on the principle of the photovoltaic effect, that is, the generation of electron-hole pairs on the absorption of incident radiation, which are collected by application of an electric field; as a result, there is the flow of current (photocurrent) in an external circuit. For a photodetector to detect light, the excitation energy should be greater than the material bandgap energy. Silicon with bandgap energy of 1.12 eV has no detection in 1.3–1.6 μm wavelength region, putting a limit on wavelengths below 1.1 μm for detection. Photodetectors for optical communication wavelengths around 1.5 or 1.3 μm are possible with the integration of a material on silicon platform for detection in this wavelength band. Materials commonly used on the silicon platform for photodetection are as follows [49, 72]:

III-V materials (bonded or epitaxial grown on silicon): Due to direct bandgap, these materials show strong absorption and emission characteristic. The bandgap of these semiconductors can be engineered by varying the relative concentration of their constituents. Bandgap engineering enables the formation of heterojunction structures for enhanced performances. However, integration of these alloys is a costly and difficult process, limiting the application of these material systems [75].

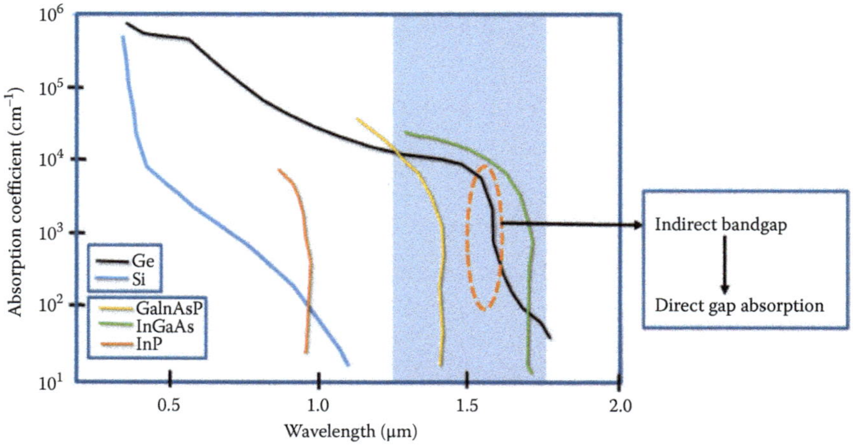

FIGURE 14.29
Absorption coefficient of III-V compounds and group IV materials.

Group IV materials (Ge-on-Si): Germanium (Ge) has strong detection in the communication wavelength (1.3 μm and 1.55 μm), as shown in Figure 14.29. Ge is compatible with CMOS process flow, its higher carrier mobility promises faster operation, making it a suitable material for photodetection [76]. However, the lattice mismatch of 4% between Ge and Si limits the thickness of the pure defectless germanium film that can be grown on silicon substrate. The resulting dislocations are sites for carrier recombination which results in reduced responsivity and large leakage currents in photodetectors.

1. *Metal–semiconductor–metal photodetector (MSM detector):* In MSM detectors, light falling on the semiconductor generates electrons-hole pairs, which are collected by the electric field applied between two Schottky contacts, that is, two metallic electrodes on a semiconductor material.
2. *P-I-N photodetector*: These detectors have an intrinsic region between the n and p doped regions where most of the photons are absorbed, carriers generated there efficiently contribute to the photocurrent.

14.7.5.1 Photo Detector Electrical Structures

Photodetectors integrated on SOI waveguide, as shown in Figure 14.30, are commonly used for light detection, where the input light travels through an optical waveguide containing the absorbing layer [77]. Epitaxial growth, flip-chip integration, and adhesive bonding are some of the integration schemes for photodetector on the SOI platform. The efficiency of a detector is dependent on the coupling mechanism of light from the waveguide to

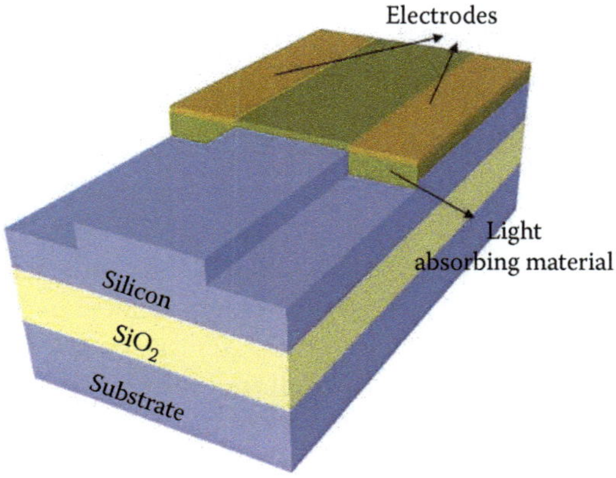

FIGURE 14.30
Photodetector integrated on SOI waveguide.

detector. Methods for coupling light from the waveguide into the detector include butt coupling, vertical coupling, and end-fire coupling.

14.7.5.2 Requirements

1. Broadband detection in the range of 1.3–1.6 μm
2. High absorption coefficient (material choice)
3. Low dark current
4. High bandwidth
   a. Low carrier transit time (material choice)
   b. Low RC constant—dependent on the electrical structure considered, such as a pin diode or MSM detector
5. High responsivity
6. Compactness

14.8 Wavelength Division Multiplexing

One of the major advancement in the optical communication technology is the wavelength division multiplexed (WDM) systems. In this technique, communication channels are allocated to different frequencies, multiplexed onto an optical link and then separated to different receivers. Using this technique, the transmission capacities of an optical link can be efficiently utilized. The excessive bandwidth of an optical interconnect can be taken advantage of with a very high-frequency single data channel; however, due to the bandwidth limitation of sender and receiver, the channel capacity is not fully utilized [78]. Wide-bandwidth channels are also affected by dispersion, restricting the transmission distance.

By combining several low transmission rate channels, the high data rate of an optical channel can be achieved. The available bandwidth of the interconnect is divided into multiple channels, each with its carrier signal. Modulated carriers are simultaneously transmitted on the link. There are international standards specifying the inter carrier spacing between two subsequent carrier signals in terms of frequency, taking into account all the possibilities of interference between adjacent channels. The essential components of a WDM system mainly include tunable laser sources and wavelength multiplexers and demultiplexer components such as switches, couplers, modulators, amplifiers, and photodetectors at the receiver end. A wavelength multiplexer (MUX) combines several wavelengths onto a single optical link at the transmission end and a demultiplexer (DEMUX) separate different wavelengths at the receiving end. Diffraction gratings and arrayed waveguide gratings (AWG) are the most common components used as optical (de)multiplexers in WDM systems.

With advancements in optical technology and the development of high performance, narrow spectral width lasers, simultaneous transmission of closely spaced channels with a negligible possibility of interference is possible. This capability of the WDM system has allowed optical carriers to be closely spaced, resulting in the dense packing of carrier signals, according to specified standard guidelines for placement. The new WDM systems is known as dense wavelength division multiplexed (DWDM) systems. Due to more closely spaced channels, DWDM has a greater overall capacity.

14.9 Performance Comparison between Electrical and Optical Interconnect

Optical interconnects (OIs) should exhibit advantages over electrical interconnects mostly in terms of power, delay, and bandwidth density to be considered a possible substitute to electrical interconnects for on-chip communication. In OIs, types of delay include delay in driving the capacitive load of the modulator, group delay while traveling through the waveguide, and delay associated with the optical transmitter and receiver circuit. In between these, receiver delay is dominant and determines the overall performance. In terms of power, the major part of power is usually associated with the optoelectronic circuitry. Delay can be reduced at the expense of power loss, for example, by increasing the optical power from the input source, reducing photodiode capacitance (decreasing its efficiency). As the waveguide length starts increasing, it starts having significant contribution in the delay of the overall link. The power dissipated in the electrical system accounts for dynamic power arising from the charging and discharging of capacitors present at wiring and load terminal and to the static power dissipated by the buffers; whereas, for optics, static power dissipation in the end-device contributes to power dissipation [79, 80].

The comparison between electrical and optical interconnect as demonstrated by Koo et al. and Cho et al. [81,82] at 22 nm technology node, are shown in Figure 14.31. As the power dissipation in optical interconnects is almost independent of length and is mostly accounted for by the optoelectronic driving circuitry, optical interconnects can be a solution to long distance interconnects (global interconnects) while, for shorter links (local interconnects) electrical interconnects are more favorable on power accounts, as shown in Figure 14.31a. The delay increases in both

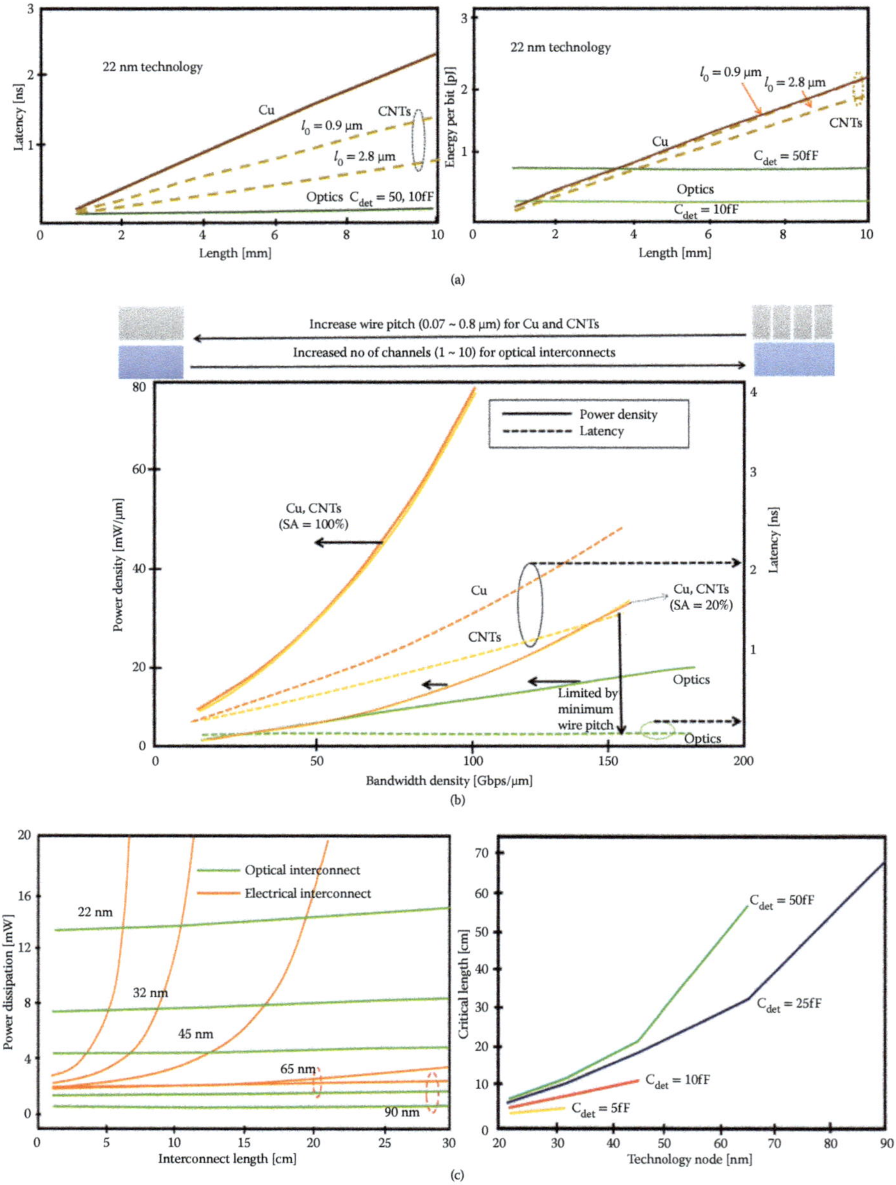

FIGURE 14.31
Comparisons for on-chip interconnects technologies. (a) Latency and energy per bit in terms of wire length, where l_0 is the mean free path of electrons in CNT and C_{det} is the detector capacitance for optics. (b) Comparison between Cu, CNTs, and optical interconnect for different switching activities (SA). (c) Critical length, beyond which optical interconnects have low power dissipation and critical length dependence on detector capacitance for different technology nodes. (K. Koo et al., *IEEE Transactions on Electron Devices*, 56, 1787–1798, 2009; H. Cho et al., *Proceedings of IEEE International Conference on Interconnect Technology*, 177–179, 2005. With permission.)

technologies with wire length. In the case of electrical interconnections, it is because of repeater insertions; whereas, in optics, latency is due to length dependence. The maximum achievable bandwidth in the case of electrical interconnect is limited by the aspect ratio. Interconnect pitch, in the case of the optical interconnect, is larger than electrical interconnect leading to smaller bandwidth density. The diffraction limit of light restricts the interconnect size, comparable to the wavelength of light. Bandwidth density in optical interconnects can be significantly improved by techniques such as wavelength division multiplexing (WDM). Figure 14.31b shows that the lowest latency is shown by optical links, followed by the CNT and Cu at all bandwidth densities. As higher bandwidth is achieved by adding more wavelengths using WDM in an optical link, there is no effect on the latency of other wavelengths. In contrast, due to increased resistance in smaller pitch wire with increased bandwidth density and more repeaters, the latency of Cu and CNT increases. Hence with more bandwidth requirements, the optical link becomes more advantageous than the electrical link. The above results depict that, beyond a certain length, optical interconnects perform better than electrical interconnects (even with delay optimization techniques.). With higher data rates and lower detector capacitance, critical length decreases to less than 10 cm, as shown in Figure 14.31c. The power-delay product is a better estimate for deciding the circuit performance. The speed of the optical interconnect is faster, but it requires a large power expenditure to get this advantage and that can mitigate this effect. Comparison between power dissipated in optical and electrical clock distribution network at the 70 nm technology node [83] is summarized in Tables 14.1 and 14.2.

TABLE 14.1

Optical Link Performance

Optical System Parameters	
Wavelength [μm]	1.55
Waveguide Core index (Si)	3.47
Waveguide cladding index (SiO_2)	1.44
Waveguide thickness [μm]	0.2
Waveguide width [μm]	0.5
Transmission loss [dB/cm]	1.3
Loss per Y-junction [dB]	0.2
Input coupling coefficient [%]	50
Photodiode capacitance [fF]	100
Photodiode responsivity [A/W]	0.95

Source: O'Connor, I. and Gaffiot, F., *Ultra Low-Power Electronics and Design,* Kluwer Academic Publishers, New York, NY, 2004. With permission.

TABLE 14.2

Electrical Link Performance

Electrical System Parameters	
Technology [μm]	0.07
V_{dd} [V]	0.9
T_{ox} [nm]	1.6
Chip size [mm^2]	400
Global wire width [μm]	1
Metal resistivity [μΩ-cm]	2.2
Dielectric constant	3
Optimal segment length [mm]	1.7
Optimal buffer size [μm]	90

Source: O'Connor, I. and Gaffiot, F., *Ultra Low-Power Electronics and Design,* Kluwer Academic Publishers, New York, NY, 2004. With permission.

14.10 Challenges with Optical Interconnects

The major challenges that are faced by optical interconnect technology in silicon platform will be in terms of power budget requirements, complexity, cost, thermal regulation, and integration capability with existing CMOS technology [84]. In terms of power consumption, there is a debate among researchers over whether optical interconnects will increase the power budget or if they will need less power as compared to electrical interconnects. Many optimistic researchers claim that the absence of resistive loss and the quantum sourcing and detecting in optics could result in lesser J/bit than conventional copper metal wires. On the contrary, pessimistic researchers doubt the potential power savings due to the power hungry nature of current optical devices unless these interconnect devices are employed for relatively long communication distances. Continuous efforts are being made in realizing efficient on-chip optical interconnects that could solve the problems associated with electrical interconnects; however, there are many challenges ahead before they can be implemented on the silicon chip. Some of the issues include the following:

1. For utilizing the advantages of optical interconnects, developments are required in the area of efficient optoelectronic devices, including light sources, modulators, sources, and photodetector.
2. Optoelectronic circuits usually require large drive voltage (~6V), which is inadmissible with CMOS electronic drivers.
3. An efficient interface circuitry is required between electronics and optics with low latency and power dissipation.

4. Integration of photonic integrated circuit in a silicon chip requires advancements in order to provide a simple cost-effective technology.
5. Coupling losses between optical components is a major source of power loss.
6. Thermal management is also one of the major challenges in order to work under the thermal constraints of a VLSI chip.
7. Low-cost optical blocks are a major concern for dense integrated circuits.

Therefore, it remains a challenge to realize dense optoelectronic integration in a silicon chip with commercialization potential, maturity, compactness, and within the budget for state-of-the-art electronic devices.

14.10.1 Recent Advancement in Graphene-Based Optical Interconnects

As the technology advances, interconnects are limiting the performance of high-speed systems. This has encouraged researcher to explore alternative interconnect materials and technologies that can replace existing conventional copper interconnect technology with less power consumption, better performance, cost effectivity, and compatibility with existing CMOS fabrication processes. Graphene, a two-dimensional allotrope of Carbon atoms, with its unique optical properties, is emerging as a promising substitute for prevalent optoelectronic, plasmonic, and nanophotonic materials [85]. It is compatible with standard CMOS processes and could be integrated with other grown technologies for high-data-rate (inter and intra-chip) optical blocks. It has tunable optical properties which can be controlled via doping or application of voltage. Due to its large thermal conductivity, high mobility, and large charge-carrier concentrations, it is considered a suitable candidate for electro-optic devices such as modulators or photodetectors [86, 87]. Several kinds of research have shown that graphene-based on-chip optical devices could meet the desired performance requirement for high-speed systems and chips.

Plasmonics is an active field of research in the direction of nanophotonics. Plasmonics is the study of the interaction of light coupled to oscillations of free electrons at the surface of a metal. These surface plasmon polaritons (SPPs) propagate along the interface of a metal-dielectric interface, providing nanoscale confinement beyond the diffraction limit for localization of light. The high loss over small propagation distance and unalterable optical properties of noble metals currently limit application of plasmonics for long distance interconnect [88]. New materials and designs are continuously being explored and graphene plasmons show some promising results in this direction with tunable properties, reduced losses, and large bandwidth operation [89].

Although optical technology at the chip level is still in an immature stage due to device size, energy, cost, integration, and architectural complexity barriers with existing CMOS technologies, the pioneering research is at its fastest pace to overcome all these impediments. Thus, for on-chip interconnects, a paradigm shift seems to be taking place in the near future from conventional metal to optical interconnects.

References

1. "ITRS Reports," *International Technology Roadmap for Semiconductors*, 2015. Available: http://www.itrs2.net/itrs-reports.html.
2. W. E. Donath, Placement and average interconnections lengths of computer logic, *IEEE Transactions Circuits Systems*, vol. CAS-26, vol. 6, no. 4, 272–277, 1979.
3. J. H. Chern, J. Huang, L. Arledge, P. C. Li, and P. Yang, Multilevel metal capacitance models for CAD design synthesis systems, *IEEE Electron Device Letters*, vol. 13, no.1, 32–33, 1992.
4. K. C. Saraswat and F. Mohammadi, Effect of scaling of interconnections on the time delay of VLSI circuits, *IEEE Transactions Electron Devices*, vol. 29, no.4, p. 645, 1982.
5. D. Miller, Rationale and challenges for optical interconnects to electronic chips, *Proceedings of the IEEE*, vol. 88, no. 6, 728–749, 2000.
6. H. B. Bakoglu and J. D. Meindl, Optimal interconnect circuits for VLSI, *IEEE Transactions Electron Devices*, vol. ED-32, no. 5, 903–909, 1985.
7. D. Miller and H. Ozaktas, Limit to the Bit-Rate capacity of electrical interconnects from the aspect ratio of the system architecture, *Journal of Parallel Distributed Computing*, vol. 41, no. 1, 42–52, 1997.
8. P. Kapur, J. P. McVittie, and K. C. Saraswat, Technology and reliability constrained future copper interconnects-Part I: Resistance modeling, *IEEE Transactions Electron Devices*, vol. 49, no. 4, 590–597, 2002.
9. W. Wu, S. Brongersma, M. Van Hove, and K. Maex, Influence of surface and grain-boundary scattering on the resistivity of copper in reduced dimensions, *Applied Physics Letters*, vol. 84, no. 15, 2838–2840, 2004.
10. K. H. Koo, Comparison Study of Future On-Chip Interconnects for High Performance VLSI Applications, PhD Diss., Department of Electrical Engineering, Stanford University, Stanford, CA, USA, 2000.
11. J. Chawla, F. Gstrein, K. O'Brien, J. Clarke and D. Gall, Electron scattering at surfaces and grain boundaries in Cu thin films and wires, *Physical Review B*, vol. 84, no. 23, 2011.
12. E. H. Sondheimer, The mean free path of electrons in metals, *Advances in Physics*, vol. 1, no. 1, 1–42, 1952.
13. P. S. Ho and T. Kwok, Electromigration in metals, *Reports on Progress Physics*, vol. 52, no. 3, 301–348, 1989.
14. C. K. Hu and J. M. E. Harper, Copper interconnections and reliability, *Materials Chemistry Physics*, vol. 52, no. 1, 5–16, 1998.

15. J. Tao, N. W. Cheung, and C. Hu, Electromigration characteristics of copper interconnects, *IEEE Electron Devices Letters*, vol. 14, no. 5, 149–251, 1993.
16. E. Ogawa, Ki-Don Lee, V. Blaschke and P. Ho, Electromigration reliability issues in dual-damascene Cu interconnections, *IEEE Transactions Reliability*, vol. 51, no. 4, 403–419, 2002.
17. K. Tu, Recent advances on electromigration in very-large-scale-integration of interconnects, *Journal of Applied Physics*, vol. 94, no. 9, 5451–5473, 2003.
18. A. Ghiti and A. G. O'Neill, The effects of grain boundary diffusion anisotropy on via electromigration failure, *Journal of Applied Physics*, vol. 81, no.7, 3064–3068, 1997.
19. C. K. Hu, R. Rosenberg, H.S. Rathore, D.B. Nguyen, and B. Agarwala, Scaling eEffect on Electromigration in On-Chip Cu Wiring, in *Proceedings of the IEEE International Interconnect Technology Conference*, pp. 267–269, 1999.
20. C. K. Hu, L. Gignac, and R. Rosenberg, Electromigration of Cu/low dielectric constant interconnects, *Microelectronics Reliability*, vol. 46, no. 2–4, 213–231, 2006.
21. M. Tada, H. Ohtake, J. Kawahara, and Y. Hayashi, Effect of material interfaces in Cu/low-κ damascene interconnects on their performance and reliability, *IEEE Transactions Electron Devices*, vol. 51, no. 11, 1867–1876, 2004.
22. J. Noguchi, Dominant factors in TDDB degradation of Cu interconnects, *IEEE Transactions Electron Devices*, vol. 52, no. 8, 1743–1750, 2005.
23. A. Vittal, L. H. Chen, S. M. Marek, K.-P. Wang, and X. Yang, Crosstalk in VLSI Interconnections, *IEEE Transactions Computer-Aided Design Integrated Circuits Systems*, vol. 18, no. 12, 1817–1824, 1999.
24. E. Lewis, An analysis of interconnect line capacitance and coupling for VLSI circuits, *Solid-State Electronics*, vol. 27, no. 8–9, 741–749, 1984.
25. L.-R. Zheng and H. Tenhunen, Wires as interconnects, in *Interconnect-Centric Design for Advanced SoC and NoC*, J. Nurmi, H. Tenhunen, J. Isoaho, and A. Jantsch, Eds. Berlin: Springer US, pp. 25–54, 2005.
26. M. R. Feldman, S. C. Esener, C. C. Guest, and S. H. Lee, Comparison between optical and electrical interconnects based on power and speed considerations, *Applied Optics*, vol. 27, no. 9, 1742–1751, 1988.
27. G. I. Yayla, P. J. Marchand, and S. C. Esener, Speed and energy analysis of digital interconnections: Comparison of on-chip, off-chip, and free-space technologies, *Applied Optics*, vol. 37, no. 2, 205–227, 1998.
28. A.K. Goel, *High-Speed VLSI Interconnections*. New York, NY: John Wiley & Sons, 1994.
29. H. Li, C. Xu, N. Srivastava, and K. Banerjee, Carbon nanomaterials for next-generation interconnects and passives: Physics, status and prospects, *IEEE Transactions Electron Devices*, vol. 56, no. 9, 1799–1821, 2009.
30. D. Areshkin, D. Gunlycke, and C. White, Ballistic transport in graphene nanostrips in the presence of disorder: Importance of edge effects, *Nano Letters*, vol. 7, no. 1, 204–210, 2007.
31. A. Srivastava, Y. Xu, and A. Sharma, Carbon nanotubes for next generation very large scale integration interconnects, *Journal of Nanophotonics*, vol. 4, no. 1, 041690, 2010.
32. B. K. Kaushik, V. R. Kumar, and A. Patnaik, *Crosstalk in Modern On-Chip Interconnects: A FDTD Approach*. Singapore: Springer, 2016.
33. S. Berber, Y. Kwon, and D. Tománek, Unusually high thermal conductivity of carbon nanotubes, *Physical Review Letters*, vol. 84, no. 20, 4613–4616, 2000.

34. J. Morris, The proof is in the nanopacking, *IEEE Nanotechnology Magazine*, vol. 2, no. 4, 25–27, 2008.
35. H. Li, W. Yin, K. Banerjee, and J. Mao, Circuit modeling and performance analysis of multi-walled carbon nanotube interconnects, *IEEE Transactions Electron Devices*, vol. 55, no. 6, 1328–1337, 2008.
36. J. Plombon, K. O'Brien, F. Gstrein, V. Dubin, and Y. Jiao, High-frequency electrical properties of individual and bundled carbon nanotubes, *Applied Physics Letters*, vol. 90, no. 6, p. 063106, 2007.
37. M. Scarselli, P. Castrucci, and M. Crescenzi, Electronic and optoelectronic nanodevices based on carbon nanotubes, *Journal of Physics Condensed Matter*, vol. 24, no. 31, 313202-1–313202-36, 2012.
38. A. Naeemi and J. D. Meindl, Electron transport modeling for junctions of zigzag and armchair graphene nanoribbons (GNRs), *IEEE Electron Device Letters*, vol. 29, no. 5, 497–499, 2008.
39. K. H. Murali, K. Brenner, Y. Yang, T. Beck, and J. D. Meindl, Resistivity of graphene nanoribbon interconnects, *IEEE Electron Device Letters*, vol. 30, no. 6, 611–613, 2009.
40. A. Naeemi and J. D. Meindl, Conductance modeling for graphene nanoribbon (GNR) interconnects, *IEEE Electron Device Letters*, vol. 28, no. 5, 428–431, 2007.
41. G. Keiser, *Optical Fiber Communications*, 2nd ed. New York: McGraw-Hill, 1991.
42. J. W. Goodman, F. J. Leonberger, S.-Y. Kung, and R. A. Athale, Optical interconnections for VLSI systems, *Proceedings of the IEEE*, vol. 72, no. 7, 850–866, 1984.
43. D. Miller, Device requirements for optical interconnects to silicon chips, *Proceedings of the IEEE*, vol. 97, no. 7, 1166–1185, 2009.
44. D. Miller, Physical reasons for optical interconnection, *International Journal of Optoelectronic*, vol. 11, no. 3, 155–168, 1997.
45. D. Miller, Optical interconnects to electronic chips, *Applied Optics*, vol. 49, no. 25, F59–F70, 2010.
46. D. Miller, Optics for low-energy communication inside digital processors: Quantum detectors, sources, and modulators as efficient impedance converters, *Optics Letters*, vol. 14, no.2, 146–148, 1989.
47. Y. I. Ismail, E. G. Friedman, and J. L. Neves, Figures of merit to characterize the importance of on-chip inductance, *IEEE Transactions VLSI Systems*, vol. 7, no.4, 442–449, 1999.
48. G. A. Keeler, B. E. Nelson, D. Agarwal, and D. A. B. Miller, Skew and jitter removal using short optical pulses for optical interconnection, *IEEE Photonics Technology Letters*, vol. 12, no.6, 714–716, 2000.
49. G. T. Reed, *Silicon Photonics: The State of the Art*, Berlin: Springer, 2008.
50. R. A. Soref and J. P. Lorenzo, Single-crystal silicon-a new material for 1.3 and 1.6 µm integrated optical components, *Electronics Letters*, vol. 21, no. 21, 953–954, 1985.
51. R. A. Soref and J. P. Lorenzo, All-silicon active and passive guided-wave components for t=1.3 and 1.6 µm, *IEEE Journal Quantum Electronics*, vol. QE-22, no. 6, 873–879, 1986.
52. B. Jalali and S. Fathpour, Silicon Photonics, *Journal of Lightwave Technology*, vol. 24, no. 12, 4600–4615, 2006.
53. B. Jalali, P. D. Trinh, S. Yegnanarayanan, and F. Coppinger, Guided-wave optics in Silicon-on-insulator technology, in *Proceedings Instruments Electronics Engineering Optoelectronics*, vol. 143, no. 5, 307–311, 1996.

54. D. Liang and J. Bowers, Recent progress in lasers on silicon, *Nature Photonics*, vol. 4, no. 8, 511–517, 2010.
55. G. Roelkens, L. Liu, D. Liang, R. Jones, A. Fang, B. Koch, and J. Bowers, III-V/silicon photonics for on-chip and intra-chip optical interconnects, *Laser Photonics Reviews*, vol. 4, no. 6, 751–779, 2010.
56. B. Garrido et al., "Silicon Photonics Course," Helios, 2016. Available at: http://www.helios-project.eu/Download/Silicon-photonics-course.
57. N. Koshida and H. Koyama, Visible electroluminescence from porous silicon, *Applied Physics Letters*, vol. 60, no. 3, p. 347, 1992.
58. Z. Zhou, B. Yin, and J. Michel, On-chip light sources for silicon photonics, *Light Science Appliations*, vol. 4, no. 11, p. 358, 2015.
59. L. Pavesi, L. Dal Negro, C. Mazzoleni, G. Franzo, and F. Priolo, Optical gain in silicon nanocrystals, *Nature*, vol. 408, no. 6811, 440–444, Nov. 2000.
60. J. Liu, X. Sun, R. Camacho-Aguilera, L. Kimerling, and J. Michel, Ge-on-Si laser operating at room temperature, *Optics Letters*, vol. 35, no. 5, p. 679, 2010.
61. Y. Vlasov and S. McNab, Losses in single-mode silicon-on-insulator strip waveguides and bends, *Optics Express*, vol. 12, no. 8, p. 1622, 2004.
62. P. D. Trinh, S. Yegnanarayanan, and B. Jalali, Integrated optical directional couplers in silicon-on-insulator, *Electronic Letters*, vol. 31, no. 24, 2097–2098, 1995.
63. A. R. Nelson, Coupling optical waveguides by tapers, *Applied Optics*, vol. 14, no. 12, 3012–3015, 1975.
64. L. B. Soldano and E. C. M. Pennings, Optical multi-mode interference devices based on self-imaging: Principles and applications, *Journal of Lightwave Technology*, vol. 13, no. 4, 615–627, 1995.
65. Y. Zhang, S. Yang, A. Lim, G. Lo, C. Galland, T. Baehr-Jones, and M. Hochberg, A compact and low loss Y-junction for submicron silicon waveguide, *Optics Express*, vol. 21, no. 1, p. 1310, 2013.
66. C. Z. Zhao, G. Z. Li, E. K. Liu, Y. Gao, and X. D. Liu, Silicon on insulator Mach–Zehnder waveguide interferometers operating at 1.3 μm, *Applied Physics Letters*, vol. 67, no. 17, 2448–2449, 1995.
67. S. Feng, T. Lei, H. Chen, H. Cai, X. Luo, and A. W. Poon, Silicon photonics: From a microresonator perspective, *Laser Photonics Reviews*, vol. 6, no. 2, 145–177, 2012.
68. Y. Inoue, A. Himeno, K. Moriwaki, and M. Kawachi, Silica-based arrayed-waveguide grating circuit as optical splitter/router, *Electronics Letters*, vol. 31, no. 9, 726–727, 1995.
69. P. Trinh, S. Yegnanarayanan, F. Coppinger, and B. Jalali, Silicon-on-insulator (SOI) phased-array wavelength multi/demultiplexer with extremely low-polarization sensitivity, *IEEE Photonics Technology Letters*, vol. 9, no. 7, 940–942, 1997.
70. A. Narasimha, Low dispersion, high spectral efficiency, RF photonic transmission systems and low loss grating couplers for silicon on-insulator nanophotonic integrated circuits, Ph.D. dissertation, University of California, Los Angeles, 2004.
71. R. A. Soref and B. R. Bennett, Electro-optical effects in silicon, *IEEE Journal of Quantum Electronics*, vol. QE-23, no. 1, 123–129, 1987.
72. D. Wood, *Optoelectronic Semiconductor Devices*. Trowbridge, UK: Prentice-Hall, p. 250, 1994.
73. V. R. Almeida, C. A. Barrios, R. Panepucci, and M. Lipson, All optical control of light on a silicon chip, *Nature*, vol. 431, no. 7012, 1081–1084, 2004.

74. G. T. Reed and C. E. Png, Silicon optical modulators, *Materials Today*, vol. 8, no. 1, 40–50, 2005.
75. D. L. Mathine, The integration of III–V optoelectronics with silicon circuitry, *IEEE Journal of Selected Topics in Quantum Electronics*, vol. 3, no. 3, 952–959, 1997.
76. L. Colace, G. Masini, and G. Assanto, Ge-on-Si approach to the detection of near-infrared light, *IEEE Journal of Quantum Electronics*, vol. 35, no. 12, 1843–1852, 1999.
77. T. K. Woodward, A. V. Krishnamoorthy, A. L. Lentine, and L. M. F. Chirovsky, Optical receivers for optoelectronic VLSI, *IEEE Journal Selected Topics in Quantum Electronics*, vol. 2, no. 1, 106–116, 1996.
78. E. A. De Souza, M. C. Nuss, W. H. Knox, and D. A. B. Miller, Wavelength-division multiplexing with femtosecond pulses, *Optics Letters*, vol. 20, 1166–1168, 1995.
79. P. Kapur and K. C. Saraswat, Optical interconnects for future high performance integrated circuits, *Physica E Low Dimensional Systems Nanostructures*, vol. 16, no. 3–4, 620–627, 2003.
80. K. Koo, H. Cho, P. Kapur, and K. C. Saraswat, Performance Comparisons Between Carbon Nanotubes, Optical, and Cu for Future High-Performance On-Chip Interconnect Applications, *IEEE Transactions Electron Devices*, vol. 54, no. 12, 3206–3215, 2007.
81. K. Koo, P. Kapur, and K. C. Saraswat, Compact Performance Models and Comparisons for Gigascale On-Chip Global Interconnect Technologies, *IEEE Transactions Electron Devices*, vol. 56, no. 9, 1787–1798, 2009.
82. H. Cho, P. Kapur, and K. C. Saraswat, The impact of technology on power for high-speed electrical and optical interconnects, in *Proceedings of the IEEE International Conference Interconnect Technology*, Burlingame, CA, 177–179, 2005.
83. I. O'Connor and F. Gaffiot, On-chip optical interconnects for low power, in *Ultra Low-Power Electronics and Design*, E. Macii, Ed. New York, NY: Kluwer Academic Publishers, 21–39, 2004.
84. M. Haurylau, C. Q. Chen, H. Chen, J. D. Zhang, N. A. Nelson, D. H. Albonesi, E. G. Friedman, and P. M. Fauchet, On-chip optical interconnect roadmap: Challenges and critical directions, *IEEE Journal of Selected Topics in Quantum Electronics*, vol. 12, no. 6, 1699–1705, 2007.
85. P. Avouris, Graphene: Electronic and photonic properties and devices, *Nano Letters*, vol. 10, no. 11, 4285–4294, 2010.
86. Z. Lu and W. Zhao, Nanoscale electro-optic modulators based on graphene-slot waveguides, *Journal of the Optical Society America B*, vol. 29, no. 6, 2012.
87. M. Liu, X. Yin, E. Ulin-Avila, B. Geng, T. Zentgraf, L. Ju, F. Wang, and X. Zhang, A graphene-based broadband optical modulator, *Nature*, vol. 474, 64–67, 2011.
88. P. R. West, S. Ishii, G. V. Naik, N. K. Emani, V. M. Shalaev, and A. Boltasseva, Searching for better plasmonic materials, *Laser Photonics Reviews*, vol. 4, no. 6, 795–808, 2010.
89. L. Ju, B. Geng, J. Horng, C. Girit, M. Martin, Z. Hao, H. A. Bechtel, X. Liang, A. Zettl, Y. R. Shen, and F. Wang, Graphene plasmonics for tunable terahertz metamaterials, *Nature Nanotechnology*, vol. 6, 630–634, 2011.

15

Electro-Thermal Considerations Dedicated to 3D Integration; Noise Coupling

Yue Ma, Olivier Valorge, J. R. Cárdenas-Valdez, J. C. Núñez–Perez, J. Verdier, Francis Calmon, and Christian Gontrand

CONTENTS

More generally, analytical substrate extraction methods, relying on the Transmission Line Matrix (TLM) method or the use of Green kernels, are proposed to model multilayer substrates on electrical and thermal points of view.

An important problem for designers of complex heterogeneous circuits combining digital and analog parts, is also to predict the perturbations coming from commutating logical gates and flowing through the substrate to reach some sensitive analog blocks. We then present, in this chapter, an application of a stochastic model based on Markovian processes; the digital switching activity is modeled as functions defined as Markov Chains. The final goal is to grasp the noise power density of such perturbations. Then formulas are derived for both the spectral lines—the discrete part and the continuous part of this spectral density. As an example, we extract an actual interference signal resulting from the modulation of parasitic injected waveforms (introduced, for instance, into some circuit supply), driven by the discrete stochastic process.

15.1 Introduction

3D Si integration seems the right way to go and compete with Moore's law (more than Moore versus more Moore). However, it is still a long way to go.

In 2010, the question was: Why 3D? Today, the questions are: When to use 3D and how? The 3D chip stacking, or multi layers circuits, are well enough known to overcome conventional 2D-IC issues using in-depth contacts or through-silicon via for signal transmission.

Mixed Signal IC's design has become a key for systems-on-chip VLSIs. Functional analog blocks like VCOs, LNA, filters, and AD/DA converters are placed on a die with high speed digital processing elements composed of a few millions digital gates. Parasitic phenomenon, commonly called crosstalk, occurs between the noisy digital and the sensitive analog part of the device. Due to many parasitic coupling mechanisms, there is a distinct possibility that the transient regimes in the digital circuitry of this kind of device will corrupt low-level analog signals and seriously compromise the achievable performance [1]. One of the main coupling phenomenon is the substrate coupling. The chip substrate acts as a collector, integrator, and distributor of coupled noise on chip. Many developments describe methods to simulate this perturbing kind of noise [2–4]. However, the majority of the models presented only considers the package bonding; whereas, many different parts of the supply lines are important in substrate noise generation: PCB, package, pad ring, on chip metal lines, and other parasitic elements of supply lines.

Nowadays, micro(nano)technology and the development of semiconductor technology enables designers to integrate multiple systems into a single chip, not only in 2D (planar), but also in 3D (in the bulk). This design technology reduces cost while improving performance and enables the possibility of a system on a chip [1–4].

The kind of silicon substrate in which 3D interconnections are processed is an important parameter and has a strong impact on 3D interconnection electrical behavior. Electrical modeling methods are dependent on the substrate type of the application. Two different kinds of P silicon substrates are typically used in CMOS/BiCMOS processes: uniform lightly doped substrate

(>1Ω.cm) and heavily doped substrate (<1Ω.cm) with lightly thinned doped (>1Ω.cm) epitaxial layer. Other semiconductor substrates are used in micro/nano technologies and could embed 3D interconnections (SOI technologies, MEMS technologies, sensor technologies, etc.)

Advancements in the field of V(U)LSI have led to more compact ICs having higher clock frequencies and lower power consumptions. The paradigms of these integrated technologies are systems-on-chip (SoC) and systems-in-package (SiP). Now, these conventional 2D planar technologies face several strategic challenges: physical limits, processing complexity, fabrication costs, etc. Consequently, technological approaches other than scaling are investigated to continue following or to get over Moore's Law. Currently, 3D integrated systems can be obtained by stacking 2D integrated circuits vertically using mature, controlled micro and nano technologies.

The electrical connections are ensured by new pre- or post-processed metallic structures redistribution layers (RDL), which distribute power and high speed signals on die top or backside surfaces. Copper pillars and mostly Through-silicon Via (TSV) are key enabling technologies for 3D integration by propagating signals through the silicon layers.

Chips can also be stacked vertically using wire-bonding and flip-chip techniques. However, the flip-chip technique solely provides an interconnection between two chips, whereas, wire-bonding only enables the connection of chip input/output pads located at their perimeter. Moreover, wire-bonding presents disadvantages in terms of surface and propagation delays depending on the application frequency clock and the wire-bond lengths. Finally, solder balls are put on the 3D systems bottom layers backsides to ensure connection with their environment. Compared with 2D classical schemes, 3D integration potential performance improvement benefits are numerous: flexible heterogeneous system combinations (logic CMOS, RF analog function, memories, sensors;) smaller form factors; reduced fabrication costs; and significantly shortened interconnect line networks, which decreases interconnect line delays and power consumption, significant obstacles in 2D VLSI systems relative to delays in transistor switching [1,2].

Many challenges are encountered with 3D integration due to its emerging technology status, notably in properly characterizing and electrically modeling the 3D interconnects. Few present-day CAD tools can design 3D architecture. Many works have been mainly dedicated to advanced TSV electrical modeling.

Indeed, different types of via play a central role in realizing high-density integrated 3D systems; it enables a large variation of shape, dimension (radius, length), dielectric thickness and filling material. Difficulty in TSV modeling is that the overall 3D interconnection electrical context must be considered. For example, current paths must be modeled if the environment comprises top/back redistribution metal lines. As with local metal layer interconnect line structures [5,6], the 3D interconnect global environment requires the modeling of the substrate coupling effects, which can no longer be neglected at high frequency. Several substrate extraction techniques have been proposed in the literature, but most of them cannot be employed for systems-on-chip with realistic dimensions or are only suitable for a particular type of structure or technology process [7–11].

We attempt, hereafter, to propose a beginning to the synthesis of unsteady electro-thermal phenomena studies in 3D layered circuits.

We solve Poisson's equation using Green Kernels [12]; here, we show that this mathematical method is an analog to the Transmission Line Matrix (TLM) [13] based on reflection-transmission of voltage and current through the layers when adapting it to embedded contacts of any shape.

Then, work on the resolution of the unsteady 3D heat equation is shown; we use the algorithms developed for the electrical modeling, particularly with the spreading impedance concept.

Finally, we show that these electrical and thermal spreading impedances can be considered entangled. This approach drives us to start an attempt to model the more general problem of noise in 3D ICs, using, again, a transfer impedance paradigm.

15.2 Mathematical Tools

Concerning the substrate itself, our intent is to replace numerical methods, such as finite element (FE) ones, with analytical algorithms such as Green kernels, since the resolution speed of the later method is much faster. Some basic recalls and concepts are first introduced. The use of discrete cosine transforms (DCT) and applying twice a FFT (Fast Fourier Transform) to accelerate calculation speed. Then, an improved model that can be applied on substrate with in-depth contacts (or via) of any shape, is shown.

15.2.1 Green Kernels

The high density integration and high system frequency make substrate noise coupling one of the most significant considerations in the design due to its great impact on the performance of ICs. The aim of this 3D substrate analysis is, first of all, to efficiently extract the Z impedance parameters between any contacts (electrode, via), which are located on or into the silicon substrate. An efficient impedance extraction tool for any two contacts could help the designer accelerate the design and optimize the final layout. Like in planar technologies, 3D interconnects can be built as an "RLCG"—divider bridge: (G//C)/ (R series L)—equivalent electrical model with a Π or a T network. Often, a simple compact model is constructed by modeling the substrate as a simple node. However, this assumption is only viable when the substrate is highly conductive in low and medium frequency domains and is not suitable for the multilayer substrates. That is the reason why, in this chapter, we propose a substrate extraction method relying on Transmission Line Matrix method (TLM) over multilayered substrate and/or Green functions to model these effects into the bulk.

15.2.2 Substrate Analysis

The most well-known pioneering paper came from Gharpurey [12]: a layered 3D substrate with 2 surface contacts.

The originality of our algorithm is that we consider and apply here all the calculations in the reciprocal domain, from the very beginning of the calculations to the end, to compute voltages and extract impedances.

Generally, the Z parameters can be defined as

$$Z_{mn} = \frac{V_m}{I_n}\bigg|_{I_{k\neq n}} = 0 \qquad (15.1)$$

In practical operation, by injecting a unitary current as an excitation source, I at the n-th point, and a unitary current excitation source, I (equivalent to a sink current I) at an m-th point and calculate the resulting voltage at the probe point, we can get the direct impedance Z_{mn} between the m and n contacts, as shown in Equation 15.1.

Under quasi-static conditions, the potential over the substrate satisfies Laplace's equation

$$\nabla^2\varphi(x,y,z) = 0 \qquad (15.2)$$

It can also be written as

$$\left(\frac{\partial^2}{\partial x^2} + \frac{\partial^2}{\partial y^2} + \frac{\partial^2}{\partial z^2}\right)\varphi(x,y,z) = 0 \qquad (15.3)$$

The analytical method is, first of all, based on the algorithm of variable separation, which leads to solving an eigen problem.

If we assume that $\varphi = X(x,x')\cdot Y(y,y')\cdot Z(z,z')$, we can rewrite Equation 15.3.

By defining

$$Z(z,z') = Z'(z,z')\cdot\cos\left(\frac{m\cdot\pi\cdot x'}{a}\right)\cdot\cos\left(\frac{n\cdot\pi\cdot y'}{b}\right) \qquad (15.4)$$

and we get a simple equation:

$$\frac{ab}{4}\cdot\left[\frac{d^2Z}{dz^2}-\gamma_{mn}^2\cdot Z\right]=-\frac{\delta(z-z')}{\varepsilon_N} \text{ with } \gamma_{mn}=\sqrt{\left(\frac{m\cdot\pi}{a}\right)^2+\left(\frac{n\cdot\pi}{b}\right)^2} = \sqrt{(\lambda)^2+(\mu)^2} \tag{15.5}$$

For $z \neq z'$, $\delta(z-z')=0$. The above Equation 15.5 has a well-known general solution

$$Z' = A\cdot e^{-\gamma_{mn}(d+z)} + B\cdot e^{\gamma_{mn}(d+z)} \tag{15.6}$$

This equation invokes a transmitted wave and a reflected one.

For instance, considering the case where both the point of charge and the point of observation are in the same dielectric layer on the surface with $z = z' = 0$, the Green function then changes to

$$\varnothing(x,y;x',y')_{z=z'=0} = (G_0)_{z=z'=0} + \left(\sum_{m=0}^{\infty}\sum_{n=0}^{\infty} f_{mn}C_{mn}\cos\left(\frac{m\pi x}{a}\right)\cos\left(\frac{m\pi x'}{a}\right) \times \cos\left(\frac{n\pi y}{b}\right)\cos\left(\frac{n\pi y'}{b}\right)\right) \tag{15.7}$$

where $C_{mn}= 0$ for $(m = n = 0)$, $C_{mn} = 2$ for $m = 0$ or $n = 0$ but $m \neq n$ and $C_{mn}= 4$ for all others m and n ($m>0$ and $n>0$).

The function f_{mn} is given by

$$f_{mn} = \frac{1}{ab\gamma\varepsilon_N}\frac{\beta_N \tanh(\gamma_{mn}d)+\Gamma_N}{\beta_N+\Gamma_N\tanh(\gamma_{mn}d)} \tag{15.8}$$

β_N and Γ_N can be computed recursively from

$$\begin{bmatrix}\beta_k\\ \Gamma_k\end{bmatrix}=\frac{1}{2}\cdot\begin{bmatrix} 1+\dfrac{\varepsilon_{k-1}}{\varepsilon_k} & \left(1-\dfrac{\varepsilon_{k-1}}{\varepsilon_k}\right)e^{2\theta_k} \\ \left(1-\dfrac{\varepsilon_{k-1}}{\varepsilon_k}\right)e^{-2\theta_k} & 1+\dfrac{\varepsilon_{k-1}}{\varepsilon_k}\end{bmatrix}\begin{bmatrix}\beta_{k-1}\\ \Gamma_{k-1}\end{bmatrix} \tag{15.9}$$

$$\text{where } \theta_k = \gamma_{mn}(d-d_k),\ \beta_0 = 1,\ \Gamma_0 = 1 \tag{15.10}$$

Then, after derivation of some formal mathematical solution from the Poisson's equation with boundary conditions, we will depict, hereafter, a more physical (and graspable) approach to the problem using a TLM with some brief recalls. Note that, in our whole algorithm, from its beginning, we work on the reciprocal space. For instance, we also take the Fourier transform of the injected currents on the discretized contacts.

In practice, then, we can choose any shape for these embedded contacts; whereas, in all of the few others works, to the best of our knowledge, all the contacts are rectangular-like, the integration over them being done in the real space.

After a bi-dimensional spatial Fourier transform in x, y direction, the Equation 15.3 above should be

$$\frac{\partial^2 \Phi}{\partial z^2} - k_x^2 \Phi - k_y^2 \Phi = 0 \tag{15.11}$$

where Φ is the spatial-frequency domain potential and k_x, k_y are the spatial-frequency variables (m^{-1}).

Since the current density $J_Z = \sigma^* E_Z$ (σ^* is the complex conductivity) and $E_Z = -\frac{\partial \varphi}{\partial z}$, we can write

$$J_z = -\sigma^* \frac{\partial \varphi}{\partial z} \tag{15.12}$$

In the spatial-frequency domain, it changes to

$$J_z(k_x, k_y, z) = -(\sigma + j\omega\varepsilon) \frac{\partial \Phi(k_x, k_y, z)}{\partial z} \tag{15.13}$$

After derivative calculations of the above equation, we obtain

$$\begin{cases} \dfrac{\partial \Phi}{\partial z} = -\dfrac{J_z}{(\sigma + j\omega\varepsilon)} \\ \dfrac{\partial J_z}{\partial z} = -(\sigma + j\omega\varepsilon)(k_x^2 + k_y^2)\Phi \end{cases} \tag{15.14}$$

These equations indicate the relationship between current density in the z direction and the potential distribution in the spatial-frequency domain.

15.2.3 Transmission Line Analogy for Multilayered Media

In its simplest form, a transmission line (TL) is a pair of conductors linking together two electrical systems (e.g., source and load) with forward and return paths; for cases where the return path is floating, a third conductor (or more) is introduced as the grounding shield. For microwaves, they are waveguides. In our case, the propagation of EM waves and their interferences through the silicon substrate are among the most serious obstacles in the steady trend toward the integration of present day microelectronics. In fact, the TL method (TLM) has been established in some cases concerning substrate surfaces; TLM can be seen as a more physical interpretation of the mathematical developments presented above.

Let us consider a plane wave through a multilayered medium, its plane of incidence being parallel to the <*x*,*y*> plane, providing the same boundary conditions as mentioned above.

The expression for the apparent impedance is given:

$$Z = \frac{V_{\text{total}}}{i_{\text{total}}} = \frac{V_F \pm V_R}{i_F \mp i_R} \tag{15.15}$$

Then, the general solution for voltage and current is

$$V(z) = V_f e^{\gamma z} + V_r e^{-\gamma z} \tag{15.16}$$

$$I(z) = Y_c \left(V_F e^{-\gamma z} - V_r e^{\gamma z} \right) \tag{15.17}$$

where the propagation coefficient γ is the square root of the product Z by Y, the linear impedance and admittance are, respectively; $\gamma = \alpha + \varphi\beta$, α: attenuation, $\beta = 360° / \lambda_{\text{si}}$.

Deriving directly from the chain matrix:

$$\begin{bmatrix} V_1 \\ I_1 \end{bmatrix} = \begin{bmatrix} \cosh \gamma L & Z_c \sinh \gamma L \\ Y_c \sinh \gamma L & \cosh \gamma L \end{bmatrix} \begin{bmatrix} V_2 \\ -I_2 \end{bmatrix} \tag{15.18}$$

and from the output branch: $V_2 = -Z_L I_2$, we get

$$\begin{bmatrix} V_1 \\ I_1 \end{bmatrix} = -I_2 \begin{bmatrix} Z_L \cosh \gamma L & Z_c \sinh \gamma L \\ Z_L Y_c \sinh \gamma L & \cosh \gamma L \end{bmatrix} \begin{bmatrix} 1 \\ 1 \end{bmatrix} \tag{15.19}$$

where $Z_c = 1/Y_c$ (square root of Z and Y ratio) is the characteristic impedance (Electric and magnetic fields modulus ratio), Z_L is the charge impedance, L is the distance from the load; we can extract an input impedance $Z_l = (V_1/I_1)$ function of the load impedance Z_{l+1} (l: layer number) as

$$Z_l = Z_c \frac{Z_{l+1} + Z_c \tanh \gamma L}{Z_c + Z_{l+1} \tanh \gamma L} \tag{15.20}$$

This well-known (TLM) equation Z_l is directly related to Equation 15.8, where $Z_c = \beta_N$ and $Z_L = \Gamma_N$ [12–14].

In fact, restarting from Equation 15.7, taking into account the limit or boundary conditions—continuity of potential, and discontinuity of the electric field if there is a surface charge at the considered layer interface—it is easy to program this iterative solution along the substrate depth. In fact, these calculations are made only one time, at each frequency.

In our simulator, using MATLAB [15], we grasp a matricial formalism, extending the impedance (Z) and the current (*T*) transmission from a layer l to its adjacent one, starting from a *k* layer:

$$\begin{bmatrix} Z_{k,\,l+1} \\ T_{k,\,l+1} \end{bmatrix} = \begin{bmatrix} A_k & -B_k \\ C_k & -D_k \end{bmatrix} \begin{bmatrix} Z_{k,l} \\ T_{k,l} \end{bmatrix} \tag{15.21}$$

where *T* addresses the derivation of the current (between two parallel regions).

The very suitable ABCD-parameters are known variously as chain, cascade, or transmission line parameters.

As mentioned earlier, all the calculations are done in the reciprocal space. For instance, the potential is derived from the (kronecker) product of (sub-) matrix Z by a DCT of the matrix of the injected current accomplished using two consecutive FFTs.

In our algorithm, the contact voltage is the mean value of the voltage of the discrete contact elements calculated via Millman's theorem, as

$$V = \frac{\sum_{n=1}^{N} j\omega C\varphi_n}{\sum_{n=1}^{N} j\omega C} = \frac{1}{N}\sum_{n=1}^{N} \varphi_n \tag{15.22}$$

where φ_n is the substrate voltage at a node and *N* is the number of contact sub-areas. So, in our modified algorithm, the contact voltage is then calculated by taking the average <Va> of the potentials of the substrate region under the injecting contact.

15.2.4 Comparison Results with FEM Method

15.2.4.1 Model via-Contact

In order to test our extractor's accuracy, we compare our simulation results with ones coming from COMSOL [16].

In Figure 15.1, a TSV of 5.1 μm a side occupies the 12 layers.

The pitch via-contact is set to 6 μm and the frequency varies from 1e8 to 1e13 Hz. We set the surface of the via as a port and input 1A current and – 1A at the other contact's entirety. The frequency RC-parameter results from the two simulators are shown in the Figure 15.2. They show good fits for the reactance and impedance. The variation between the two simulators increases

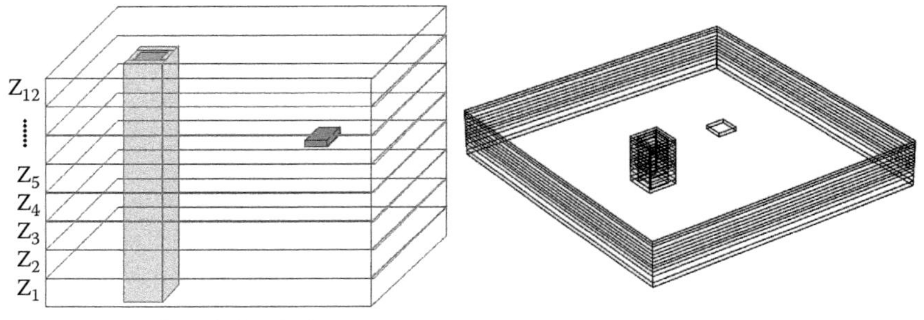

FIGURE 15.1
Via-contact contact.

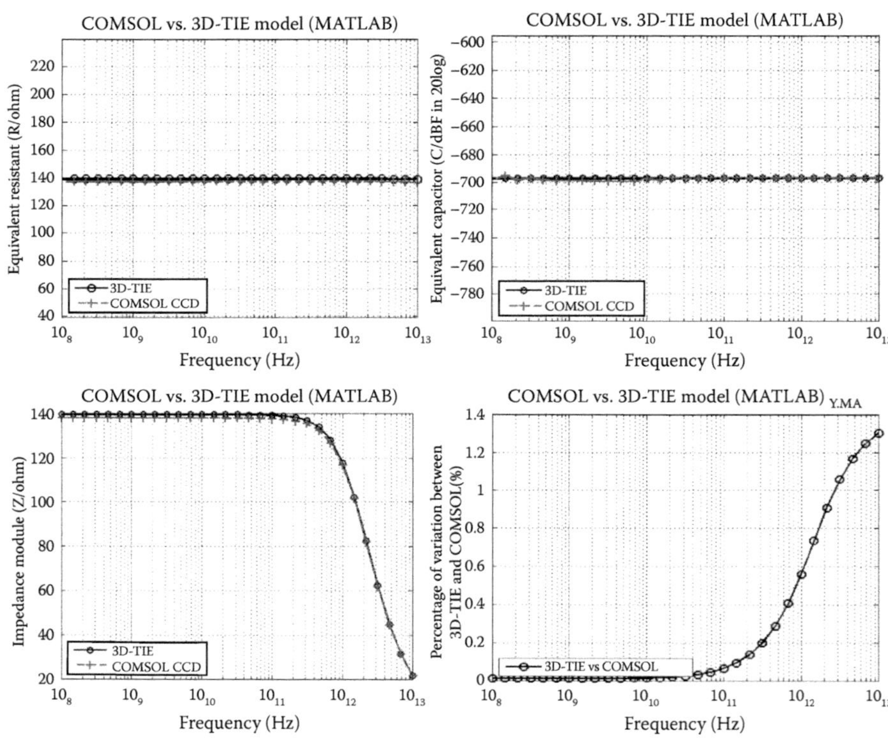

FIGURE 15.2
Via-contact: comparison between Green and FEM simulation.

versus frequency, but, even at 10 THz (far from the quasi-static regime), the variation percentage is only 1.4% maximum.

The TSV-TSV geometry is shown in Figure 15.3.

The pitch is set to 6 μm and the frequency varies from 1e8 to 1e13 Hz. We set the surface of the via as a port and input 1A current and – 1A at the other contact's entirety.

We have same types of results that those of the contact-via pair (Figure 15.4).

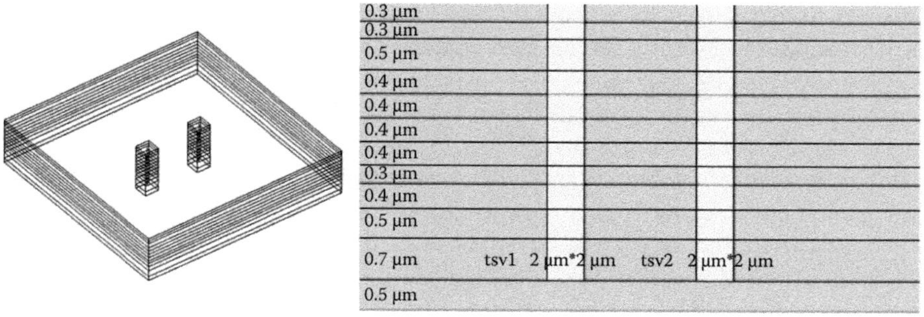

FIGURE 15.3
Via-via model.

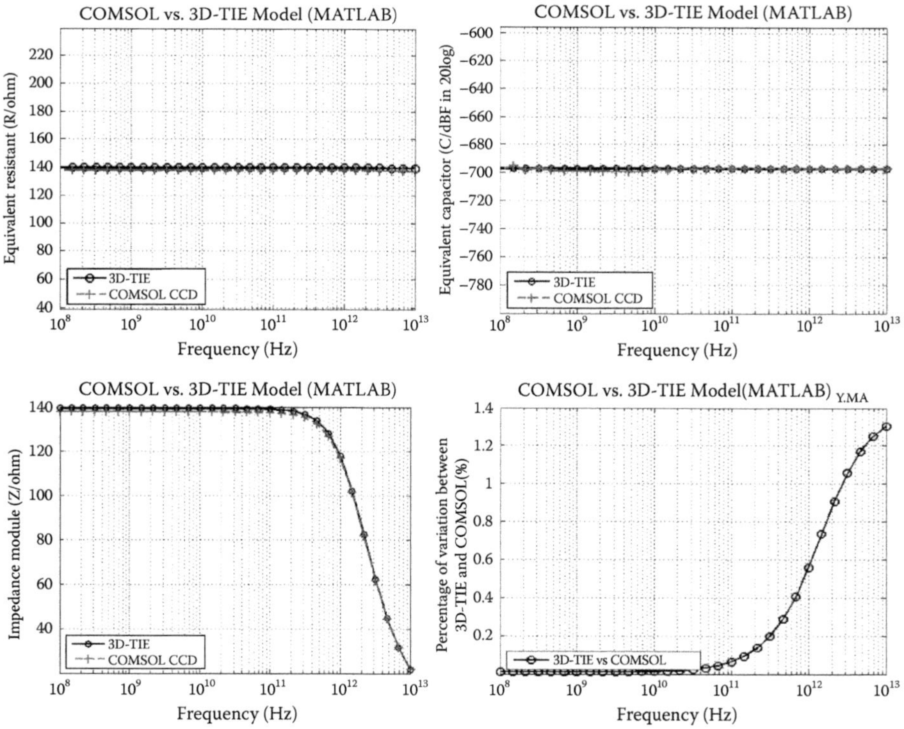

FIGURE 15.4
Via-contact: comparison between Green and FEM simulation.

15.3 Modeling Approach Validation and Test Structures' RF Behaviors

The modeling approach is validated in the frequency domain by comparing the S-parameters from the measurements performed on the test structures, for both configurations—called W1 and W2—and the simulated ones retrieved from the structures equivalent electrical models [17–21]. Some of these results are presented in Figure 15.5. The test structures RF responses are plotted with dotted blue curves; the ones of the electrical models with plain red curves.

The proposed equivalent RLCG compact models demonstrate a very good accuracy for frequencies up to 20 GHz for the coplanar waveguides (not presented here) and 10 GHz for the TSV chains. Some spikes can be observed in measurements but not in simulations at low frequency (below 1 GHz). These resonances are due to the coupling between the test structure substrate and its measurement environment. The test sample chuck is a piece of metal that acts as an excellent antenna, having uncontrolled impedance to the measurement ground. This unknown impedance can explain some unexpected poles of resonances. The spikes can be suppressed by isolating the test structure from its environment. For this purpose, the substrate must be strongly grounded or biased.

Such a pragmatic approach presents the advantage of being automatized for large circuit extraction. However, the method can be enriched by including skin effect and eddy current and yet lead to huge complexity for the automatic extraction.

15.3.1 The Simulation Platform

Figure 15.6 depicts the substrate extractor.

When getting the SPICE-compatible file from 3D-TLE, one can import it to ADS easily. An example of the generated SPICE txt file is shown Table 15.1.

15.3.2 Study of Signal Integrity in TSV Matrices

Since the 3D interconnect electrical models have been validated in low and medium frequency ranges, timing analyses are now proposed using Eye Diagrams to examine the signal integrity in the case of a 3 × 3 medium-density

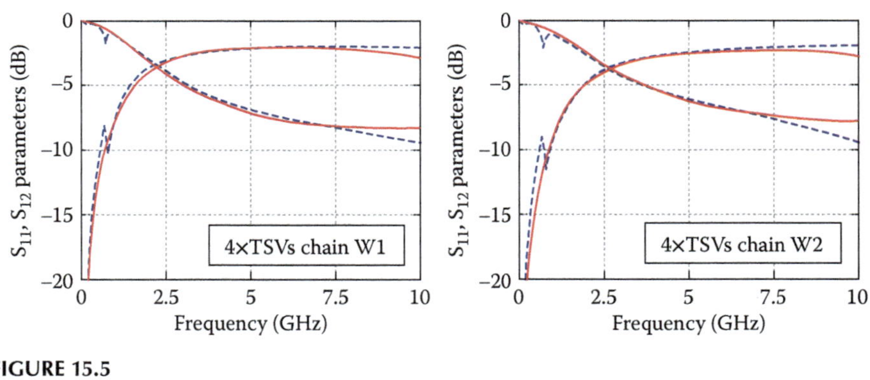

FIGURE 15.5
S-parameter comparisons up to 10 GHz between the measurements performed on 4xTSVs chain test structures and the simulations [22,23] of their electrical equivalent models for both W1 and W2 configurations.

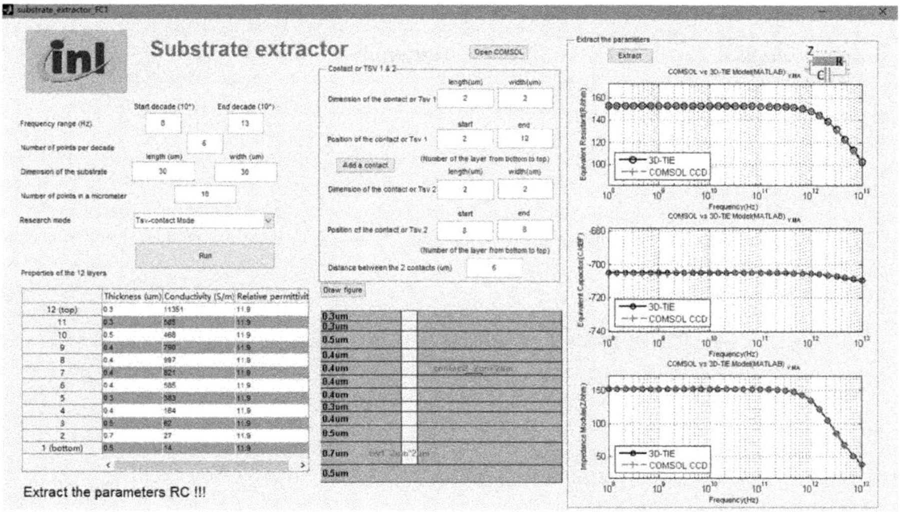

FIGURE 15.6
Substrate extractor.

TABLE 15.1
Exportable SPICE-Compatible TXT File

```
*Netlist generated from 3DIDEAS

.subckt gnd1_signal_gnd2 in1 out1 in2 out2 in3 out3

R1 in1 x1_gnd1 0.0066343
L1 x1_gnd1 x2_gnd1 6.8442e-11
L2 x2_gnd1 x3_gnd1 6.8442e-11
R2 x3_gnd1 out1 0.0066343
Cox_gnd1 x2_gnd1 0 2.0741e-13

R3 in2 x1_signal 0.0066343
L3 x1_signal x2_signal 6.8442e-11
L4 x2_signal x3_signal 6.8442e-11
R4 x3_signal out2 0.0066343
Cox_signal x2_signal 0 2.0741e-13

R5 in3 x1_gnd2 0.0066343
L5 x1_gnd2 x2_gnd2 6.8442e-11
L6 x2_gnd2 x3_gnd2 6.8442e-11
R6 x3_gnd2 out3 0.0066343
Cox_gnd2 x2_gnd2 0 2.0741e-13

K1 L1 L3 0.41668
K2 L2 L4 0.41668
Cinter1 x2_signal x2_gnd1 2.4691e-15
```

(Continued)

TABLE 15.1

Exportable SPICE-Compatible TXT File (*Continued*)

```
K3 L3 L5 0.41668
K4 L4 L6 0.41668
Cinter2 x2_gnd2 x2_signal 2.4692e-15

K5 L1 L5 0.28941
K6 L2 L6 0.28941

.ends gnd1_signal_gnd2

X1_gnd1_signal_gnd2 0 1 2 3 0 4 gnd1_signal_gnd2


.subckt pillier1_pillier2_pillier3 in1 out1 in2 out2 in3 out3

R1 in1 x1_pillier1 0.000219
L1 x1_pillier1 x2_pillier1 5.3753e-12
L2 x2_pillier1 x3_pillier1 5.3753e-12
R2 x3_pillier1 out1 0.000219

R3 in2 x1_pillier2 0.000219
L3 x1_pillier2 x2_pillier2 5.3753e-12
L4 x2_pillier2 x3_pillier2 5.3753e-12
R4 x3_pillier2 out2 0.000219

R5 in3 x1_pillier3 0.000219
L5 x1_pillier3 x2_pillier3 5.3753e-12
L6 x2_pillier3 x3_pillier3 5.3753e-12
R6 x3_pillier3 out3 0.000219

K1 L1 L3 0.22791
K2 L2 L4 0.22791
Cinter1 x2_pillier2 x2_pillier1 1.0556e-15

K3 L3 L5 0.22791
K4 L4 L6 0.22791
Cinter2 x2_pillier3 x2_pillier2 1.0556e-15

K5 L1 L5 0.11562
K6 L2 L6 0.11562

.ends pillier1_pillier2_pillier3

X1_pillier1_pillier2_pillier3 1 10 3 30 4 40 pillier1_pillier2_pillier3

.subckt gauche_centre_droite in1 out1 in2 out2 in3 out3

R1 in1 x1_gauche 0.013269
L1 x1_gauche x2_gauche 1.7262e-10
L2 x2_gauche x3_gauche 1.7262e-10
R2 x3_gauche out1 0.013269
Cox_gauche x2_gauche 0 2.7599e-13
```

```
R3 in2 x1_centre 0.013269
L3 x1_centre x2_centre 1.7262e-10
L4 x2_centre x3_centre 1.7262e-10
R4 x3_centre out2 0.013269
Cox_centre x2_centre 0 2.7599e-13

R5 in3 x1_droite 0.013269
L5 x1_droite x2_droite 1.7262e-10
L6 x2_droite x3_droite 1.7262e-10
R6 x3_droite out3 0.013269
Cox_droite x2_droite 0 2.7599e-13

K1 L1 L3 0.48924
K2 L2 L4 0.48924
Cinter1 x2_centre x2_gauche 5.2569e-15

K3 L3 L5 0.48924
K4 L4 L6 0.48924
Cinter2 x2_droite x2_centre 5.2569e-15

K5 L1 L5 0.33042
K6 L2 L6 0.33042

.ends gauche_centre_droite

X1_gauche_centre_droite 10 11 30 31 40 41 gauche_centre_droite


.subckt pillier4_pillier5_pillier6 in1 out1 in2 out2 in3 out3

R1 in1 x1_pillier4 0.000219
L1 x1_pillier4 x2_pillier4 5.3753e-12
L2 x2_pillier4 x3_pillier4 5.3753e-12
R2 x3_pillier4 out1 0.000219

R3 in2 x1_pillier5 0.000219
L3 x1_pillier5 x2_pillier5 5.3753e-12
L4 x2_pillier5 x3_pillier5 5.3753e-12
R4 x3_pillier5 out2 0.000219

R5 in3 x1_pillier6 0.000219
L5 x1_pillier6 x2_pillier6 5.3753e-12
L6 x2_pillier6 x3_pillier6 5.3753e-12
R6 x3_pillier6 out3 0.000219

K1 L1 L3 0.22791
K2 L2 L4 0.22791
Cinter1 x2_pillier5 x2_pillier4 1.0556e-15

K3 L3 L5 0.22791
K4 L4 L6 0.22791
Cinter2 x2_pillier6 x2_pillier5 1.0556e-15
```

(Continued)

TABLE 15.1

Exportable SPICE-Compatible TXT File (*Continued*)

```
K5 L1 L5 0.11562
K6 L2 L6 0.11562
.ends pillier4_pillier5_pillier6

X1_pillier4_pillier5_pillier6 11 12 31 32 41 42 pillier4_pillier5_pillier6

.subckt gnd1_end_signal_end_gnd2_end in1 out1 in2 out2 in3 out3

R1 in1 x1_gnd1_end 0.0066343
L1 x1_gnd1_end x2_gnd1_end 6.8442e-11
L2 x2_gnd1_end x3_gnd1_end 6.8442e-11
R2 x3_gnd1_end out1 0.0066343
Cox_gnd1_end x2_gnd1_end 0 2.0741e-13

R3 in2 x1_signal_end 0.0066343
L3 x1_signal_end x2_signal_end 6.8442e-11
L4 x2_signal_end x3_signal_end 6.8442e-11
R4 x3_signal_end out2 0.0066343
Cox_signal_end x2_signal_end 0 2.0741e-13

R5 in3 x1_gnd2_end 0.0066343
L5 x1_gnd2_end x2_gnd2_end 6.8442e-11
L6 x2_gnd2_end x3_gnd2_end 6.8442e-11
R6 x3_gnd2_end out3 0.0066343
Cox_gnd2_end x2_gnd2_end 0 2.0741e-13

K1 L1 L3 0.41668
K2 L2 L4 0.41668
Cinter1 x2_signal_end x2_gnd1_end 2.4691e-15
K3 L3 L5 0.41668
K4 L4 L6 0.41668
Cinter2 x2_gnd2_end x2_signal_end 2.4692e-15

K5 L1 L5 0.28941
K6 L2 L6 0.28941

.ends gnd1_end_signal_end_gnd2_end

X1_gnd1_end_signal_end_gnd2_end    12    13    32    33    43    44
gnd1_end_signal_end_gnd2_end

.end
```

TSVs regular matrix. The matrix is modeled with W1 configuration TSVs. The pitch between each TSV is 120 μm. The signal integrity is studied at low and medium frequencies, so we make the assumption that all the TSVs oxide capacitances join at a single node representing the conductive substrate. This modeling consideration is illustrated in Figure 15.5a for a 2 × 2 TSVs matrix. All the capacitances between neighbor TSVs are taken into account with inter-pillar parasitic capacitance formula. Concerning magnetic field coupling, only the TSVs proximity effects (mutual inductive couplings) are considered in these analyses.

The Eye Diagrams are produced by assuming a symmetrical test bench in the simulation, with 50 Ω output buffer equivalent resistances connected to each TSV input. The TSVs are terminated with 4fF load capacitances. The matrix SPICE netlist is generated with our 3D-TLE extractor and exported to Cadence Virtuoso Spectre for circuit timing analyses. Depending on the studied configuration, "0" logic signals or pseudo-random bit sequences, having a magnitude of 1V and a rise time of 500 ps for data rates of 500 Mb/s, are transmitted to the TSVs.

Three configurations are investigated. For the first one (A), all the TSVs are grounded except the central TSV. The second configuration (B) is a shielded arrangement for which data is transmitted through the matrix's inner TSVs while the outer ones are grounded. In the third configuration (C), data are transmitted through all the TSVs (Figure 15.7).

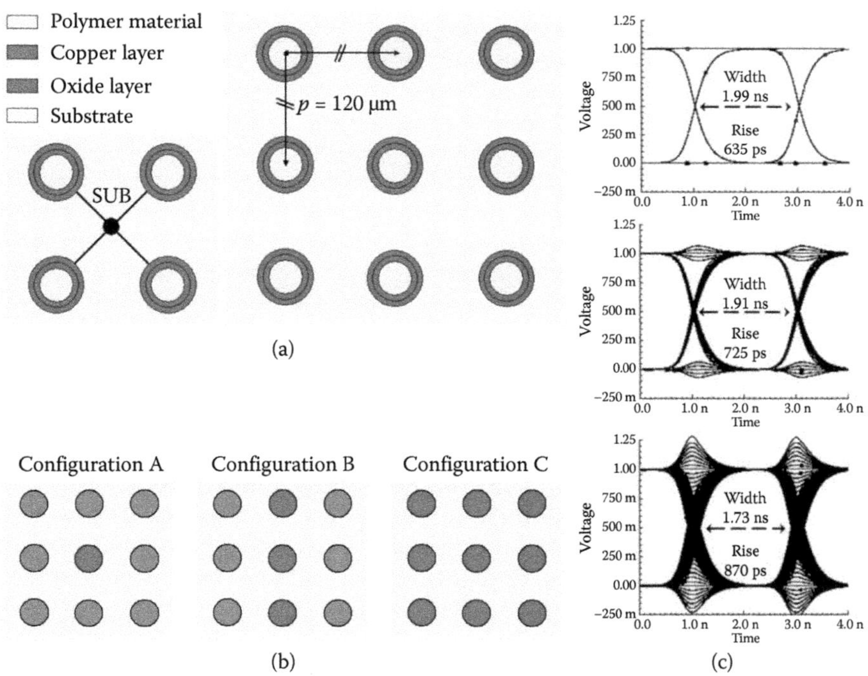

FIGURE 15.7
(a) Medium-density TSVs regular matrices. Bottom left: all the TSVs' oxide capacitances join at a single node, modeling the substrate (illustration for a 2 × 2 matrix). Right: 3 × 3 medium-density TSVs matrix considered for the timing domain analyses. (b) 3 × 3 medium-density TSV matrix configurations for generating Eye Diagrams from simulations. Color legend: gray for grounded TSV, red for signal. Configuration A: data is only transmitted through the matrix central TSV. Configuration B: shielded configuration. Configuration C: data is sent to all the TSVs. (c) Generating Eye Diagrams from simulation based on the three studied configurations. Top: A configuration. Middle: B configuration. Bottom: C configuration.

As the number of neighboring TSVs carrying data increases, the rise time and the propagation delay increase. Consequently, the eye width is shortened. In comparison to configuration A, the rise time in configuration C is 37% longer. Additionally, there appears to be a voltage overshoot linked to the crosstalk ranges from 65 mV up to 250 mV in the B and C configurations, respectively. As the noise margins decrease with technological evolution, the signal integrity in TSV matrices must be more fully quantified for cases in which signals propagate in the same direction, as in the presented analyses, or in opposite directions. The couplings between the TSVs and the substrate must also be integrated into the TSVs matrices electrical models to evaluate which couplings have the most impact on structure performances and the signal integrity.

The results show that the 3D-TLE extractor can be easily integrated with a common EDA environment by providing compact models capable of analyzing the performance of TSVs bundles for high speed data communication.

The last steps of this global work should be as follows:

1. Introducing the thermal aspect to the 3D structures, a key problem, and their electrothermal couplings. These appear, for instance, via simple formula, for example, $R = R_0(1 + \alpha_0 T)$.
2. Deal with the noise aspect of 3D. This is an old but present problem, since, for instance, the signal to noise ratio is redhibitory for nanoscale 3D structure (4 kRT, GR noise).

15.4 Heat Equation

A problem commonly encountered in the thermal analysis of electronic packages or in 3D ICs is that of thermal spreading resistance. Thermal spreading resistance occurs as heat flows by conduction between a source and a sink with different cross-sectional areas. One example is that of a multilayers chip. Another one is that of a module package with a thermal grease path between a chip and the module cap.

We are, in the case of (un)steady temperature due to one or multiple heat sources, using the principle of superposition.

A thermal spreading resistance—still in the reciprocal domain—can be extracted from discrete heat source; typical applications are cooling of device areas at the package and at die level.

We consider 3D transient heat conduction in a non-homogenous finite medium [24–26].

We can write the heat equation as follows:

$$\frac{d^2T}{dx^2}+\frac{d^2T}{dy^2}+\frac{d^2T}{dz^2}-\frac{1}{\alpha}\frac{dT}{dt}=0 \tag{15.23}$$

where $\alpha = k/(\rho c)$ is the thermal diffusivity, k is the effective isotropic thermal conductivity, and c is the isobar mass heat.

As can be seen, we get analogies with Equations 15.2 through 15.8.

In the same way that, for the Poisson's equation, we extract an associated Green Kernel, that is, if we assume $\tau = X(x,x') \cdot Y(y,y') \cdot Z(z,z')\ \tau(t,t')$, we can extend Equation 15.23, thus resulting in

$$\frac{d^2X}{dx^2}+\lambda^2X=\frac{d^2Y}{dy^2}+\mu^2Y=\frac{d^2Z}{dz^2}-\gamma^2Z=\frac{d\tau}{dt}-\alpha\left(\lambda^2+\mu^2-\gamma^2\right)\tau=0 \tag{15.24}$$

λ, μ, and γ are eigen values associated to the Green Kernel of this heat equation with

$$\lambda^2+\mu^2-\gamma^2+\frac{A^2}{\alpha}=0 \tag{15.25}$$

This last Equation in 15.25 has an evanescent solution *versus* time; for an i layer, we can write:

$$\tau_i=\tau_{i0}\exp\left(-\left(+\lambda_{ix}^2+\mu_{iy}^2-\gamma_{iz}^2\right)\cdot\alpha\cdot t\right) \tag{15.26}$$

With the reciprocal domain (so, we can take also advantage of the data compression inherent to the DCT) the Equation 15.23 becomes

$$\frac{d^2X}{dx^2}+\frac{d^2Y}{dy^2}+\frac{d^2Z}{dz^2}-\frac{1}{\alpha}\cdot\frac{d\tau}{dt}=-\lambda^2X-\mu^2\,Y+\gamma^2\,Z-\left(\frac{j}{\alpha}\right)\omega\tau=0\ \left(j^2=-1\right) \tag{15.27}$$

We can define a Fourier Coefficients Spreading (transcendental) impedance function as the following ratio:

$$R_{FCSi}=A_i/B_i \tag{15.28}$$

where A_i and B_i are the Fourier coefficients of the general *i* solution for the temperature in the direction z perpendicular to the stacked layers.

$$\mathrm{FCSi}=\mathrm{Ai}\cdot\cosh(\gamma z)+\mathrm{Bi}\cdot\sinh(\gamma z) \tag{15.29}$$

15.4.1 Boundary Conditions

The continuity of the thermal conductivity at interface layers leads to

$$k_i \frac{dT_i}{dz,i} = k_{i+1} \frac{dT_{i+1}}{dz_{i+1}} \qquad \text{(Neumann condition)} \tag{15.30}$$

We have also, at any layers interface:

$$X_i(t) = X_{i+1}(t),\ Y_i(t) = Y_{i+1}(t) \qquad \text{(Derichlet condition)} \tag{15.31}$$

$$k_i \frac{T_{i,z}}{T_{i+1,z}} = k_{i+1} \frac{R_{\mathrm{FCS}\ i,z}}{R_{\mathrm{FCS}\ i+1,\ z}} \tag{15.32}$$

cf. the analogy with the induction vector at an interface without charge:

$$\varepsilon E_i = \varepsilon_{l+1} E_{i+1} \ \text{(E: Electrical field)} \tag{15.33}$$

If Equations 15.30 and 15.31 are verified, it follows:

$$B_{iz} \cdot R_{\mathrm{FCS}\ iz} \cdot Y_{i(y)} \cdot e^{-\left(-\gamma_{iz}^2 + \mu_{iy}^2 + \lambda_{ix}^2\right)t} = -R_{\mathrm{FCS}iz} \cdot Y_{i+1(y)} \cdot e^{-\left(-\gamma_{i+1,z}^2 + \mu_{i+1,y}^2 + \lambda_{i+1,x}^2\right)} \tag{15.34}$$

$$k_{1z} \cdot R_{\mathrm{FCS}\ iz} \cdot Y_{i(y)} \cdot e^{-\left(-\gamma_{iz}^2 + \mu_{iy}^2 + \lambda_{ix}^2\right)t} = k_{1^\circ 1,z} \cdot B_{iz} \cdot R_{\mathrm{FCS}\ i+1,\ z} \cdot Y_{i+1(y)} \cdot e^{\left(-\gamma_{iz}^2 + \mu_{iy}^2 + \lambda_{ix}^2\right)t} \tag{15.35}$$

We have an analog set of Equations 15.34 and 15.35 replacing Y by X. Moreover, Equations 15.34 and 15.35 are verified if

$$\frac{B_{i,z}}{B_{i+1,z}} = \frac{R_{FCS\ i+1,z}}{R_{FCS\ i,z}} \tag{15.36}$$

$$\frac{k_{i,z}}{k_{i+1,z}} = \frac{R_{\mathrm{FCS}\ i,z}}{R_{\mathrm{FCS}\ i+1,z}} \tag{15.37}$$

Note: for different thermal diffusivities or conductivity, in each layer, but with $k_{x,i} = k_{y,i}$ (isotropy), we get

$$\gamma_{i+1z} = \left[\frac{\alpha_i}{\alpha_{i+1}} \cdot \gamma_{iz}^2 + \left(\frac{\alpha_i}{\alpha_{i+1}} - 1\right) \cdot (\lambda_{ix}^2 + \mu_{iy}^2)\right]^{\frac{1}{2}} \tag{15.38}$$

We obtain

$$A_{iz} = B_{iz} R_{\mathrm{FCS}\,iz}.(\gamma_{iz}) \tag{15.39}$$

$$R_{\mathrm{FCS}iz}(\gamma_n) = \frac{k_{i+1} R_{\mathrm{FCS}i+1,z}(\gamma_n) + k_i \tanh(\gamma_{iz} \cdot t_i)}{k_i + R_{\mathrm{FCS}i+1,z}(\gamma_n) \cdot \tanh(\gamma_{iz} \cdot t_i)} \tag{15.40}$$

Here, γ_n refers to the eigenvalue in Equation 15.25; t_i is the thickness of the ith layer.

The development above seems important because it does not have the classic form of earlier works; and we get a shape analogous to Equation 15.40, the thermal impedance (Equation 15.2.8) and the recursive electrical impedance specific to the TLM method (see Equation 15.20) also produced by Green kernels (see Equation 15.8). In addition, taking inverse Fourier transforms, we can compare with the direct calculation of time-dependent heat profile (cf. FE.M.).

Finally, in general, the real solution of Equation 15.23 takes the following expression:

$$T(x, y, x', y', z, z') = \exp\left(-\left(\lambda_{ix}^2 + \mu_{iy}^2 - \gamma_{iz}^2\right) \cdot \alpha \cdot \mathrm{t}\right)$$

$$\sum_{m=1} \sum_{n=1} \left[\frac{2\left(A_{\mathrm{N}} \mathrm{e}^{-\gamma_{mn}(\mathrm{d}+\mathrm{Z_l})} + B_{\mathrm{N}} \mathrm{e}^{\gamma_{mn}(\mathrm{d}+\mathrm{Z_l})}\right)\left(\mathrm{e}^{-\gamma_{mn}\mathrm{Z_h}} + \mathrm{e}^{\gamma \mathrm{Z_h}}\right)}{ab\Sigma_{N\gamma mn}\left(B_{\mathrm{N}} \mathrm{e}^{\gamma \mathrm{d}} - A_{\mathrm{N}} \mathrm{e}^{-\gamma \mathrm{d}}\right)} \right] \cos\left(\frac{mx}{a}\right) \cos\left(\frac{ny}{b}\right) \tag{15.41}$$

We built our simulator using the analogies between, first of all, electrical and thermal models. Particularly, if all the relevant variables calculation (voltages/temperature, current density/thermal flow, noise power/current fluctuations) made in the reciprocal domains of space and frequency are done in the main program, we do develop a single special subroutine to calculate the impedance—electrical or thermal, or transfer impedances for noise extraction—between any two points into the bulk. Typically, we take 300 points * 300 points * 12 layers.

First results are presented for temperature coming from both the FEM algorithm and from our method where the temperature source profile is applied to one or both the contacts is depicted in Figure 15.8a and the temperature map in Figure 15.8b with FEM results at left and Green's results at right.

We will go further, particularly with numerical experiences, to consolidate our algorithms. We think this modeling can be efficient for the industry concerned, for instance, with 3D electrical simulations. We can also seek hot points involving drastic transient regimes by the thermal approach. We should work on the placement of coolers via investigations on the heat spreading in three dimensions. *A minima*, the electro thermal coupling can be done by the linear simple formula as

$$R = R_0(1 + \alpha_0 T)$$

In fine, we introduce the last part of our work, designated to finalize our simulation platform: a second order electrical analysis for 3D circuits.

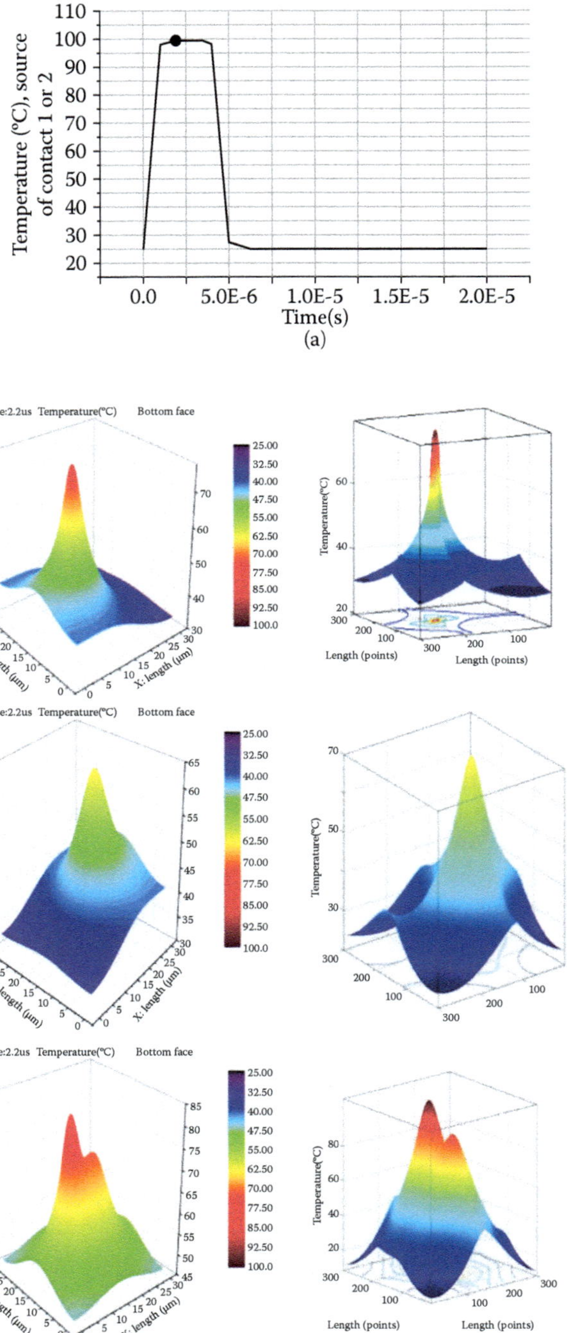

FIGURE 15.8
(a) Temperature profile and (b) temperature maps.

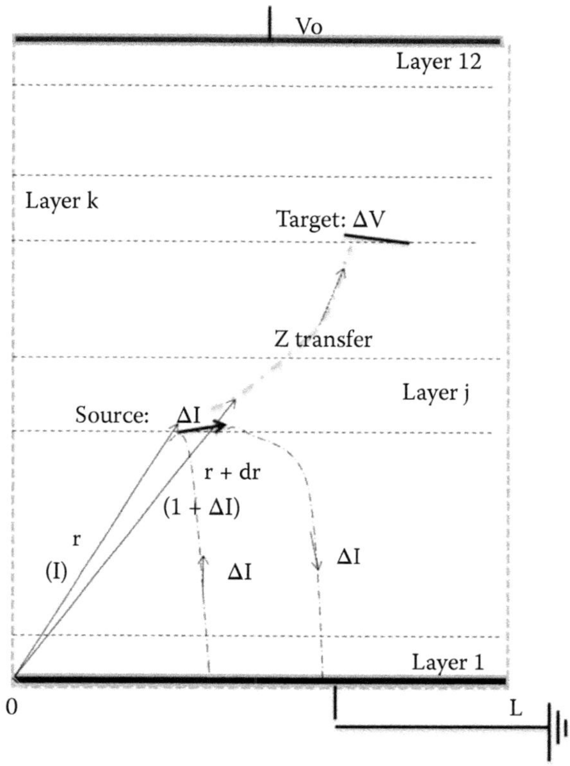

FIGURE 15.9
Schematic of the principle of the impedance field method.

15.5 3D Electrical Noise Approach

The Green method can also be dedicated to an impedance transfer method [27,28] considering signal fluctuations; injecting a small current variation at a contact creates a potential variation at any probe or contact. This way, we could analyze possible correlations between different *ab initio* independent noise sources [29], for instance considering the nano-scale.

We now consider quadratic forms. The transfer impedance method (TIM), a proper two-point internal response function, gives a linear relation between the *electric*-field response and a local current perturbation, thus, enabling one to compute the internal field noise spectra that originated from current fluctuations. This method, by generalizing the original impedance field method of Shockley et al. [27,28], has been used during the last decades for noise calculations, but in one dimension. Currently, no new insights permit the solution to this problem in 3D dimensions (even in 2D), not because of numerical difficulties, but in a sound physical point of view.

The impedance field method [27] can be stated as follows.

Each noise source at point r in the substrate produces noise in some other point (see Figure 15.9) of the bulk or the die surface (and at the electrodes).

The induced noise depends on its source(s) as the way it propagates to it(s) target(s); this propagation acts through the so-called impedance field.

In order to study the noise between two contacts embedded into the substrate, we inject a small (noise) current of complex value δI at frequency f at a point r (+/– dr) in the substrate, superimposed to the DC current between two contacts in the bulk. This small current produces a voltage V_f between the considered contacts.

Then, we can write

$$\Delta V(r+\mathrm{d}r) = Z(r+\mathrm{d}r)\delta I \tag{15.42}$$

Z(r; f) is the AC impedance.

The resulting voltage difference induced is

$$\delta V = \Delta V(r+\mathrm{d}r) - \Delta V(r) = \nabla Z \cdot \delta I \tag{15.43}$$

where the Green's function ∇Z is called the impedance field.

First of all, we calculate, as an example, the global gradient between 2 embedded contacts in the *x* and *y* direction perpendicular to the layers (see Figure 15.7). Using, at least 300 * 300 calculation points in each plane, we approximate, for the first time, the impedance derivative by its string between 2 adjacent points in *x* or *y*, since all the impedances between any two points are stocked *via* a subroutine at each frequency. The procedure is straightforward; we can do the same thing in the *z* direction, enhancing the layer number.

δIdr can be described by the current density $\delta j(r)$ in an infinitesimal volume $\mathrm{d}r^3$:

$$\delta \mathrm{I} \cdot \mathrm{d}r = \delta j(r) \cdot \mathrm{d}^3 r;\ (\delta j \text{ is the current density} - \text{unit : Am}^{-2}-) \tag{15.44}$$

Using Equation 15.31 in Equations 15.43 and 15.44 leads to

$$\delta V_f(r) = \big(Z(r,f) - Z(r-\mathrm{d}r, f)\big)\delta I = \nabla Z(r,f)\mathrm{d}r\delta I \tag{15.45}$$

$$\Delta \mathrm{V}_f = \int \nabla \mathrm{Z}(r,f) \cdot \delta j(r) \cdot \mathrm{d}r^3 \tag{15.46}$$

Equation 15.46 means that a current fluctuation in a volume $\mathrm{d}r^3$ produces a potential difference between the two contacts. Integrating over the whole domain considered, we can get the total voltage produced by all these currents.

Wanting to calculate explicitly the S_V (f) noise density, we separate the components along every direction *x*, *y*, and *z* (indexed by α). Equation 15.46 thus can be rewritten as

$$\Delta V_f = \sum_{\alpha} \left[\int \frac{\partial \mathbf{Z}(\mathbf{r},\mathbf{f})}{\partial \alpha} \,.\, \delta j_{\alpha}(\mathbf{r}) \cdot \mathbf{dr}^3 \right] \tag{15.47}$$

The spectral density of the voltage noise at the terminals then becomes

$$S_v(\mathrm{f}) = 2 \int_{-\infty}^{\infty} \overline{\delta V_f(t) \cdot \delta V_f^*(t+\tau)}\, e^{2\pi i f t} dt \tag{15.48}$$

$$S_v(f) = \sum_{\alpha} \sum_{\beta} \iint \frac{\partial Z(r,f)}{\partial \alpha} \cdot \frac{\partial Z^*(r',f)}{\partial \beta} .2. \int_{-\infty}^{\infty} \overline{\delta j_{\alpha}(r,t) . \delta j_{\beta}^*(r',t+\tau)}\; e^{2\pi f i \tau} d\tau . dr^3 \, dr'^3 \tag{15.49}$$

$$S_v(f) = \sum_{\alpha} \sum_{\beta} \iint \frac{\partial Z(r,f)}{\partial \alpha} . \frac{\partial Z^*(r',f)}{\partial \beta} .S_{j\alpha\beta}(r,r',f) dr^3 \, dr'^3 \tag{15.50}$$

$$S_v(f) = \iint \vec{\nabla} Z(r,f) \cdot Sj(r,r',f) . \vec{\nabla} Z^* \; (r',f) dr^3 \, dr'^3 \tag{15.51}$$

Supposing the current noise sources uncorrelated, we can write

$$S_{j\alpha\beta}(r,r',f) = K_{\alpha\beta}(r,f) \, .\delta(r'-r) \tag{15.52}$$

Equation 15.51 is then simplified to

$$S_v(f) = \int \vec{\nabla} Z(r,f) \cdot K(r,f) \cdot \vec{\nabla} Z^*(r,f) dr^3 \tag{15.53}$$

(cf. Salami method: stacked sausage wafers. The noise of each layer is independent of the noise of all the other layers; then, the total noise is the simple sum of the noises of each of these layers).

Figure 15.10 presents noise spectral densities relative to the band pass, in different directions.

Since the direct method gives the impedance field with respect to every equation in the electrical model, new physical noise models depending on all equations could be invoked. For example, current noise induced by temperature fluctuations in the bulk could be investigated. If temperature fluctuations occur at low frequencies, then a model for flicker noise could be developed, since this noise also contains 1/f noise.

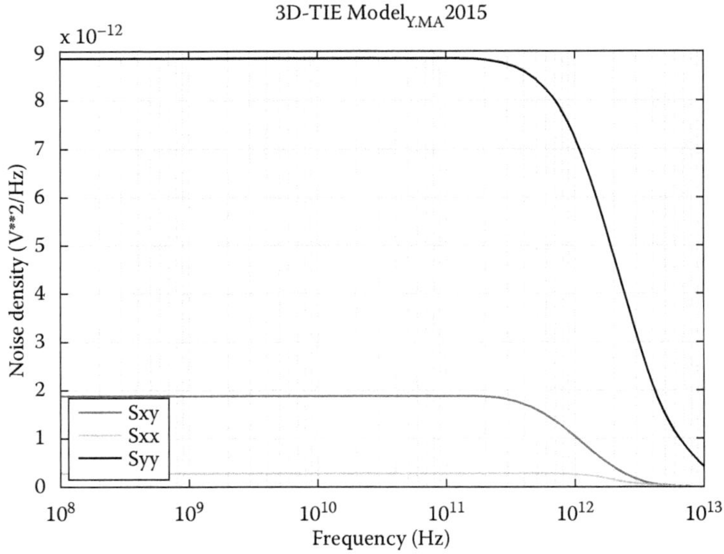

FIGURE 15.10
Noise spectral densities.

For the analysis of submicron distances, new effects can arise. For example, non-stationary effects come into account. When decreasing the length to a few mean free paths, each carrier undergoes very few collisions on its way from one terminal to another one.

Another paradigm is the space correlation of noise sources. It is shown [29] that, for small distances; the noise sources, although independent, can be spatially correlated. Moreover, noise intensities can rise during transients of the applied voltage to some sensitive areas of the circuit. Further practical investigations have to be done to implement these simulations into our simulator.

15.6 Digital Perturbations

More and more system-on-chip designs requires the integration of both very fast analog circuits and large digital blocks. Up to now, the perturbations introduced by the digital part against the analog one are the key problem and hard one for designers to handle, especially in the radio frequency (RF) range [30,31]. Substrate noise involved in the digital part will degrade the performance of analog circuits. The in-depth problem is to construct efficient models for the digital part's activity, the substrate, and their coupling in a package that could be embedded in some unique computer aided design (CAD) tool [32–34]. The well-known tool SPICE is able to model such phenomena for substrate or digital signals, but, for a complex technology with very large digital blocs, it seems too optimistic to expect SPICE to suffice. Some significant attempts [35] in modeling noise generation injected in the bulk on a "SPICE" and a process/device point of view have been proposed; but it seems accurate for relatively small circuits and simple parasitic waveforms, travelling through the substrate.

The idea described in this chapter has been proposed in [36–41]; in this work, we present its development, work out details, and finally present, for the moment, simple examples for the practical application of the methodology we did implemented.

15.6.1 Methodology

Our subjects of interest are the digital transmission/propagation systems.

For a general approach, consider M specific noise (parasitic) sources, $x_1(t)$, $x_2(t)$,.., $x_M(t)$, time-continuous signals, modulated by an encoded message c[n]. It can be written in the general form:

$$x(t) = \sum_{n=-\infty}^{\infty} g^{c[n]}(t - nT) \tag{15.54}$$

This equation means that the "symbol" c[*n*], the switching event, emitted at the instant *nT*, assigns the waveform $g^{c[n]}(t-nT)$, where *T* is the period of the digital clock.

Let s_n be a stationary Markov chain with state set $S = \{\sigma_1, \sigma_2, ..., \sigma_I\}$ and $c[n] = f(s_n)$ be a vector-valued stochastic process obtained as a memoryless function of s_n.

We assume that the Markov chain s_n is ergodic (s_n has only one class of communicating states. No degeneracy is involved in this process; a temporal average is equal to an ensemble one).

$g(t)$ is a $M \times I$ matrix, which takes the general form:

$$g(t) = \begin{bmatrix} g_1(t,\sigma_1) & g_1(t,\sigma_2) & \cdot\ \cdot & g_1(t,\sigma_I) \\ \cdot & \cdot & \cdot\ \cdot & \cdot \\ \cdot & \cdot & \cdot\ \cdot & \cdot \\ \cdot & \cdot & \cdot\ \cdot & \cdot \\ g_M(t,\sigma_1) & g_1(t,\sigma_2) & \cdot\ \cdot & g_M(t,\sigma_I) \end{bmatrix} \tag{15.55}$$

Then $c[n]$ can be written $c[n] = S_n \cdot C$, where C is a $M \times I$ matrix and where $S_n = [S_n(1), S_n(2),..., S_n(I)]$, defined by $S_n(i) = 1$ if $S_n(i) = \sigma_i$, and $S_n(i) = 0$, otherwise.

The *j*th column of *C* can be written:

$$\left[f_j(\sigma_1), f_j(\sigma_2),..., f_j(\sigma_I) \right]' \tag{15.56}$$

and the *i*th "row" of C is the value of c[*n*], when $s_n = \sigma_i$

For our demonstration's starting case for a digital transition (I = 4) and for a single noise source, ($M = 1$), we can use the simple notation, g_{ij}, where the ij indexes are the primary states, which "means" *from the i state to the j state*, where *i* and *j* take the value 0 (Low state) or 1 (High state). The digital transition signal is defined as a function $f : S \rightarrow C$ where $C \equiv \{00, 01, 11, 10\}^T$ (or$\{LL, LH, HH, HL\}^T$). Then, from s_n we could set up a Transition Probability Matrix *P* of general term *P*(*i*,*j*) (=$Pr\{s_{n+1} = s_j / s_n = s_i\}$).

Then each noise source, if necessary, will be assigned the relevant noise vector, here the simple one <$g_{00}(t), g_{01}(t), g_{10}(t), g_{11}(t)$>, at t time.

The "superposition" of all associated parasitic waveforms (which can be stored in some library), *g*, coming from the commutating gate will give the final modulated signal. Moreover, c[n] shall be modelled as a function of a Markov chain.

The very aim of our work is to compute, at any sensitive circuit node of interest, the spectral power density, called S_y(f), which will be represented as a matrix; in fact, all the development of the algorithms are based on the matricial formalism, which appears, here, concise, and very general.

Dealing with the interference noise, means to develop a second order analysis of the process, assuming it is a two order stationary one. In practice, we ought to calculate the correlation matrix function R_x (k) (associated to the *x* noise source) of the vector *g*(*t*;*c*), that is,

$$R_x[k] = E\left\{ g(t; c[n])^* \cdot g(t, c[n+k]) \right\} \tag{15.57}$$

$$(^* : \text{transpose conjugate, for the general case})$$

Let us consider a single cell (each cell will be further considered as an ideal current source or sink, which can be injected or absorbed—in power, ground, substrate, package, or board—which will be supposed as having a linear electric behavior).

The digital signal *s* state can be defined as follows (Figure 15.11):

– L → H (or: 0 → 1): with a probability = α

– H → L (or: 1 → 0): with a probability = β,

(L and H mean respectively high and low.)

Then, Figure 15.12 represents the state diagram of such a "machine."

Let us suppose the "input" *i* is, itself, a Markov Chain with a transfer probability matrix:

$$\pi = \begin{bmatrix} 1-\alpha & \alpha \\ \beta & 1-\beta \end{bmatrix} = \left(\begin{bmatrix} \pi_{00} & \pi_{01} \\ \pi_{10} & \pi_{11} \end{bmatrix} \right) \tag{15.58}$$

with: $\sum_j \pi_{ij} = 1$, i (= 0 or 1): row number; j (= 0 or 1): column number.

$$\pi_{01} = Pr\ \{j = 1,\ \text{if}\ i = 0\} = \alpha \text{ and } \pi_{10} = Pr\ \{j = 0, \text{if}\ i = 1\} = \beta$$

Then: $\pi_{00} = Pr\ \{j = 0, \text{ if } i = 0\} = 1-\alpha$, and $\pi_{11} = Pr\ \{j = 1, \text{ if } i = 1\} = 1-\beta$.

A simple digital signal that may switch from *low* to *high* with probability α, and *low* to *high* with probability β at every clock tick, is modeled in four states (see Figure 15.13). It has a state Transition Probability Matrix (TPM) representing the switching activity, expressed as an $I \times I$ Matrix P:

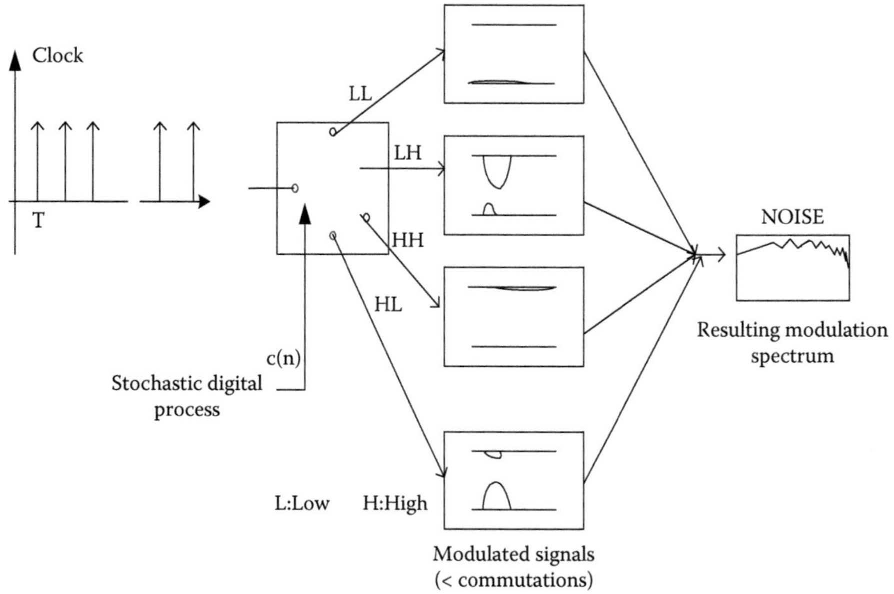

FIGURE 15.11
Block diagram of the parasitic signal propagation chain.

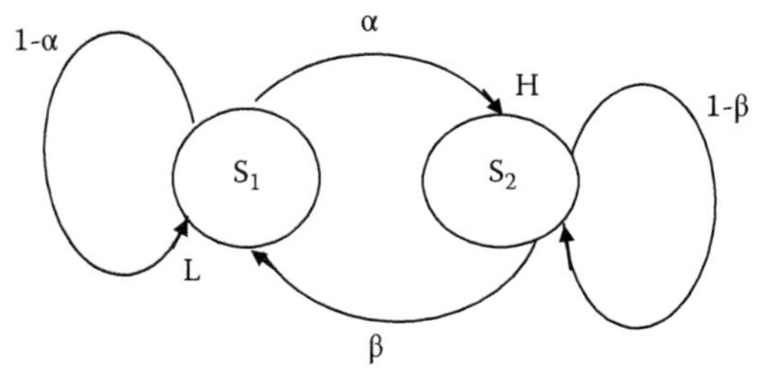

FIGURE 15.12
Binary commutating gate.

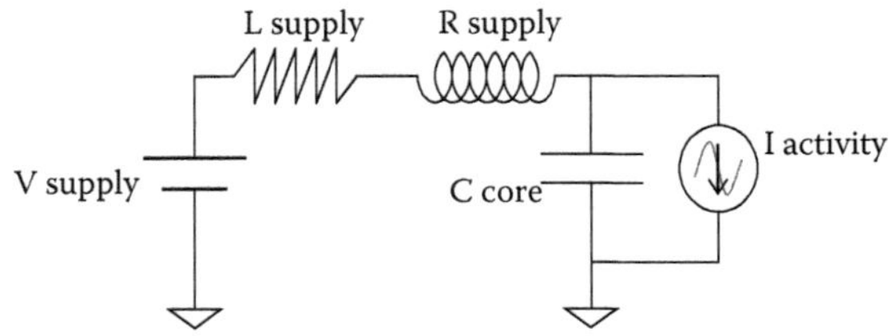

FIGURE 15.13
An RLC-like model for a supply network.

$$P = \begin{bmatrix} 1-\alpha & \alpha & 0 & 0 \\ 0 & 0 & \beta & 1-\beta \\ 1-\alpha & \alpha & 0 & 0 \\ 0 & 0 & \beta & 1-\beta \end{bmatrix} \tag{15.59}$$

Our application can draw on the communication theory [38–41] (cf. deterministic finite state sequential machine). It can be investigated as spectral properties of encoded messages. It is a direct application of the communication theory field, although it concerns, here, parasitic signal propagation and interference.

The general term of P can be understood as:

$P_{ij/kl}$ = Pr {state transition, from i to j, at n+1 time/if transition k to l at n}

Note that although the matrix P is reducible (some states do not communicate with others), it does not involve any drawback for the development of the rhythm.

A cumbersome, but not so difficult way to resolve this problem is depicted as follows.

As we said, the aim is to obtain the spectral density generated by the digital block switches; that means, to find the Fourier transform of the corresponding autocorrelation function (*cf.* second order moment).

This spectral density $w(f)$ can be split into two terms:

$w(f) = w_d(f) + w_{cont}(f)$, where f is the frequency.

The discrete (W_d) component exhibits spectral lines or jumps at the frequency of the clock and its multiples ($f = n/T$), while the continuous part (w_{cont}) can be derived by setting $z = \exp(j{*}2\pi f)$ into the z-transform of its correlation function:

$$w(z) = \sum_{k=-\infty}^{\infty} R_{cont}[k] z^{-k} \quad (R_{cont} \text{ is explicated hereafter}) \tag{15.60}$$

Moreover, the discrete random process S can be characterized by an auto-correlation matrix: $R(k) \equiv E\{S_n^T(j) S_{n+k}(i)\} = Pr\{s_{n+k} = \sigma_j, s_i = \sigma_n\}$ (T: transpose).

By using the matrix algebra tool, the whole correlation matrix (of the source codes $c[n]$) can be derived as

$R_c(k) = C^T R(k) C$,

where R(k) is the joint probability matrix between the random vectors S_n and S_{n+k}, with the "code words" $c[n] = S_n C$, where C is a matrix whom each column assigns a specific parasitic waveform. In a general point of view, a source code b[n] is encoded into a sequence of code words c[n]; the "message" c[n] is generally a non-linear transformation, namely $c[n] = f(b[n])$ and can take the general form (input-output relation): $c[n] = g_0.b_n + g_1.\, b_{n-1} + \cdots + g_N.\, b_{n-N}$; with $b[n] = (b_n, b_{n-1}, \ldots, b_{n-N})$.

Note, in our simple starting example, $N = 4$, with (see above)

$$g_0 \equiv g_{00},\ g_1 \equiv g_{01},\ g_2 \equiv g_{10},\ g_3 \equiv g_{11}$$

Finally, we make explicit the whole correlation matrix as follows:

$$R[k] = \begin{cases} \Delta P^k & k \geq 0 \\ \left(P^T\right)^{-k} \Delta & k < 0 \end{cases} \tag{15.61}$$

P^k represents the k-step TPM, (note that, for k=0, R(0) = Δ, and $E\{S_n(j) S_n(i)\} = p_i.\delta_{ij}$, where δ is the Kronecker symbol) with Δ = diag{p(1), p(2), …,p(I)}, which can be noted as

$$\Delta = \left[p_j^{\infty} \delta_{ij} \right] \tag{15.62}$$

The elements of this diagonal matrix are the stationary (cf. the infinite symbol) vector components, which are derived from the equation:

$$p = pP \ \left(\text{with } p = [p_1, p_2, p_3, p_4]\right) \tag{15.63}$$

R[k] can be as well decomposed in two "orthogonal" parts (cf. Vold's theorem [36]), a continuous and a discrete one:

$R[k] = R_{\text{cont}}[k] + R_{\text{d}}[k]$, where:

$$R_{\text{cont}}[k] = \begin{cases} \Delta H^k & k > 0 \\ \left(H^T\right)^{-k} \Delta & k < 0 \\ \Delta - \sum_{r=0}^{h-1} K_r & k = 0 \end{cases} \tag{15.64}$$

and a periodic component (cf. pure tones) with period h, given by

$$R_d[k] = \sum_{r=0}^{h-1} K_r \cdot \exp(j2\pi rk/h) \tag{15.65}$$

which is cyclic when $h > 1$.

We treat the case of arbitrary h, but, without loss of generality, we do not represent explicitly the global matrix with $h>1$, which is, itself, constructed via a cyclic form of the master matrix P. We can put in evidence the discrete and continuous part of these matrices, by transforming them by suitable permutations of row (or/and columns), so that only zeros appear in the main diagonal.

Here K_r (Equation 15.11) is the amplitude of the jumps (in our example, induced by a clock) and we have

$$K_r \equiv J_r{}^* p^T p J_r \tag{15.66}$$

(The asterisk denotes the transpose conjugate of the matrix, the hermitian one, for the general case.)

Handling the matrix tool, K_r is directly related to the eigenvalues of modulus equal to one of the TPM (p, the stationary probability vector of the TPM, is an eigen vector of P—cf. Equation 15.10.)

In our problem, the solution of Equation 15.10 gives the vector $[\alpha^*(1-\beta)/(\alpha+\beta), (\alpha^*\beta)/(\alpha+\beta), (\beta^*\alpha)/(\alpha+\beta), \beta^*(1-\alpha)/(\alpha+\beta)]$ as solution.

The matrix H is then defined as

$$\text{H} \equiv P - \sum_{r=0}^{h-1} \lambda_r J_r^* \; wpJ_r \tag{15.67}$$

where w = [111…11]′: column vector.

That implies H = P—P∞ (P∞ is the limit of P_k as k → ∞), as H has all its eigen values modulus <1 (by construction: P is split into two parts, H being the continuous one); therefore, $R_{cont}(k)$ is absolutely summable.

Here J_r has the general form:

$$J_r = \begin{bmatrix} \lambda_r^{-1} I_1 & 0 & -- & 0 \\ 0 & \lambda_r^{-2} I_1 & -- & 0 \\ ----- & ----- & --- & --- \\ 0 & 0 & -- & \lambda_r^{-h} I_h \end{bmatrix} \tag{15.68}$$

$\lambda_r = \exp(j2\ pr/h)$, with $r = 0, 1, h-1$, has a modulus equal to one.

In our example, there is only an eigen value of modulus one: λ = 1 (the others are 0—of order 2-, and 2-α−β); in that case:

$$R_d[k] = K_r \tag{15.69}$$

We substitute Equations 15.11 into 15.7 and taking into account that the limit of the matrix sum,

$M_i = \sum_{k=1}^{i} H^k z^{-k}$ converges to the matrix $M_\infty = zI - H$ (this result is easy if we use the so-called matrix formula: $\sum_{k=0}^{\infty} A^k = (I-A)^{-1}$, for any matrix A, provided all its eigen values have a modulus less than the unity).

The eigen values λ_h of H are the value null, or non-null with a modulus lesser than 1 (λ_h = 0, 0, 2-α-β).

After some cumbersome, but relatively easy calculations, the continuous part of w(z) (cf. Equation 15.7) can be written

$$w^{(z)} = (z^{-1}I - P^T)^{-1} C (zI - P)^{-1}, \text{ with } C = \Delta - P^T \Delta P \tag{15.70}$$

where $z = \exp(j * 2\pi f)$ and I is the identity matrix.

As we have been saying above, the spectral density can be considered as the sum of two parts: the continuous part and the digital part.

The spectral density of the modulated signal $x(t)$ can be formally written as

$$S_x(f) = S_x^{\text{cont}}(f) + \sum_k \delta(f - f_k^d) \cdot S_{x,k}^d \tag{15.71}$$

The continuous spectral density of the modulated signal $x(t)$ is given by

$$S_x^{\text{cont}}(f) = \frac{1}{T} \cdot G^*(f) \cdot W_x(e^{j2\pi fT}) \cdot G^T(f) \tag{15.72}$$

where $G(f)$ is the Fourier transform of $g(t)$ and W_x (cf. Equations 15.18 and 15.19) is the Fourier transform de $R_x(\tau)$, with:

$$R_x(\tau) = \frac{1}{T} \cdot \sum_{k=-\infty}^{+\infty} \int_{-\infty}^{+\infty} g^*(t) R(k) g(t + \tau - kT) \mathrm{d}t \tag{15.73}$$

Indeed, note that $R_x(\tau)$ depends on the statistics of the Markov chain s_n, depending on $R(k)$.

On the other hand, the noise signal $x(t)$ exhibits spectral lines (jumps) at frequency multiples of $f_0 = 1/T$, of amplitude

$$w_x^d(r) = C^* K_r C, \qquad r = 0, 1, \ldots, h-1 \tag{15.74}$$

Then $w_x^d(r)$, with K_r defined by Equation 15.13 becomes

$$w_x^d(r) = \left| \frac{1}{T} p J_r G_r(f_0) \right|^2 \tag{15.75}$$

with G as the Fourier transform of the matrix g(t).

The TPM P has only one eigenvalue, λ_0, equal to 1; all eigeinvalues of modulus 1 result in Dirac distributions. In our case, only λ_0 results in a train of Dirac distributions at the frequency 1/T and its harmonics.

Our problem is analogous to some modulation in digital communication. Then $g(t{:}\ c[n])$ can be straightforwardly written $c[n] \cdot f(t)$, $f(t)$ being the parasitic basic waveform vector. Finally, $g(t) \equiv f(t) \cdot C$.

Therefore the continuous part of the spectrum becomes

$$w_x^{\text{cont}}(f) = \frac{1}{T} \cdot |F(f)|^2 \cdot Wc(e^{j2\pi fT}), \text{ with } w_C(e^{j2\pi fT}) = C^* w(e^{j2\pi fT}) C \tag{15.76}$$

And the discrete one,

$$S_x(f) = \frac{1}{T^2} \cdot |F(f)|^2 \, w_c^d \tag{15.77}$$

If we now actually study the propagation of the noise, it is necessary to take into account the transfer function of the "propagation channel," said $T(f)$ (assuming linearity of the network).

We get the spectrum noise density, taking into account the network parasitic function transfer, T. The final power spectrum will be

$$S_y(f) = T(f) \cdot S_x(f) \cdot T^*(f) \tag{15.78}$$

where $S_x(f)$ is the Power spectral density of the noise sources obtained using modulation theory and $T(j2\pi f)$ is the (generally) multi input (noise sources), multi output linear (number of sensitive points) system.

15.6.2 Numerical Experiences and Discussion

As a check to illustrate the algorithm explicated above, we study the case of signals injected into the analog power grid by one switching digital cell.

Power (and ground) lines have to be efficiently modeled to predict voltage drops. Thanks to the ICEM (Integrated Circuit Emission Model) model [42,43] predicting conducted-mode emission levels on a chip and in the application board is possible by means of SPICE-based analog simulations. Each part is modeled by an RLC network (see the schematic on Figure 15.13). We can very easily derive the following relation:

$$V_{\text{core}} = V_{\text{supply}} - \frac{L_{\text{sup}} \cdot j\omega + R_{\text{sup}}}{R_{\text{sup}} C_{\text{core}} \cdot j\omega - L_{\text{sup}} C_{\text{core}} \omega^2 + 1} \cdot I_{\text{act}}(j\omega) \tag{15.79}$$

(or, in other words, the transfer function T: $T^{-1} = jC\omega + (R + jL\omega)^{-1}$).

The values used for the parasitic components that model the power grid are based on measurements and extractions. Typically, we put R = 0.5 Ω, L = 0.2 nH, C = 0.2 nF.

The resonant frequency is ten $f_0 = 8\ 10^{18}$ Hz.

The core is represented by its transient consumption of current and by its capacitance. This dynamical consumption is the sum of all currents "demanded" by the logical gates of the digital circuit during its operating. The core capacitance is also the sum of all capacitances of digital gates and interconnects. For a single gate, for instance, its capacitance is made of its drain-substrate capacitance and its output capacitance, itself made of interconnects and inputs of other gates. In this approach, it means that we consider the contribution to the supply lines voltage drop of each gate will be the same; in other words, the power bounces due to PCB, packaging, bonding wire, and other parasitic elements are more significant than the ones due to the power grid distribution. The power grid is considered as a unique voltage source. The power fluctuations do not depend on the topography of the design. In our case, and for the future developments, for a digital block, the switching activity current will be applied to one point: the middle point of this block.

As an example to validate the methodology and provide insight into some aspect of the problem, we consider a gate with only the state transition LH and HL (since the LL and HH state transitions involve a very low parasitic signal) the LH and HL parasitic signal shape will be assumed to be gaussian versus the time. (This is not so far from the reality and considering moments greater than 2 for the signal shape is not useful, for the moment, to validate our algorithm; what is more, a parasitic signal library is the best solution for an industrial circuit simulation.) We extracted the parameter mean values, hereafter, from a 90 nm CMOS process (see the beginning of this section):

LH: delay: 20 ns; standard deviation: 0.001 ns. $V_{peak} = 0.5V$
HL: delay: 60 ns; standard deviation: 0.01 ns. $V_{peak} = -0.15V$

Figure 15.14 shows the first simulations coming from the Markovian stochastic algorithm. It depicts the total power spectrum—continuous plus discrete parts—of the parasitic waveforms injected into the V_{dd} line. A periodic

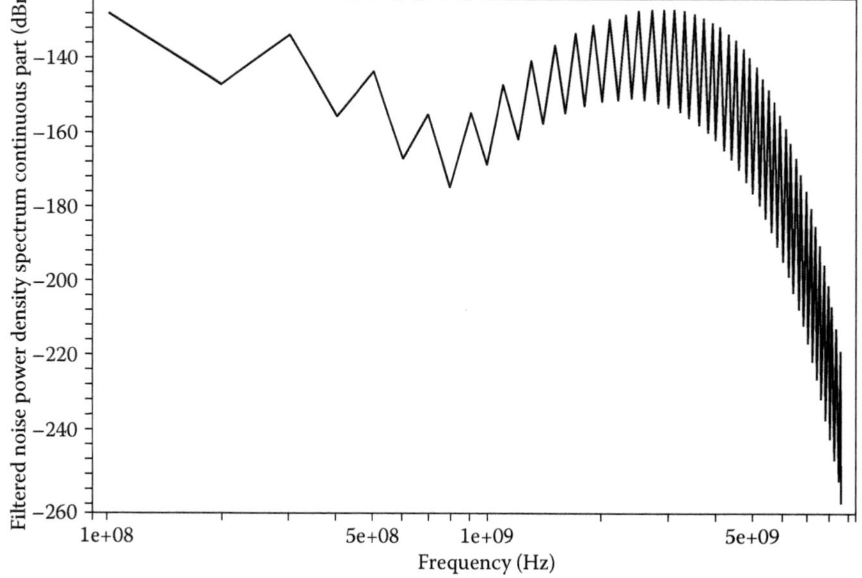

FIGURE 15.14
A typical calculated global noise spectrum of a gate commutating through a supply network.

behavior in this spectrum can be seen corresponding to the clock frequency. It indicates a global result on what can be produced by a commutating gate through a RLC-like network.

15.7 Conclusion

We propose some compact and physical modeling approach of the entanglement of 3D substrate and its interconnects. Concerning the interconnect part: it is applied essentially to redistribution lines and medium-density TSVs. It is based on parametrical extractions performed on realistic test structures and analytical formulas. Moreover, it includes the system global electrical context with the current paths and the modeling of proximity and/or substrate effects. Complete equivalent electrical models are illustrated for TSV chains structures. The modeling approach is validated through frequency analyses by comparing S-parameters measured on the test structures with the ones obtained by correspondent simulations, finally getting some equivalent electrical models. However, it implies the modeling of all the substrate coupled effects existing between the elements sharing this substrate.

Then, a substrate coupling simulation method suitable for execution in a conventional CAD environment is proposed. Moreover, to apply this approach to a realistic mixed signal design, several challenges have to be grasped. A key one concerns the way to obtain, for instance, the transient consumption current of an entire large digital part. Some CAD tools offer the possibility to simulate these currents.

Our models are tested for contact pads, coplanar waveguides, and TSV structures. Such simple approach presents the main advantage to be easily automatized; additionally, it can be improved by including skin effect and eddy current. On-going works will consist of extending the modeling approach to other 3D interconnect types, such as copper pillars, and to different types of 3D interconnect matrices.

On a deeper physico-mathematical point of view, numerical electromagnetic simulations—F.E.M.—and Green Kernels/TLM electro-thermal models of multilayer substrates are developed in 3D.

Another novelty is the ability to calculate transfer impedance, effectively considering the noise sources in two or three dimensions. The unitary concept of spreading impedance embraces the three fields in the frequency domain: electrical, thermal, and electro-thermal noise. In the near future, we should consider the study of 3D noise couplings and the possible correlations between noise sources, especially at the nano-scale. The main aim of the work, from a synthesis point of view, will be to finalize, through a simulation platform, the construction of some electro-thermal parameter sets of the first and second order for compact modeling of 3D RF circuits, an attractive multi-domain field for the industry.

Acknowledgment

This work is supported by INFIERI (Intelligent Fast Interconnected and Efficient Devices For Frontier Exploitation In Research And Industry) European Program and by UPM (Union Pour la Mediterranée.) Yue Ma thanks the China Scholarship Council (CSC).

References

1. X. Aragonès, J. L. Gonzalez, F. Moll, and A. Rubio, Noise generation and coupling mechanisms in deep-sub micron IC's, *IEEE Design & Test of Computers*, vol. 19, 27–35, September–October 2002.
2. D. K. Su, M. J. Loinaz, S. Masui, and B. A. Wooley, Experimental results and modeling techniques for substrate noise in mixed-signal integrated circuits, *IEEE Journal of Solid-State Circuits*, vol. 28, no. 4, 420–430, April 1993.
3. M. Xu, D. K. Su, D. K. Shaeffer, T. H. Lee, and B. A. Wooley, Measuring and modeling the effects of substrate noise on the LNA of a CMOS GPS receiver, *IEEE Journal of Solid-State Circuits*, vol. 36, no. 3, 473–485, March 2001.
4. M. van Heijningen, M. Baradoglu, S. Donnay, G. G. E Gielen, and H. J. De Man, Substrate noise generation in complex digital systems: Efficient modeling and simulation methodology and experimental verification, *IEEE Journal of Solid-State Circuits*, vol. 35, no. 7, 1065–1072, July 2000.
5. J. Briaire and K. S. Krisch, Principles of substrate crosstalk generation in CMOS circuits, *IEEE Transactions on Computer-Aided Design of Integrated Circuits and Systems*, vol.19, no.6, 645–653, June 2000.
6. M. Felder and J. Ganger, Analysis of ground-bounce induced substrate noise coupling in a low resistive bulk epitaxial process: Design strategies to minimize noise effects on a mixed-signal chip, *IEEE Transactions on Circuits and Systems*, vol. 46, no. 11, 1427–1436, November 1999.
7. J. Kanapka, J. Phillips, and J. White, "Fast Method for Extraction and Sparsification of Substrate Coupling," in *Proceedings of the 37th Design Automation Conference*, Los Angeles, CA, pp. 738–743, June 2000.
8. FastHenry and FastCap. Available at: http://www.fastfieldsolvers.com.
9. HFSS by Ansoft Corporation
10. SIMPLEX *Substrate Storm*.*

11. CadMOS *Seismic.**
12. M. N. Ali, R. Gharpurey, and R. G. Meyer, Numerically stable green function for modeling and analysis of substrate coupling in integrated circuits, *IEEE Transactions on Computer-Aided Design of Integrated Circuits and Systems*, vol. 17, no. 4, 305–315, April 1998.
13. C. Christopoulos, *The Transmission-Line Modeling Method: TLM*, New York, NY: IEEE Press, 1995.
14. P. S. Crovetti and F. L. Fiori, Efficient BEM-based substrate network extraction in silicon SoCs, *Microelectronics Journal*, vol. 39, 1774–1784, 2008.
15. Matlab, Available at: http://mathworks.com.
16. COMSOL, Available at: http://www.comsol.com.
17. Y. Liang and Y. Li, Closed-form expressions for the resistance and the inductance of different profiles of through-silicon vias, *IEEE Electron Device Letters*, vol. 32, 393–395, 2011.
18. Y. Eo and W. R. Eisenstadt, High-speed VLSI interconnect modeling based on S-parameter measurements, *IEEE Transactions on Components, Hybrids, and Manufacturing Technology*, vol. 16, no. 5, 555–562, August 1993.
19. A. Weisshaar, H. Lan, and A. Luoh, "Accurate closed-form expressions for the frequency-dependent line parameters of on-chip interconnects on lossy silicon Substrate, *IEEE Transactions on Advanced Packaging*, vol. 25, no. 2, 288–296, May 2002.
20. C. Xu, V. Kourkoulos, R. Suaya, and K. Banerjee, A fully analytical model for the series impedance of through-silicon vias with consideration of substrate effects and coupling with horizontal interconnects, *IEEE Transactions on Electron Devices*, vol. 58, no. 10, 3529–3540, October 2011.
21. D. H. Kim, S. Mukhopadhyay, and S. K. Lim, Fast and accurate analytical modeling of through-silicon-via capacitive coupling, *IEEE Transactions on Components, Packaging and Manufacturing Technology*, vol. 1, 168–180, 2011.
22. Advanced Design System - ADS, Available at: http://www.home.agilent.com.
23. Cadence Virtuoso Spectre Circuit Simulator, Available at: http://www.cadence.com.
24. Y. Tamene, C. Bougriou, and R. Bessaïh, Thermal behaviour of a multilayer media in transient regime, *Revue des Energies Renouvelables*, vol. 10, 397–405, 2007.
25. M. Ferraiudo and O. Manca, Heat transfer in a multi-layered thermal protection system under aerodynamic heating, *International Journal of Thermal Sciences*, vol. 53, 56–70, March 2012.
26. Y. Ma, L. Fakri Bouchet, F. Calmon, and C. Gontrand, Electro-thermal modelling for three-dimensional nanoscale circuit substrates, *IEEE Transactions on Components, Packaging and Manufacturing*, vol. 6, no. 7, 1040–1050, 2016.
27. Y. Ma, F. Calmon, J. Verdier, and C. Gontrand, A possible way to grasp electronic noise via some transfer impedance methods in three dimensional nanoscale circuit substrates, *Global Journal of Advanced Engineering Technologies*, Vol. 4, no. 4, 1051–1068, 2016.
28. K. M. Van Vliet, A. Friedmann, R. J. J. Zulstra, A. Gisolf, and A. Van Der Ziel, Noise in single injection diodes. I. A survey of methods, *Journal of Applied Physics*, vol. 46, no. 4, 1804, 1975.
29. J. P. Nougier, J. C. Vaissiere, and C. Gontrand, Two-point correlations of diffusion noise sources of hot carriers in semiconductors, *Physical Review Letters*, vol. 51, no. 6, 513–516, August 1983.

30. N. K. Verghese, D. J. Allstost, and M. A. Wolfe, Verification techniques for substrate coupling and their application to mixed-signal IC design, *IEEE Journal of Solid-State Circuits*, vol. 31, no. 3, 354, March 1996.
31. T.-J. Brazil, The modelling an simulation of high-frequency electronic circuits, *International Journal of Circuit Theory and Applications*, vol. 35, 533–546, 2007.
32. S.-J. Li, J.-F. Dai, C.-C. Chang, C.-H. Huang, and Y.-T. Tsai, Development of 3-D equivalent-circuit modelling with decoupled L-ILU factorization in semiconductor-device simulation, *International Journal of Numerical Modelling: Electronic Networks, Devices and Fields*, vol. 30, 133–148, September 1, 2005. Available at: www.interscience.wiley.com.
33. T. Sato, H. Orihara, S. Fujimori, M. Tosaka, and A. Inoue, LSI and PCB unified noise analysis CAD system, *Electronics and Communications in Japan (Part II: Electronics)*, vol. 90, 23–31, 2007.
34. E. Charbon, R. Ghapurey, R. G. Meyer, and A. Sangiovanni-Vincentelli, "Semi-Analytical Techniques for Substrate Characterization in the Design of Mixed-Signal ICs," in *IEEE ICAD*, San Jose, CA, 1996.
35. A. Papoulis and S. U. Pillai, *Probability, Random variables, and Stochastic Process*, 3rd edition, New-York, NY: McGraw-Hill, 1991.
36. A. Demir and P. Feldmann, "Modeling and Simulation of the Interference due to Digital Switching in Mixed-Signal Ics," in *Proceedings of IEEE/ACM International Conference on Computer-Aided Design*, San Jose, CA, November, 1999.
37. H. Yasuda, Calculation of power spectra of pulses sequences by means of transition matrices, *Electronics and Communications in Japan*, vol. 54-A, no. 11, 17–23, 1971.
38. G. L. Cariolaro and G. P. Tronca, Spectra of blocked coded digital signals, *IEEE Transactions on Communications*, vol. 22, 1555–1563, October 1974.
39. G. Bilardi, R. Padovani, and G. L. Pierobon, Spectral analysis of functions of Markov chains with applications, *IEEE Transactions on Communications*, vol. COM-31, 853–861, July 1983.
40. H. Kanabe, Spectral lines of codes given as functions of finite Markov chains, *IEEE Transactions on Information Theory*, vol. 37, 927–941, May 1991.
41. Demir, J. Roychowdhury, and P. Feldmann, "Modelling and Simulation of Noise in Analog/ Mixed-Signals Communications Systems," in *IEEE Custom Integrated Circuits Conference*, San Diego, CA, 1999.
42. International Electro-technical Commission, IEC 61927: EMC for components–Part 3: Integrated Circuit Electromagnetic Model (ICEM), IEC standard proposal, 2002.
43. O. Valorge, C. Andrei, B. Vrignon, F. Calmon, C. Gontrand, J. Verdier, and P. Dautriche, "Using ICEM Models for Substrate Noise Characterization in Mixed Signal IC's," in *Proceedings of the 16th International Zurich Symposium on Electromagnetic Compatibility*, Zurich, Switzerland, pp. 353–356, 2005.
44. N. K. Verghese and D. J. Allstost, Computer-aided design considerations for mixed-signals coupling in RF integrated circuits, *IEEE Journal of Solid-State Circuits*, vol. 33, no. 3, 314, March 1998.

Index

M

N

Q

R

Z